Antioxidants: Advances in Biochemistry

Antioxidants: Advances in Biochemistry

Editor: Oliver Stone

www.callistoreference.com

Callisto Reference,
118-35 Queens Blvd., Suite 400,
Forest Hills, NY 11375, USA

Visit us on the World Wide Web at:
www.callistoreference.com

ISBN: 978-1-63239-805-5 (Hardback)

The publisher's policy is to use permanent paper from mills that operate a sustainable forestry policy. Furthermore, the publisher ensures that the text paper and cover boards used have met acceptable environmental accreditation standards.

Trademark Notice: Registered trademark of products or corporate names are used only for explanation and identification without intent to infringe.

Printed in the United States of America.

Cataloging-in-publication Data

Antioxidants : advances in biochemistry / edited by Oliver Stone.
   p. cm.
Includes bibliographical references and index.
ISBN 978-1-63239-805-5
1. Antioxidants. 2. Biochemistry. 3. Chemical inhibitors. I. Stone, Oliver.
TP159.A5 A58 2017
547.23--dc23

Table of Contents

Preface

A molecule that inhibits the oxidation of other molecules is called antioxidant. The aim of this book is to present researches that have transformed this discipline and aided its advancement. This book offers a wide range of topics that deal with the interrelationship of antioxidants with biochemistry. The text provides the readers with ample examples on how antioxidants are helpful in life processes. This book aims to understand multiple branches that fall under the discipline of antioxidant biochemistry and how such concepts have practical applications. It includes some of the vital pieces of work being conducted across the world, on various topics related to this vast field. This book is a vital tool for all researching and studying this field.

This book has been an outcome of determined endeavour from a group of educationists in the field. The primary objective was to involve a broad spectrum of professionals from diverse cultural background involved in the field for developing new researches. The book not only targets students but also scholars pursuing higher research for further enhancement of the theoretical and practical applications of the subject.

It was an honour to edit such a profound book and also a challenging task to compile and examine all the relevant data for accuracy and originality. I wish to acknowledge the efforts of the contributors for submitting such brilliant and diverse chapters in the field and for endlessly working for the completion of the book. Last, but not the least; I thank my family for being a constant source of support in all my research endeavours.

Editor

Induction of Heme Oxygenase I (HMOX1) by HPP-4382: A Novel Modulator of Bach1 Activity

Otis C. Attucks[1], Kimberly J. Jasmer[3], Mark Hannink[2], Jareer Kassis[1], Zhenping Zhong[1], Suparna Gupta[1], Sam F. Victory[1], Mustafa Guzel[1], Dharma Rao Polisetti[1], Robert Andrews[1], Adnan M. M. Mjalli[1], Matthew J. Kostura[1]*

1 TransTech Pharma LLC, High Point, North Carolina, United States of America, 2 Biochemistry Department and Life Sciences Center, University of Missouri, Columbia, Missouri, United States of America, 3 Division of Biological Sciences, University of Missouri, Columbia, Missouri, United States of America

Abstract

Oxidative stress is generated by reactive oxygen species (ROS) produced in response to metabolic activity and environmental factors. Increased oxidative stress is associated with the pathophysiology of a broad spectrum of inflammatory diseases. Cellular response to excess ROS involves the induction of antioxidant response element (ARE) genes under control of the transcriptional activator Nrf2 and the transcriptional repressor Bach1. The development of synthetic small molecules that activate the protective anti-oxidant response network is of major therapeutic interest. Traditional small molecules targeting ARE-regulated gene activation (e.g., bardoxolone, dimethyl fumarate) function by alkylating numerous proteins including Keap1, the controlling protein of Nrf2. An alternative is to target the repressor Bach1. Bach1 has an endogenous ligand, heme, that inhibits Bach1 binding to ARE, thus allowing Nrf2-mediated gene expression including that of heme-oxygenase-1 (HMOX1), a well described target of Bach1 repression. In this report, normal human lung fibroblasts were used to screen a collection of synthetic small molecules for their ability to induce HMOX1. A class of HMOX1-inducing compounds, represented by HPP-4382, was discovered. These compounds are not reactive electrophiles, are not suppressed by N-acetyl cysteine, and do not perturb either ROS or cellular glutathione. Using RNAi, we further demonstrate that HPP-4382 induces HMOX1 in an Nrf2-dependent manner. Chromatin immunoprecipitation verified that HPP-4382 treatment of NHLF cells reciprocally coordinated a decrease in binding of Bach1 and an increase of Nrf2 binding to the HMOX1 E2 enhancer. Finally we show that HPP-4382 can inhibit Bach1 activity in a reporter assay that measures transcription driven by the human HMOX1 E2 enhancer. Our results suggest that HPP-4382 is a novel activator of the antioxidant response through the modulation of Bach1 binding to the ARE binding site of target genes.

Editor: Chuen-Mao Yang, Chang Gung University, Taiwan

Funding: The research described in this manuscript was funded by High Point Pharmaceuticals LLC. The funder did have a role in study design, data collection and analysis, decision to publish, and preparation of the manuscript.

Competing Interests: High Point Pharmaceuticals, LLC provided all funds for this work. TransTech Pharma, Inc., TransTech Pharma, LLC, and High Point Pharmaceuticals, LLC are affiliates. The following authors are current or former employees of TransTech Pharma, Inc. or TransTech Pharma, LLC: OA, JK, SG, SV, MG, ZZ, DP, RA, AM, and MK. These current and former employees of TransTech Pharma, Inc., and TransTech Pharma, LLC may own units or unit options in TransTech Pharma, LLC and/or High Point Pharmaceuticals, LLC. High Point Pharmaceuticals, LLC has applied for patents covering some or all of the information disclosed in this manuscript, including but not limited to the molecules, methods of treatment, and screening methods referred to in this manuscript. Two such patent applications are PCT International Application Number PCT/US2011/024311 and PCT International Application Number PCT/US2012/020459. High Point Pharmaceuticals, LLC owns patents covering HPP-4382 and other related compounds that are in development.

* Email: matthew.j.kostura@gmail.com

Introduction

The basic metabolism of a cell generates reactive oxygen species (ROS) which oxidize cellular lipids, proteins, and DNA leading to production of reactive electrophiles which can lead to deleterious consequences if not eliminated [1]. The production of ROS and reactive electrophiles is counterbalanced by a conserved, well-defined set of cellular pathways leading to increased expression of oxidative stress-responsive proteins that degrade ROS, clear reactive electrophiles and increase cellular glutathione. This adaptive program is largely controlled by two proteins: Kelch like-ECH-associated protein 1 (Keap1) and the transcription factor NFEL2L2 (Nrf2). The Keap1-Nrf2 system has evolved to respond to intracellular oxidative stress; in particular the generation of reactive electrophiles produced from oxidation of endogenous

cellular constituents as well as xenobiotics [2–4]. In the absence of cellular oxidative stress, Nrf2 levels in the cytoplasm are maintained at low basal levels by binding to Keap1 and Cullin 3, which leads to the degradation of Nrf2 by ubiquitination [2,5–9]. During periods of oxidative stress, as levels of reactive electrophilic metabolites increase, the ability of Keap1 to target Nrf2 for ubiquitin-dependent degradation is disrupted, thereby increasing Nrf2 protein levels and its transport into the nucleus, resulting in transcription of antioxidant response genes [5,6,8,10,11]. Nrf2 binds to antioxidant response elements (AREs) found in the promoters of over 200 anti-oxidant and cytoprotective genes including NAD(P)H dehydrogenase, quinone 1 (NQO1), catalase (CAT), glutamate-cysteine ligase (GCLC), aldoketoreductase family members, thioredoxin reductase (TXNRD1), and heme oxygenase-1 (HMOX1) [12]. Activation of the anti-oxidant response via the Keap1-Nrf2 pathway is considered to be protective in nearly every organ system [4,13–15].

There is, however, another mechanism by which ARE-regulated genes are controlled and that is through Bach1, a transcriptional repressor that binds to ARE promoter elements resulting in suppression of Nrf2 activity. Bach1 regulates ARE gene expression by binding to the small Maf proteins and ARE sequences that are also separately bound by Nrf2 [16–18]. Natively, Bach1 is bound by its ligand, heme, which causes it to be displaced from the ARE, exported from the nucleus and degraded [19–22]. Bach1 and its ligand coordinate the overall intracellular levels of heme and iron with anti-oxidant gene expression [23,24]. Genetic evidence indicates that Bach1 deletion leads to a significant level of protection in a wide variety of murine disease models [25–32]. These observations suggest that ARE-regulated genes may be controlled by an intracellular ligand independent of ROS generation, electrophilic reactivity or elevation of Nrf2 levels in the cell. The potential; therefore, exists to discover novel, small molecules that target Bach1 and thereby elevate expression of ARE-regulated genes.

It has been previously demonstrated that Bach1 derepression is required prior to Nrf2-dependent HMOX1 gene expression [33–34]. Based on these observations, we report the development of a cell-based screening strategy to identify compounds that specifically modulate the expression of HMOX1 in normal human lung fibroblasts. The use of endogenous HMOX1 protein expression as a readout allowed the identification of compounds that specifically derepress Bach1 and induce transcription of an Nrf2-responsive gene. The identified compounds are not electrophiles, do not deplete cellular glutathione or otherwise incite a cellular stress response. We confirmed that these compounds modulate Bach1 directly using chromatin immunoprecipitation and reporter assays.

Materials and Methods

Cell culture

Normal human lung fibroblasts (NHLF) were purchased from Lonza and maintained in FBM medium supplemented with 2% FCS plus the supplied FGM-2 SingleQuot components (insulin, hFGF-B, and antibiotic/antifungal agents). Cells were carried for a maximum of four passages and grown in large T-175 flasks (CoStar). HepG2 hepatocellular carcinoma cells were purchased from ATCC and maintained in DMEM media containing 10% FCS and antibiotics. Compounds were kept in DMSO stock and diluted to a final concentration of 1% DMSO in complete medium for treatment.

Immunofluorescence

NHLF cells were grown in either 96-well Optilux plates (Falcon; 4,000 cells per well) or 384-well Optilix plates (2,500 cells per well) and allowed to attach overnight in complete FBM medium. Cells were then treated with compound for a specified period of time depending on experiment. Following compound treatment, HMOX1 protein was detected using indirect immunofluorescence. Cells were washed in phosphate-buffered saline (PBS) containing calcium and magnesium, fixed in 4% paraformaldehyde in PBS for 10 minutes, washed twice with PBS, and then permeabilized with 0.2% Triton-X100 in PBS for 5 minutes. Afterwards, cells were blocked in a PBS solution containing 5% bovine serum albumin (BSA) and 0.05% Triton-X100. Cells were first probed with a primary mouse monoclonal antibody against human HMOX1 (Abcam) diluted in PBS containing 1% BSA, 0.01% Triton X-100 for 1 hour, washed twice, and then probed

with a secondary goat anti-mouse Alexa 488 antibody (Invitrogen) for 1 hour. Hoescht stain (Invitrogen) was included to identify cell nuclei. Stained cells were washed in PBS, and HMOX-1 was visualized using the InCell 2000 instrument (General Electric).

ROS and glutathione detection

HepG2 cells plated in 96-well Optilux plates were treated with compound for 1 hour after which 5 μM of the FITC-labeled ROS detection agent CellROX (Invitrogen) was added to the medium per manufacturer's instructions. After 15 minutes, cells were washed 3 times with PBS and then visualized live using a GE InCell 2000 imager. Glutathione was determined using the GSH/GSSG-Glo Assay (Promega). Briefly, cells grown in 96-well tissue culture plates were exposed to compound for 4 hours after which cells were lysed with the provided Total Glutathione Reagent and luminescence was determined using a SpectraMax 384 plate reader (Molecular Devices). Percent ROS or glutathione was calculated using fluorescence intensity; a 3-sigma increase in signal over control (solvent only) was deemed positive.

Gene silencing

Silencing RNA for Nrf2 (SI03246614), Keap1 (SI03246439), and Bach1 (SI04364269) genes were purchased from Qiagen. NHLF cells were plated in complete medium at 4000 cells/well in 96-well culture plates (BD Falcon) one day prior to silencing. A 4X solution of siRNA (80 nM) and SiLentFect transfection lipid (6.75 μl/ml) (BioRad, cat# 170-3360) in serum-free FBM media was prepared and incubated at room temperature for 20 minutes. The siRNA solution was then diluted 1:4 directly into NHLF cells plated in complete FBM. Cells were incubated for 48 hours prior to compound treatment. Sequences for siRNA were as follows: Nrf2: Sense 5′ GGAUUAUUAUGACUGUUAA 3′, antisense 5′ UUAACAGUCAUAAUAAUCC 3′; Keap1: sense 5′ AGGA-UGCCUCAGUGUUAAA 3′, antisense 5′ UUUAACACUGA-GGCAUCCU 3′; Bach1: sense 5′ GGAGUAGUGUGGAGC-GAGATT 3′, antisense 5′ UCUCGCUCCACACUACUCCTA 3′.

QuantiGene II mRNA detection

Gene expression was determined using the QuantiGene II system from Affymetrix following the manufacturer's protocol. Briefly, NHLF cells were grown in 96-well CoStar tissue culture plates (4,000 cells per well) and either subjected to siRNA gene silencing or directly treated with compounds for 51hours in 100 μL complete medium per well. Cells were then lysed by adding 50 μL Lysis Buffer (provided). Following the provided protocol, a portion of the RNA-containing lysate (5–10 μl) was hybridized at 54 degrees C overnight to RNA specific magnetic capture beads in the presence of blocking buffers, proteinase K and preordered mRNA probe sets specific for the genes of interest: HMOX1, Nrf2, Keap1, Bach1, and GAPDH. With the aid of a magnetic plate holder, capture beads containing the hybridized mRNA were washed and incubated with provided labeling probes. The amount and intensity of the labeled beads were determined using a Luminex xMAP cytometric scanner (BioRad). Results were tabulated and plotted using JMP software.

Chromatin immunoprecipitation

NHLF cells were grown on 150 mm BD Falcon Integrid dishes. Cells were either treated with siRNA (see above) for 48 hours and/or treated with compound for six hours. Cells were cross-linked by adding formaldehyde to a final concentration of 1% and rocked for 10 minutes at room temperature. Cross-linking was stopped by adding glycine to a final concentration of 125 mM and rocked at room temperature for 5 minutes. Cells were washed three times with ice-cold 1X PBS, scraped into 1 ml of phosphate buffered saline (PBS) containing protease inhibitors (1x G-Biosciences Protease *Arrest*, 200 µM Na_3VO_4, and 1 mM PMSF) and collected by centrifugation (700xg for 4 min). Cell pellets were resuspended in 1 ml cell lysis buffer [5 mM Pipes pH 8.0, 85 mM KCl, 0.5% NP-40] containing protease inhibitors and incubated for 10 min on ice. Nuclei were pelleted by centrifugation (5000 rpm for 5 min) and resuspended in 350 µl nuclear lysis buffer [50 mM Tris pH 8.1, 10 mM EDTA, 1% SDS] containing protease inhibitors. After 10 minutes on ice, the samples were sonicated using the following protocol: 2×30 seconds at 30% power, 2×30 seconds at 35% power, 2×30 seconds at 40% power, 2×30 seconds at 45% power. Samples were centrifuged at maximum speed for 10 minutes at 4°C and the supernatants transferred to new tubes and diluted 5-fold in ChIP dilution buffer [0.01% SDS, 1.1% Triton X-100, 1.2 mM EDTA, 16.7 mM Tris pH 8.1, 167 mM NaCl] plus protease inhibitors. Samples were pre-cleared with protein A agarose slurry containing 10 mg/ml of E. Coli tRNA for 30 min at 4°C. For the total input control, 20% of the total supernatant was saved and frozen at −80°C. The remainder was equally divided among four tubes and incubated with rotation overnight at 4°C with: no antibody, 2 µg Nrf2 antibodies (H-300, Santa Cruz sc-13032), 4 µg Bach1 antibodies (2 µg of R&D Systems AF5776 and 2 µg of C-20, Santa Cruz sc-14700), or 2 µg Pol II antibodies (CTD4H8, Santa Cruz sc-47701). Immune complexes incubated with protein A agarose slurry containing tRNA for 1 hr at 4°C with rotation. Beads were collected by centrifugation at 4000 RPMs for 5 minutes. Beads were washed consecutively for 5 minutes on a rotating platform with 1 ml of each of the following solutions: low salt wash buffer [0.1% SDS, 1% Triton X-100, 2 mM EDTA, 20 mM Tris pH 8.1, 150 mM NaCl]; high salt wash buffer [0.1% SDS, 1% Triton X-100, 2 mM EDTA, 20 mM Tris pH 8.1, 500 mM NaCl]; LiCl wash buffer [0.25 M LiCl, 1% NP40, 1% deoxycholate, 1 mM EDTA, 10 mM Tris pH 8.0]; followed by a wash in TE Buffer. After each wash, beads were collected by centrifugation at 4000 RPMs for 5 minutes and supernatant was discarded. Complexes were eluted by adding 250 µl of elution buffer [1% SDS, 0.1 M NaHCO3] to pelleted beads and vortexed for 30 minutes. Samples were centrifuged at 14,000 rpm for 3 minutes and supernatant transferred to clean tubes. Elution was repeated and combined. Formaldehyde cross-links were reversed by adding 1 µl of 10 mg/ml RNase and NaCl to a final concentration of 0.3 M and incubation at 65°C for 4–5 hours. To precipitate DNA, 2.5 volumes of 100% ethanol was added and the samples incubated overnight at −20°C. DNA was pelleted by centrifugation at max speed for 30 minutes at 4°C. The DNA was resuspended in 100 µl of water and 2 µl of 0.5 M EDTA, 4 µl 1 M Tris pH 6.5 and 1 µl of 20 mg/ml Proteinase K were added to each sample and incubated overnight at 45°C. DNA was purified using Thermo Scientific GeneJet PCR purification kit and eluted from the column in 50 µl of sterile dH_2O. All chromatin immunoprecipitations were quantified using quantitative PCR.

Quantitative PCR and data analysis

All quantitative PCR was carried out on an Applied Biosystems 7500 Real Time PCR System. Quantitative PCR was conducted in triplicate in an Applied Biosystems MicroAmp Optical 96-well Reaction Plate with a 25 µl reaction volume containing 12.5 µl of Thermo Scientific Maxima SYBR Green/ROX qPCR Master Mix, 2 µl of purified DNA and a final primer concentration of 0.15 µM for both forward and reverse primers. Primer were

ordered from Sigma and sequences were as follows: HMOX1 EN2 ARE Sense: 5′-CACGGTCCCGAGGTCTATT-3′, REV: 5′-TAGACCGTGACTCAGCGAAA- 3′ and HMOX1 Promoter FOR: 5′-CAGAGCCTGCAGCTTCTCAGA-3′ REV 5′-GGAAACAAAGTCTGGCCATAGGAC-3′. Quantitative PCR was represented as % Input. The DNA used in each sample was representative of .8% of the total chromatin collected (20% total chromatin x 4% used for each qPCR replicate). This is a dilution factor of 125. For this reason, the Input was adjusted for dilution by subtracting $Log_2(125)$ from the raw Ct value. Percent Input was calculated for each sample by the following calculation: 125*2^(Adjusted Input – Ct(IP sample)). Procedure for calculating %Input from raw Ct values was obtained from Invitrogen. Error was reported as the standard deviation of %Input value triplicates.

Western blotting

NHLF cells transfected with siRNA molecules were lysed in High Salt ELB lysis buffer [1 M Tris pH 8.0, 1% NP-40, 250 mM NaCl, 5 mM EDTA] supplemented with protease and phosphatase inhibitors (1x G-Biosciences Protease Arrest, 200 µM Na_3VO_4, and 1 mM PMSF). One-half volume of 3x Sample buffer [6.7% SDS, 160 mM Tris-HCl pH 6.8, .005% Bromophenol Blue dye, 8.3% glycerol, 15% 2-BME) was added to the lysates. Lysates were sonicated using a Fisher Scientific Sonic Dismembrator (Model 500) at 35% power for 30 seconds on ice and then boiled for 10 minutes. Lysates were separated via SDS-PAGE on a 12.5% Bis-Tris polyacrylamide gel and transferred onto nitrocellulose membrane. After blocking overnight at 4° in 5% non-fat dry milk in PBS/0.1%Tween-20, blots were probed with the appropriate primary antibody for Keap1 (Cell Signaling), Nrf2 (Santa Cruz Biotechnology), or β-Tubulin (Sigma Aldrich). Blots were then probed with an appropriate horseradish peroxidase-conjugated secondary antibody (αMouse: Jackson-Immuno Research, αRabbit: Santa Cruz Biotechnology). Immunodetection was performed using Millipore Western HRP substrate and developed in a Fujifilm Intelligent Dark Box using LAS-3000 software.

Bach1 luciferase assay

Single DNA strand bearing three copies of the human Maf-recognition element (MARE) core motifs, 5′-CTAGCTGCT-GAGTCATGCTGAGTCATGCTGAGTCATC 3′, and its complementary strand, 5′-TCGAGATGACTCAGCATGACTCAG-CATGACTCAGCAG 3′, were synthesized and annealed through standard procedures. The generated DNA fragment was then subjected to NheI and XhoI digestion and cloned into the pGL3-Luc basic vector that had also been digested with the same restriction enzymes. The clone, pGL-MARE-Luc, was confirmed via DNA sequencing before being used in the luciferase reporter assay. A FLAG tag was introduced to the N-terminus of the human Bach1 gene by cloning the gene into a pFLAG-CMV-6c vector (Sigma). Cysteine-to-Alanine substitutions (C435A, C461A, C492A and C646A) in the CP motifs were achieved through site-directed mutagenesis using the QuikChange II Site Directed Mutagenesis Kit from Agilent Technologies. HepG2 cells in 100 mm cell culture dishes were transfected with pGL-MARE-Luc plasmid DNA along with plasmid carrying the human Bach1 gene or the empty vector pFLAG-CMV-6c using Fugene6 (Promega). Transfected cells were trypsinized and re-plated into 96-well plates 20–24 hours after transfection. Compounds were added to cells 5–6 hours later, and then incubated overnight. The transfected and compound-treated cells were then gently washed with PBS followed by the addition of Luciferase substrate (Steadyliteplus, PerkinElmer). The cells were incubated for 15–

30 minutes at room temperature to allow complete cell lysis before determining luminescent levels using in an Envision plate reader.

Results

Screening HMOX1 protein expression in NHLF cells

Normal human lung fibroblasts (NHLF) cells grown in 384-well Optilux plates were treated with candidate compounds and incubated for 18 hours prior to fixation and staining with Hoescht dye and anti-HMOX1 antibody as described in *Materials and Methods*. Figure 1A provides representative data on performance of the assay; Cobalt Protoporphyrin IX (CoPP) was used as an internal positive control. The mean expression level and confidence limits of HMOX1 protein were estimated from the raw pixel intensities using JMP software (SAS Institute). Control charting was used to determine the relative ability of a compound to induce HMOX1. The global mean and variance of HMOX1 expression was estimated for all wells of the plate tested. From that, lower and upper confidence limits representing 3SD units above and below the mean are plotted. Values above the upper confidence limit indicate a well with a potentially active compound. As shown in Figure 1B, NHLF cells have a very low level of basal HMOX1 expression. Treatment with the positive control CoPP results in induction of HMOX1 protein as measured by specific immunofluorescence. Based on this method of compound activity classification, we identified a class of thiol-reactive (electrophilic) HMOX1 inducing compounds, exemplified by HPP-1014. In addition, a separate class of non-electrophilic yet potent HMOX inducing compounds, represented by HPP-4382, was discovered. The relative potency of the compounds was established using NHLF cells giving the rank order of potency as HPP-4382>HPP-1014>CoPP (Figures 1B, 1C). This rank order was maintained in HepG2 cells (Figure S1 in Data S1).

HPP-4382 is not an electrophile, is not affected by N-acetylcysteine, and does not increase ROS

Chemical induction of Nrf2-dependent gene activation is often described as being driven by compounds with electrophilic groups. The chemical reactivity of these groups leads to alkylation of reactive thiols and generation of ROS. A key test of chemical reactivity is to incubate the compounds with a thiol-containing reductant. If the compound is reactive, a thiol-containing adduct will be formed that is detectable using mass spectrometry. Using this methodology, the chemical reactivity of HPP-4382 was compared to the electrophile bardoxolone-methyl (CDDO-Me) (Figure S2 in Data S2). Solutions of HPP-4382 and CDDO-Me were exposed to the thiol-containing reductants N-acetylcysteine (NAC), cysteine and dithiothreitol. CDDO-Me reacted with thiol groups as determined by detection of specific adducts by LC-MS (Tables S1 and S2 and Figure S3 in Data S2). Similar results are observed with HPP-1014 (data not shown). In contrast, no thiol-containing HPP-4382 adducts were detected, demonstrating that HPP-4382 is not thiol-reactive.

To assess thiol reactivity in cells, the ability of NAC to block HMOX1 induction was determined. NAC has been shown to suppress induction of Nrf2-dependent gene activation by electrophilic compounds, an attribute of both its chemical reactivity and its ability to maintain cellular glutathione levels. To test this premise, NHLF cells were treated with either CDDO-Me, the electrophilic compound HPP-1014, CoPP, or HPP-4382 in the presence or absence of 5 mM NAC. Both CoPP and HPP-4382 induced HMOX1 expression in the presence of NAC whereas induction of HMOX1 by both CDDO-Me and HPP-1014 was inhibited (Figure 2A).

Thiol-reactive electrophilic compounds often increase ROS levels in cells, as a consequence of their ability to deplete

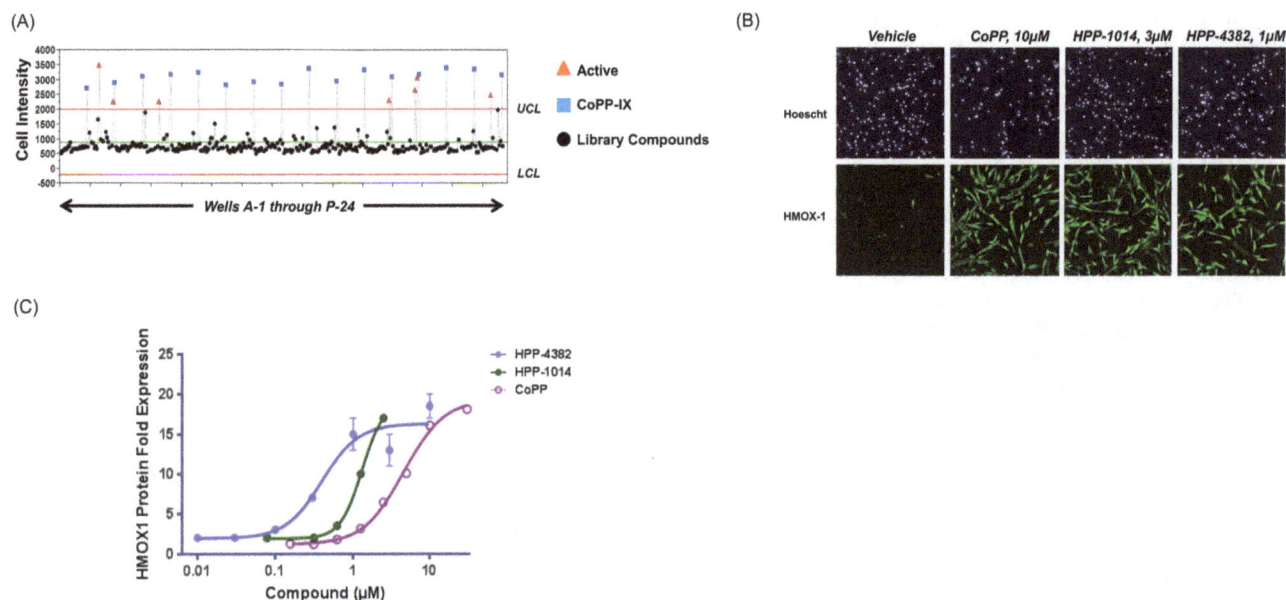

Figure 1. Identification of molecules that induce HMOX1 expression. (A) Human lung fibroblast cells were plated in 384-well Optilux plates and screened with compound libraries at 15 µM for 18 hours. Cells were then fixed, permeabilized, and probed with anti-HMOX1 antibody. Fluorescence intensity of HMOX1 staining was quantified with a GE InCell imager. Control charts were prepared using the statistical software JMP. HMOX1-staining intensities greater than the upper confidence limit were deemed hits. (B) Representative images of cells expressing HMOX1 following compound treatment. NHLF cells were cultured in 96-well Optilux plates as described in *Materials and Methods*. Cells were treated with indicated compound at selected concentrations for 18 hours after which HMOX1 expression was determined by immunofluoresence and quantified on a GE InCell imager. (C) Potency of CoPP, HPP-1014, and HPP-4382 were determined in NHLF cells. Cells were treated for 18 hours, after which they were fixed, permeabilized, and HMOX1 expression determined via immunofluoresence captured on a GE InCell imager.

(A)

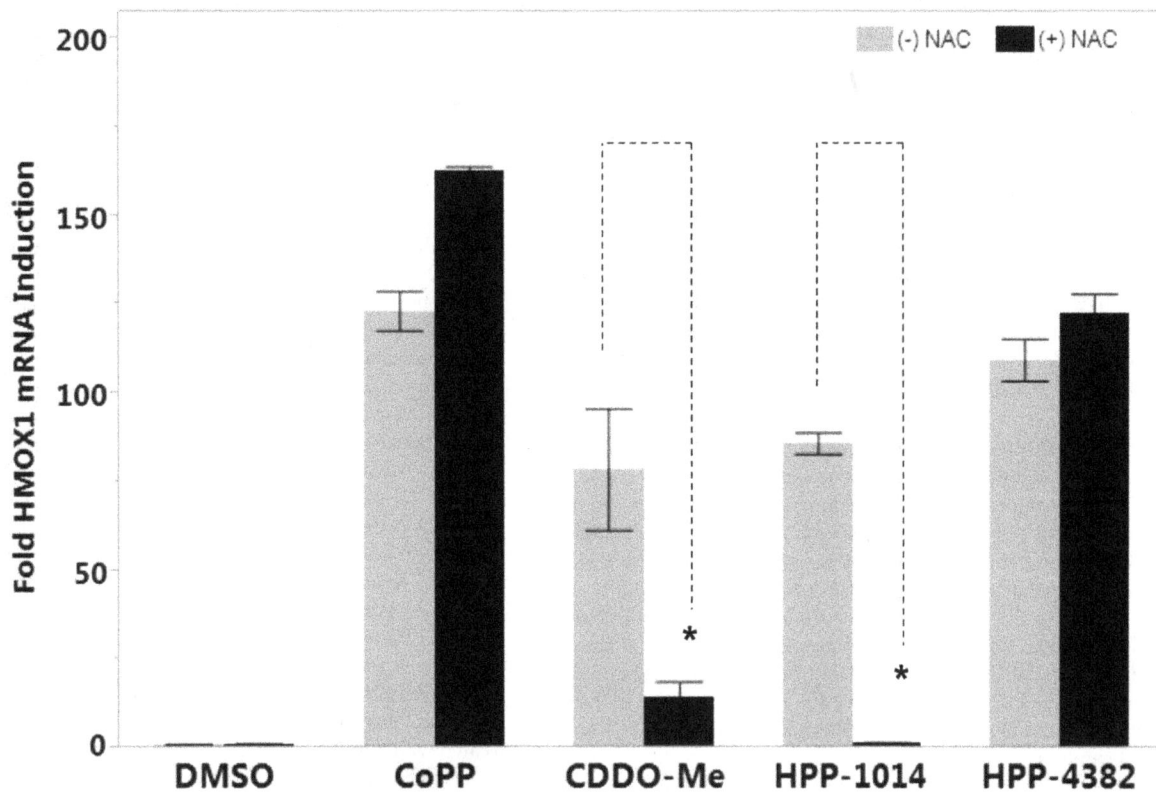

(B)

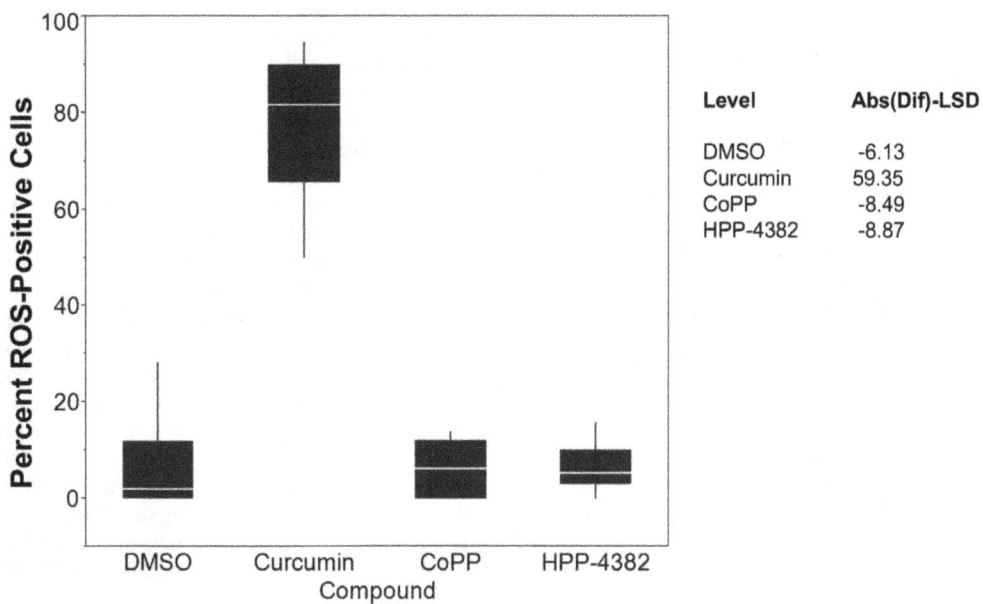

Figure 2. HPP-4382 is not a thiol-reactive electrophile. (A) Effect of N-acetylcysteine (NAC) on HMOX1 induction by HPP compounds. NHLF cells were pretreated with 5 mM NAC for one hour prior to treating with compounds for a further 5 hours (CoPP, 3 µM; CDDO-Me, 0.1 µM; HPP-1014, 3 µM; HPP-4382, 3 µM). Cells were then lysed and HMOX1 mRNA was detected using the Quantigene II method as described in *Materials and Methods*. *$p<0.05$, (B) Induction of reactive oxygen species (ROS) by HPP compounds in HepG2 cells. Cells attached to Optilux plates were treated with compound for 1 hour after which the FITC-labeled ROS-detecting agent CellROX was added for 15 minutes. The number and intensity of ROS-stained cells were captured with a GE InCell imager and the percentage of cells expressing ROS above a set threshold were determined; positive values show pairs of means that are significantly different. All samples in duplicate.

glutathione. Levels of ROS were measured in HepG2 cells following exposure to either HPP-4382 or curcumin, a highly reactive electrophilic compound. ROS levels, as measured by the proportion of cells that stained positive for CellROX, increased from an average of 8.1% in cultures treated with DMSO to 78% in cultures treated with curcumin. In contrast, at the highest tested dose of HPP-4382 (3 µM), ROS levels did not increase above background (6.4%; Figure 2B).

HPP-4382 does not deplete cellular levels of glutathione

Increased cellular ROS is often accompanied by a decrease in cellular glutathione levels. Glutathione was measured in NHLF cells following a 4-hour treatment with buthionine sulphoximine (BSO, an inhibitor of gamma-glutamylcysteine synthetase), electrophilic compounds including bardoxolone, sulforaphane and HPP-1014, and non-electrophilic compounds, including CoPP and HPP-4382. Glutathione levels were markedly reduced in cells treated with BSO (48%, $p<.0001$) or with the electrophilic compounds. However, neither CoPP nor HPP-4382 reduced cellular glutathione. In fact, cellular glutathione levels were significantly increased by HPP-4382 (129%, $p=.0007$) within four hours (Figure 3). Extended treatment of NHLF cells with all compounds revealed a recovery of cellular glutathione with all compounds except BSO (data not shown). The combination of a lack of ROS generation and increased levels of cellular glutathione suggest that HPP-4382 induces HMOX1 in a manner distinct from electrophilic activators of Nrf2.

HPP-4382 induction of HMOX1 is Nrf-2 dependent

To determine if induction of HMOX protein expression by HPP-4382 remained dependent on Nrf2 despite being independent of ROS production, RNAi was used to reduce the expression of Nrf2, Keap1 and Bach1. NHLF cells were treated with siRNA to each gene, resulting in reduced mRNA levels for each gene by 73%, 72%, and 73%, respectively, as determined by the QuantiGene II mRNA plex (Figure 4A). Silencing of Nrf2 significantly decreased baseline levels of HMOX1, whereas silencing of Bach1 resulted in a 50-fold increase in expression of HMOX1 mRNA (Figure 4B). Keap1 silencing, which stabilizes Nrf2 protein levels (see below), only minimally elevated HMOX1 gene expression (approximately 3-fold). In NHLF cells transfected with siRNA against Nrf2 and subsequently treated with either the thiol-reactive compounds CDDO-Me or HPP-1014; or the non-reactive compounds CoPP or HPP-4382, there was a marked reduction in HMOX1 expression. (Figure 4C). Thus, maximal induction of HMOX1 by HPP-4382 is independent of ROS but is still dependent on the presence of Nrf2.

HPP-4382 alters the balance of Nrf2 and Bach1 bound to the HMOX1 E2 ARE independent of Nrf2 and Keap1

Transcription of the HMOX1 gene is controlled, in part, through the binding of either Nrf2 or Bach1 to an ARE, termed HMOX1 E2, located approximately 9 kbp from the transcription start site. Chromatin immunoprecipitation was used to monitor Nrf2 and Bach1 occupancy at the HMOX1 E2 ARE. Under basal

conditions, no significant differences in Nrf2 occupancy were observed at the HMOX1 E2 ARE. Following treatment of NHLF cells with either HPP-4382 or CDDO-Me, a 2- to 3-fold increase in Nrf2 occupancy was observed at the ARE. Under basal conditions, Bach1 occupancy was markedly higher than Nrf2 occupancy at the HMOX1 E2 ARE. HPP-4382, but not CDDO-Me, significantly reduced Bach1 occupancy at the HMOX1 E2 ARE (Figure 5A).

To provide insight into the mechanism whereby HPP-4382 is able to both increase occupancy of Nrf2 and decrease occupancy of Bach1, siRNA was used to reduce steady-state levels of either Nrf2 or Keap1 (Figure 5B). While siRNA knockdown of Nrf2 markedly decreased steady-state levels of Nrf2 protein, siRNA knockdown of Keap1 increased steady-state levels of Nrf2 protein as lack of Keap1-mediated degradation results in accumulation of Nrf2. Knockdown of Nrf2 decreased occupancy by Nrf2 at the HMOX1 E2 ARE (Figure 5C). HPP-4382 increased occupancy by Nrf2 in cells treated with control siRNA. In cells treated with both anti-Nrf2 siRNA and HPP-4382, HMOX1 E2 ARE occupancy by Nrf2 was also increased relative to the levels observed in cells treated with anti-Nrf2 siRNA only, but not to the level observed in cells treated with HPP-4382 without Nrf2 silencing (Figure 5C). Occupancy of the phosphorylated form of RNA polymerase II at the promoters of these genes paralleled Nrf2 occupancy at the corresponding ARE (data not shown).

Bach1 occupancy of the HMOX1 E2 ARE was reduced to approximately 50% of untreated control cells by the anti-Nrf2 siRNA. Bach1 occupancy of the HMOX1 E2 ARE was reduced to a greater extent, about 25% of untreated controls, in cells treated with both anti-Nrf2 siRNA and HPP-4382 (Figure 5C). Thus, reduction in Bach1 occupancy by HPP-4382 is not dependent on the presence of Nrf2. Instead, HPP-4382 reduces Bach1 occupancy of the ARE even when steady-state levels of Nrf2 are reduced by siRNA.

While anti-Nrf2 siRNA molecules decrease steady-state levels of Nrf2, anti-Keap1 siRNA molecules have the opposite effect of increasing steady-state levels of Nrf2 (Figure 5B). Thus the ability of siRNA-mediated knockdown of Keap1 to perturb occupancy by Nrf2 and Bach1 at the HMOX1 E2 ARE was determined. In general, Keap1 siRNA alone resulted in a modest increase in Nrf2 occupancy at the HMOX1 E2 ARE while Keap1 siRNA in combination with HPP-4382 resulted in a further increase of Nrf2 occupancy. Importantly, in the presence of anti-Keap1 siRNA, HPP-4382 was still able to decrease Bach1 occupancy to the same extent as treatment with HPP-4382 only (Figure 5D). Taken together, these results suggest that HPP-4382 induces changes in Bach1 occupancy regardless of steady-state levels of Nrf2.

The ability of HPP-4382 to alter occupancy of Nrf2 and Bach1 at the HMOX1 E2 ARE was compared to HPP-1014 and to CoPP (Figure 5D). HPP-1014 is expected to act through Keap1 to stabilize Nrf2, while CoPP is a mimetic of heme, a known ligand for Bach1 that reduces its steady-state levels [20]. Nrf2 occupancy at the HMOX1 E2 ARE was increased by HPP-1014 while Bach1 occupancy was only slightly reduced by HPP-1014 in the absence of anti-Keap1 siRNA. No reduction of Bach1 occupancy by HPP-1014 was observed in the presence of anti-Keap1 siRNA. In

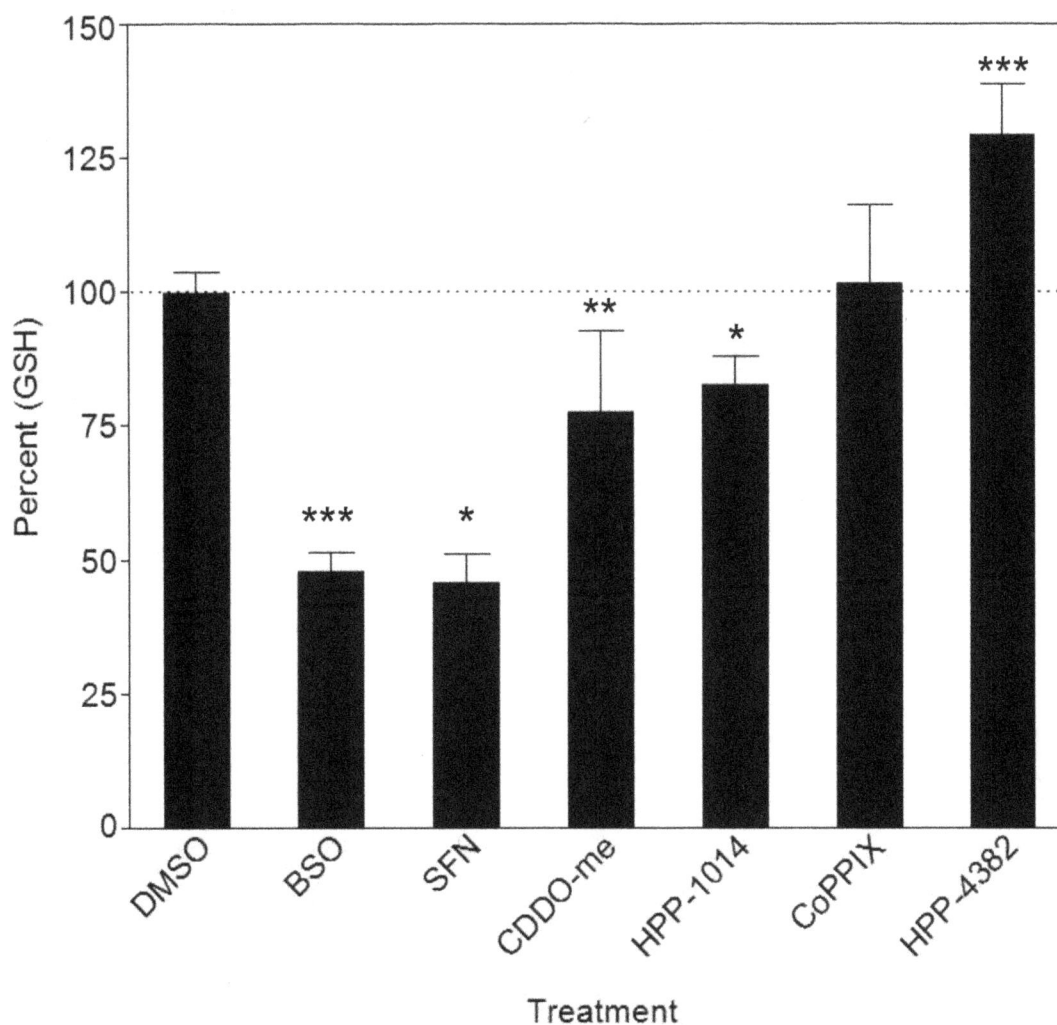

Figure 3. HPP-4382 increases cellular glutathione levels: NHLF cells grown in 96-well Optilux plates were treated with compounds for 4 hours (BSO, 200 μM; Sulphorafane, 10 μM; CDDO-Me, 0.1 μM; HPP-1014, 10 μM; CoPP, 10 μM; HPP-4382 3 μM) and glutathione levels were determined using the GSH/GSSG-Glo Assay Kit (Promega). Positive values show pairs of means that are significantly different. All samples in duplicate, error bars represent standard deviation compared to DMSO. *, $p < 0.05$; **, $p < 0.01$, ***; $p < 0.001$.

contrast, both HPP-4382 and CoPP markedly reduced Bach1 occupancy at the HMOX E2 ARE either in the absence or presence of anti-Keap1 siRNA. Thus, the pattern of altered Nrf2/Bach1 occupancy induced by HPP-4382 does not resemble the pattern induced by an electrophile but closely resembles the pattern induced by CoPP.

Heme binding motifs are required for promoter activity by HPP-4382

The ChIP experiments demonstrated the ability of HPP-4382 to elicit the removal of Bach1 from the HMOX1 promoter independent of Nrf2 steady-state levels. This suggests that HPP-4382 acts directly to modulate binding of Bach1 to AREs. To more fully explore the effects of compound on Bach1 activity, a luciferase reporter assay using the HMOX1 E2 ARE as a target was developed. Expression of HMOX1 E2-dependent luciferase expression was determined in HepG2 cells co-transfected with an HMOX1 E2-dependent reporter plasmid and a plasmid expressing FLAG-tagged wildtype Bach1 (Figure 6A). Luciferase expression was markedly lower in cells expressing Bach1, indicating

effective repression of HMOX1 E2-dependent transcription by Bach1 (Figure 6B). To determine the ability of test compounds to activate HMOX E2-dependent expression in the presence of Bach1, cells were treated with CoPP, CDDO-Me, or HPP-4382. All three compounds were able to induce luciferase expression, demonstrating their ability to overcome Bach1 repression of HMOX E2-dependent transcription (Figure 6C).

The ability of hemin and CoPP to derepress Bach1 has been related to the presence of 4 CP motifs spanning the bZIP domain of the protein [21]. Mutation of these CP motifs markedly reduced heme binding to Bach1 and abrogated the ability of hemin to derepress Bach1. To probe the role of these CP motifs in modulation of Bach1 by HPP-4382, a mutant Bach1 protein containing Cysteine to Alanine substitutions at CP motifs 4 through 7 was constructed (Figure 6A). FLAG-hBach1-AP4-7 was more effective at repression of HMOX1 E2 ARE-dependent transcription than wild-type Bach1 (Figure 6B). Nonetheless, CDDO-Me and HPP-1014 still were able to activate HMOX1 E2-dependent luciferase expression in the presence of mutant Bach1 proteins, indicating that the CP motifs in Bach1 are not critical for efficient derepression of Bach1 by an electrophile

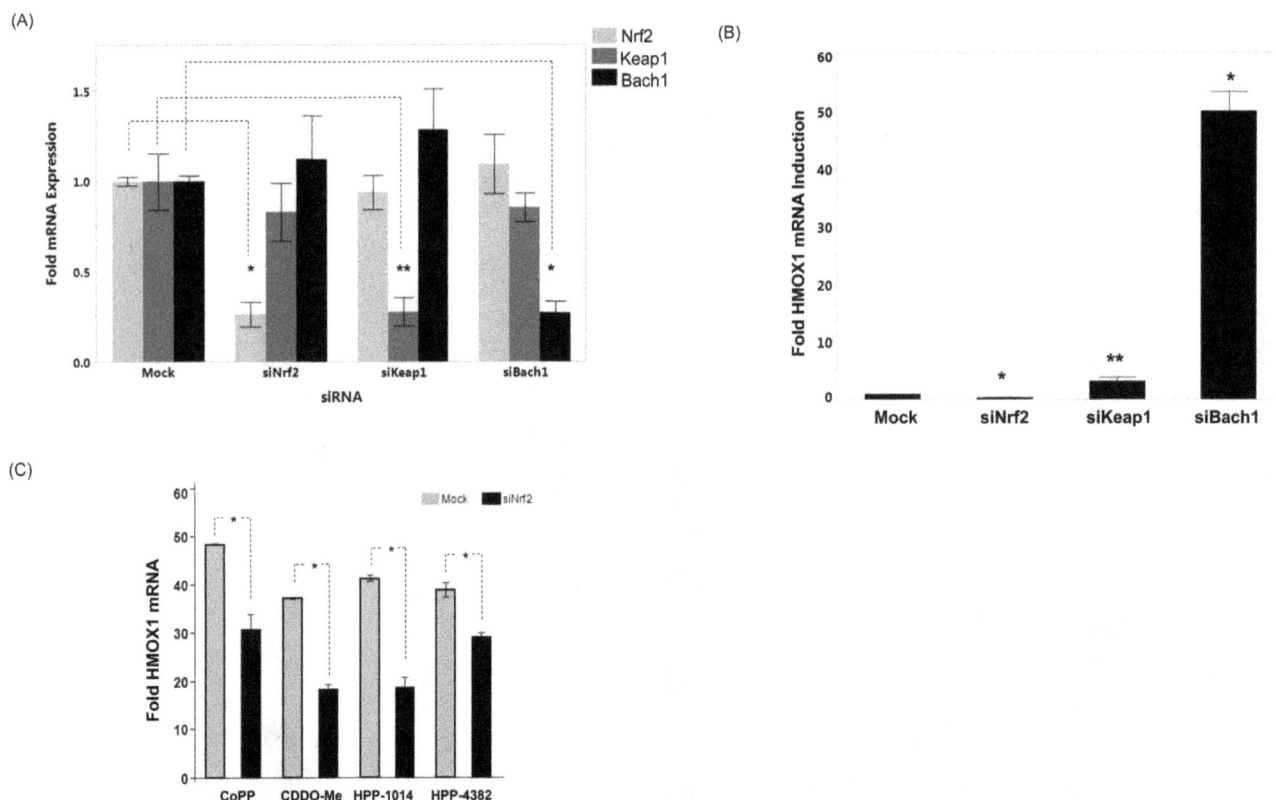

Figure 4. HMOX1 expression by HPP-4382 requires Nrf2. (A) NHLF cells were exposed to 20 nM per well each of Nrf2, Keap1 or Bach1 silencing RNA (or mock) for 48 hours as described in *Material and Methods*. Cells were then lysed and probed for transcription of Nrf2, Keap1, or Bach1 using the QuantiGene II RNA plex (Affymetrix). RNA expression was normalized to total GAPDH and expressed as fold induction over DMSO vehicle for the same gene. *, $p<0.0001$; **, $p<0.05$; all others not significant ($p>0.1$). (B) NHLF cells exposed to Nrf2, Keap1, or Bach1 siRNA or lipid-only vehicle for 48 hours were lysed and probed for transcription of HMOX1 using the QuantiGene II RNA plex. Detected HMOX1 RNA per well was normalized to total GAPDH in same well and shown as fold induction over DMSO;. *, $p<0.0001$; **, $p<0.05$. (C) HMOX1 gene expression in NHLF cells treated with Nrf2 siRNA. After silencing for 48 hours, cells were exposed to vehicle (DMSO) or compounds for 5 hours. HMOX1 and GAPDH RNA expression was determined using QuantiGene II mRNA quantitation. All samples were performed in duplicate, error bars represent standard deviation. *, $p<0.05$.

(Figure 6D). In contrast, the ability of both CoPP and HPP-4382 to induce HMOX1 E2 ARE-dependent luciferase expression in the presence of the mutant Bach1 protein was sharply inhibited. The failure of CoPP to derepress FLAG-hBach1-AP4-7 is in line with observations demonstrating that these CP motifs are required for derepression of Bach1 by heme, and are essential components of a metalloporphyrin binding site in Bach1. That these CP motifs are also required for the ability of HPP-4382 to derepress Bach1 indicates a requirement for this metalloporphyrin binding site in Bach1 for induction of HMOX gene expression by HPP-4382.

Discussion

In light of the widespread role of oxidative stress in the pathology of diverse human diseases and the ability of the Nrf2-dependent antioxidant response gene network to protect against oxidative stress, considerable effort has been directed towards discovering compounds that can increase the activity of Nrf2. Currently, all described small molecule inducers of Nrf2 activity are reactive electrophiles [13,35,36]. Typically, such compounds are not considered pharmaceutically acceptable as they can present safety and toxicity liabilities. Two such molecules, bardoxolone (CDDO) and dimethyl fumarate (DMF), have recently completed clinical trials. Both compounds are chemically reactive alkylating electrophiles. The intrinsic chemical promiscu-

ity of Bardoxolone results in alkylation of a large number of proteins [37]. As a consequence, bardoxolone has a complicated pharmacological and toxicological profile with significant clinical safety problems. Similarly, dimethyl fumarate (DMF) is an electrophile that rapidly reacts with glutathione [38–40]. DMF, however, has not shown the same toxicities in humans as seen with bardoxolone. Given the rather divergent toxicology and adverse event profiles seen with bardoxolone and DMF, we conclude that induction of Nrf2 can be advantageous, but that the electrophilic character of the molecule is crucial and thus sets significant limitations on the safety and efficacy of such compounds.

An alternative approach to regulating Nrf2-dependent gene expression is through targeting the transcriptional repressor Bach1. Bach1 is a member of the BTB and CNC transcriptional regulator family that, like Nrf2, binds to ARE sequences as heterodimeric complexes with small Maf proteins [18] A major physiological role for Bach1 is in iron homeostasis through regulation of the expression of heme oxygenase-1 (HMOX1), ferroportin (FPN1) and Ferritin (FTH) genes [23,24,33,41]. Elevation of intracellular hemin leads to induction of HMOX1 enzyme activity. Consequently, hemin is converted to carbon monoxide, bilirubin and free iron. As hemin levels are reduced, Bach1 is resynthesized and repression of HMOX1 and other genes is restored. Thus Bach1 coordinates the overall intracellular levels

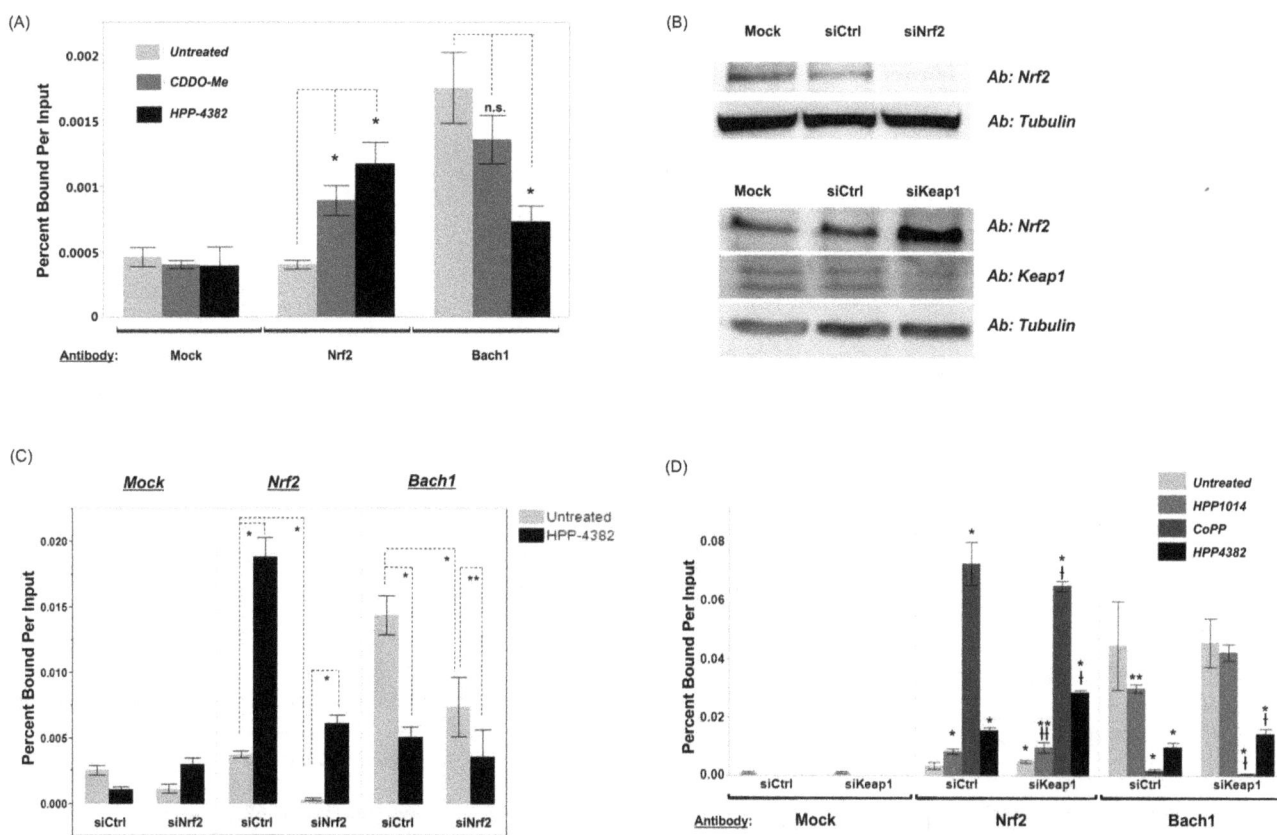

Figure 5. HPP-4382 alters occupancies of Nrf2 and Bach1 on the HMOX1 E2 promoter. (A) NHLF Cells were treated with 0.1 μM CDDO-Me or 1 μM HPP-4382 for 6 hours after which they were crosslinked with 1% formaldehyde in media, washed, and collected to be processed for chromatin immunoprecipitation as described in *Materials and Methods*. Precleared nuclear lysates were incubated with antibodies against Nrf2 or Bach1. Immune complexes were than isolated with E.coli tRNA/Protein A agarose beads, and the obtained purified DNA with subjected to qPCR using primers for HMOX1 E2 promoter. *, $p<0.05$ compared to the untreated sample of same antibody; n.s. = not significant. (B) NHLF cells were exposed to 20 nM Nrf2, Keap1, or control siRNA for 48 hours. Cells were lysed and separated via SDS-PAGE then Western blotted with antibodies against Nrf2, Keap1, or tubulin. (C) Cells transfected with either Nrf2 or control siRNA were subjected to chromatin immunoprecipitation after treatment with 1 μM HPP-4382 for 6 hours. Precleared nuclear lysates were probed with antibodies against Nrf2 or Bach1; a third set was not probed (mock). *, $p<0.01$; **, $p<0.05$. (D) Cells transfected with either Keap1 or control siRNA were subjected to chromatin immunoprecipitation after treatment with either 10 μM HPP-1014, 10 μM CoPP, or 1 μM HPP-4382 for 6 hours. Precleared nuclear lysates were probed with antibodies against Nrf2 or Bach1; a third set was not probed (mock). All samples were performed in triplicate, error bars represent standard deviation. *, $p<0.01$; **, $p<0.05$ compared to untreated siCtrl for same antibody probe. +, $p<0.01$; ++, $p<0.05$ compared to untreated siKeap1 for same antibody probe.

of hemin and iron metabolizing genes with anti-oxidant gene expression [19,21,22,24].

The pharmacology of Bach1 modulation by heme and its metalloporphyrin mimetics has been examined in a variety of settings. Cobalt Protoporphyrin (CoPP) has been shown to have considerable pharmacological benefit in models of diabetes-linked vascular and renal damage [42–45], Ang II mediated hypertension [46,47], renovascular hypertension [48], arterial thrombosis [49] and other oxidative stress-mediated pathologies. Inhibition of Bach1 itself has been suggested to be of benefit in diseases such as non-alcoholic steatohepatitis [50] and insulin resistance [51]. However, CoPP and most metalloporphyrins have limited bioavailability and therefore are unsuitable in most clinical settings. Thus, identifying molecules that can mimic the ability of metalloporphyrins to modulate Bach1 activity directly may have a high degree of therapeutic utility in a number of clinical settings without the potential liabilities of an electrophilic molecule.

Herein, we report the characterization of a novel molecule, HPP-4382, which induces HMOX1 in a manner distinct from other Nrf2 activators. We also demonstrated the ability of HPP-4382 to induce other Phase 2 genes, including NQO1 and

TXNRD1 (Table S3 in Data S3). HPP-4382 does not have electrophilic properties, as determined by its structure and lack of chemical reactivity with common thiol-containing compounds such as N-acetylcysteine. To further characterize HPP-4382, we screened a selection of alternative genes for expression: two markers of endoplasmic- or general cellular stress, HSPA6 and GADD45A, and ICAM1, a target of NF-κB. Using these orthogonal measures of cellular pathway analysis, we confirmed that HPP-4382, in contrast to bardoxolone, did not induce significant cellular stress at high doses as measured by HSPA6 expression and that the mechanism of HPP-4382 activity does not appear dependent on NF-κB as ICAM1 expression is not induced by HPP-4382 (data not shown).

It has been demonstrated that CoPP and hemin induce HMOX1 in an Nrf2-dependent manner through inhibition of Bach1 binding to HMOX1 promoter elements [21,22,33,34]. Our data suggest that HPP-4382 functions to induce HMOX1 in a similar manner. We confirmed the role of Nrf2 in the regulation of HMOX1 gene expression by HPP-4382 using genetic silencing of Nrf2. Knockdown of Nrf2 expression resulted in reduced induction of HMOX1 by HPP-4382, CoPP and bardoxolone,

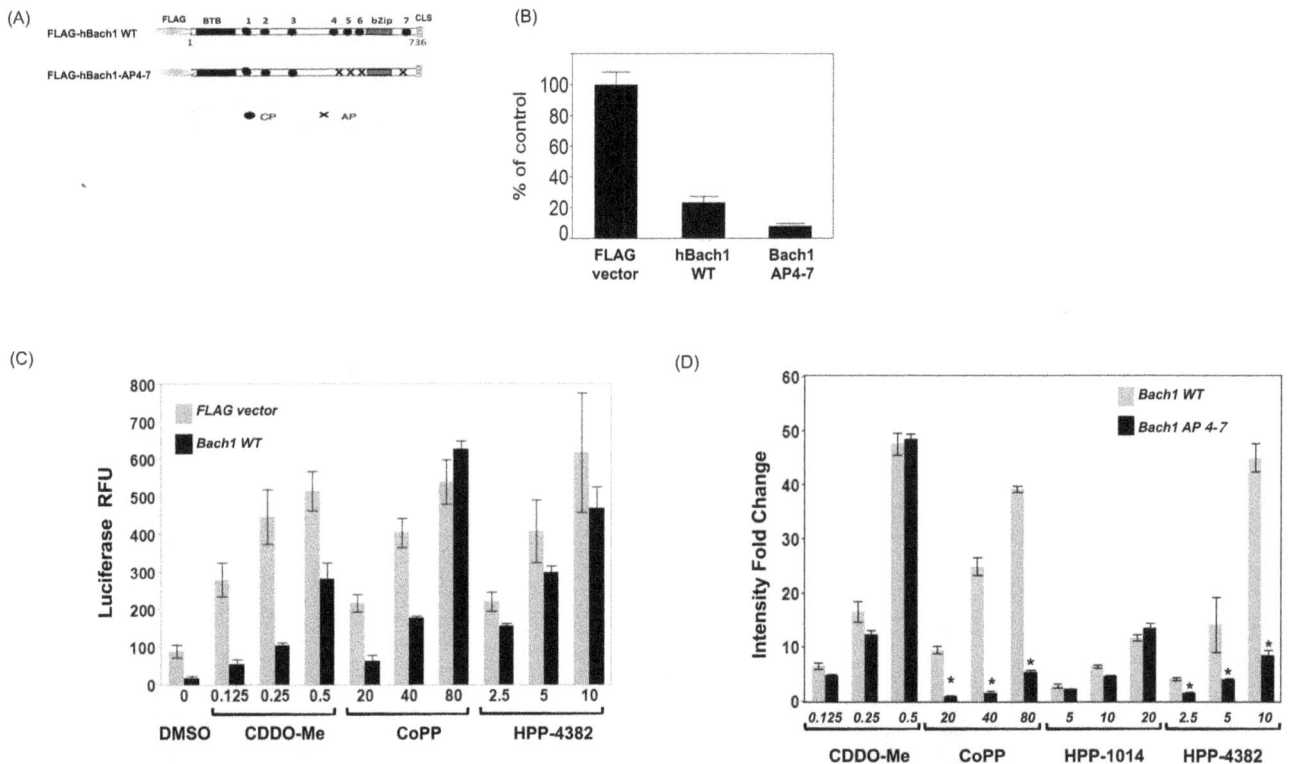

Figure 6. Heme binding motifs are required for activity of both CoPP and HPP-4382 on the HMOX E2 promoter. (A) Schematic representation of pFLAG-Bach1 (WT) and pFLAG-Bach1 (AP4-7) used in these experiments. (B) HepG2 cells were transfected with pGL-MARE-Luc plasmid DNA (containing the HMOX1 E2 promoter) plus a plasmid carrying either pFLAG-Bach1 (WT), pFLAG-Bach1 (AP4-7), or pFLAG-only for 24 hours. Cells were then transferred to 96-well plates and allowed to recover for 6 hours. After washing, Luciferase substrate was added for 30 minutes and fluorescence was measured on an Envision reader. (C, D) HepG2 cells were transfected and replated in 96-well plates as described in B, but treated with compounds at indicated concentrations (μM) overnight prior to determining luciferase activity. In (D), data is reported as fluorescence intensity fold over DMSO-treated cells in each set of transfection *, p<0.0001 compared to Bach1-WT expressing cells at same compound doses. Each sample was performed in quadruplicate, error bars represent standard deviation.

consistent with the well-characterized role of Nrf2 as a critical activating transcription factor for HMOX1.

Pharmacological elevation of Nrf2 protein levels without concomitant derepression of Bach1 fails to induce HMOX1 [33]. Similarly, genetic silencing of Keap1 is insufficient to maximally activate HMOX1 gene expression in Keap1 null mice [52]. These data indicates the clear need for Bach1 derepression for HMOX1 gene expression. We probed this hypothesis in NHLF cells by silencing the three key components of the regulatory pathway. First, Bach1 silencing is sufficient to maximally induce HMOX1 mRNA expression, consistent with published results. On the other hand, Keap1 silencing resulted in significantly less HMOX1 induction in the absence of compound. Our results are consistent with the suggestion that Bach1 represents a dominant layer of control on HMOX1 expression in NHLF cells.

We further probed the ability of HPP-4382 to modulate transcription factor binding to the HMOX1 promoter via chromatin immunoprecipitation. In these experiments, HPP-4382 was compared to the electrophile CDDO-Me (Bardoxolone). Both compounds increased binding of Nrf2 at the HMOX E2 enhancer and binding of RNA polymerase II to the HMOX promoter, consistent with the ability of these compounds to activate HMOX1 transcription in an Nrf2-dependent manner. However, only HPP-4382, but not CDDO-Me, resulted in robust decreases in binding of Bach1 to the HMOX1 E2 enhancer

element, suggesting that HPP-4382 has a mode of action distinct from that of CDDO-Me. To test this idea further, we altered steady-state levels of Nrf2 by gene silencing and measured occupancy of Bach1 at the HMOX1 E2 enhancer. In the presence of anti-Nrf2 siRNA, which significantly reduced steady state levels of Nrf2, Bach1 occupancy of the HMOX1 E2 enhancer was decreased by HPP-4382. In the converse experiment, when steady-state levels of Nrf2 were increased by gene silencing of Keap1, HPP-4382 was also able to decrease occupancy of Bach1 at the HMOX1 E2 enhancer. Thus, the ability of HPP-4382 to decrease binding of Bach1 to the HMOX1 E2 enhancer is independent of steady-state levels of Nrf2.

To further examine the mechanism by which HPP-4382 modulates Bach1, we created reporter assays controlled by the ARE element found in HMOX1-E2 and which is known to be regulated by Bach1. In addition, we created a modified Bach1 that is unable to respond to hemin and hemin mimetics, including CoPP. In these assays, both wild-type Bach1 and FLAG-hBach1-AP4-7 efficiently repressed basal levels of luciferase expression. CDDO-Me was able to derepress both the mutant and wild-type Bach1 proteins, resulting in increased levels of ARE-dependent gene expression. However, while CoPP efficiently derepressed the wild-type Bach1 protein, CoPP did not affect the repressive action of the mutant Bach1 protein. Similarly, HPP-4382 was able to overcome repression of ARE-dependent gene expression by wild-type Bach1 protein but not mutant Bach1 protein. Taken together, the results from the ChIP and derepression assays provide supporting evidence that HPP-4382 interferes with the ability of Bach1 to bind DNA. However, while heme has been reported to induce nuclear export and subsequent cytoplasmic degradation of Bach1, HPP-4382 does not appear to alter the steady-state levels or nuclear-cytoplasmic distribution of Bach1 (data not shown), suggesting that HPP-4382 may not fully mimic the action of heme as a ligand of Bach1. Nevertheless, the non-electrophilic character of HPP-4382 and the fact that an intact heme binding site in Bach1 is required for modulation of Bach1 activity indicates that HPP-4382 represents a first-in-class compound that is able to activate the anti-oxidant response gene network by specific modulation of Bach1 activity. We believe that this type of compound will provide therapeutic benefit in a variety of disease settings without the toxicities associated with electrophilic inducers of Nrf2 activity.

References

1. Marnett LJ, Riggins JN, West JD (2003) Endogenous generation of reactive oxidants and electrophiles and their reactions with DNA and protein. J Clin Invest 111: 583–593.
2. Itoh K, Wakabayashi N, Katoh Y, Ishii T, O'Connor T, et al. (2003) Keap1 regulates both cytoplasmic-nuclear shuttling and degradation of Nrf2 in response to electrophiles. Genes Cells 8: 379–391.
3. Kobayashi M, Li L, Iwamoto N, Nakajimi-Takagi Y, Kaneko H, et al. (2009) The Antioxidant Defense System Keap1-Nrf2 Comprises a Multiple Sensing Mechanism for Responding to a Wide Range of Chemical Compounds. Mol Cell Biol 29: 493–502.
4. Motohashi H, Yamamoto M (2004) Nrf2-Keap1 defines a physiologically important stress response mechanism. Trends Mol Med 10: 549–557.
5. Zhang DD, Hannink M (2003) Distinct cysteine residues in Keap1 are required for Keap1-dependent ubiquitination of Nrf2 and for stabilization of Nrf2 by chemoprotective agents and oxidative stress. Mol Cell Biol 23: 8137–8151.
6. Zhang DD, Lo SC, Cross JV, Templeton DJ, Hannink M (2004) Keap1 is a redox-regulated substrate adapter protein for a Cul3-dependent ubiquitin ligase complex. Mol Cell Biol 24: 10941–10953.
7. Kobayashi A, Kang MI, Okawa H, Ohtsuji M, Zenke Y, et al. (2004) Oxidative stress sensor Keap1 functions as an adaptor for Cul3-based E3 ligase to regulate proteasomal degradation of Nrf2. Mol Cell Biol 24: 7130–7139.
8. Kobayashi A, Kang MI, Watai Y, Tong KI, Shibata T, et al. (2006) Oxidative and electrophilic stresses activate Nrf2 through inhibition of ubiquitination activity of Keap1. Mol Cell Biol 26: 221–229.
9. Lo S-C, Li X, Henzl MT, Beamer LJ, Hannink M (2006) Structure of the Keap1:Nrf2 interface provides mechanistic insight into Nrf2 signaling. EMBO J 25: 3605–3617.
10. Yamamoto T, Suzuki T, Kobayashi A, Wakabayashi J, Maher J, et al. (2008) Physiological significance of reactive cysteine residues of Keap1 in determining Nrf2 activity. Mol Cell Biol 28: 2758–2770.
11. Wakabayashi N, Dinkova-Kostova AT, Holtzclaw WD, Kang MI, Kobayashi A, et al. (2004) Protection against electrophile and oxidant stress by induction of the phase 2 response: fate of cysteines of the Keap1 sensor modified by inducers. Proc Natl Acad Sci U S A 101: 2040–2045.
12. Malhotra D, Portales-Casamar E, Singh A, Srivastava S, Arenillas D, et al. (2010) Global mapping of binding sites for Nrf2 identifies novel targets in cell survival response through ChIP-Seq profiling and network analysis. Nucleic Acids Res 38: 5718–5734.
13. Copple IM (2012) The Keap1-Nrf2 cell defense pathway–a promising therapeutic target? Adv Pharmacol San Diego Calif 63: 43–79.
14. Sykiotis GP, Bohmann D (2010) Stress-activated cap'n'collar transcription factors in aging and human disease. Sci Signal 3(112): re3.
15. Nguyen T, Yang CS, Pickett CB (2004) The pathways and molecular mechanisms regulating Nrf2 activation in response to chemical stress. Free Radic Biol Med 37: 433–441.
16. Dhakshinamoorthy S, Jain AK, Bloom DA, Jaiswal AK (2005) Bach1 competes with Nrf2 leading to negative regulation of the antioxidant response element (ARE)-mediated NAD(P)H:quinone oxidoreductase 1 gene expression and induction in response to antioxidants. J Biol Chem 280: 16891–16900.
17. Warnatz HJ, Schmidt D, Manke T, Piccini I, Sultan M, et al. (2011) The BTB and CNC homology 1 (BACH1) target genes are involved in the oxidative stress response and in control of the cell cycle. J Biol Chem 286: 23521–23532.
18. Oyake T, Itoh K, Motohashi H, Hayashi N, Hoshino H, et al. (1996) Bach proteins belong to a novel family of BTB-basic leucine zipper transcription factors that interact with MafK and regulate transcription through the NF-E2 site. Mol Cell Biol 16: 6083–6095.
19. Suzuki H, Tashiro S, Hira S, Sun J, Yamazaki C, et al. (2004) Heme regulates gene expression by triggering Crm1-dependent nuclear export of Bach1. EMBO J 23: 2544–2553.
20. Zenke-Kawasaki Y, Dohi Y, Katoh Y, Ikura T, Ikura M, et al. (2007) Heme Induces Ubiquitination and Degradation of the Transcription Factor Bach1. Mol Cell Biol 27: 6962–6971.
21. Ogawa K, Sun J, Taketani S, Nakajima O, Nishitani C, et al. (2001) Heme mediates derepression of Maf recognition element through direct binding to transcription repressor Bach1. EMBO J 20: 2835–2843.
22. Hira S, Tomita T, Matsui T, Igarashi K, Ikeda-Saito M (2007) Bach1 a heme-dependent transcription factor reveals presence of multiple heme binding sites with distinct coordination structure. IUBMB Life 59: 542–551.
23. Marro S, Chiabrando D, Messana E, Stolte J, Turco E, et al. (2010) Heme controls ferroportin1 (FPN1) transcription involving Bach1, Nrf2, and a MARE/ARE sequence motif at position -7007 of the FPN1 promoter. Haematologica 95: 1261–1268.
24. Hintze KJ, Katoh Y, Igarashi K, Theil EC (2007) Bach1 repression of ferritin and thioredoxin reductase1 is heme-sensitive in cells and in vitro, and coordinates expression with heme oxygenase1 beta-globin and NADP(H) quinone (oxido) reductase1. J Biol Chem 282: 34365–34371.
25. Watari Y, Yamamoto Y, Brydun A, Ishida T, Mito S, et al. (2008) Ablation of the bach1 gene leads to the suppression of atherosclerosis in bach1 and apolipoprotein E double knockout mice. Hypertens Res Off J Jpn Soc Hypertens 31: 783–792.
26. Iida A, Inagaki K, Miyazaki A, Yonemori F, Ito E, et al. (2009) Bach1 deficiency ameliorates hepatic injury in a mouse model. Tohoku J Exp Med 217: 223–229.
27. Inoue M, Tazuma S, Kanno K, Hyogo H, Igarashi K, et al. (2011) Bach1 gene ablation reduces steatohepatitis in mouse MCD diet model. J Clin Biochem Nutr 48: 161–166.
28. Tanimoto T, Hattori N, Senoo T, Furonaka M, Ishikawa N, et al. (2009) Genetic ablation of the Bach1 gene reduces hyperoxic lung injury in mice: role of IL-6. Free Radic Biol Med 46: 1119–1126.
29. Yano Y, Ozono R, Oishi Y, Kambe M, Yoshizumi M, et al. (2006) Genetic ablation of the transcription repressor Bach1 leads to myocardial protection against ischemia/reperfusion in mice. Genes Cells 11: 791–803.
30. Harusato A, Naito Y, Takagi T, Yamada S, Mizushima K, et al. (2009) Inhibition of Bach1 ameliorates indomethacin-induced intestinal injury in mice. J Physiol Pharmacol 60 (Suppl 7): 149–154.
31. Harusato A, Naito Y, Takagi T, Uchiyama K, Mizushima K, et al. (2011) Suppression of indomethacin-induced apoptosis in the small intestine due to Bach1 deficiency. Free Radic Res 45: 717–727.
32. Mito S, Ozono R, Oshima T, Yano Y, Watari Y, et al. (2008) Myocardial protection against pressure overload in mice lacking Bach1 a transcriptional repressor of heme oxygenase-1. Hypertension 51: 1570–1577.
33. Reichard JF, Motz GT, Puga A (2007) Heme oxygenase-1 induction by NRF2 requires inactivation of the transcriptional repressor BACH1. Nucleic Acids Res 35: 7074–7086.
34. Reichard JF, Sartor MA, Puga A (2008) BACH1 is a specific repressor of HMOX1 that is inactivated by arsenite. J Biol Chem 283: 22363–22370.
35. Kansanen E, Bonacci G, Schopfer FJ, Kuosmanen SM, Tong KI, et al. (2011) Electrophilic nitro-fatty acids activate NRF2 by a KEAP1 cysteine 151-independent mechanism. J Biol Chem 286: 14019–14027.
36. Kensler TW, Wakabayashi N, Biswal S (2007) Cell survival responses to environmental stresses via the Keap1-Nrf2-ARE pathway. Annu Rev Pharmacol Toxicol 47: 89–116.
37. Yore MM, Kettenbach AN, Sporn MB, Gerber SA, Liby KT (2011) Proteomic Analysis Shows Synthetic Oleanane Triterpenoid Binds to mTOR. PLoS ONE 10.1371/journal.pone0022862.
38. Linker RA, Lee DH, Ryan S, van Dam AM, Conrad R, Bista P, et al. (2011) Fumaric acid esters exert neuroprotective effects in neuroinflammation via activation of the Nrf2 antioxidant pathway. Brain J Neurol 134: 678–692.

Supporting Information

Data S1   HMOX1 activation in HepG2 cells.

Data S2   Comparison of electrophilic reactivity towards reduced glutathione.

Data S3   Expression of Phase II genes in NHLF cells.

Author Contributions

Conceived and designed the experiments: OA KJ MH RA SG MG JK DP SV ZZ AM MK. Performed the experiments: OA KJ SG MG JK SV ZZ. Analyzed the data: OA KJ JK ZZ. Contributed reagents/materials/analysis tools: SG MG SV. Wrote the paper: JK MK.

39. Schmidt TJ, Ak M, Mrowietz U (2007) Reactivity of dimethyl fumarate and methylhydrogen fumarate towards glutathione and N-acetyl-l-cysteine—Preparation of S-substituted thiosuccinic acid esters. Bioorg Med Chem 15: 333–342.

40. Scannevin RH, Chollate S, Jung MY, Shackett M, Patel H, et al. (2012) Fumarates Promote Cytoprotection of Central Nervous System Cells against Oxidative Stress via the Nuclear Factor (Erythroid-Derived 2)-Like 2 Pathway. J Pharmacol Exp Ther 341: 274–284.

41. Sun J, Hoshino H, Takaku K, Nakajima O, Muto A, et al. (2002) Hemoprotein Bach1 regulates enhancer availability of heme oxygenase-1 gene. EMBO J 21: 5216–5224.

42. Vanella L, Sodhi K, Kim DH, Puri N, Maheshwari M, et al. (2013) Increased heme-oxygenase 1 expression decreases adipocyte differentiation and lipid accumulation in mesenchymal stem cells via upregulation of the canonical Wnt signaling cascade. Stem Cell Res Ther 4: 28.

43. Elmarakby AA, Faulkner J, Baban B, Saleh MA, Sullivan JC (2012) Induction of hemeoxygenase-1 reduces glomerular injury and apoptosis in diabetic spontaneously hypertensive rats. Am J Physiol Renal Physiol 302: F791–800.

44. Burgess A, Li M, Vanella L, Kim DH, Rezzani R, et al. (2010) Adipocyte heme oxygenase-1 induction attenuates metabolic syndrome in both male and female obese mice. Hypertension 56: 1124–1130.

45. Kruger AL, Peterson SJ, Schwartzman ML, Fusco H, McClung JA, et al. (2006) Up-regulation of heme oxygenase provides vascular protection in an animal model of diabetes through its antioxidant and antiapoptotic effects. J Pharmacol Exp Ther 319: 1144–1152.

46. Vera T, Kelsen S, Stec DE (2008) Kidney-specific induction of heme oxygenase-1 prevents angiotensin II hypertension. Hypertension 52: 660–665.

47. Stec DE, Vera T, McLemore GR Jr, Kelsen S, Rimoldi JM, et al. (2008) Heme oxygenase-1 induction does not improve vascular relaxation in angiotensin II hypertensive mice. Am J Hypertens 21: 189–193.

48. Botros FT, Schwartzman ML, Stier CT, Goodman AI, Abraham NG (2005) Increase in heme oxygenase-1 levels ameliorates renovascular hypertension. Kidney Int 68: 2745–2755.

49. Johns DG, Zelent D, Ao Z, Bradley BT, Cooke A, et al. (2009) Heme-oxygenase induction inhibits arteriolar thrombosis in vivo: effect of the non-substrate inducer cobalt protoporphyrin. Eur J Pharmacol 606: 109–114.

50. Inoue M, Tazuma S, Kanno K, Hyogo H, Igarashi K, et al. (2011) Bach1 gene ablation reduces steatohepatatis in mouse MCD diet model. J Clin Biochem Nutr 48(2): 161–166.

51. Kondo K, Ishigaki Y, Gao J, Yamada T, Imai J, et al. (2013) Bach1 deficiency protects pancreatic β-cells from oxidative stress injury. Am J Physiol Endocrinol Metab 305(5): E641–E648.

52. Reisman SA, Yeager RL, Yamamoto M, Klaassen CD (2009) Increased Nrf2 activation in livers from Keap1-knockdown mice increases expression of cytoprotective genes that detoxify electrophiles more than those that detoxify reactive oxygen species. Toxicol Sci 108, 35–47.

2

Altered Antioxidant System Stimulates Dielectric Barrier Discharge Plasma-Induced Cell Death for Solid Tumor Cell Treatment

Nagendra K. Kaushik*[᠑], Neha Kaushik[᠑], Daehoon Park, Eun H. Choi*

Plasma Bioscience Research Center, Kwangwoon University, Seoul, Korea

Abstract

This study reports the experimental findings and plasma delivery approach developed at the Plasma Bioscience Research Center, Korea for the assessment of antitumor activity of dielectric barrier discharge (DBD) for cancer treatment. Detailed investigation of biological effects occurring after atmospheric pressure non-thermal (APNT) plasma application during *in vitro* experiments revealed the role of reactive oxygen species (ROS) in modulation of the antioxidant defense system, cellular metabolic activity, and apoptosis induction in cancer cells. To understand basic cellular mechanisms, we investigated the effects of APNT DBD plasma on antioxidant defense against oxidative stress in various malignant cells as well as normal cells. T98G glioblastoma, SNU80 thyroid carcinoma, KB oral carcinoma and a non-malignant HEK293 embryonic human cell lines were treated with APNT DBD plasma and cellular effects due to reactive oxygen species were observed. Plasma significantly decreased the metabolic viability and clonogenicity of T98G, SNU80, KB and HEK293 cell lines. Enhanced ROS in the cells led to death via alteration of total antioxidant activity, and $NADP^+$/NADPH and GSH/GSSG ratios 24 hours (h) post plasma treatment. This effect was confirmed by annexin V-FITC and propidium iodide staining. These consequences suggested that the failure of antioxidant defense machinery, with compromised redox status, might have led to sensitization of the malignant cells. These findings suggest a promising approach for solid tumor therapy by delivering a lethal dose of APNT plasma to tumor cells while sparing normal healthy tissues.

Editor: Shama Ahmad, University of Colorado, Denver, United States of America

Funding: This work was supported by the National Research Foundation of Korea (NRF) grant funded by the Korea government (MSIP) (NRF-2010-0027963) and Kwangwoon University in 2014. The funders had no role in study design, data collection and analysis, decision to publish, or preparation of the manuscript.

Competing Interests: The authors have declared that no competing interests exist.

* Email: kaushik.nagendra@kw.ac.kr (NKK); ehchoi@kw.ac.kr (EHC)

᠑ These authors contributed equally to this work.

Background

Cancer is the foremost cause of increasing human death in economically developed countries [1]. Chemotherapy [2] and photodynamic therapy [3] are frequently applied in cancer therapy to eradicate tumor cells for maximum treatment efficacy, but they also cause side effects that influence normal healthy cells. The use of radiotherapy is only 40% effective if used prior to surgery [4]. Although medical science has progressively improved treatment techniques to cure cancer, treatment approaches are still imperfect [5] due to inadequate drug distribution, dose limiting toxicity, and poor cancer cell selectivity. Nevertheless, even with many advances in chemotherapy and radiotherapy, survival rates have persistently decreased over the past years. Hence, a new cancer treatment modality is required to improve survival rates.

The use of non-thermal atmospheric-pressure plasma has recently expanded into biomedical fields (a research area called 'plasma medicine') [6]. Plasma sources usually contain a mixture of charged particles, radicals (e.g., reactive oxygen species (ROS)) and other reactive molecules (e.g., hydrogen peroxide, nitric oxide) as well as photons (UV). Free radicals play a big role in cellular redox signaling pathways, but high levels of ROS can have adverse

effects on cells and lead to activation of cellular apoptotic pathways. Recently, our group reported valuable effects of non-thermal plasma on cancer cell death [7]. Several reports on the application of plasma for treatment of cancer were limited to a few types of cancer targets [8–16], which is not sufficient to establish non-thermal plasma effects on every type of cancer. Different types of cancer cell lines may have different responses to the same treatment therapies. Plasma-induced cancer cell death seems to be dependent on cellular ROS pathways [17]. Some researchers claim that ROS induced by anticancer drugs produce a shift in cellular antioxidant machinery [18,19] and in mitochondrial membrane potential, which is related to induction of programmed cell death (apoptosis) in cancer cells [20,21].

Herein, we report on APNT plasma interaction with three tumor cell lines, human glioblastoma cells (T98G), thyroid carcinoma cells (SNU80) and oral carcinoma cells (KB) and a non-malignant embryonic cells (HEK293). It is crucial to explore the interactions between the production of plasma-induced reactive species and cellular responses. While plasma–mediated oxidative stress may bring about harmful or beneficial cellular responses, one should examine carefully the plasma-dependent effects within target cells by comparing the effects on cancer and normal cells [22]. Previously, we reported that plasma-induced cell death in T98G brain cancer cells and have the least toxic effect on non-malignant HEK293 cells [23]. This additional study was designed to explore the role of ROS sensitive antioxidant machinery against the APNT DBD plasma induced oxidative stress in different cancer cells.

Materials and Methods

Human cell lines

The human cancer cell lines glioblastoma (T98G), thyroid carcinoma (SNU80), oral carcinoma (KB) and non-malignant embryonic cells (HEK293) were acquired from the KCLB (Korean Cell Line Bank, Seoul, Korea). For the plasma-cell interaction, these cells were maintained in Dulbecco's Modified Eagle Medium (Hyclone, USA) supplemented with 10% fetal bovine serum (Hyclone, USA) and 1% penicillin-streptomycin (PS) at 37°C in a humidified atmosphere of 5% CO_2.

Experimental device specifications and plasma treatment

Atmospheric pressure non-thermal (APNT) DBD plasma was designed and used to provide uniform treatment for biomedical purposes. Our plasma system primarily consisted of a high-voltage power supply, electrodes and dielectrics. We used this device for the treatment of human cells in an ambient air environment. For treatment, the distance was at fixed approximately 5 mm between source and sample. Throughout plasma exposure, cells were attached to the bottom of tissue culture plates and were covered in 2 mm of growth medium. Plasma was applied for 30–240 sec (s) with 80 V input voltage. A cost- effective transformer for neon light operated at 60 Hz was used for high-voltage power supply. The upper electrode is made up of silver, and down electrode facing the sample is made of stainless steel mesh. The width of the metal mesh for DBD plasma generation was approximately 80 mm, the thickness of mesh was 1 mm and the space between the two adjacent metal grids was 1 mm. They were separated by 1.8 mm-thick glass and tightly wrapped using insulating paste. The power supply was provided by slidacs, a neon trance invertor. The operational temperature of the plasma device was of 24–35°C during treatment time. Electrical power (5.7 W) was delivered to the upper and lower electrodes to produce the APNT DBD plasma. During 240 s plasma exposure, the pH of the media increased to 8.1 and the temperature of the cell culture media increased to 34.5°C (data not shown in figures). All plasma treatments were given under ambient air conditions and the room temperature recorded by infrared camera was 26°C. For APNT DBD plasma treatment, we used 5 mL of cell suspension with of 10^5 cells/mL on the cell culture dish (SPL, Korea) and incubated cells for approximately 20–24 hours (h) to reach confluence.

Cells viability and survival assay

For evaluation of mitochondrial viability, cells were seeded in a cell culture dish (SPL, Korea) under similar conditions as described previously (experimental device specifications section). Cells were treated with plasma for 30, 60, 120 and 240 s. A control group without plasma treatment was included in each assay. MTT [3-(4,5-dimethyl-2-thiazolyl)-2,5-diphenyl-2H-tetra-zolium bromide] was used to assess cell viability. The MTT assay is a novel method of quantifying metabolically viable cells through their ability to reduce a soluble yellow tetrazolium salt to blue-purple formazan crystals [24]. After incubation times of 24, 48 and 72 h, 20 μL/well of MTT solution (5 mg/mL; Sigma-Aldrich, Korea) was added to each well of the 96-well plate. The absorbance was measured at 540 nm after 3 h incubation using

a microplate reader (Biotek, VT, USA). All assay results are reported as percentage (%) viability, which is directly proportional to the number of metabolically active cells. Percentage (%) viability was calculated as:

% Viability =

Optical density in sample well/Optical density in control well

$\times 100$

For quantitative comparison, we performed cell counts using trypan blue dye (Sigma Aldrich, Korea) and a haemocytometer.

Tumor proliferation test by clonogenic survival assay

The clonogenic assay or colony formation assay is the most well-known cell survival assay and is based on the ability of a single cell to grow into a colony. The assay tests each cell in the population for its ability to undergo "unlimited" division. Because a single portion of seeded cells retains the capacity to produce colonies before or after treatment, cells were diluted appropriately to allow formation of colonies in 10 days. Briefly, after harvesting with 0.05% trypsin, 150–400 (depending on the exposure time) cells were plated for 20–24 h before plasma treatment in DMEM at 37°C [25]. Cultured cells were treated with plasma for 30, 60, 120 and 240 s. After plasma exposure cells were incubated in the dark in a humidified, 5% CO_2 atmosphere at 37°C. After 8–10 days, cells were fixed with 1% crystal violet (Sigma-Aldrich, Korea) as described previously [26]. This assay provides evidence of limited tumor growth rate.

Evaluation of cellular morphology

Scanning electron microscopic (JSM 7001F, JEOL, Tokyo, Japan) analyses were performed to examine the morphology of the cells. Briefly, after 24 h of plasma treatment, plasma exposed cells were fixed in 1 mL of Karnovsky's fixative (2% paraformaldehyde and 2% glutaraldehyde) overnight as described in previous reports [27]. SEM sample preparation involved dehydration of the material in hexamethyldisilazane (HMDS), followed by mounting and coating on glass with carbon tape and examining using a FE-SEM. Images and cell size (length and width) were recorded with PC/SEM software (Jeol Serving Advance Technology).

ROS detection assays

Intracellular oxidative stress ensues when an imbalance exist that favors the production of various types of ROS, such as superoxide ($O_2^{\bullet-}$), hydroxyl radical ($OH^{\bullet}$) and hydrogen peroxide (H_2O_2) over antioxidant defenses [28]. To explore intracellular reactive oxygen species (ROS), we used two types of approaches. In the first scheme, fluorochrome probe 2′,7′-dichlorodihydro-fluorescein diacetate (H_2DCFDA; Invitrogen, USA) was used to detect total ROS. Briefly, after APNT plasma treatment, cells were incubated with 10 μM of H_2DCFDA for 30 minutes (min) at 37°C in the dark. ROS fluorescence was measured using a microplate reader (Biotek, VT, USA) with excitation at 485 nm and emission at 528 nm.

In the second scheme, to ensure that the cell death upon exposure to plasma was connected with intracellular H_2O_2 generation, the level of intracellular H_2O_2 in cells was evaluated. H_2O_2 is highly stable and has strong oxidizing capacity and is therefore considered a strong ROS. Generally, superoxide radical can react with ambient water to form H_2O_2. Consequently, it can damage mitochondrial cellular components and cause cell death.

(a)

(b)

(c)

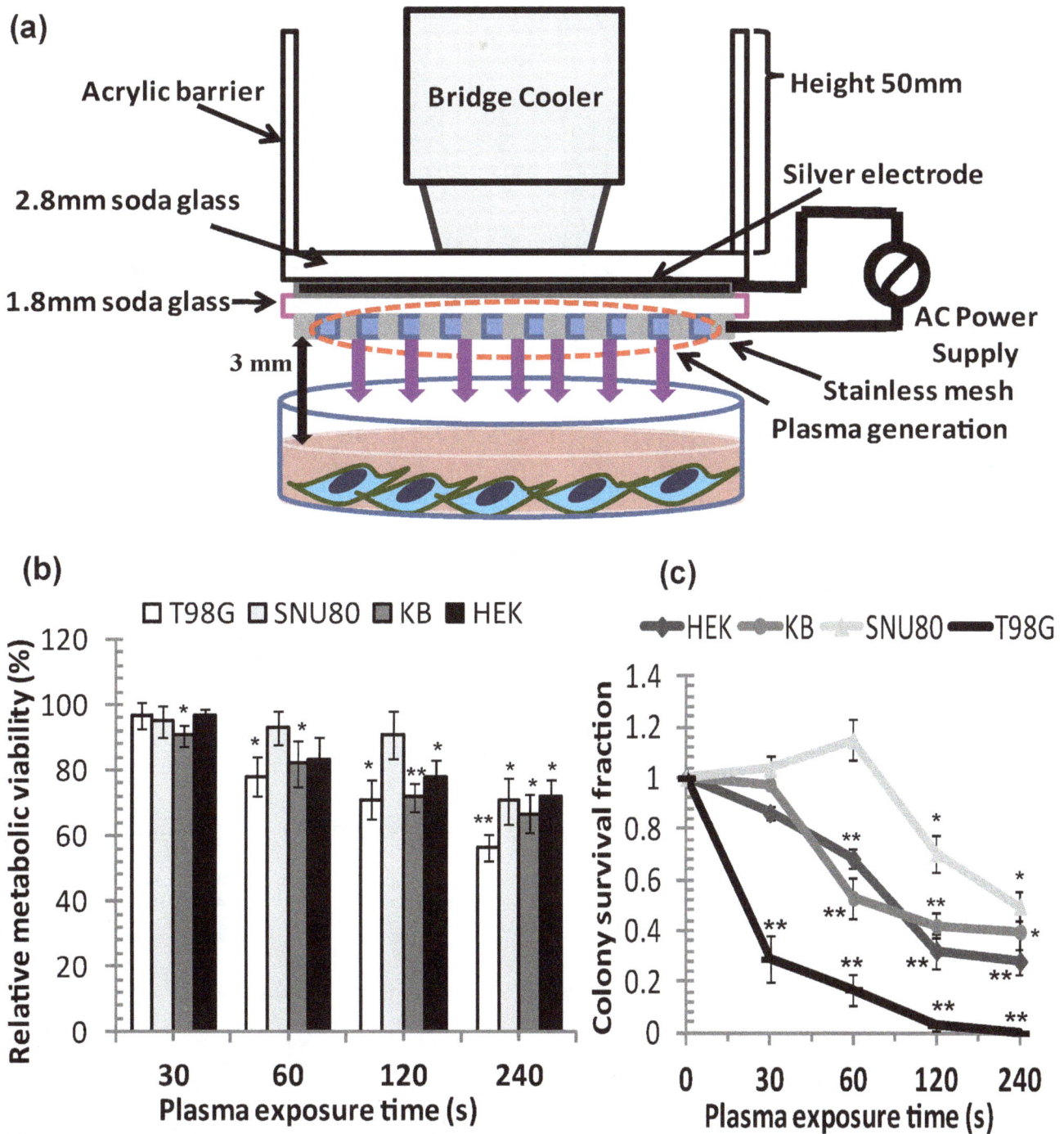

Figure 1. APNT DBD plasma shows growth inhibitory effect in cancer cells. (a) Schematic diagram of an atmospheric pressure non-thermal dielectric barrier discharge (APNT DBD) plasma device. (b) Metabolic viability (%) of cells after plasma treatment was compared after 24 h incubation. (c) Colony forming capacity and clonogenic survival of exponentially growing T98G, SNU80, KB and HEK293 cells. Results from four independent experiments are shown as mean ± SD, and Student's t-test was performed to controls ($*p<0.05$ and $**p<0.01$).

ADHP (10-acetyl-3,7-dihydroxyphenoxazine) was used for H_2O_2 detection. ADHP reacts with H_2O_2 to produce highly fluorescent resorufin. All steps of detection were performed according to the manufacturer's (Hydrogen Peroxide Cell-Based Assay Kit; Cayman chemicals, USA) instructions. Resorufin fluorescence was measured using microplate reader (Biotek, VT, USA) with excitation at 540 nm and emission at 600 nm.

Total glutathione assay

Reduced glutathione (GSH) and oxidized glutathione (GSSG) plays key roles in cellular redox systems. It is essential to examine intracellular glutathione levels in experiments, because fluctuations in the GSH/GSSG ratio are related to human disease therapy, aging and other cell signaling activities. The level of GSSG reflects cell health and oxidative stress. Briefly, 24 h after plasma treatment, cells were washed with PBS and total cell extract was

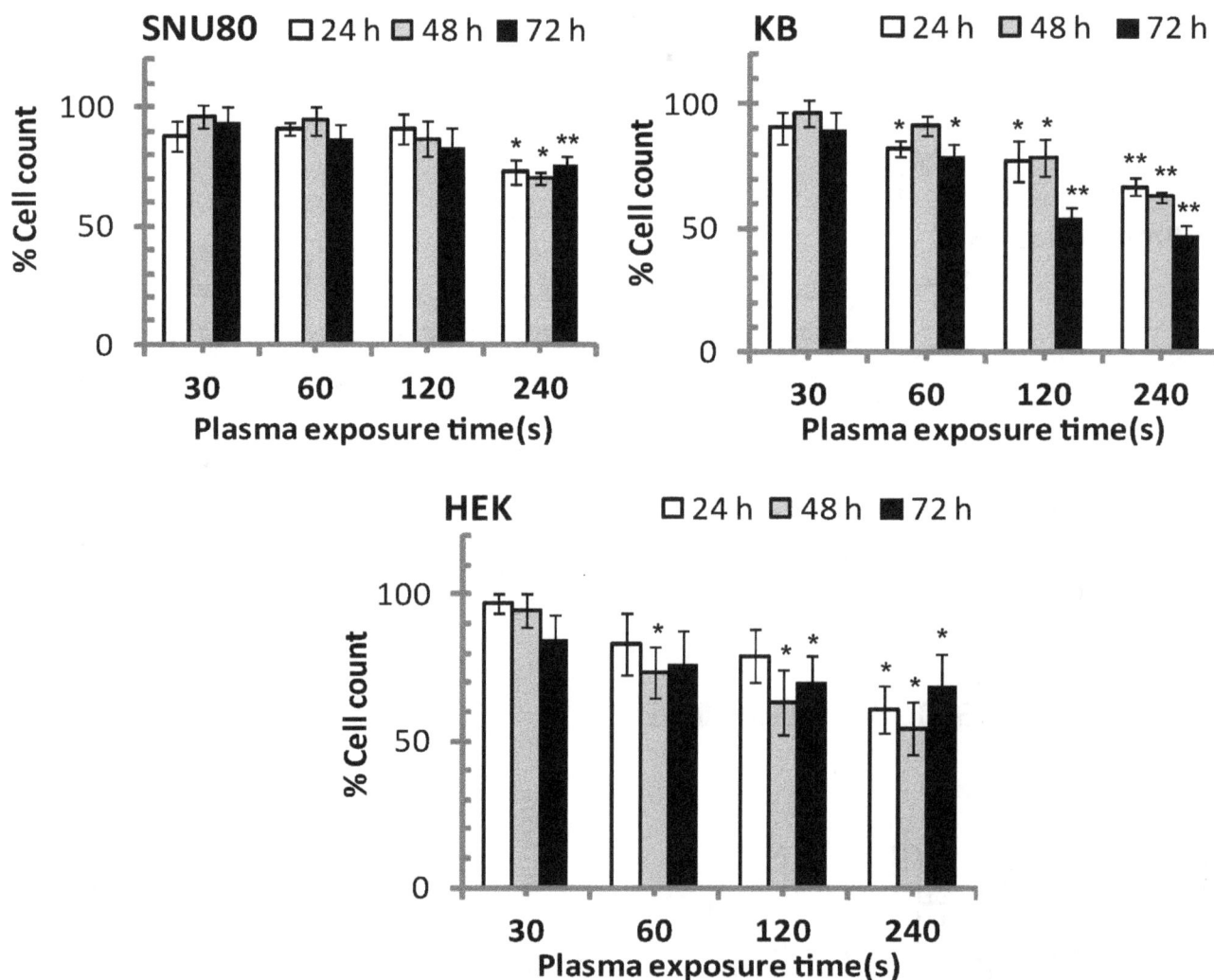

Figure 2. The cell counts (relative to control) showed exposure/incubation time-dependent death rate. KB cells underwent more severe loss than SNU80 and HEK293 by APNT DBD plasma treatment. Results from four independent experiments are shown as mean ± SD, and Student's *t*-test was performed to controls (*$p<0.05$ and **$p<0.01$).

prepared separately for GSH and GSSG quantification in the lysis reagent provided with the kit. GSH, GSSG and GSH/GSSG were determined by a luminescence based biochemical method using the GSH/GSSG-Glo Assay Kit (Promega, Korea) following the manufacturer's instructions. Luminescence was detected using a microplate reader (Biotek, VT, USA).

NADP$^+$/NADPH quantification assay

NADPH is intricately involved in protecting against ROS toxicity, allowing the renewal of GSH. Nicotinamide adenine dinucleotide phosphate (NADP) is an enzymatic cofactor involved in many redox reactions where it cycles between the reduced (NADPH) and oxidized (NADP) forms. Cell samples were prepared 24 h after treatment with plasma according to manufacturer's instructions provided in the NADP/NADPH Quantification Kit (Sigma-Aldrich, Korea). Sample supernatants were collected and analyzed immediately for NADP$^+$ and NADPH. There was no need to purify samples because this assay is specific for NADPH and NADP$^+$. NADP$_{total}$ and NADPH were quantified at 450 nm using a microplate reader (Biotek, VT, USA).

Assessment of cell antioxidant activity

Mammalian cells have developed complex antioxidant systems to stabilize ROS and to reduce injury. Samples of cells 24 h after plasma treatment were prepared according to the manufacturer's instructions provided in the Antioxidant Assay Kit (Cayman Chemicals, USA) and analyzed for antioxidant activity. Absorbance was detected using a microplate reader (Biotek, VT, USA) at 405 or 750 nm.

Caspase 3/7 activity assay

Caspases, a family of cysteine proteases, are the central regulators of apoptosis. Caspases are also necessary for other biological purposes, including cell proliferation, and differentiation. The most studied members of this cysteine protease family include executioner caspase-3 and caspase-7, which play a central role in cell apoptosis and differentiation [29]. To measure apoptosis, the Caspase-Glo 3/7 Assay Kit (Promega, Korea) was used according to the manufacturer's instructions. Luminescence was measured using a luminometer (Biotek, VT, USA).

a

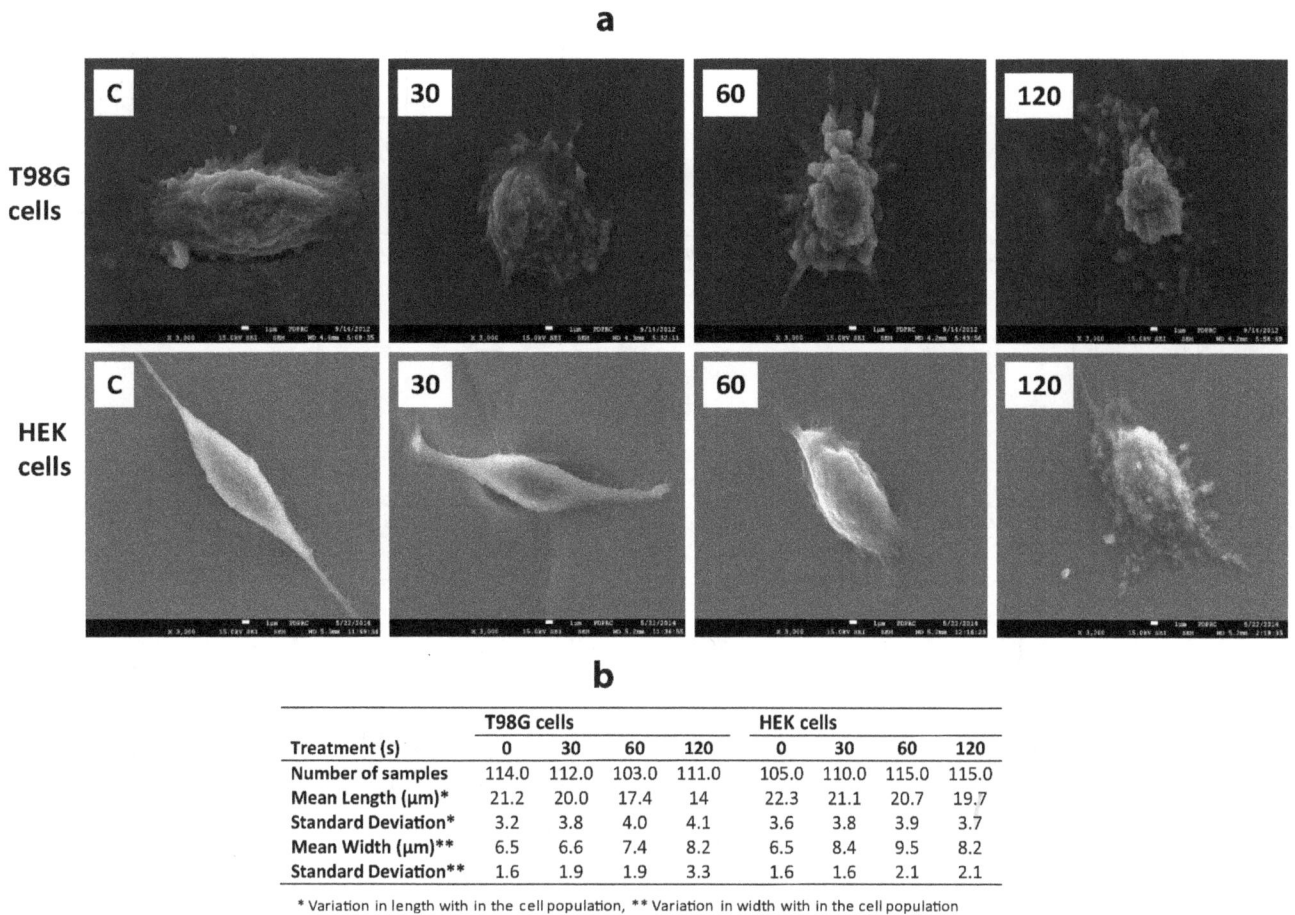

b

Treatment (s)	T98G cells				HEK cells			
	0	30	60	120	0	30	60	120
Number of samples	114.0	112.0	103.0	111.0	105.0	110.0	115.0	115.0
Mean Length (µm)*	21.2	20.0	17.4	14	22.3	21.1	20.7	19.7
Standard Deviation*	3.2	3.8	4.0	4.1	3.6	3.8	3.9	3.7
Mean Width (µm)**	6.5	6.6	7.4	8.2	6.5	8.4	9.5	8.2
Standard Deviation**	1.6	1.9	1.9	3.3	1.6	1.6	2.1	2.1

* Variation in length with in the cell population, ** Variation in width with in the cell population

Figure 3. APNT plasma effects on morphological structure of T98G cancer and HEK293 cells. (a) Cell morphology analyzed by scanning electron microscope (SEM). Cells have blebbing and clear changes in morphology on their outer surface 24 h after plasma treatment. (b) Summary of cellular morphological parameters (length and width).

Mitochondrial membrane potential (ΔΨm) analysis

Mitochondria serve a major role in cell apoptosis induced by many stimulating factors, and the drop of mitochondrial membrane potential (ΔΨm) is an earlier event during apoptosis [30]. The mitochondrial membrane potential was monitored by Mito Flow (Cell Technology Inc, USA). The Mito Flow assay utilizes a cationic dye to visualize the change of mitochondrial membrane potential (MMP). It is a cell permeable, rhodamine-based dye. Membrane potential driven accumulation of the dye within the inner membrane region of healthy functioning mitochondria results in a strong red-orange fluorescence. In the apoptotic cells the dye does not accumulate in the mitochondria, therefore these cells exhibit a lower fluorescence signal. A total of 2×10^5 cells/mL were treated with the plasma for 0, 30, 60, 120 and 240 s. 5 µL of 20X MitoFlow dye was added before the cells were harvested and incubated for 30 min at 37°C. Then, the cells were collected, rinsed with dilution buffer and analyzed for emission at 488 nm by FACS analysis (BD FACSVerse, NJ, USA instrument and FACS suite software).

Cell apoptosis assay (Annexin-V and PI staining)

Apoptosis is a key method by which cancer cells die after treatment [31]. To evaluate plasma induced cell death, annexin V/PI staining was performed followed by flow cytometry. Cells were seeded and treated for 120 s because this was shown to exert maximum plasma effect. Briefly, 24 h after plasma treatment, cells were collected and subjected to annexin V/PI staining using the EzWay AnnexinV-FITC Apoptosis Detection Kit (Koma Biotech Inc, Seoul, Korea) according to the manufacturer's protocol by FACS analysis (BD FACSVerse NJ, USA). Actinomycin D (5 µg/mL, Cayman chemicals, USA) was used as a positive control reagent for apoptosis activation (data not shown).

Statistical analyses

All result values were expressed as the mean ± standard deviation (S.D.) of four independent tests. Statistical analysis was performed using Student's t-test. Statistical significance was recognized at *$p < 0.05$ and **$p < 0.01$.

Results

APNT plasma reduced cell metabolic viability and colony forming capacity

The APNT DBD plasma device was used for cell treatment (**Figure 1a**) under conditions similar to those described previously (experimental device specifications and plasma treatment section). In our study we examined the viability of T98G glioma, SNU80 thyroid cancer, KB oral cancer and HEK293 non-malignant cells. MTT assay results indicated that the APNT plasma has a greater inhibitory effect on cancer cells than on normal cells in a dose/

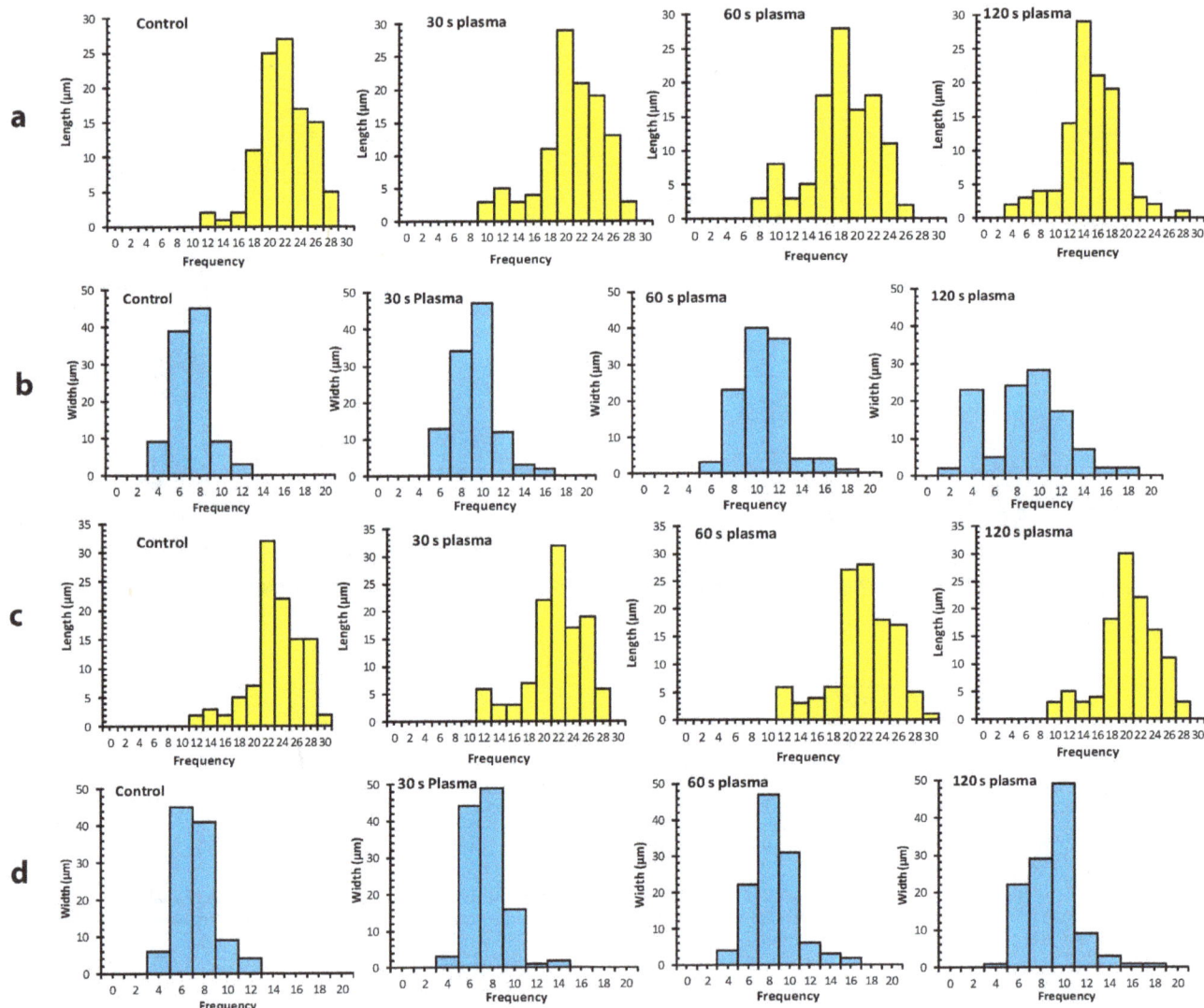

Figure 4. Analyses of the size variability of T98G and HEK293 cells. (a) and (b) show the frequency distribution of length and width, respectively, in the T98G cell population. (c) and (d) shows the frequency distribution of length and width, respectively, in the HEK293 cell population.

incubation time-dependent manner. We observed that the cells exposed to 30 and 60 s plasma treatments showed a lesser effect than those exposed to a 120 and 240 s treatments. However, plasma did not induce any inhibitory effect on SNU80 cells ($p>$ 0.05) at treatments as much high as 120 s. A significant inhibitory effect was seen after 60 s plasma exposure of cells, as shown by inhibition of cell viability up to 22.1% ($p<0.05$), and 17.9% ($p<$ 0.05) respectively in T98G, and KB cells, with a range of viability of 77.9%–83.1% ($p<0.05$). We found that after 120s exposure, up to 29%, 28.2% and 26% cells died in T98G, KB and HEK293 population, respectively, and their viability was 71% ($p<0.05$), 71.8% ($p<0.05$) and 74% ($p<0.05$), respectively, after treatment. Our data clearly show that a higher dose (240 s) of plasma exposure results in a severe decrease in metabolic activity and 56.4%–70.7% viability ($p<0.05$) of all four types of cells (**Figure 1b**).

To confirm the changes in metabolic viability of cells, we performed clonogenic assay. **Figure 1c** demonstrates the effect of the APNT plasma on the colony forming capacity. APNT plasma treatment resulted in a decline in the viability of cells, which resulted in the reduction in the number of colonies formed after plasma exposure. The colony survival fraction of T98G, KB and HEK293 cells was found to be drastically decreased and directly depended on plasma exposure time. Even after 60 s plasma exposure, a highly significant decline ($p<0.01$) in colony survival fraction up to 0.2, 0.52 and 0.68, respectively was detected in T98G, KB and HEK293 cells. SNU80 cells were not significantly affected ($p>0.05$) by a plasma treatment of up to 60 s. A significant, drastic decline in the survival fraction ($p<0.01$) was observed after a 120 s plasma treatment with a survival fraction of 0.04, 0.4 and 0.321, respectively in T98G, KB and HEK293 cells. After a 240 s plasma exposure, we found significant inhibition ($p<$ 0.05) in the clonogenic capacity of all four cell lines compared to untreated controls. This shows that these treatments exert an inhibitory effect on the colony formation capabilities of all cancer cells at all doses and also indicates that plasma treatment causes cell death as reflected by decreased clonogenicity.

(a)

(b)

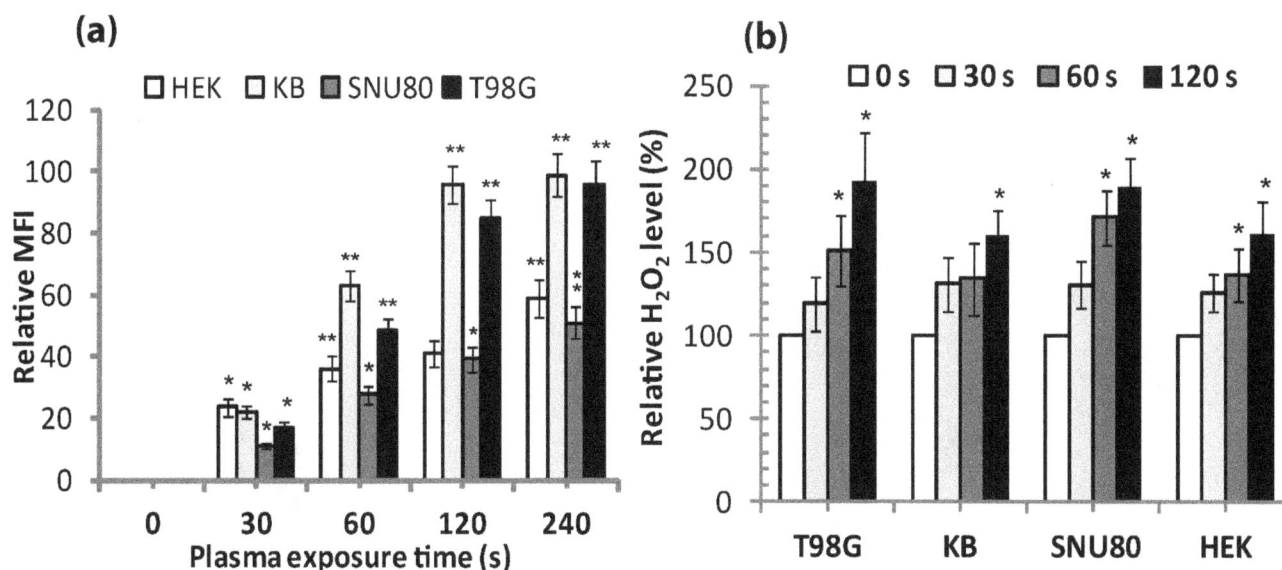

Figure 5. Induction of ROS in APNT DBD plasma treated cells. (a) T98G, SNU80, KB and HEK293 cells were treated with the oxidation-sensitive fluorescent probe 2,7-dichlorodihydrofluorescein diacetate (H_2DCFDA) for detection of total ROS, (b) detection of H_2O_2 level (in µM) in cells observed at 24 h after exposure. In (a) and (b), all fluorescence levels were expressed as fluorescence intensity (FL intensity). Results from four independent experiments are shown as mean ± SD, and Student's t-test was performed to controls (*$p<0.05$ and **$p<0.01$).

Survival Rate

We quantitated cells by counting cell number using a hemocytometer during 3 day incubation. The ratio of test cell and control cells is shown in **Figure 2**. The cell number ratio was significantly decreased after 120 s plasma exposure for cancer cells and for non-malignant HEK293 cells. The results of this study clearly demonstrate that cell number is reduced in an exposure time dependent fashion. DBD plasma had inhibitory effect on the growth of T98G cells in an exposure time-dependent manner. Plasma treatment for 30, 60, 120 and 240 s decreased the T98G cell count to 73% ($p<0.05$), 66% ($p<0.05$), 58% ($p<0.01$), and 20% ($p<0.01$), respectively, at 72 h incubation [25]. The SNU80, KB and HEK293 cells count were significantly decreased ($p<0.05$) at a 240 s plasma-dose, but not after a 30 s plasma dose ($p>0.05$). In addition, the KB cell survival percentage was significantly decreased ($p<0.05$) after a 120 s plasma dose at all incubation time, which could be predicted from increased intracellular and extracellular ROS levels or the failure of cell antioxidant systems. Interpreted together, the results from the MTT, clonogenic, survival rate assays clearly demonstrate that the plasma sensitized cells in a dose dependent manner, which may be due to an increase in mitotic (linked to cytogenetic damage) or interphase (apoptosis) death.

The cell morphology study also revealed that the morphology of the cells was affected by APNT plasma treatments. **Figure 3** shows the differences in morphology of the treated T98G and HEK293 cells, compared to the untreated controls. **Figure 4** shows the frequency distribution of cell size (length and width) and images of cells analyzed using the JEOL PC/SEM software system. We found significant ($p<0.01$) size variability between control and plasma treated (≥60 s dose) T98G cancer cell populations. HEK293 cells only showed significant ($p<0.01$) morphological variations at a dose of 120 s. The membrane of T98G cells membrane started blebbing and leaking inner components after a ≥60 s dose of plasma exposure and these cells ultimately died via apoptotic pathway.

Intracellular ROS accumulation

Figure 5 shows ROS measurement in plasma exposed cells. Enhanced levels of different ROS lead to increased oxidative stress that result in DNA, lipid, and protein destruction in cells. Therefore, to determine if plasma enhanced oxidative stress, cells were labeled with the oxidation sensitive probe (H_2DCFDA) that is capable of being oxidized to fluorescent product (DCF) by reactive oxygen species and other peroxide radicals generated from ROS. **Figure 5a** displays total ROS production in cells at 24 h following plasma treatment. Plasma increased the DCF fluorescence by 1.2 to 2-folds in cells compared with untreated controls and this increase depended on plasma dose. However, a less significant change ($p>0.01$) in ROS production was observed in cells treated with a low plasma dose (30 s) at 24 h. A 240 s plasma dose significantly increased ($p<0.01$) the intracellular ROS by 2-fold in T98G and KB cell lines. Plasma treatment of SNU80 and HEK293 cells significantly increased ($p<0.05$) the intracellular ROS by 1.2 to1.5-folds.

We further examine the levels of intracellular H_2O_2 in the cells as described earlier (ROS detection assays section). It is renowned that H_2O_2 is a cytotoxic agent whose levels must be minimized by the action of antioxidant defense system. **Figure 5b** shows increased fluorescence for H_2O_2 observed in cells in plasma dose-dependent fashion at 24 h. A significantly higher intracellular level of H_2O_2 was observed in plasma treated cancer cells than non-malignant HEK293 cells. A 60 s plasma dose significantly increased ($p<0.05$) the resorufin fluorescence in T98G, SNU80 and HEK293 cells. However, A 120 s plasma dose significantly increased ($p<0.05$) the resorufin fluorescence by 1.6 to 2-fold at in all cells compared to untreated controls.

Instabilities occurs in cellular redox status by APNT plasma

GSH has a central role in maintaining the cellular redox homeostasis. Some previous reports have also claimed that many diseases such as cancer, neurodegenerative disorders, cystic fibrosis, and Crohn's disease occurs if there is a disturbance in

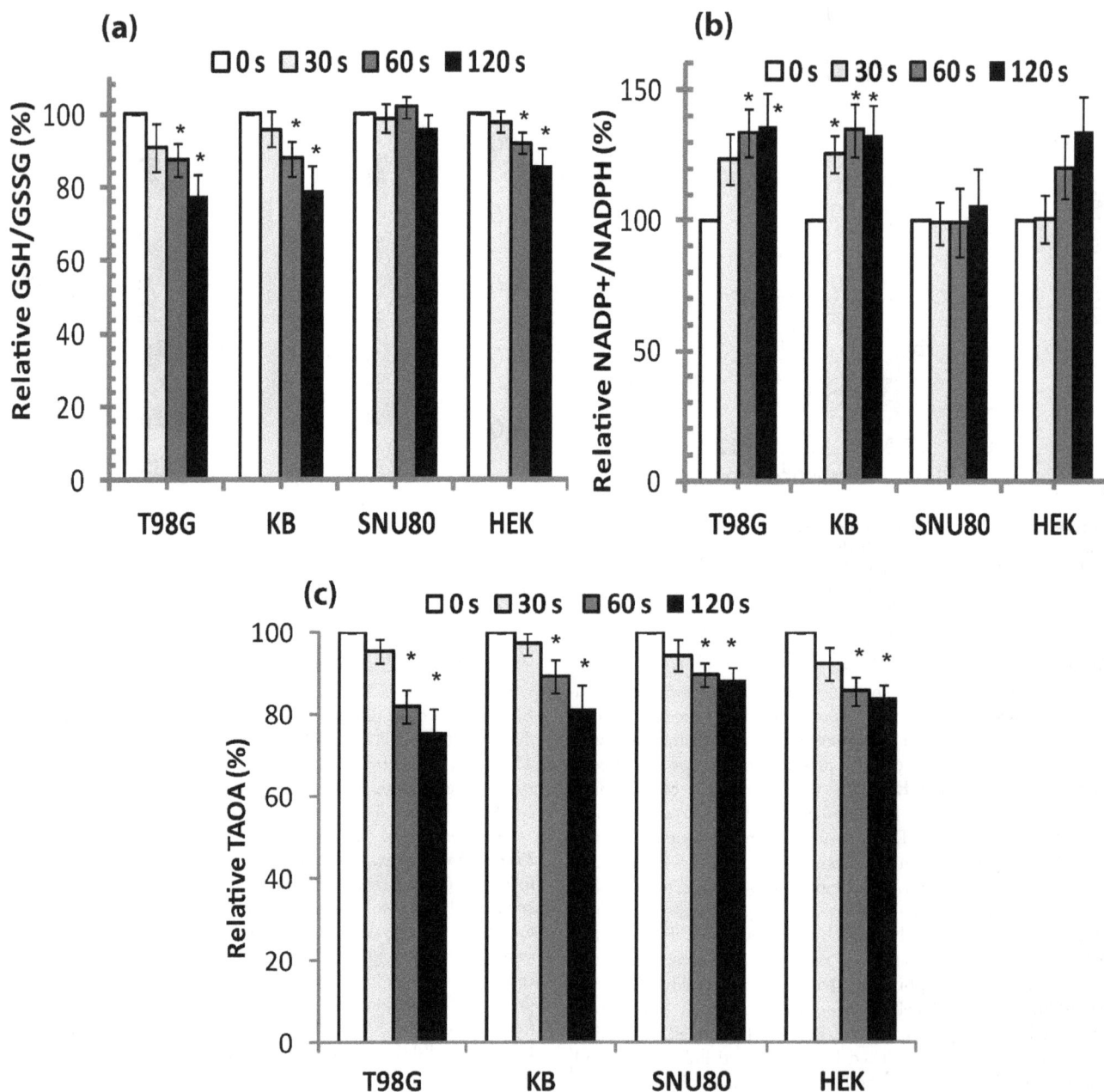

Figure 6. Changes in redox indicators due to APNT plasma exposure. (a) Detection of GSH/GSSG levels in cells. (b) NADP$^+$/NADPH ratio in cells levels of NADP$^+$ and NADPH were measured using a standard prepared for NADPH. The ratio of NADP$^+$ and NADPH was plotted as a function of treatment time. (c) Total antioxidant activity (TAOA) was assessed in APNT plasma treated cells. Results from four independent experiments are shown as mean $\pm$ SD, and Student's t-test was performed to controls (*$p < 0.05$ and **$p < 0.01$).

GSH [32,33]. We assessed total cellular glutathione (both reduced and oxidized) in our study. **Figure 6a** shows a significant decrease ($p < 0.05$) in GSH and an increase in GSSG in T98G, HEK293 and KB cells, 24 h after 60 and 120 s plasma treatments. In contrast, changes in GSH and GSSG levels inside plasma treated SNU80 cells were not significant ($p > 0.05$) 24 h after treatment. A reduction in GSH and a simultaneous increase in GSSG (which was more prominent after 60 and 120 s plasma treatments) suggest a compromised redox state in both T98G and KB cells. The ratio of GSH/GSSG, a measurement of redox status, was found to decrease significantly ($p < 0.05$) in T98G and KB cells after 60 and 120 s plasma treatments (**Figure 6a**). The ratio of GSH/GSSG

in HEK293 cells also decreased significantly only after a 60 s plasma dose. Plasma significantly decreased the level of GSH/GSSG by 22.2%, 21% and 14% in T98G, KB and HEK293 cells, respectively, after a 120 s plasma treatment.

It has been reported that NADPH is vital for GSH synthesis. Therefore, we conducted a study to determine the ratio of NADP$^+$/NADPH in cells. The ability of cell to neutralize H$_2$O$_2$ by glutathione peroxidase (GPx) depends on the regeneration of NADPH from NADP$^+$ by the GSH synthesis pathway. We found that treatment with plasma may interrupt regeneration of NADPH in cells. After 24 h, the level of NADPH was reduced significantly (>25%, $p < 0.05$) in T98G and KB cells treated with plasma for 60

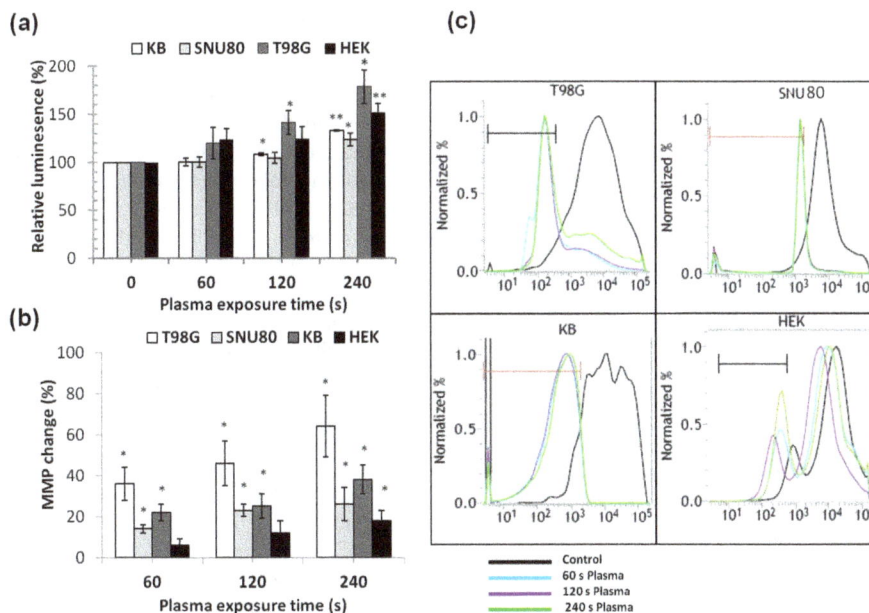

Figure 7. Involvement of caspase activation and loss in mitochondrial membrane potential during apoptosis. (a) APNT plasma induced activation of caspase-3/7 of human glioblastoma (T98G) and a non-malignant (HEK293) cell lines. (b) APNT plasma affects mitochondrial membrane potential of T98G, SNU80, KB and HEK293 cells. (c) Flow cytometric plot of mitochondrial membrane potential in cells, using Mito Flow rhodamine dye. Results from four independent experiments are shown as mean ± SD, and Student's t-test was performed to controls (*$p < 0.05$ and **$p < 0.01$).

and 120 s plasma compared with untreated controls. In contrast, plasma treatment did not significant change this ($p > 0.05$) in SNU80 and HEK293 cells.

Figure 6b shows that the ratio of NADP$^+$/NADPH (an indicator of redox status) was increased up to 20%–35% in T98G, KB and HEK293 cell lines compared with untreated control cells 24 h after plasma treatment. Conversely, a 30 s plasma dose did not significantly change the NADP$^+$/NADPH ratio ($p > 0.05$) in all four cells when compared to untreated controls. No significant change in the ratio of NADP$^+$/NADPH

($p > 0.05$) was observed in HEK293 and SNU80 cells at up to a 120 s plasma dose. The increase in the NADP$^+$/NADPH ratio clearly directs the compromised regeneration of NADPH from NADP$^+$ under these experimental conditions. For this reason, failure to regenerate NADPH during treatment could lead to enhanced intracellular oxidative stress levels.

Reduction in the antioxidant system activity in cells

Because an increase in ROS has been demonstrated in plasma exposed cells, we measured the total antioxidant events in cells. Oxidative stress ensues when antioxidant mechanisms are overwhelmed by generation of excessive reactive oxygen and nitrogen species that damage membrane lipids, proteins and nucleic acids [34]. We observed that the plasma resulted in a decrease in antioxidant activity in cells. **Figure 6c** shows a significant decrease in the endogenous antioxidant activity level ($p < 0.05$) in all four type of cells after 60 and 120 s plasma doses compared with the untreated control group. However, there was no significant decrease ($p > 0.05$) in antioxidant activity in cells after 30 s plasma treatment. Plasma decreased the antioxidants activity by 24.5%, 18.6%, 11.7% and 15.8% of T98G, KB, SNU80 and HEK293 cells, respectively, after 120 s plasma dose. Data from this experiment shows that these cells used their antioxidant system (both enzymatic and non-enzymatic) to reduce the load of ROS induced by the plasma treatment.

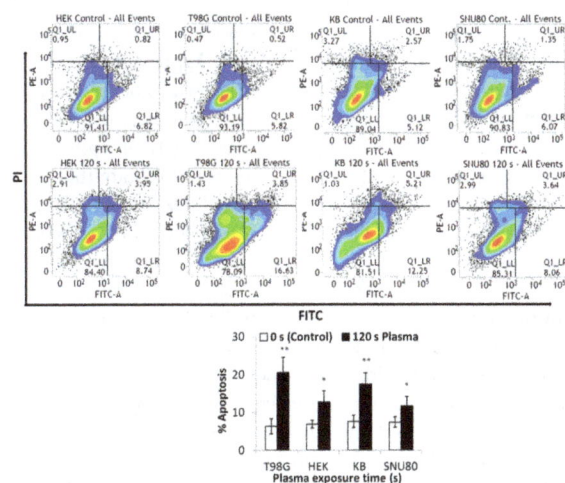

Figure 8. Analysis of APNT induced cell death (apoptosis). Flow cytometry data of Annexin V and PI staining of human T98G, SNU80, KB and HEK293 cells after plasma treatment. Apoptosis of each cell was evaluated after 24 h incubation post plasma treatment for 120 s. Results from four independent experiments are shown as mean ± SD, and Student's t-test was performed to controls (*$p < 0.05$ and **$p < 0.01$).

Caspase activation and loss of mitochondrial membrane potential are involved in APNT plasma-induced apoptosis

To evaluate the influence of plasma on caspase 3/7 activity, we used three cancer cell lines (T98G, KB, and SNU80) and a non-malignant cell line (HEK293). Caspase-3 and -7 play an important role in the cleavage of cellular constituents during apoptosis. **Figure 7a** shows the caspase activity in both control and plasma treated cells at 24 h post-treatment. In untreated cells, the detected level of caspase activity was related to the fraction of apoptotic cells generated in the normally growing population due to natural aging. In treated cells, caspase-3/7 activity increased over basal levels. Plasma exposure for 120 and 240 s significantly increased ($p<0.05$) the caspase -3/7 activity by 41.3% and 78.6%, respectively, in T98G cells. Plasma exposure for up to 60 s did not produce a significant ($p>0.05$) change in caspase-3/7 activity in all four cell lines. KB cells was also affected significantly ($p<0.05$) at 120 s plasma exposure, with an increase of caspase activity by 8.2%. Plasma exposure for 240 s significantly increased ($p<0.01$) the caspase activity in HEK293 cells by 52%. SNU80 plasma treated cells showed significant increases ($p<0.05$) in caspase activity only at a 240 s plasma dose. We observed a significant change ($p<0.05$) in caspase activity in all cell lines at a higher dose (240 s) of plasma.

Because mitochondria are important for both intrinsic and extrinsic apoptosis pathways, mitochondrial membrane potential was measured. An enhanced ROS level often induces a membrane permeability change, which is an early event in cell apoptosis [35]. APNT plasma generates different reactive oxygen species that can disrupt membrane polarization. Following 24 h of incubation with similar treatment conditions, cells were incubated with Mito Flow dye for 30 min and then analyzed by FACS analysis. When gated cells were examined for dye uptake, those that had been treated with plasma clearly differed in fluorescence compared with the untreated control. **Figures 7b and 7c** demonstrate the MMP change in cancer and normal cells. The MMP of 60 and 120 s plasma treated T98G, KB and SNU80 cells show significant changes ($p<0.05$) when compared to HEK293 cells. Plasma treatment for 60, 120 and 240 s induced a significant MMP change ($p<0.05$) of 36%, 46% and 64%, respectively, in T98G glioma cells (**Figure 7b**). Plasma treatment for 240 s induced significant change ($p<0.05$) in MMP of all four type of cell lines. However, plasma treatment for up to 120 s did not result in a significant ($p>0.05$) change in the MMP of HEK293 cells. **Figure 7c** shows a band shift phenomenon in cells observed after 24 h incubation. Plasma exposure resulted in membrane depolarization within the cells in a dose-dependent fashion, which is an indication of early/late apoptosis.

Apoptosis

Programmed cell death (apoptosis) is well known in multicellular organisms and, can be recognized by morphologic characteristics (such as membrane blebbing), cellular size and volume reduction, caspase activity, chromatin shortening etc [36]. Because cell survival was dramatically reduced in all cells and to confirm the observed MMP and caspase assay results, we next evaluated whether the plasma induced apoptosis in cells. Cell populations were evaluated by flow cytometer after staining with fluorescein isothiocyanate (FITC)-labeled annexin V (green fluorescence) and with the the nonvital dye propidium iodide (PI). This allowed discrimination between intact, early apoptotic, late apoptotic and necrotic cells [37]. **Figure 8** shows a flow cytogram with four quadrants for measuring intact, early apoptotic, late apoptotic and necrotic cells. A plasma exposure of 120 s induced significant ($p<0.05$) apoptosis in T98G, KB, SNU80 and HEK293 cells when measured 24 h after treatment. This effect was more prominent ($p<0.01$) in T98G and KB cancer cells. As shown in **Figure 8**, plasma treatment increased annexin V-FITC binding 1 to 2-fold; from the negative control of 6.34%–7.42% to 20.5% ($p<0.01$), 12.69% ($p<0.05$), 17.6% ($p<0.01$), and 11.7% ($p<0.05$) in T98G, HEK293, KB and SNU80 cell line, respectively. These results can be correlated with the results from the caspase activation and mitochondrial membrane potential change experiments.

Discussion

Numerous reports have revealed that atmospheric non-thermal plasmas can encourage cancer cell apoptosis in a dose-dependent manner and that this can be related to DNA damage resulting from the generation of ROS [38]. Plasma action seems to be lethal for cells and modulation in the cellular metabolic activity was observed (by MTT assay) up to 24 h post-treatment. This effect was found to be dependent on the plasma dose. Current results of anticancer studies clearly demonstrate that APNT DBD plasma treatments sensitize cells by increasing cell death and failure of endogenous antioxidant system to counteract ROS burden in treated cells. In this study, we have demonstrated through the evaluation of metabolic viability, clonogenicity, cell count, ROS, and H_2O_2 that plasma exposure has a significant inhibitory effect on cancer cell growth, and it generated a large amount of oxidative stress in cells. Our results confirmed that intracellular ROS level was increased significantly by APNT DBD plasma treatment. The increased oxidative stress level and a resulting loss of MMP were reported to be distinctive phenomena during mitochondria-dependent apoptosis [39]. MMP loss also induces apoptosis by causing the release of cytochrome c from the inner mitochondrial space to the cytosol [40]. It has also been reported that cytochrome c release can activate initiator caspase-9, which eventually activates executioner caspases like caspase-3 via cleavage [41]. We conclude that APNT plasma effectively promotes activation of caspases and a loss of MMP that results in the decline of cellular viability (**Figure 7**).

We also found that plasma treatment showed no or less effect on SNU80 cells. This anomalous behavior is due to the anaplastic condition of these cells. SNU80 is an anaplastic thyroid carcinoma (ATC) cell line obtained from Korean thyroid carcinoma patients, which divides rapidly and has little or no resemblance to normal cells. The mechanism of its carcinogenesis is unclear, and they are highly resistant to chemotherapy and radiotherapy. No effective therapeutic regimen has been identified for ATC. We are now focusing on novel therapeutic approaches to treat SNU80-like cell lines using various strategies with plasma treatment and will report on these experiments in the future.

Based upon our findings in this report, decreased viability of cancer cells strongly suggests that oxidative stress induced by plasma resulted in the observed cell death (**Figure 1, 2**). GSH, one of the major redox state regulators, is known to be involved in elimination of ROS [42]. To test this hypothesis, we analyzed antioxidant markers and showed decreased levels of both GSH and NADPH. That revealed a compromised redox state that may be responsible for the cell death (**Figures 5, 6** and **8**). Results from the current study emphasize the need for a deeper understanding of the role that plasma induced reactive oxygen species play in tumor suppression. Our study suggests that the failure of the antioxidant system to neutralize ROS is responsible for ROS elevation and cell death. Excess ROS and H_2O_2, which are not reduced by antioxidant enzymes inside the mitochondria, may also damage mitochondria and can cross the mitochondrial membrane into the cytosol and cause cell damage. These findings offer insight into the mechanism underlying APNT plasma-mediated apoptosis for future studies that may provide insight into methods to develop better plasma-therapeutic approaches that are selective for various types of cancer cells.

Author Contributions

Conceived and designed the experiments: NKK NK EHC. Performed the experiments: NKK NK DP. Analyzed the data: NKK NK EHC.

Contributed reagents/materials/analysis tools: NKK EHC. Contributed to the writing of the manuscript: NKK NK EHC.

References

1. World Health Organization (2008) The Global Burden of Disease: 2004 Update. Geneva: World Health Organization.

2. Wong HL, Bendayan R, Rauth AM, Li Y, Wu XY (2007) Chemotherapy with anticancer drugs encapsulated in solid lipid nanoparticles. Adv Drug Delivery Rev 59: 491–504.

3. Chiaviello A, Postiglione I, Palumbo G (2011) Targets and mechanisms of photodynamic therapy in lung cancer cells: a brief overview. Cancers 3: 1014–1041.

4. Baskar R, Lee KA, Yeo R, Yeoh KW (2012) Cancer and radiation therapy: current advances and future directions. Int J Med Sci 9: 193–199.

5. Depan D, Shah J, Misra RDK (2011) Controlled release of drug from folate-decorated and graphene mediated drug delivery system: Synthesis, loading efficiency, and drug release response. Mater. Sci. Eng. C 31: 1305–1312.

6. Ninomiya K, Ishijima T, Imamura M, Yamahara T, Enomoto H, et al. (2013) Evaluation of extra- and intracellular OH radical generation, cancer cell injury, and apoptosis induced by a non-thermal atmospheric-pressure plasma jet. J Phys D: Appl Phys 46: 425401.

7. Panngom K, Baik KY, Nam MK, Han JH, Rhim H, et al. (2013) Preferential killing of human lung cancer cell lines with mitochondrial dysfunction by nonthermal dielectric barrier discharge plasma. Cell Death Dis 4: e642.

8. Keidar M, Walk R, Shashurin A, Srinivasan P, Sandler A, et al. (2011) Cold plasma selectivity and the possibility of a paradigm shift in cancer therapy. Br J Cancer 105: 1295–1301.

9. Volotskova O, Hawley TS, Stepp MA, Keidar M (2012) Targeting the cancer cell cycle by cold atmospheric plasma. Sci Rep 2: 636.

10. Arndt S, Wacker E, Li YF, Shimizu T, Thomas HM, et al. (2013) Cold atmospheric plasma, a new strategy to induce senescence in melanoma cells. Exp Dermatol 4: 284–289.

11. Fridman G, Shereshevsky A, Jost MM, Brooks AD, Fridman A, et al. (2007) Floating electrode dielectric barrier discharge plasma in air promoting apoptotic behavior in melanoma skin cancer cell lines. Plasma Chem Plasma P 27: 163–176.

12. Bundscherer L, Wende K, Ottmüller K, Barton A, Schmidt A, et al. (2013) Impact of non-thermal plasma treatment on MAPK signaling pathways of human immune cell lines. Immunobiology, 218: 1248–1255.

13. Chang JW, Kang SU, Shin YS, Kim KI, Seo SJ, et al. (2014) Non-thermal atmospheric pressure plasma inhibits thyroid papillary cancer cell invasion via cytoskeletal modulation, altered MMP-2/-9/uPA Activity. PLoS ONE 9: e92198.

14. Wang M, Holmes B, Cheng X, Zhu W, Keidar M, et al. (2013) Cold atmospheric plasma for selectively ablating metastatic breast cancer cells. PLoS ONE 8: e73741.

15. Utsumi F, Kajiyama H, Nakamura K, Tanaka H, Mizuno M, et al. (2013) Effect of indirect nonequilibrium atmospheric pressure plasma on anti-proliferative activity against chronic chemo-resistant ovarian cancer Cells *In Vitro* and *In Vivo*. PLoS ONE 8: e81576.

16. Kang SU, Cho JH, Chang JW, Shin YS, Kim KI, et al. (2014) Nonthermal plasma induces head and neck cancer cell death: the potential involvement of mitogen-activated protein kinase-dependent mitochondrial reactive oxygen species. Cell Death Dis 2: e1056.

17. Ma Y, Ha CS, Hwang SW, Lee HJ, Kim GC, et al. (2014) Non-thermal atmospheric pressure plasma preferentially induces apoptosis in p53-mutated cancer cells by activating ROS stress-response pathways. PLoS ONE 9(4): e91947.

18. Trachootham D, Alexandre J, Huang P (2009) Targeting cancer cells by ROS-mediated mechanisms: a radical therapeutic approach? Nat Rev Drug Discov 7: 579–591.

19. Fuchs-Tarlovsky V (2013) Role of antioxidants in cancer therapy. Nutrition 1: 15–21.

20. Scarlett JL, Sheard PW, Hughes G, Ledgerwood EC, Ku HH, et al. (2000) Changes in mitochondrial membrane potential during staurosporine-induced apoptosis in Jurkat cells. FEBS Lett 475: 267–72.

21. Zamzami N, Marchetti P, Castedo M, Decaudin D, Macho A, et al. (1995) Sequential reduction of mitochondrial transmembrane potential and generation of reactive oxygen species in early programmed cell death. J Exp Med 2: 367–377.

22. Ma R, Feng H, Li F, Liang Y, Zhang Q, et al. (2012) An evaluation of anti-oxidative protection for cells against atmospheric pressure cold plasma treatment. Appl PhysLett 100: 123701.

23. Kaushik NK, Attri P, Kaushik N, Choi EH (2013) A preliminary study of the effect of DBD plasma and osmolytes on T98G brain cancer and HEK non-malignant cells. Molecules 18: 4917–4928.

24. Mosmann T (1983) Rapid colorimetric assay for cellular growth and survival: application to proliferation and cytotoxicity assays. J Immunol Methods 65: 55–63.

25. Kaushik NK, Uhm HS, Choi EH (2012) Micronucleus formation induced by dielectric barrier discharge plasma exposure in brain cancer cells. Appl Phys Lett 100: 084102.

26. Kaushik NK, Kim YH, Han YG, Choi EH (2013) Effect of jet plasma on T98G human brain cancer cells. Curr Appl Phys 13: 176–180.

27. Passey S, Pellegrin S, Mellor H (2007) Scanning electron microscopy of cell surface morphology. Bonifacino JS, Dasso M, Harford JB, Lippincott-Schwartz J, Yamada KM, editors. Current Protocols in Cell Biology. New Jersey: John Wiley and Sons, Inc. 37: 4.17.1–13.

28. Finkel T, Holbrook NJ (2000) Oxidants, oxidative stress and the biology of ageing. Nature 408: 239–247.

29. Vickers CJ, González-Páez GE, Wolan DW (2013) Selective Detection of caspase-3 versus caspase-7 using activity-based probes with key unnatural amino acids. ACS Chem Biol 8: 1558–1566.

30. Liu C, Yin L, Chen J, Chen J (2014) The apoptotic effect of shikonin on human papillary thyroid carcinoma cells through mitochondrial pathway. Tumor Biol 35: 1791–1798.

31. Brown JM, Attardi LD (2005) The role of apoptosis in cancer development and treatment response. Nat Rev Cancer 5: 231–237.

32. Valko M, Leibfritz D, Moncol J, Cronin MT, Mazur M, et al. (2007) Free radicals and antioxidants in normal physiological functions and human disease. Int J Biochem Cell Biol 39: 44–84.

33. Townsend DM, Tew KD, Tapiero H (2003) The importance of glutathione in human disease. Biomed Pharmacother 57: 145–155.

34. Halliwell B, Gutteridge JM (1995) The definition and measurement of antioxidants in biological systems. Free Radical Biol Med 18: 125–126.

35. Green DR, Reed JC (1998) Mitochondria and Apoptosis. Science 281: 1309–1312.

36. Kroemer G, Galluzzi L, Vandenabeele P, Abrams J, Alnemri ES, et al. (2009) Classification of cell death: recommendations of the nomenclature committee on cell death. Cell Death Differ 16: 3–11.

37. Vermes I, Haanen C, Steffens-Nakken H, Reutelingsperger C (1995) A novel assay for apoptosis. Flow cytometric detection of phosphatidylserine expression on early apoptotic cells using fluorescein labeled annexin V. J Immunol Meth 184: 39–51.

38. Vandamme M, Robert E, Lerondel S, Sarron V, Ries D, et al. (2012) ROS implication in a new antitumor strategy based on non-thermal plasma. Int J Cancer. 130: 2185–94.

39. Vaux DL, Korsmeyer SJ (1999) Cell death in development. Cell 96: 245–254.

40. Van Loo G, Saelens X, Van Gurp M, MacFarlane M, Martin SJ, et al. (2002) The role of mitochondrial factors in apoptosis: a Russian roulette with more than one bullet. Cell Death Differ 9: 1031–1042.

41. Green DR (2005) Apoptotic pathways: ten minutes to dead. Cell 121: 671–674.

42. Armstrong JS, Steinauer KK, Hornung B, Irish JM, Lecane P, et al. (2002) Role of glutathione depletion and reactive oxygen species generation in apoptotic signaling in a human B lymphoma cell line. Cell Death Differ 9: 252–63.

3

L-Carnitine Protects against Carboplatin-Mediated Renal Injury: AMPK- and PPARα-Dependent Inactivation of NFAT3

Yuh-Mou Sue[1,9], Hsiu-Chu Chou[2,9], Chih-Cheng Chang[3,4], Nian-Jie Yang[3,4], Ying Chou[3,4], Shu-Hui Juan[3,4]*

1 Department of Nephrology, Taipei Medical University-Wan Fang Hospital, Taipei, Taiwan, **2** Department of Anatomy, School of Medicine, College of Medicine, Taipei Medical University, Taipei, Taiwan, **3** Graduate Institute of Medical Sciences, Taipei Medical University, Taipei, Taiwan, **4** Department of Physiology, School of Medicine, College of Medicine, Taipei Medical University, Taipei, Taiwan

Abstract

We have previously shown that carboplatin induces inflammation and apoptosis in renal tubular cells (RTCs) through the activation of the nuclear factor of activated T cells-3 (NFAT3) protein by reactive oxygen species (ROS), and that the ROS-mediated activation of NFAT3 is prevented by N-acetyl cysteine and heme oxygenase-1 treatment. In the current study, we investigated the underlying molecular mechanisms of the protective effect of L-carnitine on carboplatin-mediated renal injury. Balb/c mice and RTCs were used as model systems. Carboplatin-induced apoptosis in RTCs was examined using terminal-deoxynucleotidyl-transferase-mediated dUTP nick end labeling. We evaluated the effects of the overexpression of the peroxisome-proliferator-activated receptor alpha (PPARα) protein, the knockdown of PPARα gene, and the blockade of AMPK activation and PPARα to investigate the underlying mechanisms of the protective effect of L-carnitine on carboplatin-mediated renal injury. Carboplatin reduced the nuclear translocation, phosphorylation, and peroxisome proliferator responsive element transactivational activity of PPARα. These carboplatin-mediated effects were prevented by L-carnitine through a mechanism dependent on AMPK phosphorylation and subsequent PPARα activation. The activation of PPARα induced cyclooxygenase 2 (COX-2) and prostacyclin (*PGI2*) *synthase* expression that formed a positive feedback loop to further activate PPARα. The coimmunoprecipitation of the nuclear factor (NF) κB proteins increased following the induction of PPARα by L-carnitine, which reduced NFκB transactivational activity and cytokine expression. The in vivo study showed that the inactivation of AMPK suppressed the protective effect of L-carnitine in carboplatin-treated mice, indicating that AMPK phosphorylation is required for PPARα activation in the L-carnitine-mediated protection of RTC apoptosis caused by carboplatin. The results of our study provide molecular evidence that L-carnitine prevents carboplatin-mediated apoptosis through AMPK-mediated PPARα activation.

Editor: Shree Ram Singh, National Cancer Institute, United States of America

Funding: This study was funded by grants from Wan Fang Hospital and Taipei Medical University (100TMU-WFH-06 and 101TMU-WFH-09) and the National Science Council, Taiwan (NSC100-2320-B-038-016-MY2(1-2) and NSC102-2320-B-038-030-MY3(1-3)). The funders had no role in study design, data collection and analysis, decision to publish, or preparation of the manuscript.

Competing Interests: The authors have declared that no competing interests exist.

* Email: juansh@tmu.edu.tw

⑨ These authors contributed equally to this work.

Introduction

The quaternary ammonium compound, L-carnitine (L-trimethyl-3-hydroxy-ammoniabutanoate), is synthesized in cells from lysine and methionine precursors [1], and is required for the transport of fatty acids from the cytosol into the mitochondria during lipid catabolism. It has been sold as the nutritional supplement vitamin Bt, and has been used as a growth factor for mealworms. In cells, L-carnitine induces antioxidant proteins, including endothelial nitric oxide synthase, heme oxygenase-1 (HO-1), and super oxide dismutase (SOD) [2], and protects against lipid peroxidation in phospholipid membranes and oxidative stress in cardiomyocytes and endothelial cells [3]. In addition, L-carnitine protects renal tubular cells (RTCs) from gentamicin-induced apoptosis through prostaglandin (PG) I2-mediated acti-

vation of the peroxisome-proliferator-activated receptor (PPAR) α protein [4].

The second-generation platinum-containing anticancer drug, carboplatin (*cis*-diammine-1,1-cyclobutanedicarboxylate platinum II), is used to treat lung, ovarian, and head and neck cancers [5]. The antitumor action of carboplatin is mediated by the alkylation of DNA, which can lead to cell death in tumor cells. Carboplatin is more water-soluble and has fewer adverse effects than its analog, cisplatin, and has equivalent DNA-damaging activity as cisplatin at similar toxic doses [6]. Cisplatin is a potent chemotherapy agent used to treat various malignant cancers, but the doses are limited due to its detrimental effects in renal tubular function and a decline in the glomerular filtration rate [7,8]. Because carboplatin has fewer toxic adverse effects than cisplatin, increased doses of carboplatin are commonly used in the clinic in order to achieve optimal antitumor effects. However, the predominant dose-limiting toxicities of carboplatin are bone marrow suppression and ototoxicity caused by free-radical oxidative injury [9]. Using both gain- and loss-of-function strategies, we previously showed that the activation of the transcription factor, nuclear factor of activated T cells-3 (NFAT3), induces RTC apoptosis, and that NFAT3-mediated apoptosis in RTCs is blocked by HO-1 gene therapy and N-acetyl cysteine (NAC) treatment [10]. The antioxidant activities of L-carnitine warrant further investigation to determine whether it might provide protection against carboplatin-mediated renal injury.

The ligand-activated transcription factors, PPARα and PPARγ, form a heterodimer with the retinoid X receptor, and bind to peroxisome proliferator responsive elements (PPREs) in target genes [11,12]. The activities of PPARα and PPARγ are also regulated by phosphorylation [13,14]. We have previously shown that the activation of PPARα by adenosine-monophosphate-activated protein kinase (AMPK) is dependent on the adiponectin-induced activation of HO-1 and cyclooxygenase (COX)-2 [15,16]. In addition, PGI2 might be a ligand of PPARα and PPARδ [17,18]. Garrelds et al. (1994) reported that PGI2 expression significantly increased in rat peritoneal leukocytes after a short-term (4 d) L-carnitine treatment [18]. Recent studies have also revealed that the L-carnitine-induced expression of PGI2 can induce the vasodilation of subcutaneous arteries in humans [19,20]. Therefore, investigations of the mechanism by which the interwoven relations of PGI2 and PPARα are involved in protection of L-carnitine in carboplatin-challenged RTCs are warranted.

The activation of PPARα has been shown to play a beneficial role in preventing various diseases by inhibiting the NFκB-induced expression of inflammatory mediators, including vascular cell adhesion molecule-1, interleukin (IL)-6, endothelin-1, and tissue factor, in a broad range of cells, including endothelial cells, smooth muscle cells, and macrophages [21–23]. The activation of PPARα by fibrates inhibits the IL-1-induced secretion of IL-6 in human aortic smooth muscle cells [24]. By contrast, the aorta of PPARα-null mice undergoes an exacerbated response to lipopolysaccharide, demonstrating that the anti-inflammatory effect of fibrates on the vascular wall requires PPARα activation in vivo [25]. In addition, PPARα ligands also regulate hepatic inflammation, and fibrates reduce serum levels of acute-phase proteins, such as C-reactive protein (CRP) and fibrinogen [26]. Evidence from clinical trials also supports the role of PPARα ligands in suppressing inflammation. Fenofibrate treatment reduces the plasma concentrations of fibrinogen, IL-6, CRP, interferon-γ, and tumor necrosis factor (TNF)-α in patients with hyperlipidemia and atherosclerosis [24,27]. Additionally, PGI2 and PPARα have been shown to protect against ischemia-reperfusion injury through the suppression of inflammation [28].

In our current study, we evaluated the protective effects of L-carnitine on carboplatin-mediated renal injury in vitro and in vivo. We also investigated the mechanisms underlying the PPARα-dependent suppression of carboplatin-mediated NFAT3 activation and inflammation.

Materials and Methods

Cell culture and reagents

We used the rat renal proximal tubular epithelial cell line, NRK-52E, for the in vitro RTC model in our study. NRK-52E epithelial cell lines are composed of differentiated, anchorage-dependent, nontumorigeic cells that undergo density-dependent inhibition of proliferation [29]. The widely used NRK-52E rat

kidney cell lines have been characterized with the morphological and kinetic properties of kidney tubule epithelial cells [30]. The NRK-52E cells were purchased from the Bioresource Collection and Research Center (Hsinchu, Taiwan) and were cultured in Dulbecco's modified Eagle medium (DMEM) supplemented with 10% fetal bovine serum (FBS) and an antibiotic and antifungal solution. The NRK-52E cell monolayers were grown until confluence was reached. The DMEM, FBS, and other tissue culture reagents were obtained from Life Technologies (Gaithersburg, MD, USA). The L-carnitine was purchased from Sigma-Tau (Rome, Italy). All of the other chemicals were of reagent grade, and were purchased from Sigma-Aldrich (St. Louis, MO, USA).

Plasmid construction and expression analysis of the PPAR and the NFκB enhancers

A pBV-luc plasmid containing the prototypic sequence of the PPAR response element, 5′-AGGTCAAAGGTCA-3′, from the acyl-CoA oxidase gene promoter was provided by Dr. Vogelstein of Johns Hopkins University [31]. The NFκB-luciferase reporter plasmid, which contains the multimeric NFκB regulatory element, (TGGGGACTTTCCGC)₅, was purchased from Stratagene (La Jolla, CA, USA). The RTC cells were transfected with these vectors using the LipofectAMINE 2000 (Invitrogen, Carlsbad, CA, USA) transfection reagent. After transfection for 4 h, the medium was replaced with complete medium, and the transfected cells were incubated for an additional 20 h. The transfected cells were treated with carboplatin for 2 h. The luciferase activity of the cell lysates were recorded using the Dual Luciferase Assay Kit (Promega, Madison, WI, USA) in a TD-20/20 luminometer (Turner Designs, Sunnyvale, CA, USA). The luciferase activity of

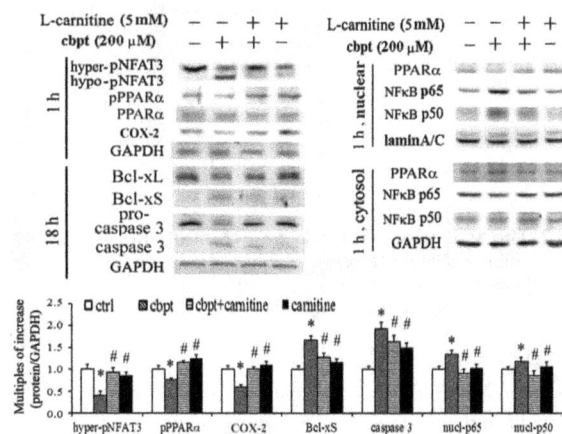

Figure 1. The protective effect of L-carnitine on the cytotoxicity of carboplatin in RTCs. The RTCs were pretreated with L-carnitine (5 mM) for 24 h, followed by treatment using 200 µM carboplatin for 1 h or 18 h, and the levels of PPARα, COX-2, Bcl-xL, Bcl-xS, and cleaved caspase-3 in the cell lysates were examined using a western blot analysis. The cell lysates of the samples with 1 h of carboplatin challenge were partitioned into cytosolic and nuclear factions. Band intensities were normalized based on GAPDH band intensity using densitometry. The bar chart shows the normalized intensities of each protein band. Lamin A/C and GAPDH were used as internal controls for the nuclear fraction and whole-cell lysate, respectively. Comparisons were subjected to ANOVA followed by Bonferoni's post-hoc tests. Results are expressed as the mean ± SD (n = 4). Data from a representative experiment are shown. Significant difference (*$P < 0.05$ vs. the control; #$P < 0.05$ vs. the Cbpt-treated group).

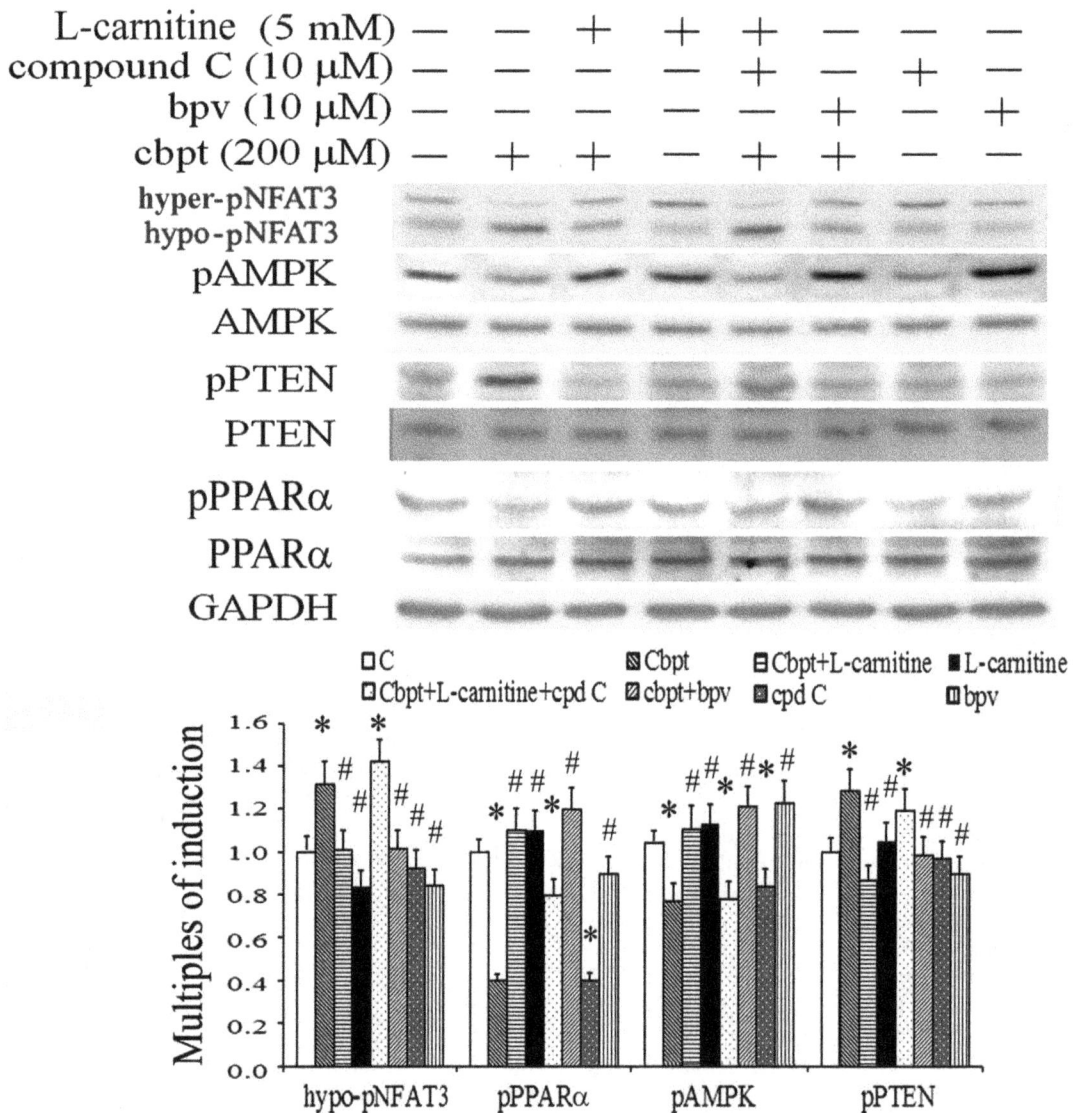

Figure 2. The activation of PPARα by L-carnitine in carboplatin-induced RTC injury. RTCs were pretreated with L-carnitine for 24 h or with the PTEN inhibitor BPV or the AMPK inhibitor compound C for 1 h, followed by carboplatin challenge for 20 min. The levels of NFAT3, AMPK, pPTEN, and pPPARα in the cell lysates were analyzed by conducting western blotting, using GAPDH as an internal control. The data represent the mean ± SD of the results of 3 independent experiments (*P<0.05 vs. the control; #P<0.05 vs. the Cbpt-treated group).

the reported plasmids was normalized to that of the empty reporter plasmid and the pRL-TK Renilla luciferase plasmid.

Small interfering RNA-mediated gene silencing of PPARα

The PPARα small interfering (si) RNA duplexes, 5′-GAA-CAUCGAGUGUCGAAUATT-3′ and 5′-GACUACCAGUA-CUUAGGAATT-3′ were purchased from Ambion (Austin, TX, USA). The RTCs were seeded in 6-well plates, and were transfected for 24 h using 100 pmol of the PPARα siRNAs, or the scrambled siRNAs in 100 μL of siPORTNeoFX. The expression of PPARα and other relevant proteins was analyzed using western blotting.

Co-immunoprecipitation and western blot analysis of cytosolic and nuclear fractions of cell lysates

The PPARα protein was immunoprecipitated in samples containing 200 μg of total protein using 2 μg of an anti-PPARα

antibody and 20 μg of protein-A-plus-G agarose beads to determine whether the p65 and/or p50 proteins coprecipitated with the PPARα protein. The precipitates were washed 5 times with a lysis buffer and once with phosphate-buffered saline (PBS). The washed pellet was resuspended in a sample buffer containing 50 mM Tris, 100 mM bromophenol blue, and 10% glycerol at pH 6.8, and incubated at 90°C for 10 min. The precipitated proteins were released from the agarose beads during gel electrophoresis. The RTCs were cultured in 10-cm² dishes. The RTCs were pretreated using 5 mM L-carnitine for 24 h, and were harvested after carboplatin challenge for the indicated time points. The cell lysates were partitioned into cytosolic and nuclear fractions using the NE-PER nuclear extraction reagents (Pierce, Rockford, IL, USA) and protease inhibitors.

The western blotting procedure has been described elsewhere [15]. The following antibodies were used in the western blot analysis at the dilutions indicated: antibodies against the NFAT3,

PTEN, PPARα, Bcl-xL, Bcl-xS, NFκB-p65, NFκB-p50, PGIS, lamin A/C (1:1000; Santa Cruz Biotechnology, Dallas, TX, USA), pPPARα-Ser21 (1:500; ABR Affinity Bioreagents, Rockford, IL, USA), cleaved caspase-3, COX-2 (1:500; Cayman Chemical, Ann Arbor, MI, USA), GAPDH (1:2000; Ab Frontier, Seoul, Korea), AMPK and phospho-AMPK proteins (1:500; Millipore, Burlington, MA, USA). Aliquots of the nuclear and cytosolic fractions containing 50 μg of total protein were separated on a 10% acrylamide gel using sodium dodecyl sulfate-polyacrylamide gel electrophoresis. The protein bands in the acrylamide gel were electrophoretically transferred to a Hybond-P membrane (GE Healthcare Life Sciences, Waukesha, WI, USA), and the membranes were probed using the various primary antibodies. Band intensities in the western blots were normalized based on GAPDH (control) band intensity using an IS-1000 digital imaging system (ARRB, Victoria, Australia).

Analysis of gene expression using a reverse-transcription polymerase chain reaction

A previously described method was used to obtain the total RNA for the analysis of gene expression using a reverse-transcription polymerase chain reaction (RT-PCR), with minor modifications (Pang et al., 2008). Sequences of the primer pairs used for the amplification of each gene were as follows: 5′-TGCCTCAGCCTCTTCTCATT-3′ and 5′-CCCATTTGGG-AACTTCTCCT-3′ for the TNFα gene (108 bp); 5′-AGGTATC-CATCCATCCCACA-3′ and 5′-GCCACAGTTCTCAAAGCA-CA-3′ for the ICAM-1 gene (209 bp); 5′-ATGCAGTTAAT-GCCCCACTC-3′ and 5′-TTCCTTATTGGGGTCAGCAC-3′ for the MCP-1 gene (167 bp); and 5′-AACTTTGGCATTGTG-GAAGG-3′ and 5′-TGTTCCTACCCCCAATGTGT-3′ for the GAPDH gene (223 bp). In each experiment, 5 μg of total RNA from the extracts of RTCs was used. The total cDNA in each RT-PCR sample was normalized to that of the GAPDH samples. The PCR products were separated on a 2% agarose gel and quantified using an electrophoresis image analysis system (Eastman Kodak, Rochester, NY, USA).

Animals and treatments

All animal study procedures were conducted in accordance with the Taipei medical university animal care and use rules (licenses No. LAC-101-0102) and an Association for Assessment and Accreditation of Laboratory Animal Care approved protocol. Eight-week-old male Balb/c mice weighing 20 to 25 g were obtained from the Research Animal Center at National Taiwan University (Taipei, Taiwan). The animals were housed in a central facility, were subjected to a 12-h light–dark cycle, and were given regular rat chow and tap water. The mice were separated into the control, carboplatin, carboplatin+L-carnitine, carboplatin+L-carnitine+compound C (an AMPK inhibitor), compound C groups, carboplatin+compound C, and L-carnitine, with 12 mice in each group except for the compound C group with 16 mice. Compound C (10 mg/kg) was intraperitoneally injected 1 h before the L-carnitine was administered. The L-carnitine (50 mg/kg) or compound C was given 2 days before a single dose of carboplatin (75 mg/kg) was intraperitoneally injected. Within the 4-day period of carboplatin challenge, L-carnitine and compound C were given every 2 days.

At the end of the treatment period, animals were anaesthetized intramuscularly with a combination of ketamine (8 mg/100 g body weight), xylazine (2 mg/100 g) and atropine (0.16 mg/100 g). Mice's blood samples were collected to measure the serum levels of creatinine and urea nitrogen using Fuji Dri-Chem slides (Fujifilm, Tokyo, Japan). The kidneys were harvested by performing a laparotomy, and tissue samples of the renal cortex were snap-frozen in dry ice before being stored at −80°C. The kidney tissue samples were fixed in 10% formalin, and embedded in paraffin. Serial 5-μm sections were prepared from the paraffin-embedded samples from the control and carboplatin-treated groups, and the sections were stained with hematoxylin and eosin (H&E) for histological analysis that was performed by a pathologist in a single-blind fashion. Frozen sections (5-μm) were also prepared for terminal deoxynucleotidyl transferase dUTP nick end labeling (TUNEL) 4 d after carboplatin treatment.

Statistical analysis

The data are expressed as the mean ± the standard deviation (SD), and represent the results of at least 3 experiments. The means of the experimental and control groups were compared using a one-way analysis of variance, or the Bonferroni method was used for the post-hoc analysis. A value of $P<0.05$ was considered to indicate a statistically significant difference.

Results

Protection of L-carnitine in carboplatin-mediated inflammation and apoptosis in RTCs

Because L-carnitine has antioxidant and antiapoptotic properties, the molecular mechanism of the protective effect of L-carnitine in carboplatin-challenged RTCs was examined using a western blot analysis. Our preliminary data showed that treatments using 5–40 mM L-carnitine were effective in prevention of carboplatin-mediated apoptosis in RTCs. Therefore, we used a low dose of 5 mM L-carnitine in our in vitro experiments. Carboplatin increased the levels of proteins involved in apoptotic and inflammatory signaling in RTCs, including hypo-pNFAT3, Bcl-xS, caspase 3, and p65/p50, and the levels of these proteins were significantly reduced in L-carnitine-treated cells. We observed that the protective effect of L-carnitine coincided with increased levels of PPARα and COX-2, whereas the levels of PPARα and COX-2 were reduced in cells treated using carboplatin alone (Figures 1).

Signaling pathway of PPARα phosphorylation by L-carnitine in rescuing carboplatin-mediated changes in NFAT3 and PPARα activation in RTCs

We have previously shown that NFAT3 is activated by ROS in carboplatin-challenged RTCs. We investigated the signaling pathways involved in the L-carnitine-induced NFAT3 inactivation but PPARα activation in RTCs treated with carboplatin by using a western blot analysis. As shown in Figure 2, carboplatin caused NFAT3 activation (hypo-phosphorylation) and PPARα inactivation (de-phosphorylation), which are correlated with the opposite phosphorylation status of PTEN and AMPK; increased phosphorylation of PTEN but reduced that of AMPK, compared with the phosphorylation of these proteins in the control cells. Nevertheless, L-carnitine reversed the activation of NFAT3 and PPARα caused by carboplatin through altering the phosphorylation of PTEN and AMPK.

The involvement of AMPK and PTEN by L-carnitine in the prevention of the dephosphorylation of NFAT3 and PPARα caused by carboplatin was verified using the AMPK inhibitor, compound C, and the PTEN inhibitor, BPV. Compound C inactivated AMPK, resulting in the elimination of L-carnitine-mediated PPARα phosphorylation in carboplatin-treated RTCs. By contrast, BPV inactivated PTEN, which coincided with increased AMPK and PPARα phosphorylation in carboplatin-

(A)

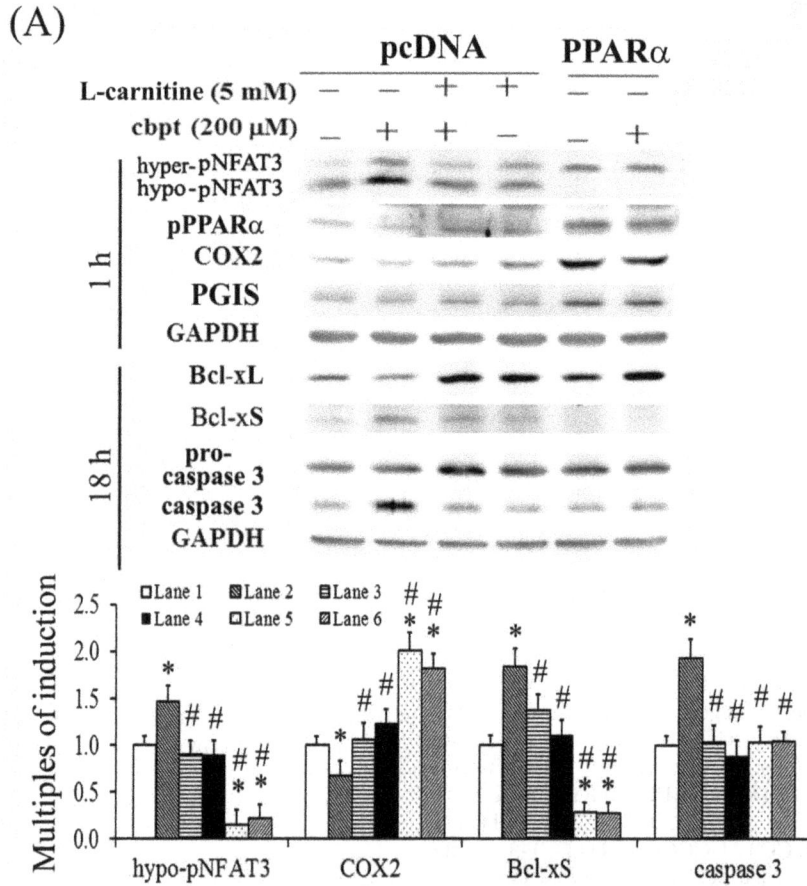

(B)

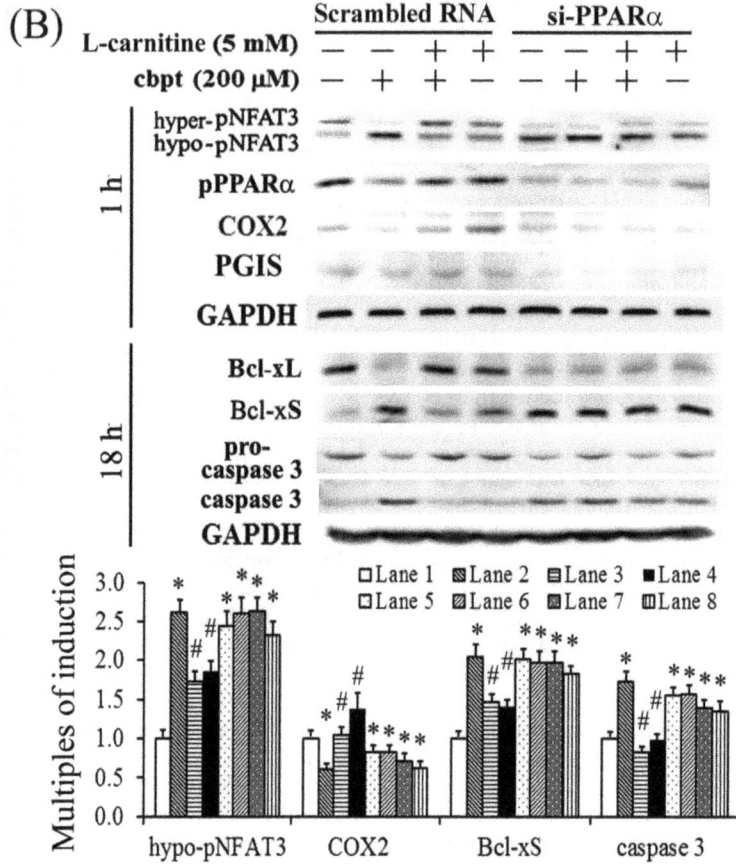

Figure 3. Overexpression of PPARα mimics the protective effect of L-carnitine in carboplatin-induced RTC apoptosis. Cells with PPARα overexpression (A) or PPARα silence (B) were pretreated with 5 mM L-carnitine for 24 h, followed by a 200-μM carboplatin challenge for 1 h and 18 h, and the levels of NFAT3, PPARα, Bcl-xL, Bcl-xS, and cleaved caspase-3 were analyzed by western blotting. The bar chart in each panel shows the normalized intensities of each protein band with GAPDH band using densitometry. Data were derived from the results of 4 independent experiments, and are presented as the mean ± SD (*$P < 0.05$ vs. the control; #$P < 0.05$ vs. the Cbpt-treated group).

treated RTCs. This suggests that L-carnitine activates PPARα through an AMPK/PTEN-dependent mechanism, resulting in the inactivation of NFAT3, which might be beneficial for the protection against carboplatin-mediated renal injury.

PPARα overexpression mimics the protective effect of L-carnitine in carboplatin-mediated apoptosis and oxidative insult

We intend to unravel the causal relationship between PGI2 and PPARα signaling and their influence in the L-carnitine-mediated protection of carboplatin-treated RTCs. The essential role of PPARα in the antiapoptotic effect of L-carnitine in carboplatin-challenged RTCs was examined using both gain- and loss-of-function experimental designs. As shown in Figure 3A, the effect of PPARα overexpression mimicked the protective effect of L-carnitine with regard to the reduced ratio of Bcl-xS to Bcl-xL. Similar to the adverse effect of carboplatin, PPARα knockdown increased the ratio of Bcl-xS to Bcl-xL (Figure 3B), indicating the critical role of PPARα activation in the protective effect of L-carnitine. In addition, PPARα overexpression increased COX-2/PGIS expression in RTCs, whereas PPARα knockdown reduced COX-2/PGIS expression.

We also examine the causal relationship among PPARα, COX-2, and PGIS in the L-carnitine-mediated protection of carboplatin-induced RTC apoptosis by assessing the effects of the PPARα agonist Wy14643, the PPARα antagonist GW6471, the PGI2 analog beraprost, and the selective COX-2 inhibitor NS398. As shown in Figure 4A, Wy14643 and beraprost mimicked the effect of L-carnitine in reducing the activation of NFAT3 and caspase 3, whereas GW6471 and NS-398 produced the opposite effect. In addition, Wy14643 induced COX-2 expression, which is consistent with the effects of PPARα overexpression and PPARα knockdown (Figure 3A and B). A PPRE-driven luciferase assay was used to examine whether the inhibitor and/or the analog of COX-2/PGIS affected the transactivational activity of PPAR. As shown in Figure 4B, carboplatin reduced the transactivational activity of PPAR. This reduction in PPARα transactivational activity was significantly reversed by treatment using additional L-carnitine or beraprost, whereas NS398 reduced the effect of L-carnitine in carboplatin-challenged RTCs. These data suggest that COX-2 and PGI2 (a feedback loop) are involved in the L-carnitine-induced increase in PPARα transactivational activity.

Transactivational activity of NFκB is regulated by L-carnitine through PPARα binding of p65 and p50

The mechanism underlying the anti-inflammatory effect of L-carnitine in carboplatin-treated RTCs was examined using genetic and pharmacological approaches. As shown in Figure 5, PPARα overexpression and treatments using agonists of PPARα and PGI2 reduced the carboplatin-mediated nuclear translocation of the NFκB-p65 and NFκB-p50 protein complexes, whereas the PPARα antagonist and COX-2 inhibitor eliminated the protective effect of L-carnitine in carboplatin-treated RTCs. The results of the NFκB-driven luciferase assay (Figure 6A) showed that carboplatin increased NFκB transactivational activity, and that treatment with additional L-carnitine or the overexpression of

PPARα reversed the carboplatin-mediated increase in NFκB-transactivational activity in RTCs.

The mechanism underlying the PPARα-mediated reduction in NFκB activation was examined using immunoprecipitation. As shown in Figure 6B, both PPARα overexpression and L-carnitine-induced PPARα activation increased the amount of p65 and p50 that coprecipitated with PPARα, and reduced the transactivation activity of NFκB (Figures 5A and 6A). We examined the expression of ICAM, MCP-1, and TNFα in RTCs treated with L-carnitine and carboplatin to investigate the relationship between the levels of these cytokines and the regulation of NFκB transcriptional activity. The results of the RT-PCR assays showed that the L-carnitine treatment significantly reduced the expression of these inflammatory cytokines in carboplatin-treated RTCs relative to that in RTCs treated with carboplatin alone (Figure 6C).

Essential roles of AMPK and PPARα activation in the L-carnitine-mediated protection of carboplatin-mediated renal injury in mice

The protective effect of L-carnitine in vitro was verified in carboplatin-challenged Balb/c mice, and the essential role of AMPK/PPARα activation in exerting protective effect of L-carnitine in carboplatin-induced renal injury was examined by using compound C, an AMPK inhibitor. As shown in Figure 7A, mice challenged with carboplatin showed severe condense nuclei (blue arrow), drop out of some epithelial cells (blue circle) and vacuolization of some proximal epithelial cells (red arrow). By contrast, the carboplatin-induced histological damage was milder in the mice treated with L-carnitine. Additional compound C treatment suppressed the protective effects of L-carnitine with regard to RTC structural integrity and histological changes with decreased number of epithelial cell nuclei in mice treated with carboplatin. Additionally, cell apoptosis in the kidney sections was evaluated by TUNEL assay with nuclei of cells stained with DAPI. The brightly stained nuclei produced by TUNEL were detected in the renal cortex of carboplatin-treated mice, but occurred rarely in those of the control and L-carnitine-treated mice. Most of the TUNEL-labeled nuclei were concentrated in proximal RTCs. These results demonstrate that carboplatin-induced apoptosis in mouse RTCs is inhibited by L-carnitine by approximately 75%. Additional compound C treatment reverted the protective effect of L-carnitine, increasing RTC apoptosis by approximately 65% compared with that of mice treated with both L-carnitine and carboplatin. The western blot analysis in Figure 7B showed that carnitine-mediated AMPK and PPARα activation in renal tissues was reverted in mice following compound C treatment, suggesting that L-carnitine exerts its protective effect on carboplatin-mediated renal injury through a mechanism that is dependent on the activation of AMPK and PPARα. Furthermore, in the renal functional assessment shown in Figure 7C, mice subjected to carboplatin insult showed increased serum levels of urea and creatinine, which suggests renal dysfunction. This can be significantly alleviated in mice additionally treated with L-carnitine, suggesting a marked prevention of renal function associated with carboplatin toxicity. However, levels of serum urea and creatinine in mice additionally treated with compound C

(A)

(B)

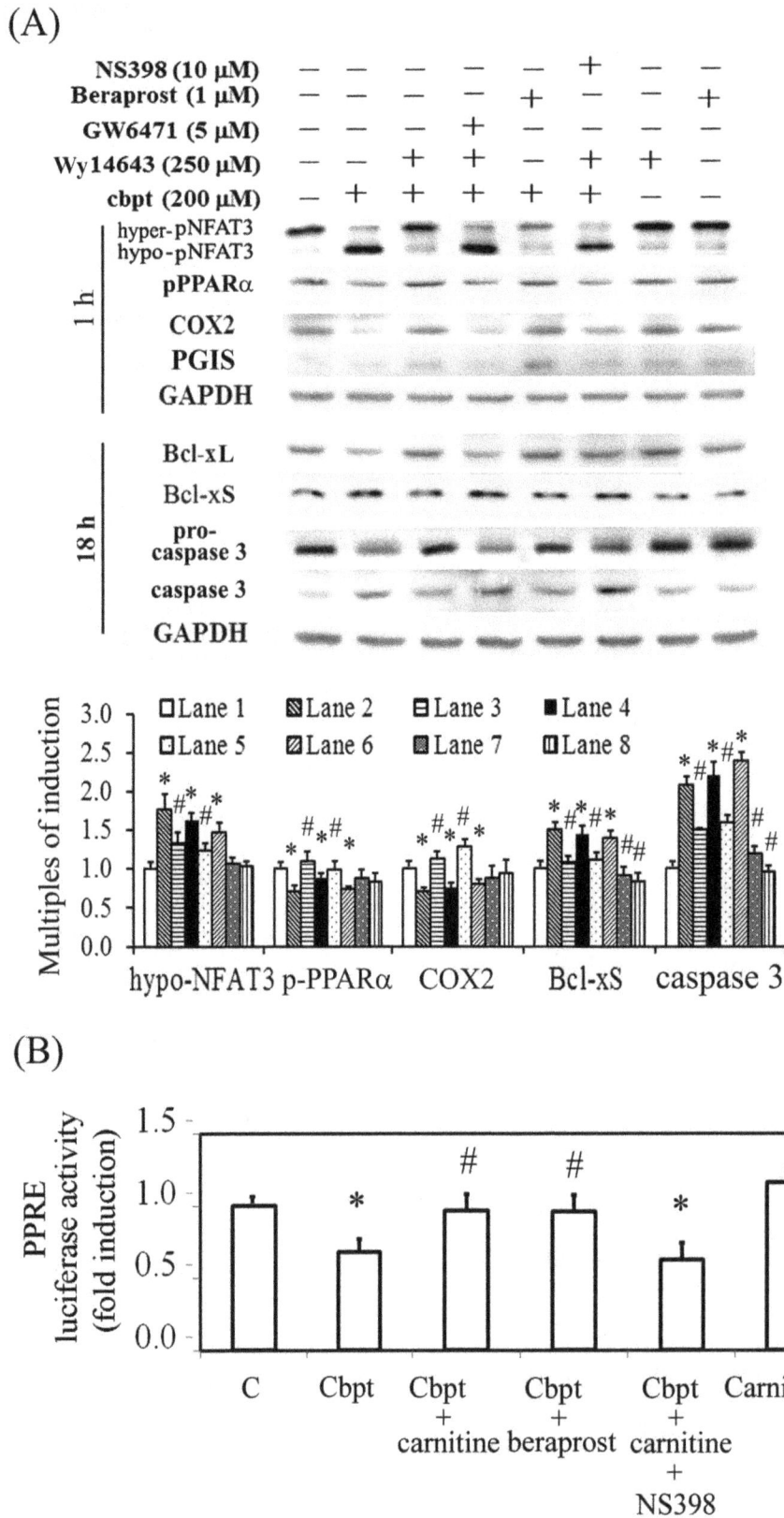

Figure 4. Causal relationship of PPARα, COX-2, and PGIS for the protective effect of L-carnitine in carboplatin-induced RTC apoptosis. (A) RTCs were pretreated using various antagonists/inhibitors and agonists of PPARα, COX-2, and/or PGI2 for 1 h or 5 mM L-carnitine for 24 h, followed by a 200-μM carboplatin challenge for 1 h and 18 h, and the levels of NFAT3, PPARα, Bcl-xL, Bcl-xS, and cleaved caspase-3 were analyzed by western blotting. Band intensities were normalized based on GAPDH band intensity using densitometry. The bar chart in each panel shows the normalized intensities of each protein band. (B) The RTCs were transfected with the PPRE-driven reporter plasmids overnight, and

pretreated with 5 mM L-carnitine for 24 h, followed by a 200-µM carboplatin challenge for 2 h. The reporter luciferase activity was calculated by normalizing the intrinsic activity of samples based on the intensity of the pGL3-promoter vector and the transfection efficiency. Data were derived from the results of 4 independent experiments, and are presented as the mean ± SD (*$P<0.05$ vs. the control; #$P<0.05$ vs. the Cbpt-treated group).

were significantly higher than those observed in the carboplatin-L-carnitine group. These results suggest that L-carnitine protects against carboplatin-mediated renal injury through an AMPK/PPARα-dependent pathway, in which AMPK-dependent PPARα activation mediates the protective effect of L-carnitine.

Discussion

We have previously shown that oxidative stress induces NFAT3 activation (hypophosphorylation) in carboplatin chemotherapy, resulting in inflammation and apoptosis in RTCs, and that carboplatin-induced NFAT3 activation is reduced by NAC treatment and HO-1 overexpression [10,32]. The results of our current study provide both in vitro and in vivo evidence that L-carnitine inhibits carboplatin-mediated renal injury by suppressing the activation of NFAT3 and NFκB through AMPK-mediated PPARα activation, and that PPARα activation is essential to the protective effect of L-carnitine in RTCs.

Both L-carnitine and propionyl-L-carnitine (PLC), a carnitine derivative, induce endothelium-dependent vasodilation, and have been used for the treatment of cardiovascular diseases [33,34]. Nitric oxide production in the aorta of hypertensive rats is enhanced by L-carnitine through PI3 and Akt kinases [35], and PLC promotes prostaglandin synthesis in subcutaneous arteries in humans [20]. We showed that the antioxidant and anti-inflammatory properties of L-carnitine protected RTCs from carboplatin-mediated injury. Although L-carnitine has been shown to induce SOD, catalase, and glutathione peroxidase [2], the antioxidant properties of L-carnitine were not addressed in our study.

Although Chen et al. [4] demonstrated that L-carnitine protects against gentamicin-mediated renal injury though a PGI2-PPARα pathway, we showed that PPARα regulated COX-2 and PGIS expression, which also increased PPARα transactivational activity, and the roles of PGI2, COX-2 and PGIS were confirmed in both gain- and loss-of-function experiments (Figure 3). This is in agreement with our previous finding that adiponectin-mediated COX-2 induction through a PPARα-dependent mechanism, and COX-2 exerted an anti-inflammatory effect of adiponectin in hepatocytes subjected to iron challenge [15]. In addition, a PGI2 agonist (beraprost) enhanced PPRE-driven transactivational activity (Figure 4B) and PPARα nuclear translocation, suggesting the existence of a positive feedback mechanism in the regulation of PPARα activation that has not been previously reported. Likewise, beraprost caused COX-2 upregulation (Figure 4A) through the positive loop of PGI2 in PPARα activation. Different to the concept of COX-2 as a pro-inflammatory molecule, we demonstrated that COX-2-mediated PGIS activation can directly or indirectly potentiate the PPARα activation, resulting in the anti-inflammation by suppressing NFκB activation in carboplatin-challenged RTC treated with L-carnitine. This finding is agreed by the study that COX-2 might induce anti-inflammatory effect by generating an alternative set of prostaglandins [36]. Additionally, nuclear factor erythroid-2-related factor-2 (Nrf2) has been proved to confer protection against oxidative stress [37]. Furthermore, COX-2-dependnet electophile oxo-derivative molecules have been shown to modulate the anti-inflammatory action via activation of Nrf2-dependent antioxidant response element (ARE) [38]. We also showed that AMPK regulates PPARα phosphorylation in L-carnitine-mediated protection in RTCs challenged with carboplatin (Figure 2), which is consistent with our previous finding that

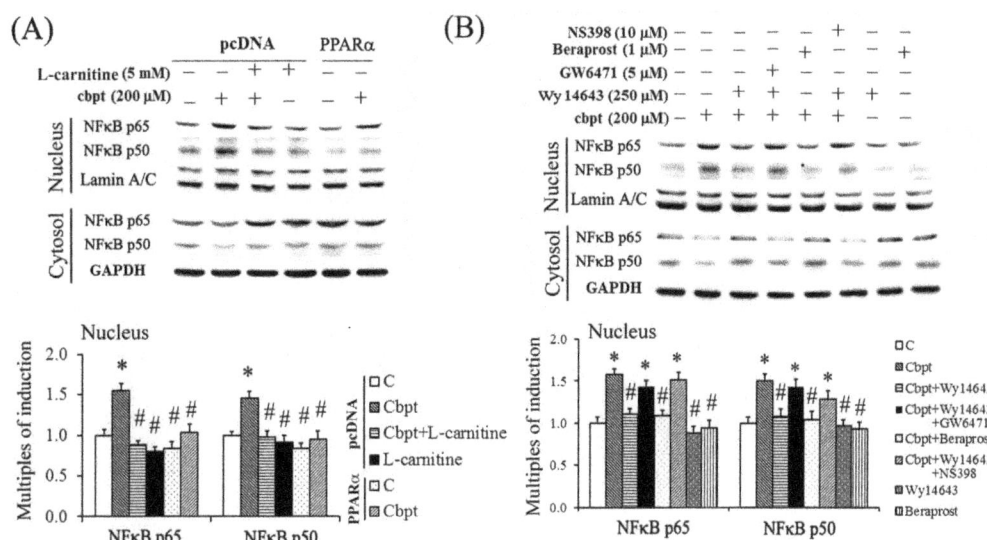

Figure 5. The inhibition of the carboplatin-mediated nuclear translocation of p65 and p50 by PPARα activation. The RTCs were (A) transfected with PPARα overnight and (B) pretreated using various antagonists/inhibitors and agonists of PPARα, COX-2, and/or PGI2 for 1 h, followed by carboplatin challenge for 1 h. The cell lysates were fractionated, and the cytosolic and nuclear distributions of p65 and p50 were analyzed using western blotting. Band intensities were normalized based on the GAPDH band intensity using densitometry. The bar chart in each panel shows the normalized intensities of each protein band. The data were derived frm the results of 3 independent experiments, and are presented as the mean ± SD (*$P<0.05$ vs. the control; #$P<0.05$ vs. the Cbpt-treated group).

(A)

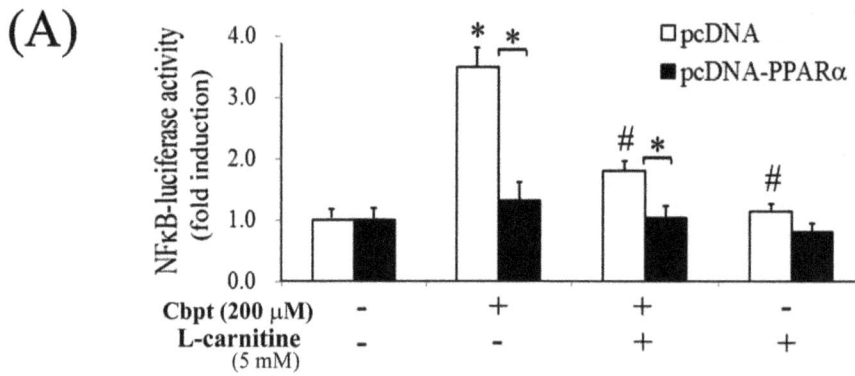

(B)

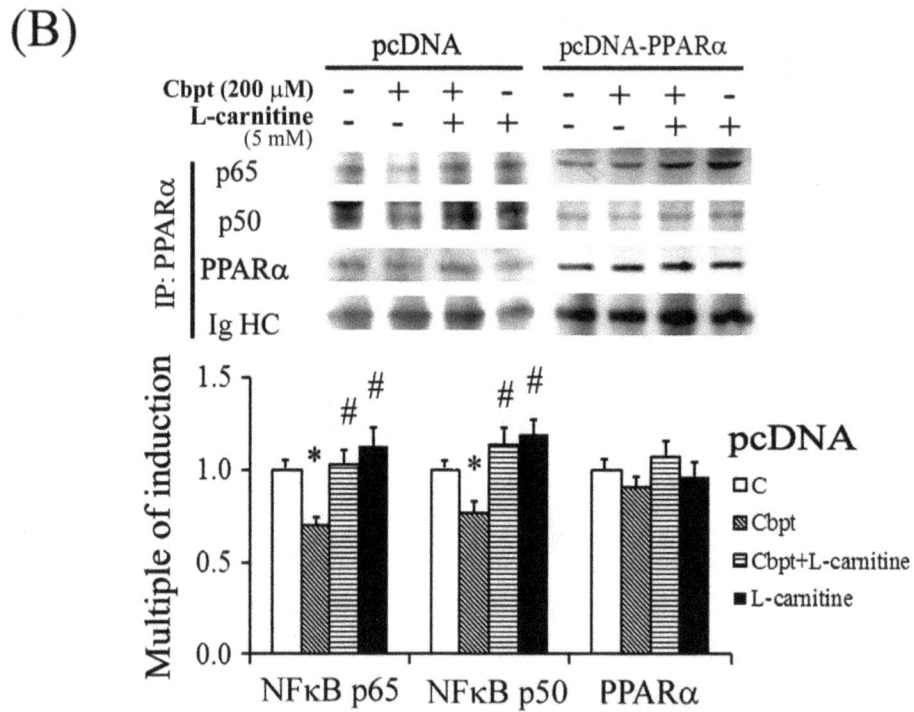

(C)

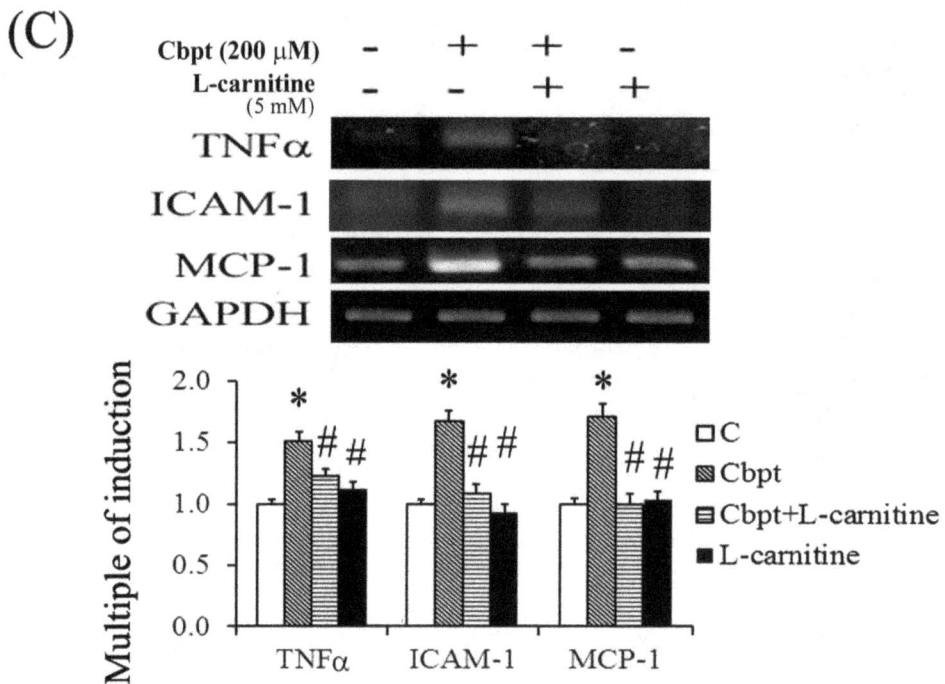

Figure 6. The inhibition of the carboplatin-mediated increase in NFκB activation by PPARα overexpression. (A) The RTCs were treated as described in Figure 3D, except the cells were transfected with the NFκB-luciferase vector. Luciferase activity was measured, and the data were processed as described in the Methods section. The data were derived from 3 independent experiments, and are presented as the mean ± SD (*P< 0.05 vs. the control). (B) Cells were transfected with pcDNA or pcDNA-PPARα overnight, followed by L-carnitine or carboplatin treatment. The PPARα protein was immunoprecipitated using an anti-PPARα antibody, and the precipitates were probed using anti-p65 and anti-p50 antibodies. The data are representative of the results of 3 independent experiments. (C) The L-carnitine-mediated downregulation of inflammatory cytokines was analyzed using RT-PCR. The RTCs were pretreated using L-carnitine for 24 h, followed by carboplatin challenge for 4 h, to investigate the effect of L-carnitine on the carboplatin-mediated production of TNFα, MCP1, and ICAM1. The expression of GAPDH was used as an internal control. Comparisons were subjected to ANOVA followed by Bonferoni's post-hoc tests. The data are presented as the mean ± SD of the results of 4 independent experiments (*P<0.05 vs. the control).

adiponectin protects hepatocytes from iron-overload-mediated apoptosis and inflammation through the AMPK-PPARα-mediated inductions of HO-1 and COX-2 [15,16].

To gain insight into the mechanism by which PPARα is activated by L-carnitine, we focused our investigation on the relationship between AMPK and PTEN in PPARα activation. We demonstrated that a reciprocal relationship exists between AMPK and PTEN in the L-carnitine-mediated activation of PPARα in carboplatin-treated RTCs. These findings are consistent with those of a previous study, which showed that metformin, an

insulin-sensitizing drug, suppressed the expression of PTEN through an AMPK-dependent mechanism in preadipocyte 3T3-L1 cells [39]. By contrast, AMPK knockdown eliminated the effects of metformin on reducing PTEN, suggesting that AMPK is involved in the regulation of PTEN. Additionally, we observed that L-carnitine stimulated the phosphorylation of PPARα through an AMPK/PTEN-dependent pathway. Although a previous study suggested that the phosphorylation of PTEN at S370 and S380 inhibits certain PTEN functions, how these phosphorylation events affect enzymatic activity is unclear [40].

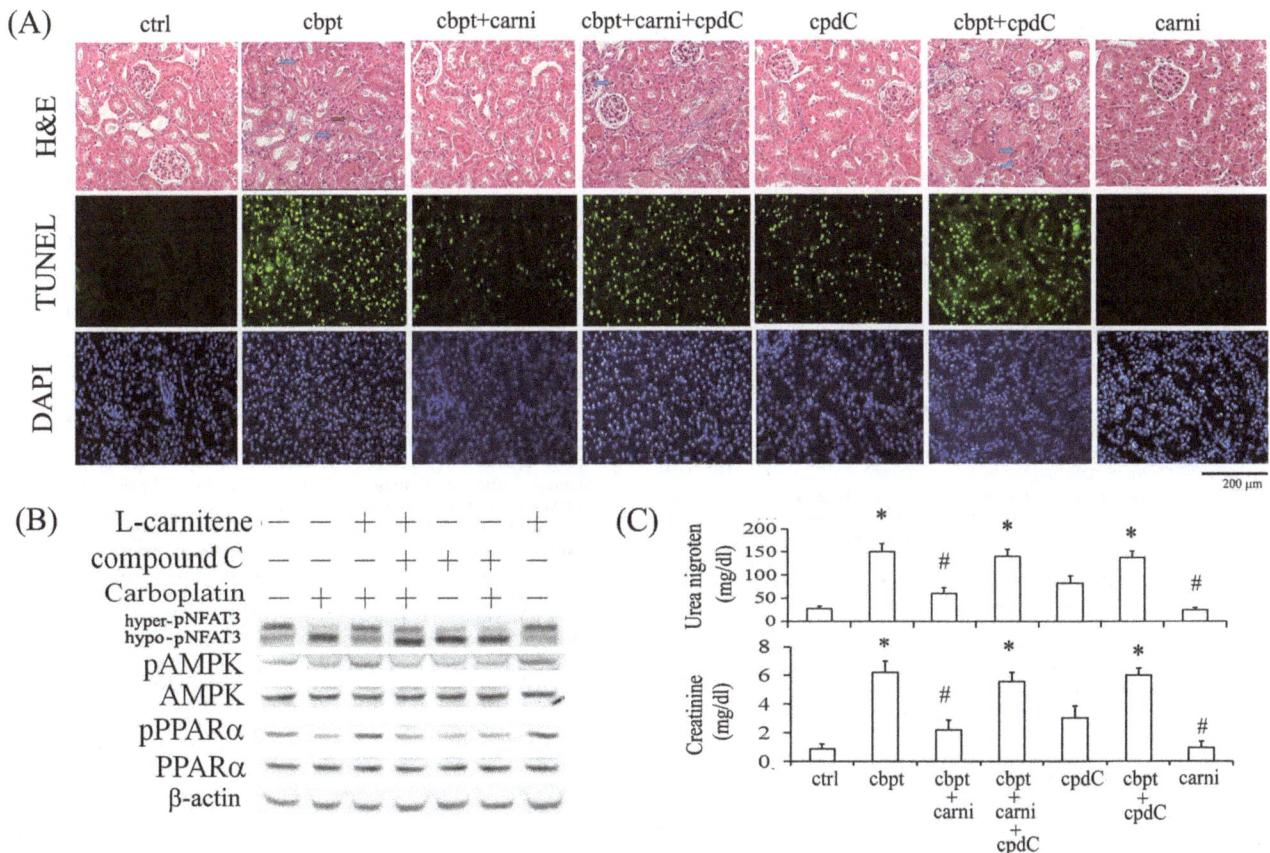

Figure 7. The inhibition of the protective effect of L-carnitine on carboplatin-mediated changes in renal structure and function in mice treated with an AMPK inhibitor. The kidneys were dissected and sectioned for (A) histological examination and TUNEL assay. Representative photographs of H&E staining are shown on top panel. Blue arrow: severe condense nuclei, red arrow: vacuolization of some proximal epithelial cells and blue circle: drop out of some epithelial cells. Apoptotic cells in the kidneys of experimental animals were detected in vivo using TUNEL staining. (Second panel) The TUNEL-labeled nuclei were visible as bright spots in the cortical sections of untreated and treated mouse kidneys. (Third panel) The identical fields were stained using DAPI to confirm the positions of the TUNEL-labeled cell nuclei. (B) The levels of NFAT3, AMPK and PPARα in the kidney extracts after mice with various treatments were evaluated using a western blot analysis using β-actin as an internal control. (C) The serum levels of urea nitrogen and creatinine in the 7 groups were measured at 4 days after carboplatin challenge. Comparisons were subjected to ANOVA followed by Bonferoni's post-hoc tests. The results are expressed as the mean ± SD (n = 10; *P<0.05 vs. the control; #P<0.05 vs. the Cbpt-treated group).

We observed that carboplatin increased PTEN phosphorylation at S380, and that the carboplatin-mediated phosphorylation of PTEN is inhibited by the PTEN inhibitor BPV, which rescued the carboplatin-mediated reduction of AMPK phosphorylation. Thus, how PTEN phosphorylation affects PTEN enzymatic activity largely remains unclear. Future investigations are warranted to unravel the intricate relationship between AMPK and PTEN in L-carnitine-mediated PPARα activation.

The results of this study also demonstrated that L-carnitine reduces NFκB transactivational activity and then the production of TNFα, ICAM1, and MCP-1 in carboplatin-treated RTCs. A previous study showed that PPARγ modulates NFκB activity by interacting with the RelA/p65 subunit of NFκB in cells treated using TNFα, and this interaction was disrupted by exposure to cigarette smoke [41]. Likewise, in our immunoprecipitation experiments, we observed that PPARα directly interacted with NFκB (Figure 6B), and that PPARα-overexpression reduced NFκB transactivational activity in carboplatin-treated RTCs (Figure 6A). Additionally, both genetic and pharmacological activation of PPARα confirmed its involvement in the modulation of NFκB activation and NFκB-related inflammation through its effect on the nuclear translocation of p65 and p50 (Figure 5). Our findings are consistent with those of a previous study that showed that fibrates inhibit vascular inflammatory response through a PPARα-dependent reduction in NFκB and AP-1 transactivation [22]. Collectively, our results show that L-carnitine, a naturally occurring compound, can prevent the renal toxicity induced by carboplatin chemotherapy. The AMPK-induced activation of PPARα is essential to the protective effect of L-carnitine in carboplatin-mediated renal injury.

Acknowledgments

We acknowledge Prof. Chi-Long Chen's suggestion in evaluating the histological slides.

Author Contributions

Conceived and designed the experiments: YMS SHJ. Performed the experiments: HCC CCC NJY YC. Analyzed the data: HCC CCC NJY YC. Contributed reagents/materials/analysis tools: HCC YMS SHJ. Contributed to the writing of the manuscript: SHJ.

References

1. Steiber A, Kerner J, Hoppel CL (2004) Carnitine: a nutritional, biosynthetic, and functional perspective. Mol Aspects Med 25: 455–473.
2. Miguel-Carrasco JL, Monserrat MT, Mate A, Vazquez CM (2010) Comparative effects of captopril and l-carnitine on blood pressure and antioxidant enzyme gene expression in the heart of spontaneously hypertensive rats. Eur J Pharmacol 632: 65–72.
3. Pisano C, Vesci L, Milazzo FM, Guglielmi MB, Fodera R, et al. Metabolic approach to the enhancement of antitumor effect of chemotherapy: a key role of acetyl-L-carnitine. Clin Cancer Res 16: 3944–3953.
4. Chen HH, Sue YM, Chen CH, Hsu YH, Hou CC, et al. (2009) Peroxisome proliferator-activated receptor alpha plays a crucial role in L-carnitine anti-apoptosis effect in renal tubular cells. Nephrol Dial Transplant 24: 3042–3049.
5. Fujiwara K, Sakuragi N, Suzuki S, Yoshida N, Maehata K, et al. (2003) First-line intraperitoneal carboplatin-based chemotherapy for 165 patients with epithelial ovarian carcinoma: results of long-term follow-up. Gynecol Oncol 90: 637–643.
6. Alberts DS (1995) Carboplatin versus cisplatin in ovarian cancer. Semin Oncol 22: 88–90.
7. Santana-Davila R, Szabo A, Arce-Lara C, Williams CD, Kelley MJ, et al. (2014) Cisplatin versus carboplatin-based regimens for the treatment of patients with metastatic lung cancer. an analysis of Veterans Health Administration data. J Thorac Oncol 9: 702–709.
8. Ardizzoni A, Boni L, Tiseo M, Fossella FV, Schiller JH, et al. (2007) Cisplatin-versus carboplatin-based chemotherapy in first-line treatment of advanced non-small-cell lung cancer: an individual patient data meta-analysis. J Natl Cancer Inst 99: 847–857.
9. Husain K, Scott RB, Whitworth C, Somani SM, Rybak LP (2001) Dose response of carboplatin-induced hearing loss in rats: antioxidant defense system. Hear Res 151: 71–78.
10. Lin H, Sue YM, Chou Y, Cheng CF, Chang CC, et al. (2010) Activation of a nuclear factor of activated T-lymphocyte-3 (NFAT3) by oxidative stress in carboplatin-mediated renal apoptosis. Br J Pharmacol 161: 1661–1676.
11. Evans RM, Barish GD, Wang YX (2004) PPARs and the complex journey to obesity. Nat Med 10: 355–361.
12. MacAulay K, Blair AS, Hajduch E, Terashima T, Baba O, et al. (2005) Constitutive activation of GSK3 down-regulates glycogen synthase abundance and glycogen deposition in rat skeletal muscle cells. J Biol Chem 280: 9509–9518.
13. Shalev A, Siegrist-Kaiser CA, Yen PM, Wahli W, Burger AG, et al. (1996) The peroxisome proliferator-activated receptor alpha is a phosphoprotein: regulation by insulin. Endocrinology 137: 4499–4502.
14. Juge-Aubry CE, Hammar E, Siegrist-Kaiser C, Pernin A, Takeshita A, et al. (1999) Regulation of the transcriptional activity of the peroxisome proliferator-activated receptor alpha by phosphorylation of a ligand-independent trans-activating domain. J Biol Chem 274: 10505–10510.
15. Lee FP, Jen CY, Chang CC, Chou Y, Lin H, et al. (2010) Mechanisms of adiponectin-mediated COX-2 induction and protection against iron injury in mouse hepatocytes. J Cell Physiol 224: 837–847.
16. Lin H, Yu CH, Jen CY, Cheng CF, Chou Y, et al. (2010) Adiponectin-mediated heme oxygenase-1 induction protects against iron-induced liver injury via a PPARalpha dependent mechanism. Am J Pathol 177: 1697–1709.
17. Lim H, Dey SK (2002) A novel pathway of prostacyclin signaling-hanging out with nuclear receptors. Endocrinology 143: 3207–3210.
18. Garrelds IM, Elliott GR, Zijlstra FJ, Bonta IL (1994) Effects of short- and long-term feeding of L-carnitine and congeners on the production of eicosanoids from rat peritoneal leucocytes. Br J Nutr 72: 785–793.
19. Bueno R, Alvarez de Sotomayor M, Perez-Guerrero C, Gomez-Amores L, Vazquez CM, et al. (2005) L-carnitine and propionyl-L-carnitine improve endothelial dysfunction in spontaneously hypertensive rats: different participation of NO and COX-products. Life Sci 77: 2082–2097.
20. Cipolla MJ, Nicoloff A, Rebello T, Amato A, Porter JM (1999) Propionyl-L-carnitine dilates human subcutaneous arteries through an endothelium-dependent mechanism. J Vasc Surg 29: 1097–1103.
21. Marx N, Sukhova GK, Collins T, Libby P, Plutzky J (1999) PPARalpha activators inhibit cytokine-induced vascular cell adhesion molecule-1 expression in human endothelial cells. Circulation 99: 3125–3131.
22. Delerive P, De Bosscher K, Besnard S, Vanden Berghe W, Peters JM, et al. (1999) Peroxisome proliferator-activated receptor alpha negatively regulates the vascular inflammatory gene response by negative cross-talk with transcription factors NF-kappaB and AP-1. J Biol Chem 274: 32048–32054.
23. Neve BP, Corseaux D, Chinetti G, Zawadzki C, Fruchart JC, et al. (2001) PPARalpha agonists inhibit tissue factor expression in human monocytes and macrophages. Circulation 103: 207–212.
24. Staels B, Koenig W, Habib A, Merval R, Lebret M, et al. (1998) Activation of human aortic smooth-muscle cells is inhibited by PPARalpha but not by PPARgamma activators. Nature 393: 790–793.
25. Delerive P, Gervois P, Fruchart JC, Staels B (2000) Induction of IkappaBalpha expression as a mechanism contributing to the anti-inflammatory activities of peroxisome proliferator-activated receptor-alpha activators. J Biol Chem 275: 36703–36707.
26. Kockx M, Gervois PP, Poulain P, Derudas B, Peters JM, et al. (1999) Fibrates suppress fibrinogen gene expression in rodents via activation of the peroxisome proliferator-activated receptor-alpha. Blood 93: 2991–2998.
27. Madej A, Okopien B, Kowalski J, Zielinski M, Wysocki J, et al. (1998) Effects of fenofibrate on plasma cytokine concentrations in patients with atherosclerosis and hyperlipoproteinemia IIb. Int J Clin Pharmacol Ther 36: 345–349.
28. Chen HH, Chen TW, Lin H (2009) Prostacyclin-induced peroxisome proliferator-activated receptor-alpha translocation attenuates NF-kappaB and TNF-alpha activation after renal ischemia-reperfusion injury. Am J Physiol Renal Physiol 297: F1109–1118.
29. de Larco JE, Todaro GJ (1978) Epithelioid and fibroblastic rat kidney cell clones: epidermal growth factor (EGF) receptors and the effect of mouse sarcoma virus transformation. J Cell Physiol 94: 335–342.
30. Best CJ, Tanzer LR, Phelps PC, Merriman RL, Boder GG, et al. (1999) H-ras-transformed NRK-52E renal epithelial cells have altered growth, morphology, and cytoskeletal structure that correlates with renal cell carcinoma in vivo. In Vitro Cell Dev Biol Anim 35: 205–214.
31. He TC, Chan TA, Vogelstein B, Kinzler KW (1999) PPARdelta is an APC-regulated target of nonsteroidal anti-inflammatory drugs. Cell 99: 335–345.
32. Sue YM, Cheng CF, Chou Y, Chang CC, Lee PS, et al. (2011) Ectopic overexpression of haem oxygenase-1 protects kidneys from carboplatin-mediated apoptosis. Br J Pharmacol 162: 1716–1730.
33. Ning WH, Zhao K (2013) Propionyl-L-carnitine induces eNOS activation and nitric oxide synthesis in endothelial cells via PI3 and Akt kinases. Vascul Pharmacol.

34. Ferrari R, Merli E, Cicchitelli G, Mele D, Fucili A, et al. (2004) Therapeutic effects of L-carnitine and propionyl-L-carnitine on cardiovascular diseases: a review. Ann N Y Acad Sci 1033: 79–91.

35. Herrera MD, Bueno R, De Sotomayor MA, Perez-Guerrero C, Vazquez CM, et al. (2002) Endothelium-dependent vasorelaxation induced by L-carnitine in isolated aorta from normotensive and hypertensive rats. J Pharm Pharmacol 54: 1423–1427.

36. Gilroy DW, Colville-Nash PR, Willis D, Chivers J, Paul-Clark MJ, et al. (1999) Inducible cyclooxygenase may have anti-inflammatory properties. Nat Med 5: 698–701.

37. Groeger AL, Cipollina C, Cole MP, Woodcock SR, Bonacci G, et al. (2010) Cyclooxygenase-2 generates anti-inflammatory mediators from omega-3 fatty acids. Nat Chem Biol 6: 433–441.

38. Itoh K, Wakabayashi N, Katoh Y, Ishii T, Igarashi K, et al. (1999) Keap1 represses nuclear activation of antioxidant responsive elements by Nrf2 through binding to the amino-terminal Neh2 domain. Genes Dev 13: 76–86.

39. Lee SK, Lee JO, Kim JH, Kim SJ, You GY, et al. (2011) Metformin sensitizes insulin signaling through AMPK-mediated PTEN down-regulation in preadipocyte 3T3-L1 cells. J Cell Biochem 112: 1259–1267.

40. Odriozola L, Singh G, Hoang T, Chan AM (2007) Regulation of PTEN activity by its carboxyl-terminal autoinhibitory domain. J Biol Chem 282: 23306–23315.

41. Caito S, Yang SR, Kode A, Edirisinghe I, Rajendrasozhan S, et al. (2008) Rosiglitazone and 15-deoxy-Delta12,14-prostaglandin J2, PPARgamma agonists, differentially regulate cigarette smoke-mediated pro-inflammatory cytokine release in monocytes/macrophages. Antioxid Redox Signal 10: 253–260.

4

The Potential of Mycelium and Culture Broth of *Lignosus rhinocerotis* as Substitutes for the Naturally Occurring Sclerotium with Regard to Antioxidant Capacity, Cytotoxic Effect, and Low-Molecular-Weight Chemical Constituents

Beng Fye Lau[1], Noorlidah Abdullah[1]*, Norhaniza Aminudin[1,2], Hong Boon Lee[3,4], Ken Choy Yap[5], Vikineswary Sabaratnam[1]

1 Mushroom Research Centre and Institute of Biological Sciences, Faculty of Science, University of Malaya, Kuala Lumpur, Malaysia, 2 University of Malaya Centre for Proteomics Research (UMCPR), Faculty of Medicine, University of Malaya, Kuala Lumpur, Malaysia, 3 Drug Discovery Laboratory, Cancer Research Initiatives Foundation (CARIF), Subang Jaya, Selangor, Malaysia, 4 Department of Pharmacy, Faculty of Medicine, University of Malaya, Kuala Lumpur, Malaysia, 5 Advanced Chemistry Solutions, Kuala Lumpur, Malaysia

Abstract

Previous studies on the nutritional and nutraceutical properties of *Lignosus rhinocerotis* focused mainly on the sclerotium; however, the supply of wild sclerotium is limited. In this investigation, the antioxidant capacity and cytotoxic effect of *L. rhinocerotis* cultured under different conditions of liquid fermentation (shaken and static) were compared to the sclerotium produced by solid-substrate fermentation. Aqueous methanol extracts of the mycelium (LR-MH, LR-MT) and culture broth (LR-BH, LR-BT) demonstrated either higher or comparable antioxidant capacities to the sclerotium extract (LR-SC) based on their radical scavenging abilities, reducing properties, metal chelating activities, and inhibitory effects on lipid peroxidation. All extracts exerted low cytotoxicity (IC_{50}>200 µg/ml, 72 h) against selected mammalian cell lines. Several low-molecular-weight compounds, including sugars, fatty acids, methyl esters, sterols, amides, amino acids, phenolics, and triterpenoids, were identified using GC-MS and UHPLC-ESI-MS/MS. The presence of proteins (<40 kDa) in the extracts was confirmed by SDS-PAGE and SELDI-TOF-MS. Principal component analysis revealed that the chemical profiles of the mycelial extracts under shaken and static conditions were distinct from those of the sclerotium. Results from bioactivity evaluation and chemical profiling showed that *L. rhinocerotis* from liquid fermentation merits consideration as an alternative source of functional ingredients and potential substitute for the sclerotium.

Editor: Michael P. Bachmann, Carl-Gustav Carus Technical University-Dresden, Germany

Funding: This study was supported by the Postgraduate Research Fund PV097/2011A from the University of Malaya (http://um.edu.my) [BFL NA NHA] and UM High Impact Research Grant HIR/MOHE/ASH/01CH-23001-G000008 from the Ministry of Education, Malaysia [NA VS]. The funders had no role in study design, data collection and analysis, decision to publish, or preparation of the manuscript.

Competing Interests: Ken Choy Yap is an employee of Advanced Chemistry Solutions, Kuala Lumpur, Malaysia.

* Email: noorlidah@um.edu.my

Introduction

The different morphological/developmental stages of a mushroom (i.e., the fruiting body, mycelium, and sclerotium) contain bioactive components with health-promoting effects. Only a handful of mushrooms are known to form sclerotia in their life cycles. One representative from the Polyporaceae family is *Lignosus rhinocerotis* [as 'rhinocerus'] (Cooke) Ryvarden (synonym: *Polyporus rhinocerus*), which is located throughout tropical regions. It is also popularly referred to as the "tiger's milk mushroom" ("*cendawan susu rimau*" in Malay) by the local and indigenous communities in Malaysia. Previous chemical investigations on *L. rhinocerotis* focused mainly on its proximate composition [1] and other nutritional attributes, such as fatty acids, vitamins, minerals, and β-glucans [2]; in particular, the physicochemical and functional properties of the sclerotial dietary fibres have been extensively investigated [3]. Among the bioactive components in *L. rhinocerotis*, the water-soluble, polysaccharide-protein complexes and β-glucans have been thoroughly studied for anti-tumour [4] and immunomodulatory effects [5]. On the other hand, little information on the low-molecular-weight constituents is available even though the use of *L. rhinocerotis* as folk medicine for overall wellness and cancer treatment [6] might be attributed to the presence of secondary metabolites with antioxidative (reduction of oxidative stress) and/or cytotoxic effects against cancer cells.

Wild-growing *L. rhinocerotis* make up the main source of these mushrooms; however, supply is limited due to their rarity [2,6]. Because of this, attempts have been made to domesticate this highly prized mushroom. Abdullah et al. [7] reported that solid-substrate fermentation of the mycelium on agroresidues yielded the fruiting body and sclerotium. In addition, liquid fermentation for the production of mycelium in bioreactors [8] as well as flasks under shaken [2] and static [9] conditions has been documented. Despite the advantages conferred by liquid fermentation for the production of fungal biomass and metabolites [10], the economic potential of the mycelium and culture broth of *L. rhinocerotis* as sources of nutraceuticals has been overlooked due to continued reliance and emphasis on the naturally occurring sclerotium. This is supported by the fact that previous studies on the mushroom's bioactivities focused solely on the sclerotium [4–5,11–13]. Indeed, the sclerotium is a compact mass of hardened mycelium; however, it is not known if the mycelium can substitute for the sclerotium with respect to bioactivities and chemical constituents. Besides, the chromatographic fingerprints of the extracts of *L. rhinocerotis* from different morphological/developmental stages have not been reported. Consequently, the chemical nature of many bioactive, low-molecular-weight compounds in the extracts remains unidentified [11,13]. Extensive studies were directed at bioactivity screening and metabolite production, but comparative studies on mushroom mycelia from different culture conditions of liquid fermentation (e.g., shaken and static conditions), which could produce varying amounts of active constituents and affect the bioactivities, has received lesser attention [14]. Aside from the mutagenicity and genotoxicity studies by Chen et al. [8], bioactivities of mycelium and culture broth of *L. rhinocerotis* have not been evaluated. In this study, we focused on the comparative analyses of bioactivities and chemical profiling of *L. rhinocerotis* from different morphological/developmental stages (mycelium and sclerotium) and culture conditions (shaken and static cultures) of liquid fermentation. The potential of the mycelium and culture broth as substitutes for the sclerotium is discussed.

Materials and Methods

Mushroom cultivation

The axenic culture of *L. rhinocerotis* (KUM61075) was obtained from the Mushroom Research Centre, University of Malaya. The sclerotium of *L. rhinocerotis* was produced by solid-substrate fermentation of mycelium on agroresidues according to the method of Abdullah et al. [7]. Harvested sclerotium was washed with distilled water and dried in the oven at 40°C for 3−5 days. The glucose-yeast extract-malt extract-peptone (GYMP, Oxoid, Hampshire, UK) medium was used for liquid fermentation [2]. Flasks were inoculated with mycelial plugs and incubated at 25°C under static conditions or placed on a reciprocal shaker at 150 rpm. After 15 days, the cultures were harvested; mycelium was filtered off from the culture broth and repeatedly washed with distilled water. Mycelium and culture broth were freeze-dried and kept in air-tight containers at −20°C.

Preparation of aqueous methanol extracts

Mushroom samples were ground to a fine powder using a Waring blender. The powdered mycelium and sclerotium as well as the freeze-dried culture broth were soaked in 80% (v/v) methanol (analytical grade) in water at a ratio of 1:20 (w/v) for 3 days. The extract was then decanted and filtered through Whatman No. 1 filter paper, and the residues were re-extracted twice. The filtrates were combined, and excess solvent was removed under pressure at 40°C using a rotary evaporator,

producing five brownish extracts: LR-MH (mycelium from shaken conditions); LR-MT (mycelium from static conditions); LR-BH (culture broth from shaken conditions); LR-BT (culture broth from static conditions); and LR-SC (sclerotium). The extracts were kept at −20°C prior to analyses. A summary of the different cultivation techniques, culture conditions of liquid fermentation, and extraction procedures involved is depicted in Figure 1.

Evaluation of antioxidant capacity of the extracts

The antioxidant capacity of *L. rhinocerotis* extracts was evaluated based on methods previously reported (below); hence, only the necessary modifications will be indicated. Standards including quercetin dihydrate, 6-hydroxy-2,5,7,8-tetramethylchromane-2-carboxylic acid (Trolox), ferrous sulphate heptahydrate ($FeSO_4 \cdot 7H_2O$), and disodium ethylenediamine tetraacetate (Na_2EDTA) were obtained from Sigma-Aldrich (St. Louis, USA), while 1,1,3,3-tetraetoxypropane (TEP) (the tetraethylacetal of malondialdehyde [MDA]) was purchased from Merck (Darmstadt, Germany). Other chemicals and solvents used were of analytical grade. All extracts were dissolved in 50% (v/v) methanol in water to produce stock solutions of 20 mg/ml and diluted to desired concentrations for the following assays:

2,2-Diphenyl-1-picrylhydrazyl (DPPH) free-radical-scavenging activity. The ability of the extracts to scavenge DPPH free radicals was measured according to methods of Kong et al. [15]. The results were expressed in terms of IC_{50} values (the concentration of extract required to produce 50% inhibition).

2,2'-Azino-bis(3-ethylbenzthiazoline-6-sulphonic acid) (ABTS) radical-scavenging activity. The ABTS radical-scavenging activity of the extracts was evaluated based on the method of Re et al. [16]. The results were expressed as mmol Trolox equivalent/g extract.

Ferric-reducing antioxidant power (FRAP) assay. The FRAP assay was performed according to Benzie and Strain [17] with modifications, in which 10 μl of extracts were mixed with 300 μl of freshly prepared FRAP reagent. The results were expressed as μmol $FeSO_4 \cdot 7H_2O$ equivalent/g extract.

Cupric ion-reducing antioxidant capacity (CUPRAC) assay. The CUPRAC assay was performed based on the method by Ribeiro et al. [18]. The results were expressed as μmol Trolox equivalent/g extract.

Metal-chelating activity. The ability of the extracts to chelate metal ions was analysed using the method by Jimenez-Alvarez et al. [19] with modifications. Briefly, 50 μl of extracts and 50 μl of 100 μM $FeCl_2$ were mixed. After 20 min of incubation, 50 μl of 100 μM ferrozine were added to the mixture. The results were expressed as μmol Na_2EDTA equivalent/g extract.

Inhibition of lipid peroxidation. The inhibitory effect of the extracts against lipid peroxidation was determined based on a method to measure thiobarbituric-acid-reactive substances (TBARS) in $FeSO_4$-induced lipid peroxidation in egg yolk homogenates [20] with minor modifications. The concentration of $FeSO_4$ used was 20 mM. The results were expressed as TEP equivalent/g extract.

Cell culture

The following cell lines were purchased from the American Type Culture Collection (ATCC, Manassas, VA, USA): A549 (human lung carcinoma); Caco-2; HCT 116; HT-29 (human colorectal carcinoma); Chang Liver (HeLa derivative); HEK-293 (human embryonic kidney); Hep G2 (human hepatocellular carcinoma); HL-60 (human acute promyelocytic leukemia); MCF7, MDA-MB-231 (human breast adenocarcinoma); MCF 10A (human breast epithelial); NRK-52E (rat kidney epithelial);

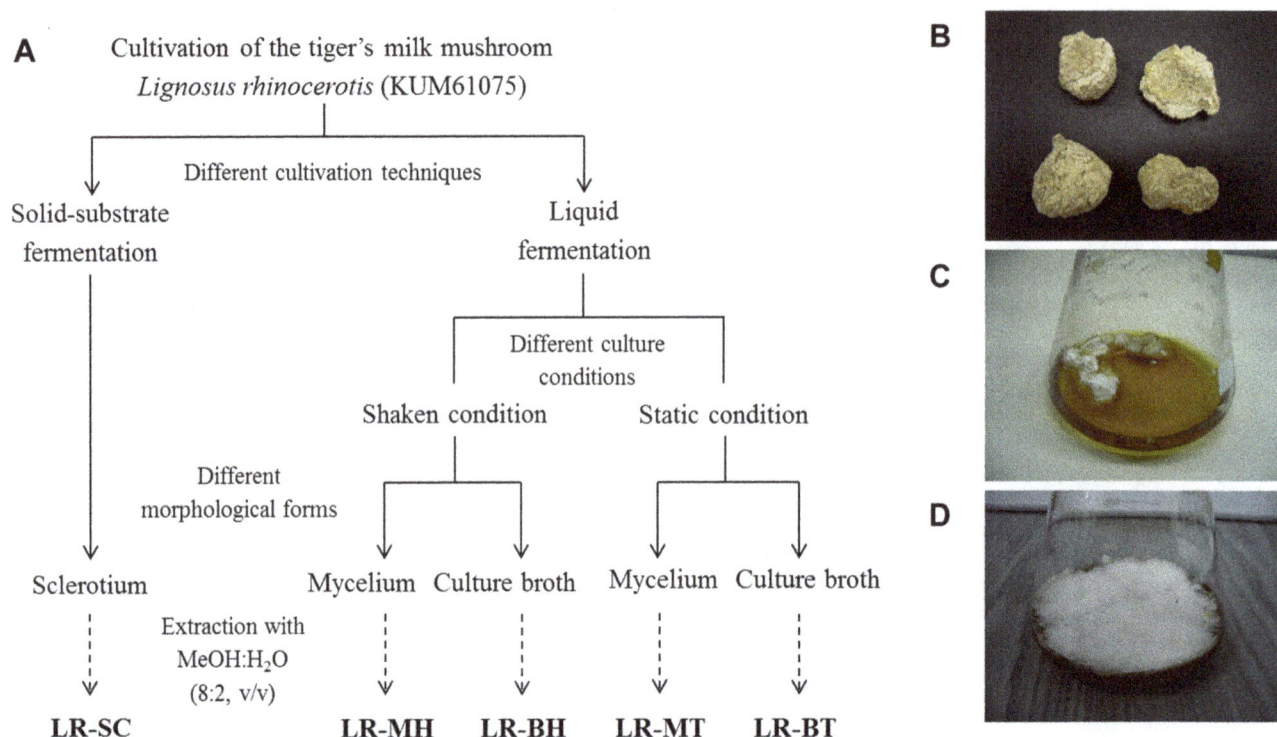

Figure 1. Overview of experimental design. (A) Cultivation of *Lignosus rhinocerotis* and extraction of low-molecular-weight compounds using aqueous methanol. Extracts were prepared from the mycelium (LR-MH, shaken cultures; LR-MT, static cultures), culture broth (LR-BH, shaken cultures; LR-BT, static cultures), and sclerotium (LR-SC). The different developmental/morphological forms of *L. rhinocerotis*: (B) sclerotium from solid-substrate fermentation, (C) mycelial pellet in shaken cultures, and (D) mycelial pellicle in static cultures of liquid fermentation.

PC-3 (human prostate adenocarcinoma); RAW 264.7 (mouse leukemic monocyte macrophage); Vero (African green monkey kidney epithelial); WRL 68 (HeLa derivative); and 4T1 (mouse mammary gland carcinoma). The HSC-2 (human oral squamous carcinoma) line was obtained from the Human Science Research Resources Bank (Japan), and HK1 (human nasopharyngeal carcinoma) was a gift from Professor Tsao at the University of Hong Kong. The OKF6 (immortalised human oral epithelial) and NP 69 (immortalised human nasopharyngeal epithelial) lines were obtained from the BWH Cell Culture and Microscopy Core at the Harvard Institutes of Medicine (USA) and University of Hong Kong Culture Collections, respectively.

Cell culture media and supplements were purchased from Gibco Invitrogen (Life Technologies, USA) unless otherwise stated. The A549, HT-29, HCT 116, HL-60, MCF7, PC-3, and 4T1 lines were maintained in RPMI-1640 media; Chang Liver, HEK-293, Hep G2, MDA-MB-231, NRK-52E, RAW 264.7, Vero, and WRL 68 were maintained in DMEM; while Caco-2, HK1, and HSC-2 were grown in MEM. All media were supplemented with 10% (v/v) heat-inactivated fetal bovine serum (FBS) and 100 units/ml of penicillin/streptomycin. The NP 69 and OKF6 lines were cultured in keratinocyte serum-free media (Keratinocyte-SFM, Invitrogen) supplemented with L-glutamine, human epidermal growth factor (hEGF, 0.1 ng/ml), bovine pituitary extract (BPE, 50 µg/ml), and Ca^{2+} (final concentration,

Table 1. Antioxidant capacity of *L. rhinocerotis* extracts.

Extracts	*DPPH (IC$_{50}$, mg/ml)	TEAC (mmol trolox equiv./g extract)	FRAP (µmol FeSO$_4$·7H$_2$O equiv./g extract)	CUPRAC (µmol Trolox equiv./g extract)	Metal chelating (µmol Na$_2$EDTA equiv./g extract)	Inhibition of lipid peroxidation (mmol MDA/g extract)
LR-MH	0.94±0.01 a	143±13.42 a	71.25±1.91 a	350.41±5.15 a	40.44±0.07 a	1.51±0.08 a
LR-MT	3.72±0.11 b	128.43±9.25 a	21.21±1.04 b	214.33±11.66 b, c	31.97±1.68 b	1.55±0.05 a
LR-BH	4.23±0.08 c	186.67±7.54 a, b	67.02±3.00 a	274.78±7.34 d	26.76±0.50 c	1.89±0.09 a
LR-BT	6.87±0.06 d	223.05±8.26 b	85.73±4.02 c	268.01±5.61 a, d	59.43±0.36 d	1.48±0.04 b
LR-SC	3.60±0.10 b	162.93±24.63 a, b	23.01±1.31 b	192.53±15.86 b	57.95±0.14 d	1.59±0.02 a

The extracts were dissolved in 50% (v/v) methanol in water for the antioxidant assays. Results were expressed as mean ± SE of at least three independent experiments (n = 3–5) performed in triplicates. The different letters (a–d) within a column represent means with significance difference ($p<0.05$).
*Quercetin dihydrate (IC$_{50}$: 0.091 mg/ml) was used as the positive control in the DPPH free radical scavenging assay.

Table 2. Cytotoxic effect of *L. rhinocerotis* extracts.

Cell lines and extracts	Cell viability (%) after 72 h of incubation (Mean ± SE)									
	LR-MH		LR-MT		LR-BH		LR-BT		LR-SC	
Concentration (µg/ml)	20	200	20	200	20	200	20	200	20	200
A549	98.1±2.37	94.3±5.03	82.5±2.46	67.0±2.70	101.0±1.10	99.5±2.28	105.2±4.85	89.2±2.94	98.5±5.33	80.0±1.73
Caco-2	99.2±2.42	91.6±3.02	97.1±2.07	79.3±2.89	95.0±0.94	79.9±3.25	98.4±5.46	84.1±1.16	98.6±3.17	76.9±1.56
Chang Liver	90.4±0.89	99.0±1.78	101.1±1.23	93.0±1.13	103.4±0.77	95.5±0.97	102.5±2.07	98.0±1.35	100.6±3.89	88.8±0.71
HCT 116	110.0±4.09	92.2±1.68	93.7±0.64	81.0±3.38	111.2±5.9	107.7±1.24	111.6±0.33	104.3±0.14	92.2±4.10	90.5±1.74
HEK-293	99.9±3.32	85.5±3.25	102.3±1.69	76.6±1.21	93.8±3.79	70.7±2.20	94.2±6.77	69.8±5.49	94.9±6.23	48.1±0.17
Hep G2	104.1±0.63	100.1±1.16	99.8±0.44	88.6±2.22	102.1±2.53	90.0±2.54	100.1±2.05	103.9±1.17	88.5±3.79	70.6±2.23
HK1	91.2±5.70	89.4±5.08	92.0±1.10	93.0±1.88	92.8±4.40	86.5±9.41	94.6±1.58	93.1±5.29	95.4±0.44	88.3±7.21
HL-60	101.0±1.63	83.3±0.84	108.1±1.97	96.4±1.87	119.6±3.86	100.9±2.11	124.2±3.03	101.1±1.31	106.2±0.43	91.0±2.43
HT-29	111.4±2.75	110.8±7.96	106.2±1.64	71.6±3.87	114.1±2.17	108.7±3.09	124.0±5.33	113.1±1.61	115.5±2.63	105.3±4.94
HSC2	98.2±3.71	90.4±0.94	101.3±2.09	94.5±1.53	90.9±0.88	91.1±1.27	95.0±1.25	97.7±2.11	92.6±1.34	95.8±1.87
MCF7	91.6±5.72	92.1±1.01	92.7±4.88	85.7±3.03	83.3±2.95	77.5±2.51	104.7±0.08	95.6±3.50	114.6±3.39	108.8±3.12
MCF 10A	110.3±3.64	81.4±1.95	102.1±2.57	37.1±4.34	100.6±6.95	107.3±3.23	102.3±6.54	83.2±4.69	93.8±3.95	94.0±2.22
MDA-MB-231	109.1±0.19	89.8±1.80	109.7±3.44	77.5±0.17	102.6±0.71	85.0±3.24	102.2±2.40	82.1±1.41	97.5±3.57	73.0±4.33
NP 69	81.2±0.87	65.4±0.20	79.6±0.74	62.1±2.29	79.6±0.68	67.5±1.97	79.1±0.82	77.5±0.69	60.3±0.42	56.01±0.38
NRK-52E	107.9±5.02	84.2±2.82	102.4±1.29	75.5±2.21	113.9±1.60	97.6±1.23	112.9±3.40	113.9±1.33	109.2±3.95	102.2±0.71
OKF6	93.7±0.76	68.3±0.16	89.0±1.11	65.6±2.83	105.6±2.19	103.1±1.69	105.7±0.48	105.4±0.57	78.0±1.86	70.7±0.67
PC3	91.2±3.29	89.4±2.93	92.0±0.64	93.0±1.09	92.8±2.54	86.5±5.43	94.6±0.91	93.1±3.05	95.4±0.25	88.3±4.16
RAW 264.7	99.9±4.66	88.8±3.82	97.4±2.00	75.4±0.58	103.9±1.03	107.8±4.26	102.3±3.73	94.7±3.22	93.1±3.64	64.1±2.19
Vero	101.3±1.85	94.0±2.24	85.8±1.00	84.6±1.66	118.7±1.43	117.2±2.55	123.3±4.21	132.7±0.64	90.6±2.57	80.6±3.15
WRL 68	104.3±3.68	105.8±3.32	91.8±2.89	83.7±0.68	98.5±0.11	102.7±0.42	97.8±0.19	104.7±1.21	94.7±1.79	90.4±1.81
4T1	98.9±4.35	92.7±4.07	89.7±0.76	69.9±1.16	84.2±3.35	68.8±2.23	87.6±5.44	86.9±6.07	72.6±2.87	61.1±1.52

The extracts were dissolved in 50% (v/v) dimethyl sulphoxide (DMSO) in water and diluted in media for the MTT assay. The final concentration of DMSO in the well was less than 0.5% (v/v) and this did not affect cell viability. Results were expressed as mean ± SE of three independent experiments ($n = 3$) performed in triplicates.

Table 3. Chemical composition of *L. rhinocerotis* extracts.

Extracts	Total sugars (mg glucose/g extract)	Total proteins (mg protein/g extract)	Total phenolic content (mg GAE/g extract)
LR-MH	185.8±3.91 a	3.4±0.08 a	18.8±1.49 a
LR-MT	413.9±41.6 b	4.5±0.15 b	7.9±2.12 b
LR-BH	267.9±22.6 c	3.7±0.13 c	15.3±1.12 a, c
LR-BT	267.9±22.6 d	1.9±0.11 d	11.8±1.14 b, c
LR-SC	118.1±16.51 a, d	7.4±0.07 e	13.2±2.41 c

Results were expressed as mean ± SD of triplicate measurements. The different letters (a–d) within a column represent means with significance difference ($p < 0.05$).

0.3 mM). The MCF 10A line was grown in serum-free mammary epithelial growth media (MEGM BulletKit, Lonza, USA). The basal medium was supplemented with BPE (50 μg/ml), hydrocortisone (0.5 μg/ml), hEGF (10 ng/ml), insulin (5 μg/ml), and cholera toxin (100 ng/ml). Cells were cultured in a 5% CO_2 incubator at 37°C in a humidified atmosphere.

Evaluation of cytotoxic effect of the extracts

The effect of the extracts on cell viability was evaluated using the 3-(4,5-dimethylthiazol-2-yl)-2,5-diphenyltetrazolium bromide (MTT) (Sigma-Aldrich, USA) assay, as previously described [21]. All extracts were dissolved in 50% (v/v) dimethyl sulphoxide (DMSO) in water to produce stock solutions of 50 mg/ml, which were further diluted with culture media to desired concentrations. Cells ($3-5 \times 10^3$ cells/well) were seeded and allowed to attach overnight prior to treatment with extracts at final concentrations of

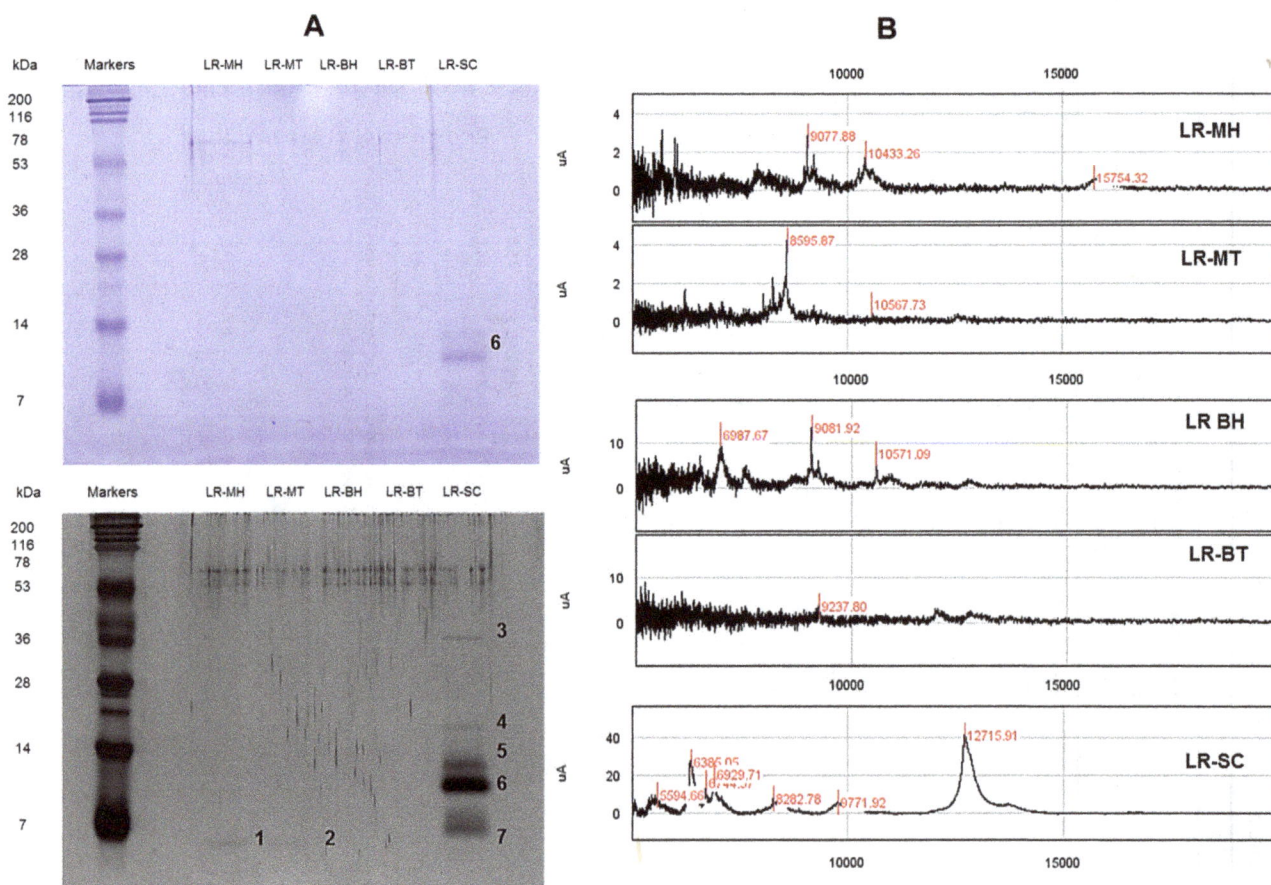

Figure 2. Protein profiling. (A) Electrophoretic analysis of proteins in the extracts of *Lignosus rhinocerotis* and visualisation by Coomassie Brilliant Blue (top) and silver staining (bottom). Molecular weight (MW) of the bands was estimated from the plot of log MW vs. relative migration distance (R_f) based on the values obtained from the bands of the marker (7–200 kDa). The estimated sizes of the bands were as follows: 1, 2 (4.0 kDa), 3 (38.0 kDa), 4 (14.0 kDa), 5 (9.5 kDa), 6 (8.0 kDa), and 7 (4.7 kDa). (B) Representative SELDI-TOF-MS spectra of the low-molecular-weight proteins (5–20 kDa) in the extracts. The x-axis represents the m/z values, and the y-axis represents the intensity of the signals (μA). Peaks with signal/noise ratios (S/N) >5 were automatically detected.

Table 4. Chemical constituents in LR-MH and LR-MT based on GC-MS analysis.

R_T (min)	Compounds	Molecular formula	Molecular weight	Area (%)	
				LR-MH	LR-MT
12.12	Methyl β-D-galactopyranoside	$C_7H_{14}O_6$	194.18	8.16	ND
12.97	Arabinitol	$C_5H_{12}O_5$	152.15	ND	4.46
13.51	Pyrrolo[1,2-a]pyrazine-1,4-dione, hexahydro-	$C_7H_{10}N_2O_2$	154.17	8.36	10.78
14.91	Hexadecanoic acid, methyl ester	$C_{17}H_{34}O_2$	270.45	1.23	6.50
15.30	n-Hexadecanoic acid	$C_{16}H_{32}O_2$	256.42	6.14	7.10
16.56	9,12-Octadecadienoic acid (Z,Z)-, methyl ester	$C_{19}H_{34}O_2$	294.47	2.63	11.23
16.97	9,12-Octadecadienoic acid (Z,Z)-	$C_{18}H_{32}O_2$	280.45	11.73	14.52
17.19	Octadecanoic acid	$C_{18}H_{36}O_2$	284.48	ND	1.25
17.39	Hexadecanamide	$C_{16}H_{33}NO$	255.44	ND	1.40
18.35	11,13-Eicosadienoic acid, methyl ester	$C_{21}H_{38}O_2$	322.53	ND	0.53
18.70	Cyclopentadecanone, 2-hydroxy-	$C_{15}H_{28}O_2$	240.38	ND	0.23
18.93	9-Octadecenamide, (Z)-	$C_{18}H_{35}NO$	281.48	4.76	3.54
21.51	9,12-Octadecadienoic acid (Z, Z)-, 2-hydroxy-1-(hydroxymethyl)ethyl ester	$C_{21}H_{38}O_4$	354.52	13.26	16.98
22.60	2,3-Dihydroxypropyl elaidate	$C_{21}H_{40}O_4$	356.54	0.28	0.27
25.69	Ergosterol	$C_{28}H_{44}O$	396.65	ND	1.30

Area (%) was determined based on the TIC of LR-MH (Figure S1) and LR-MT (Figure S2). Identification of the compounds was based on mass spectral analysis. R_T, retention time; ND, not detected.

20 and 200 μg/ml. The percentage of cell viability after 72 h of incubation was determined by the following equation:

Cell viability (%) =

(absorbance of treated cells/absorbance of untreated cells)

× 100

Determination of chemical composition of the extracts

Total sugars were determined using the phenol-sulphuric assay [22]. D-glucose (Merck) was used as the standard. Protein content was analysed using the Pierce Coomassie Plus (Bradford) Protein Assay (ThermoScientific, Massachusetts, USA) according to the manufacturer's protocol, with bovine serum albumin as the standard. The level of phenolics was estimated using the Folin-Ciocalteu reagent [23] with gallic acid (Sigma-Aldrich, USA) as the standard.

Table 5. Chemical constituents in LR-BH and LR-BT based on GC-MS analysis.

R_T (min)	Compounds	Molecular formula	Molecular weight	Area (%)	
				LR-BH	LR-BT
3.61	2-Furancarboxaldehyde, 5-methyl-	$C_6H_6O_2$	110.11	0.89	0.65
4.46	Benzeneacetaldehyde	C_8H_8O	120.15	1.24	0.51
6.70	1,4:3,6-Dianhydro-α-D-glucopyranose	$C_6H_8O_4$	144.13	2.08	ND
11.16	β-D-glucopyranose, 1,6-anhydro	$C_6H_{10}O_5$	162.14	0.80	ND
13.68	Pyrrolo[1,2-a]pyrazine-1,4-dione, hexahydro-	$C_7H_{10}N_2O_2$	154.17	15.60	20.12
15.19	Pyrrolo[1,2-a]pyrazine-1,4-dione, hexahydro-3-(2-methylpropyl)-	$C_{11}H_{18}N_2O_2$	210.27	ND	4.18
15.23	n-Hexadecanoic acid	$C_{16}H_{32}O_2$	256.42	2.86	ND
16.95	9,12-Octadecadienoic acid (Z,Z)-	$C_{18}H_{32}O_2$	280.45	2.63	ND
17.36	Hexadecanamide	$C_{16}H_{33}NO$	255.44	2.50	ND
18.93	9-Octadecenamide, (Z)-	$C_{18}H_{35}NO$	281.48	7.43	3.19
19.25	Pyrrolo[1,2-a]pyrazine-1,4-dione, hexahydro-3-(phenylmethyl)-	$C_{14}H_{16}N_2O_2$	244.29	ND	1.07

Area (%) was determined based on the TIC of LR-BH (Figure S3) and LR-BT (Figure S4). Identification of the compounds was based on mass spectral analysis. R_T, retention time; ND, not detected.

Table 6. Chemical constituents in LR-SC based on GC-MS analysis.

R_T (min)	Compounds	Molecular formula	Molecular weight	Area (%) LR-SC
12.87	D-glucopyranoside, methyl	$C_7H_{14}O_6$	194.18	32.17
13.55	Pyrrolo[1,2-a]pyrazine-1,4-dione, hexahydro	$C_7H_{10}N_2O_2$	154.17	3.51
15.28	n-Hexadecanoic acid	$C_{16}H_{32}O_2$	256.42	3.31
16.43	Oleic acid	$C_{18}H_{34}O_2$	282.46	0.24
16.94	9,12-Octadecadienoic acid (Z,Z)-	$C_{18}H_{32}O_2$	280.45	4.71
17.15	Octadecanoic acid	$C_{18}H_{36}O_2$	284.49	1.04
18.91	9-Octadecenamide, (Z)-	$C_{18}H_{35}NO$	281.48	1.56
25.68	Ergosta-4,7,22-trien-3β-ol	$C_{28}H_{44}O$	396.65	5.31

Area (%) was determined based on the TIC of LR-SC (Figure S5). Identification of the compounds was based on mass spectral analysis. R_T, retention time; ND, not detected.

Electrophoretic analysis of proteins

Sodium dodecyl sulphate-polyacrylamide gel electrophoresis (SDS-PAGE) of the extracts was carried out using 16% (w/v) separating and 5% (w/v) stacking gels in a vertical slab gel apparatus (C.B.S. Scientific Company, Inc., California, USA), as previously reported [21]. Bands were visualised by Coomassie Brilliant Blue R-250 (Sigma-Aldrich) and silver staining.

Chromatographic and mass-spectrometric analyses

SELDI-TOF-MS. For the surface-enhanced-laser-desorption-ionisation time-of-flight mass spectrometry (SELDI-TOF-MS) analysis, extracts were spotted on the reverse-phase or hydrophobic H50 ProteinChip arrays and analysed using a ProteinChip SELDI System PSC 4000 (Bio-Rad Laboratories, Inc., California, USA), as previously described [9,21].

GC-MS. The gas chromatography-mass spectrometry (GC-MS) analysis was performed using a 6890 N gas chromatograph (Agilent Technologies, Inc., California, USA) equipped with a 5975 Mass Selective Detector. The HP-5 MS (5% phenylmethyl-siloxane) capillary column (30.0 m×25 mm×25 μm) was initially set at 70°C, increased to 300°C, and then held for 10 min. Helium was used as the carrier gas at flow rate of 1 ml/min. The total ion chromatogram (TIC) was auto-integrated by ChemStation. Chemical constituents were identified by comparison with the accompanying spectral database (NIST 2011, Mass Spectral Library, USA) and literature, where applicable.

UHPLC-ESI-MS/MS. The analysis was performed using a Flexar FX15 ultra high-performance liquid chromatograph (UHPLC, PerkinElmer, Inc., Massachusetts, USA) coupled with an AB SCIEX 3200 QTrap hybrid linear ion trap triple-quadruple mass spectrometer equipped with a turbo ion spray source. Chromatographic separation was achieved on a Phenomenex Aqua C18 (5 μm, 50 mm×2 mm) column. Mobile phase A was composed of water with 0.1% (v/v) formic acid and 5 mM ammonium formate, whereas the mobile phase B consisted of acetonitrile containing 0.1% (v/v) formic acid and 5 mM ammonium formate. Elution was performed by means of a linear gradient from 10−90% B (0−8 min) held for 3 min, returned to 10% B in 0.1 min, and then re-equilibrated for 4 min before the next injection. Ionisation was achieved via electrospray ionisation on the AB Sciex Turbo V source with an ionisation temperature of 500°C and purified nitrogen gas (99%) as the collision gas via nebulisation. Collision energy was set at 35 eV for mass-fragmentation purposes. Full scan with MS/MS data collection analyses was performed in negative mode. Data analysis, processing, and interpretation were carried out using the AB SCIEX Analyst 1.5 and Advanced Chemistry Development, Inc., (ACD/Labs, Ontario, Canada) MS Processor software. Marker-View Software (AB SCIEX, Massachusetts, USA) was used for principal component analysis (PCA). The following parameters were used for PCA: retention time (R_T) range: 0−15 min, R_T tolerance: 0.5 min, mass range: m/z 100−1000, mass tolerance: 0.01 Da, and noise threshold: 5.

Statistical analysis

Analysis was performed in triplicates. Results were expressed as means ± standard deviation (SD) or standard error (SE). The data were statistically analysed using the IBM SPSS Statistics Version 19 software (SPSS Inc., New York, USA). All mean values were analysed by one-way analysis of variance (ANOVA) followed by Tukey's Honestly Significant Difference ($p < 0.05$) to detect significance between groups.

Results and Discussion

Yields of mushroom samples

Liquid fermentation is routinely used for production of mushroom mycelia and metabolites. The yields (g/100 ml) of the freeze-dried culture broth (LR-BH: 2.94, LR-BT: 3.30) were higher than that of mycelium (LR-MH: 0.62, LR-MT: 0.59) regardless of culture conditions of liquid fermentation. Although noted as a slow-growing species [7], the yield of the mycelium of *L. rhinocerotis* was higher than the maximum yield of the mycelium of an edible mushroom (*Agaricus brasiliensis*) cultured by shaken flasks (1.02 g/100 ml) [24] and an medicinal mushroom (*Ganoderma lucidum*) cultured in static flasks (1.25 g/100 ml) [14]; however, the culture conditions, such as media composition and aeration, were different, and these could have affected fungal growth. On the other hand, the yield of sclerotium (on a dry-weight basis) was 1.3−2.0 g/g substrate [7].

Yields of mushroom extracts

A mixture of methanol and water was used to widen the spectrum of extracted low-molecular-weight constituents, especially compounds with higher polarity, such as phenolic compounds [24]. The yields (w/w) of the aqueous methanol extracts (in descending order) were LR-BT (75.1%) > LR-BH (69.9%) > LR-MT (21.0%) > LR-MH (14.7%) > LR-SC (2.7%). The higher yields of LR-BH and LR-BT, compared to other extracts, indicated that constituents in the culture broths were readily

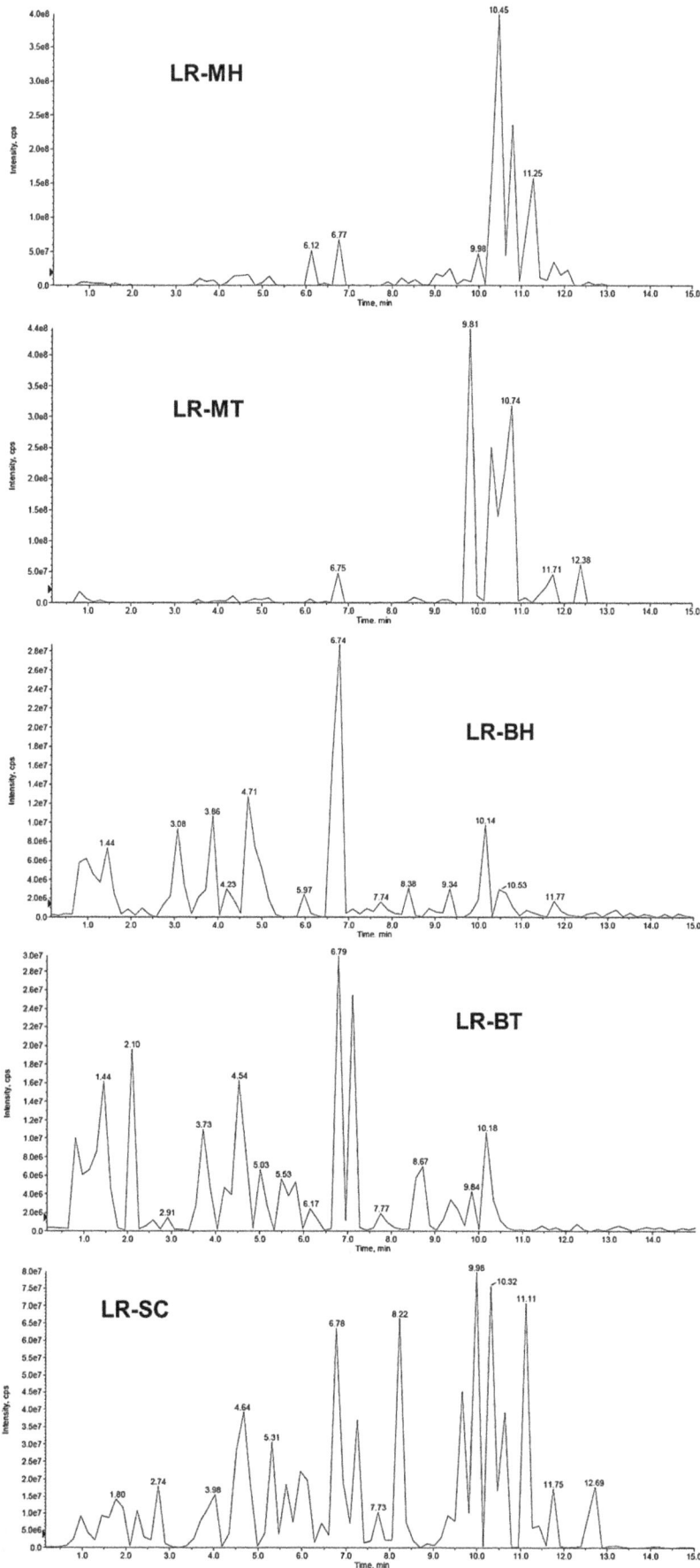

Figure 3. The UHPLC-ESI-MS TIC (negative mode) of the extracts of *Lignosus rhinocerotis.* The profiles of the extracts of the mycelium (LR-MH, LR-MT), culture broth (LR-BH, LR-BT), and sclerotium (LR-SC) were different.

soluble in aqueous methanol. Our results were consistent with previous findings where the yield of culture broth was higher than that of mycelia and/or fruiting bodies [25,26]. The yield of LR-SC was the lowest, as the sclerotium of *L. rhinocerotis* was reported to predominantly consist of carbohydrates [1,2], such as dietary fibres, that are insoluble in the extraction solvent used in this study.

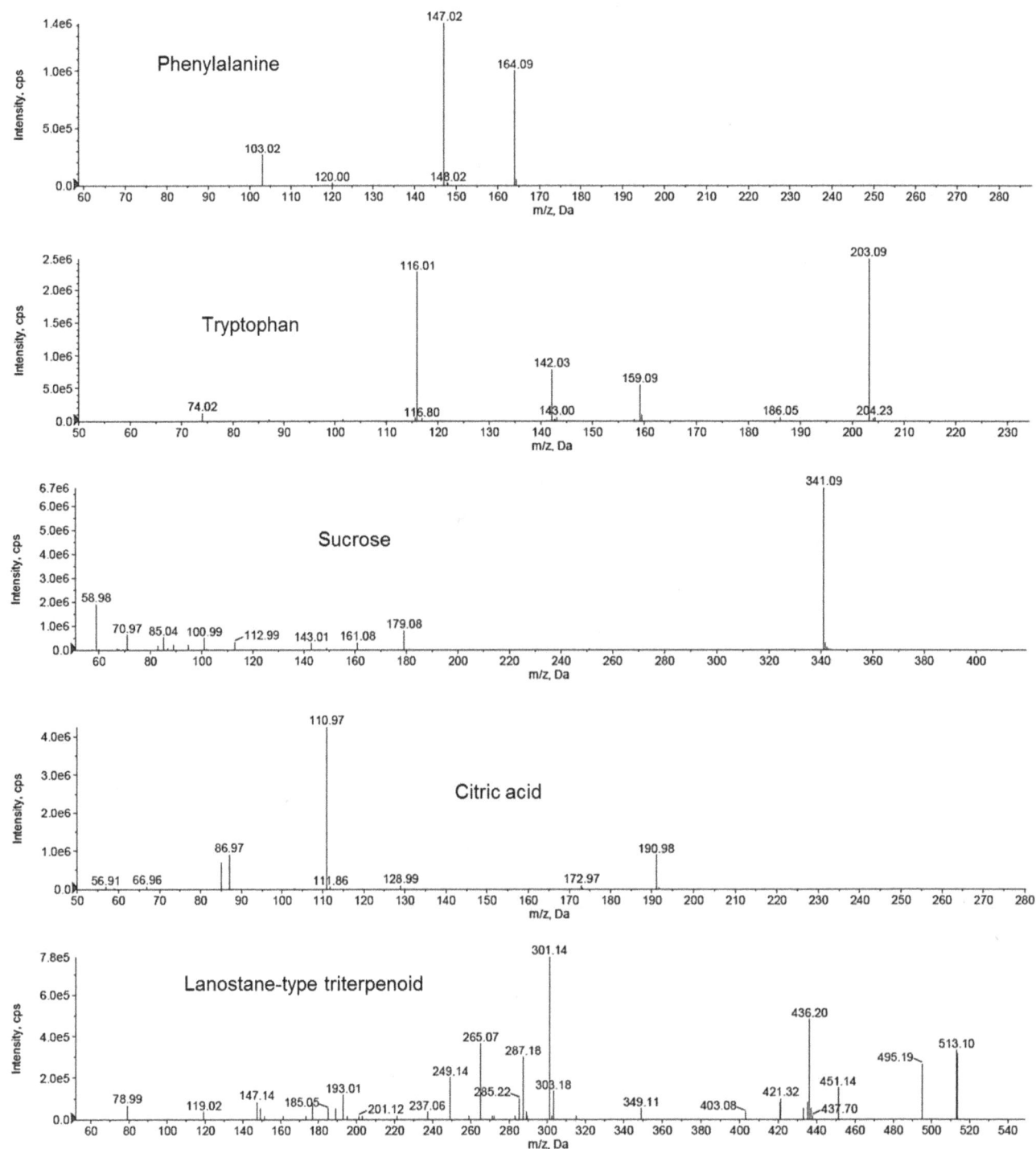

Figure 4. The MS/MS fragmentation (in negative mode) of selected low-molecular-weight compounds in the extracts of *Lignosus rhinocerotis.* Collision energy was set at 35 eV. The compounds were tentatively identified based on their mass fragmentation patterns.

Table 7. Chemical constituents in LR-MH and LR-MT based on UHPLC-ESI-MS/MS.

R_T (min)	[M-H]$^-$	Mass fragments, MS/MS	Suggested identification	Reference
LR-MH				
1.13	164	147, 120, 103	Phenylalanine	Ying et al. [36] Lu et al. [35] MassBank
1.61	241	197, 167, 141	2-(2-amino-3-imidazol-5-ylpropanoylamino)-3-hydroxypropanoic acid (Histidylserine)	MassBank
3.87	497	451, 433, 333, 225	Lanostane-type triterpenoid	Yang et al. [33] Liu et al. [34]
4.51	451	433, 333, 225	Lanostane-type triterpenoid	Yang et al. [33] Liu et al. [34]
6.12	345	201, 171	Derivative of 9,10-dihydroxy-12Z-octadecenoic acid	MassBank
LR-MT				
0.80	341	179, 161, 143, 113, 101, 85, 71, 59	Sucrose	Brudzynski and Miotto [37] Taylor et al. [38]
6.11	345	201, 171	Derivative of 9,10-dihydroxy-12Z-octadecenoic acid	MassBank

R_T, retention time.

Comparative antioxidant capacity

Antioxidants confer protection against cellular damage caused by oxidative stress and thus potentially ameliorate diseases, such as cancer, diabetes, and cardiovascular and neurodegenerative disorders. The medicinal properties of *L. rhinocerotis* might be partially associated with its antioxidant capacity. In this study, several assays based on different antioxidant mechanisms were employed to assess the antioxidant capacity of the extracts. The free radical-scavenging activities, reducing properties, metal-chelating activities, and inhibitory effects on lipid peroxidation by the extracts of *L. rhinocerotis* are shown in Table 1. Overall, the antioxidant capacity of the mycelium and culture broth of *L. rhinocerotis* was found to be either higher or comparable to that of the sclerotium; however, the relative potency of the five extracts, in different assays, was not consistent. For radical scavenging, the extracts exhibited varying degrees of DPPH free-radical-scavenging activities with extracts of the mycelium, and sclerotium (IC$_{50}$: 0.9−3.6 mg/ml) showed stronger scavenging activities than those of the culture broth (IC$_{50}$: 4.2−6.9 mg/ml). The ability of the extracts to quench the ABTS radicals was comparable, but the activity decreased in the order of culture broth > sclerotium > mycelium. The reducing properties of the extracts were measured using the FRAP and CUPRAC assays. In the FRAP assay, LR-BH, LR-BT, and LR-MH showed higher reducing properties (67.0−85.7 µmol FeSO$_4$·7H$_2$O equivalent/g extract) than other extracts. The reducing properties of the extracts as measured by the CUPRAC assay revealed a trend consistent with the FRAP assay, in that LR-MH, LR-BH, and LR-BT also exhibited higher activities (268.0−350.4 µmol Trolox equivalent/g extract). Through the Fenton reactions, hydroxyl radicals generated by transition metals could stimulate lipid peroxidation. By stabilising transition metals, chelating agents might impair the production of free radicals. The metal-chelating activity of the extracts ranged from 26.8−59.4 µmol Na$_2$EDTA equivalent/g extract. The LR-BT exhibited the highest ferrous-chelating activity, comparable to LR-SC. On the other hand, the level of MDA was taken as an indicator of lipid peroxidation, where a lower concentration of MDA reflects a higher inhibitory potential. The inhibitory

Table 8. Chemical constituents in LR-BH and LR-BT based on UHPLC-ESI-MS/MS.

R_T (min)	[M-H]$^-$	Mass fragments, MS/MS	Suggested identification	Reference
LR-BH				
0.80	377	341, 221, 179, 161, 97, 87	Hexose-based compound	MassBank
1.29	227	183	Phenolic	-
1.45	241	197, 181, 169, 140	Derivative of emodin	MassBank
4.20	497	451, 433, 225	Lanostane-type triterpenoid	Yang et al. [33] Liu et al. [34]
4.84	451	433, 333, 225, 207, 81	Lanostane-type triterpenoid	Yang et al. [33] Liu et al. [34]
4.99	497	451, 433, 333, 225	Lanostane-type triterpenoid	Yang et al. [33] Liu et al. [34]
LR-BT				
1.29	203	159, 143, 116. 74	Tryptophan	Ying et al. [36] MassBank
1.45	241	197, 181, 169, 140	Derivative of emodin	Von Wright et al. [42] MassBank
4.20	497	451, 433, 333, 225	Lanostane-type triterpenoid	Yang et al. [33] Liu et al. [34]
5.01	497	451, 433, 333, 225	Lanostane-type triterpenoid	Yang et al. [33] Liu et al. [34]

R_T, retention time.

Table 9. Chemical constituents in LR-SC based on UHPLC-ESI-MS/MS.

R_T (min)	[M-H]$^-$	Mass fragments, MS/MS	Suggested identification	Reference
0.96	191	172, 111, 87	Citric acid	John and Shahidi [39] MassBank
3.87	451	433, 333, 225, 207, 143	Lanostane-type triterpenoid	Yang et al. [33] Liu et al. [34]
7.73	513	495, 451, 436, 301, 265, 249, 193	Lanostane-type triterpenoid	Yang et al. [33] Liu et al. [34]
8.22	495	451, 301, 285, 193, 149	Lanostane-type triterpenoid	Yang et al. [33] Liu et al. [34]
10.31	564	504, 279, 224, 153	Lanostane-type triterpenoid	Yang et al. [33] Liu et al. [34]
10.63	504	279, 224, 153	Lanostane-type triterpenoid	Yang et al. [33] Liu et al. [34]

R_T, retention time.

potentials of the extracts against $FeSO_4$-induced lipid peroxidation were comparable to each other except for LR-BT, in which the MDA level was significantly lower ($p<0.05$).

Previous investigations found that no firm conclusions regarding the relative antioxidant capacity of mushroom samples from different morphological/developmental stages and cultivation techniques, such as the fruiting body, mycelium, culture broth, and/or sclerotium. A direct comparison of values obtained from antioxidant capacity evaluation assays performed in different laboratories is not possible due to the differences in methodologies used. In addition, comparative analyses on the antioxidant capacity of mushrooms from different morphological/developmental stages are scarce, and, in most cases, findings are inconsistent. For instance, according to Reis et al. [27], the

fruiting bodies of several cultivated mushrooms generally revealed higher antioxidant properties than the corresponding mycelia. In a separate report on *A. brasiliensis*, Carvajal et al. [24] found that mycelial extracts exhibited stronger ABTS radical-scavenging and ferrous ion-chelating abilities but weaker DPPH free-radical scavenging and inhibition of lipid peroxidation than the fruiting body. Wong et al. [28] observed that the mycelial extract (consisting of both mycelium and culture broth) of *Hericium erinaceus* showed stronger reducing capacity than the fruiting bodies as determined by the FRAP assay, but the extract's ability to scavenge DPPH free radicals was lower. When comparisons are made between fruiting bodies and mycelia, other factors such as mushroom strain, cultivation techniques, culture conditions, and postharvest processing should be considered. As indicated, most

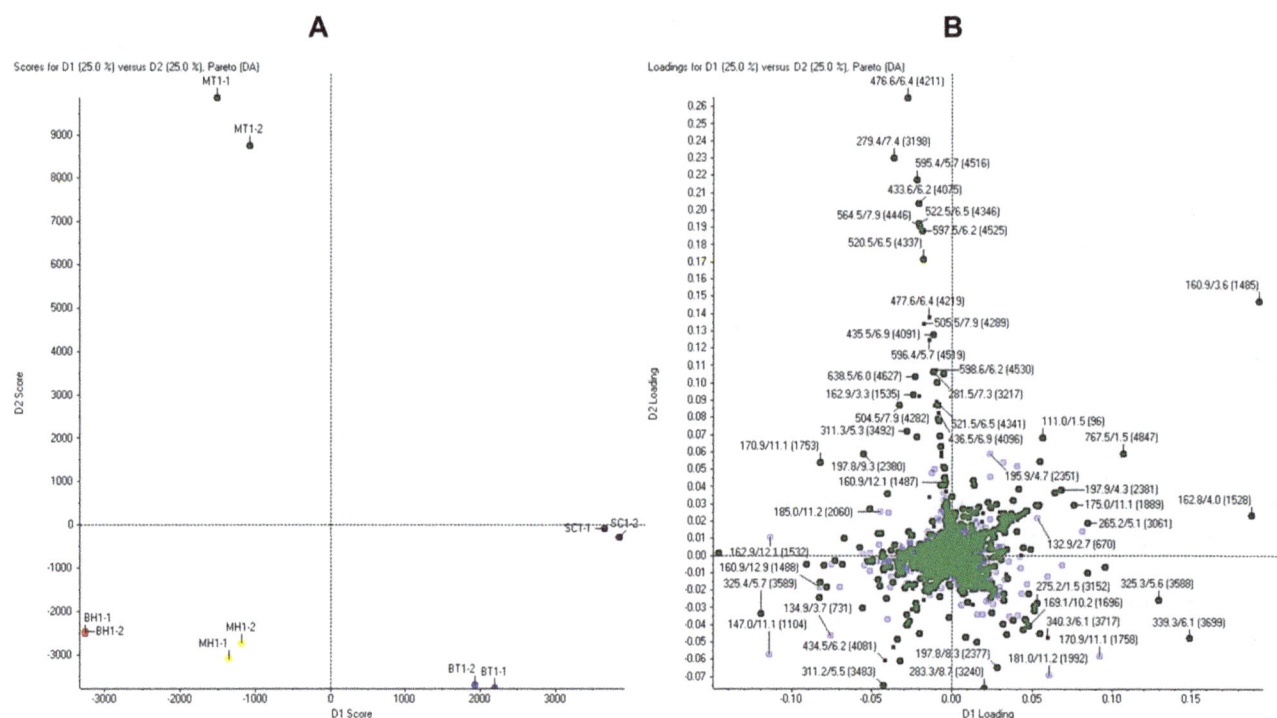

Figure 5. The UHPLC-ESI-MS (m/z 100–1000) principal component analysis of the extracts of *Lignosus rhinocerotis*. Duplicate analysis of the extracts of mycelium from shaken (MH1, MH2) and static (MT1, MT2) conditions, culture broth from shaken (BH1, BH2) and static (BT1, BT2) conditions, and sclerotium (SC1, SC2) were performed. (A) Score plot revealed that mycelia from shaken and static conditions were distinct from the sclerotium. (B) Loading plot with multiple ions common to all extracts (centre) and marker ions far from the centre, e.g. m/z 161, 325, 339, and 766, were characteristic of individual extracts. The identification of the compounds is warranted for determining biomarkers for *L. rhinocerotis* from different morphological/developmental stages.

studies focused on the comparison between fruiting bodies and mycelia, and sclerotia received lesser attention. One plausible explanation for this is that very few sclerotia-producing mushrooms are commercially available. Since mushroom sclerotia are, in general, predominantly carbohydrates, the antioxidant capacity of different types of sclerotial polysaccharides have been extensively studied, e.g. the water- and alkaline-soluble polysaccharides from *Pleurotus tuber-regium* [29] and *Inonotus obliquus* [30]; however, antioxidants in the form of low-molecular-weight constituents remain poorly investigated.

Comparative cytotoxic effect

Earlier, Chen et al. [8] reported that the mycelium of *L. rhinocerotis* did not provoke mutagenicity and genotoxicity; however, its cytotoxicity in mammalian cells was not evaluated. In light of this, extracts were screened for cytotoxicity against a panel of 21 mammalian cell lines using the MTT assay. According to the U.S. National Cancer Institute, crude extracts with IC_{50} values less than 20 µg/ml, after an incubation period of 48−72 h, are considered active [31]. As shown in Table 2, cellular viability of most cells was maintained above 70% following treatment with 20 µg/ml of extracts; hence, the extracts were considered non-cytotoxic. At higher concentration (200 µg/ml), some of the non-tumourigenic cells, usually used as models of normal cells in cytotoxicity evaluation, were found to be more susceptible than the corresponding solid tumours. For instance, NP 69 and OKF6 were observed to be more susceptible to the extracts than HK1 and HSC-2, respectively; hence, this implied non-selective cytotoxicity of the extracts against these cell lines.

Our results also indicated that the extracts of the mycelium and culture broth of *L. rhinocerotis* showed mild cytotoxic effects against most cell lines, comparable to the sclerotium extract. LR-MT (200 µg/ml) was noted to exert relatively strong cytotoxicity against MCF 10A (cell viability: 37.1%) compared to MCF7 (85.7%) and MDA-MB-231 (77.5%); however, other extracts (LR-MH, LR-BH, LR-BT, LR-SC) did not affect the viability of MCF 10A (>80%), and this is likely to indicate presence of cytotoxic metabolites that might be found only in LR-MT. This is the first attempt to screen for cytotoxicity in the extracts of *L. rhinocerotis* from liquid fermentation (i.e., mycelium and culture broth). Earlier, the cytotoxic effects of alcoholic extracts of *L. rhinocerotis* sclerotium were studied, albeit there was slight variation in the methodology used (e.g., solvent and extraction techniques, cell lines, and duration of treatment). In a previous study by Eik et al. [11], an ethanol extract of the sclerotium of the *L. rhinocerus* TM02 cultivar was reported to exert low toxicity (IC_{50}: 282.1 µg/ml) against PC-12 cells (rat pheochromocytoma) after 48 h treatment. Similarly, an aqueous methanol extract of a wild-type *L. rhinocerotis*, prepared using a pressurised liquid extraction method, showed weak cytotoxicity (IC_{50}: 600 µg/ml) against HCT 116 and no effect (IC_{50}>2000 µg/ml) against CCD-18Co (human colon fibroblast) cells after 24 h treatment [13]. Therefore, the alcoholic extracts of the sclerotium of *L. rhinocerotis* (including LR-SC from this study), in general, were non-cytotoxic (IC_{50}> 20 µg/ml) against mammalian cells. On the other hand, the cold aqueous extracts showed relatively high cytotoxicity based on previous findings. Cytotoxic components in the sclerotium of *L. rhinocerotis* were suspected to be mainly heat-labile protein/peptide(s) [21] and/or high-molecular-weight, protein-carbohydrate complexes [12], rather than the low-molecular-weight constituents.

Chemical composition

Table 3 shows the chemical characterisation of the extracts of *L. rhinocerotis*. The extracts contained relatively low concentrations of sugars and proteins. In aqueous methanol, the solubility of polysaccharides and proteins is low, but simple compounds (e.g., sugars, amino acids, and peptides) can be dissolved [24]. Although the level of sugars in LR-SC was the lowest, its protein content was significantly higher than others. The concentration of phenolics in the extracts ranged from 7.9−18.8 mg gallic acid equivalent/g extract. Interestingly, the mycelium and culture broth from shaken cultures contained significantly higher phenolics than their counterparts from static cultures.

Protein profiling

The higher protein content in LR-SC compared to other extracts was confirmed with further protein profiling (Figure 2). Results from SDS-PAGE showed that LR-SC was characterised by a single band (approximately 8 kDa) that could be visualised after Coomassie blue staining. Silver staining, a more sensitive visualisation technique, revealed the presence of other proteins, presumably those of lower abundance. These include a faint band (approximately 5 kDa) common to LR-MH and LR-MT and additional bands (5−40 kDa) in LR-SC. Compared to our previous work [21], LR-SC lacked most of the proteins present in the cold aqueous extract of *L. rhinocerotis*. The SELDI-TOF-MS analysis was performed to detect low-molecular-weight proteins that might have been resolved poorly on the gel. The number of peaks in the extracts and their intensities were low. Most peaks were in the range of 15 kDa or less. The SELDI-TOF-MS spectrum of LR-SC was different from that of LR-MH, LR-MT, LR-BH, and LR-BT; however, the profile showed some resemblance to that of the cold aqueous extract of *L. rhinocerotis*, as previously reported [21].

Identification of chemical constituents by GC-MS

By using GC-MS, several low-molecular-weight compounds composed of sugars and their derivatives, fatty acids and their methyl esters, cyclic peptides, sterols, and amides in the extracts of *L. rhinocerotis* were identified (Tables 4−6, Figures S1−S5). The LR-MH and LR-MT were characterised by the presence of 9,12-octadecadienoic acid (Z,Z) (linoleic acid) (11.7−14.5%), its methyl ester (2.6−11.2%), and a derivative of its ethyl ester, 9,12-octadecadienoic acid (Z,Z)-,2-hydroxy-1-(hydroxymethyl)ethyl ester (2-monolinolein) (13.3−17.0%). Both extracts also contained n-hexadecanoic acid (palmitic acid) (6.1−7.1%) and its methyl ester (1.2−6.5%); however, hexadecanamide (palmitic amide) was detected only in LR-MT. Some compounds were found only in LR-MT, such as arabinitol, octadecanoic acid (stearic acid), and ergosterol.

On the other hand, pyrrolo[1,2-a]pyrazine-1,4-dione, hexahydro or cyclo(leucyloprolyl) (15.6−20.1%) and 9-octadecanamide (7.4−3.2%) were the major compounds in LR-BH and LR-BT. Another two cyclic peptides were present only in LR-BT. These were pyrrolo[1,2-a]pyrazine-1,4-dione, hexahydro-3-(2-methylpropyl)- or cyclo(D-phenylalanyl-L-prolyl) (4.2%) and pyrrolo[1,2-a]pyrazine-1,4-dione,hexahydro-3-(phenylmethyl)- or cyclo(phenylalanylprolyl) (1.1%). Several compounds identified from the mycelial extracts (e.g., n-hexadecanoic acid and hexadecanamide) were also found in LR-BH. In addition, sugars and their derivatives, such as 1,4:3,6-dianhydro-α-D-glucopyranose and β-D-glucopyranose, 1,6-anhydro (levoglucosan) were detected.

The LR-SC was characterised having methyl D-glucopyranoside, a glucoside, as the major component (45.3%) as well as ergosta-4,7,22-trien-3β-ol (5.3%), linoleic acid (4.7%), cyclo(leucyloprolyl) (3.5%), and palmitic acid (3.1%) as minor components. Minor amounts of oleic acid (0.2%) were detected. Our findings showed that the major volatile constituents in the extracts of the mycelium, culture broth, and sclerotium of *L. rhinocerotis* were different. The abundance of fatty acids in the extracts of *L. rhinocerotis* was consistent with a previous report by Lau et al. [2].

Identification of chemical constituents by UHPLC-ESI-MS/MS

The extracts of *L. rhinocerotis* were also analysed using UHPLC-ESI-MS/MS. The TICs of the extracts are shown in Figure 3. The nature/class of the compounds was determined based on their mass fragmentation patterns (Figure 4) and comparison with literature and databases (e.g., MassBank [http://www.massbank.jp]). Triterpenoids, amino acids, sugars, organic acids, and phenolics were tentatively identified (Tables 7–9). These represent some common metabolites found in most culinary/medicinal mushrooms.

Lanostane-type triterpenoids with high degrees of oxidation have been previously isolated from *Ganoderma* spp. and other polypores including *Inonotus obliquus*, *Wolfiporia cocos*, *Taiwanofungus camphoratus*, and *Laetiporus sulphurous* [32]; hence, their presence in the extracts of *L. rhinocerotis* (a polypore), is not entirely surprising. In negative mode, the triterpenoids were reported to produce two types of molecular ions, i.e., [M-H]$^-$ and [2M-H]$^-$; fragmentation typically begins with prominent losses of H_2O or CO_2 before cleavage takes place on the ring skeleton [33]. A compound (LR-SC, $R_T = 7.73$ min) produced a deprotonated molecular ion at m/z of 513, and further losses of H_2O and CO_2 yielded fragments at m/z 495 and 451, respectively. This fragmentation pattern is similar to ganoderic acid AM$_1$, D, and ganoderenic acid B, which can be found in *G. lucidum* [33,34]. Another compound with an m/z of 497 and fragments at m/z 451 and 433 might possibly have structures similar to ganoderic acid B, D, G, and K, which were reported to form a prominent [M-H-H_2O]$^-$ ion at m/z 497. Other compounds with an m/z of 495 and fragments at m/z 451, 301, and 193 in the extracts were also suspected to be lanostane-type triterpenoids since they possessed fragments considered to be characteristics of ganoderic acids.

Two amino acids having hydrophobic side chains were identified from the extracts. Their mass fragmentation patterns were in agreement with previous reports [35,36]. Phenylalanine (LR-MH, $R_T = 1.13$ min) exhibited a deprotonated molecular ion ([M-H]$^-$) at m/z 164 and a mass fragment at m/z 147, possibly corresponding to the further loss of an amino group ($-NH_2$). Tryptophan (LR-BT, $R_T = 1.29$ min) gave a deprotonated molecular ion at m/z 203. Further loss of a carboxyl group (CO_2) produced a fragment at m/z 159. Identification of free amino acids in the extracts of *L. rhinocerotis* corroborates previous findings on the presence of amino acids in the aqueous alcohol extract of mushrooms [24]. Hexoses (6-C sugars) are characterised by m/z fragments at 179, 161, 143, 113, and 89 [37]. A compound (LR-MT, $R_T = 0.80$ min) with an m/z of 341 was determined to be sucrose based on postulated cleavage of the glycosidic bond to produce fragments at m/z 179 and 161 [37,38]. Another compound (LR-BH, $R_T = 0.80$ min, LR-MH) with an m/z of 341 had a constant loss of 162 units, consistent with the loss of a hexose moiety to produce a fragment at m/z 179.

Previous studies have revealed that mushrooms are rich in phenolic compounds [24,27]. In this study, however, very few phenolics were identified in the extracts (data not shown), and this corroborated the low phenolic content (Table 3). A compound in LR-SC ($R_T = 0.96$ min) with m/z 199 and mass fragment at m/z 111 was tentatively identified as citric acid, in accordance with the

literature [39]. Organic acids are commonly found in mushroom fruiting bodies [40]. Citric acid is an important intermediate in the Krebs cycle, which is one of the major cellular energy-yielding pathways. The lack of organic acids in the mycelium and culture broth might be because that these were used to support rapid vegetative growth in the mycelia, as proposed by Pinto et al. [41]. A compound present in LR-BH and LR-BT ($R_T = 1.45$) produced a [M-H]$^-$ ion at m/z 241 and fragments at 197, 181, 169, and 140. Its fragmentation pattern closely resembled that of 1,3,8-trihydroxy-6-methylanthraquinone (emodin), which was previously reported to be present in a wild mushroom (*Dermocybe sanguinea*) [42]. The compound was deduced to be a type of anthraquinone based on the similarity of fragmentation patterns.

Mass signals from the UHPLC-ESI-MS/MS for LR-MH, LR-MT, LR-BH, LR-BT, and LR-SC were subjected to PCA. In the score plot (Figure 5A), LR-SC and LR-BT (positive region) were separated from LR-MH, LR-BH, and LR-MT (negative region) by the first principal component. The LR-MT (positive region) could be distinguished from LR-MH and LR-BH (negative regions) by the second principal component. Extracts from shaken cultures (LR-MH and LR-BH) were clustered together. Some of the compounds in the mycelia might have been secreted into the culture broth, and hence, at the harvest time (day 15), the chemical profile of the intracellular (LR-MH) and extracellular (LR-BH) constituents were comparable. The results also showed that chemical constituents in the mycelium under shaken and static conditions were distinct from those of the sclerotium. A loading plot (Figure 5B) was generated to identify the variables that contributed to the differences in the extracts. It was found that several marker ions are far from the centre of the loading plot, suggesting that the concentrations of these compounds in the extracts were highly varied.

Bioactivities in relation to chemical constituents

The considerable variation in the chemical profiles, as described above, might be the main reason for differences in the antioxidant capacities between the mycelium, culture broth, and sclerotium of *L. rhinocerotis*. Mushroom extracts are good sources of phenolic compounds, and the correlation between phenolics and antioxidant capacity implies the possible roles of these compounds as antioxidants [43]. Due to the low phenolic content in the extracts of *L. rhinocerotis*, the roles of other compounds present in the extracts and have been reported to exhibit antioxidant capacities, such as triterpenoids [44], organic acids [24], proteins [45], ergosterol, sterol derivatives, and fatty acids [46], should be considered. As suggested by Carvajal et al. [24], synergistic effects of the antioxidant compounds in the extracts should not be ruled out.

Extensive work has been done to identify low-molecular-weight cytotoxic compounds from medicinal mushrooms and their possible modes of action [47]. The cytotoxic and apoptotic effects of triterpenoids, such as ganoderic acids from *G. lucidum* [32,34] and inotodiol from the sclerotium of *I. obliquus* [48], have been documented. Other classes of potentially cytotoxic metabolites are fatty acids, their conjugated forms, and sterols. The most abundant fatty acid in the extracts of *L. rhinocerotis* was linoleic acid, followed by palmitic and steric acids. Previously, it was reported that linoleic acid did not exert growth inhibition against the testosterone-dependent MCF-7aro cell [49]. Palmitic acid, on the other hand, has been shown to induce apoptosis in human leukemic cells (MOLT-4) [50], and stearic acid was reported to inhibit colony-forming abilities of human cancer cells [51].

The potential of mycelium and culture broth as a substitute for sclerotium

The aqueous methanol extracts, composed of low-molecular-weight compounds, of the mycelium and culture broth of *L. rhinocerotis* showed comparable bioactivities to the sclerotium. In the antioxidant capacity assays, LR-BT was the most potent extract with respect to its ABTS radical scavenging activity, ferric and cupric ion reducing capacities, ferrous ion chelating potential, and inhibitory effect on lipid peroxidation. This indicated that, in terms of antioxidant capacity (Table 1), the sclerotium is not superior compared to the mycelium and culture broth. Secondly, results from the MTT assay showed that all extracts were non-cytotoxic ($IC_{50} > 200$ μg/ml) against a panel of mammalian cell lines. This implied that *L. rhinocerotis* from different morphological/developmental stages (i.e., mycelium and sclerotium) do not contain low-molecular-weight, cytotoxic compounds in abundance. It should be noted that in this study, an exhaustive extraction using aqueous methanol was employed; hence, the resulting extracts would contain lower proportions of non-polar constituents than extracts prepared from other solvents, such as hexane, chloroform, dichloromethane and/or ethyl acetate. A more detailed investigation (e.g., successive extraction using solvents of increasing polarity and/or fractionation of the aqueous methanol extracts) is warranted should bio-prospecting of cytotoxic metabolites from *L. rhinocerotis* be desired.

According to Lau et al. [2], the proximate composition and some nutritional attributes of the mycelium were comparable to those of the sclerotium. This has provided a basis for considering the mycelium an alternative to the sclerotium. The extensive chemical profiling by GC-MS, UHPLC-ESI-MS/MS, SDS-PAGE, and SELDI-TOF-MS in this investigation provided insight into the nature of different low-molecular-weight compounds in *L. rhinocerotis*; nevertheless, further confirmation of these compounds would require additional chemical investigation which is currently in progress. Previously, we found that protein profiles of *L. rhinocerotis* cultured in a stirred tank reactor and static cultures were different [9]. Our results here demonstrated that culture conditions also affected the composition of low-molecular-weight compounds and their bioactivities. The strong antioxidant capacity of LR-BT and cytotoxicity of LR-MT against MCF 10A might be due to compounds produced specifically during static conditions. The chemical basis for this observation has yet to be elucidated, but the lack of aeration and spatial homogeneity as well as the merging of growth phases in static cultures might have effects on the biosynthesis of secondary metabolites [9]. In fact, it has been reported that production of microbial secondary metabolites is enhanced in stressed conditions. For instance, oxygen limitation has been shown to enhance the production of ganoderic acid by *G. lucidum* in submerged cultures [52]. Several workers have investigated the correlation between culture conditions and bioactivities [14,25]. The effects of culture conditions on the production of chemical constituents and their bioactivities require further investigation.

Mycelium, culture broth, and sclerotium represent mushroom samples from different cultivation techniques. Some of the advantages of liquid fermentation over solid-substrate fermentation, such as shorter time, greater quality control, and lesser contamination, might favour large-scale production of mycelium and culture broth as substitutes for either the cultivated or wild sclerotia for use in formulation of nutraceuticals. The diversity in the chemical constituents between mycelium, culture broth, and sclerotium, as demonstrated by the chromatographic and mass-spectrometric analyses, warrants future work pertaining to the metabolomics of mushrooms from different morphological/developmental stages and culture conditions. Regarding our results from bioactivity evaluation and chemical profiling, *L. rhinocerotis* from liquid fermentation merits further consideration as a source of functional ingredients.

Supporting Information

Figure S1   GC-MS TIC of LR-MH.

Figure S2   GC-MS TIC of LR-MT.

Figure S3   GC-MS TIC of LR-BH.

Figure S4   GC-MS TIC of LR-BT.

Figure S5   GC-MS TIC of LR-SC.

Acknowledgments

Technical assistance from the staff and postgraduate students of the Mycology and Proteomics E1.1 laboratories (Faculty of Science, University of Malaya) and the Drug Discovery Laboratory (CARIF) is appreciated. The SELDI-TOF-MS analysis was facilitated by access to the Medical Biotechnology Laboratory (Faculty of Medicine, University of Malaya). The author B. F. Lau is grateful for the SLAB/SLAI-Bright Sparks fellowship supported by the University of Malaya and the Ministry of Education, Malaysia.

Author Contributions

Conceived and designed the experiments: BFL N. Abdullah. Performed the experiments: BFL KCY. Analyzed the data: BFL N. Abdullah N. Aminudin HBL KCY. Contributed reagents/materials/analysis tools: N. Abdullah N. Aminudin HBL VS. Contributed to the writing of the manuscript: BFL N. Abdullah N. Aminudin HBL.

References

1. Wong KH, Cheung PCK, Wu JZ (2003) Biochemical and microstructural characteristics of insoluble and soluble dietary fibre prepared from mushroom sclerotia of *Pleurotus tuber-regium*, *Polyporus rhinocerus*, and *Wolfiporia cocos*. J. Agric. Food Chem. 51: 7197–7202.

2. Lau BF, Abdullah N, Aminudin N (2013) Chemical composition of the tiger's milk mushroom, *Lignosus rhinocerotis* (Cooke) Ryvarden, from different developmental stages. J. Agric. Food Chem. 61: 4890–4897.

3. Wong KH, Cheung PCK (2008) Sclerotia: emerging functional food derived from mushrooms. In: Cheung PCK (Ed.), Mushrooms as Functional Foods. John Wiley and Sons, Inc., NJ, 111–146.

4. Lai CKM, Wong KH, Cheung PCK (2008) Anti-proliferative effects of sclerotial polysaccharides from *Polyporus rhinocerus* Cooke (Aphyllophoromycetideae) on different kinds of leukemic cells. Int. J. Med. Mushrooms 10: 255–264.

5. Wong KH, Lai CKM, Cheung PCK (2011) Immunomodulatory activities of mushroom sclerotial polysaccharides. Food Hydrocolloids 25: 150–158.

6. Lee SS, Chang YS, Noraswati MNR (2009) Utilisation of macrofungi by some indigenous communities for food and medicine in Peninsular Malaysia. Forest Ecol. Manag. 257: 2062–2065.

7. Abdullah N, Dzul Haimi MZ, Lau BF, Annuar MSM (2013) Domestication of a wild medicinal sclerotial mushroom, *Lignosus rhinocerotis* (Cooke) Ryvarden. Ind. Crop. Prod. 47: 256–261.

8. Chen TI, Zhuang HW, Chiao YC, Chen CC (2013) Mutagenicity and genotoxicity effect of *Lignosus rhinocerotis* mushroom mycelium. J. Ethnopharmacol. 149: 70–74.

9. Lau BF, Aminudin N, Abdullah N (2011) Comparative SELDI-TOF-MS profiling of low-molecular-mass proteins from *Lignosus rhinocerus* (Cooke) Ryvarden grown under stirred and static conditions of liquid fermentation. J. Microbiol. Methods 87: 56–63.

10. Tang YJ, Zhu LW, Li HM, Li DS (2007) Submerged culture of mushrooms in bioreactors: challenges, current state-of-the-art, and future prospects. Food Technol. Biotechnol. 45: 221–229.

11. Eik LF, Naidu M, Sabaratnam V (2012) *Lignosus rhinocerus* (Cooke) Ryvarden: a new potential medicinal mushroom that can stimulate neurite outgrowth at low concentrations of extract. Mushroom Science XVIII: Proceedings of the 18th Congress of the International Society for Mushroom Science, 630–634.

12. Lee ML, Tan NH, Fung SY, Tan CS, Ng ST (2012) The anti-proliferative activity of sclerotia of *Lignosus rhinocerus* (tiger's milk mushroom). Evidence-based Complement. Altern. Med., doi:10.1155/2012/697603.

13. Zaila CFS, Zuraina MYF, Norfazlina MN, Lek Mun L, Nurshahirah N, et al. (2013) Anti-proliferative effect of *Lignosus rhinocerotis*, the tiger's milk mushroom, on HCT 116 human colorectal cancer cells. The Open Conference Proceedings Journal 4: 65–70.

14. Xu JW, Xu YN, Zhong JJ (2010) Production of individual ganoderic acids and expression of biosynthetic genes in liquid static and shaking cultures of *Ganoderma lucidum*. Appl. Microbiol. Biotechnol. 85: 941–948.

15. Kong KW, Mat-Junit S, Aminudin N, Ismail A, Abdul Aziz A (2012) Antioxidant activities and polyphenolics from the shoots of *Barringtonia racemosa* (L.) Spreng in a polar to apolar medium system. Food Chem. 134: 324–332.

16. Re R, Pellegrini N, Proteggente A, Pannala A, Yang M, et al. (1999) Antioxidant activity applying an improved ABTS radical cation decolourisation assay. Free Radic. Biol. Med. 26: 1231–1237.

17. Benzie IFF, Strain JJ (1996) The ferric-reducing ability of plasma (FRAP) as a measure of "antioxidant power": the FRAP assay. Anal. Biochem 239: 70–76.

18. Riberiro JP, Magalhaes LM, Reis S, Lima JL, Segundo MA (2011) High-throughput total cupric ion-reducing antioxidant capacity of biological samples determined using flow injection analysis and microplate-based methods. Anal. Sci. 27: 483–488.

19. Jimenez-Alvarez D, Giuffrida F, Vanrobaeys F, Golay PA, Cotting C, et al. (2008) High-throughput methods to assess lipophilic and hydropholic antioxidant capacity of food extracts *in vitro*. J. Agric. Food Chem 6: 3470–3477.

20. Daker N, Abdullah N, Vikineswary S, Goh PC, Kuppusamy UR (2008) Antioxidant from maize and maize fermented by *Marasmiellus* sp. as a stabiliser of lipid-rich foods. Food Chem. 107: 1092–1098.

21. Lau BF, Abdullah N, Aminudin N, Lee HB (2013) Chemical composition and cellular toxicity of ethnobotanical-based hot and cold aqueous preparations of the tiger's milk mushroom (*Lignosus rhinocerotis*). J. Ethnopharmacol. 150: 252–262.

22. Masuko T, Minami A, Iwasaki N, Majima T, Mishimura S, et al. (2005) Carbohydrate analysis by a phenol-sulphuric acid method in microplate format. Anal. Biochem. 339: 69–72.

23. Slinkard K, Singleton VL (1977) Total phenol analysis: automation and comparison with manual methods. Am. J. Enol. Viticult. 28: 49–55.

24. Carvajal AESS, Koehnlein EA, Soares AA, Eler GJ, Nakashima AT, et al. (2012) Bioactivities of fruiting bodies and submerged culture mycelia of *Agaricus brasiliensis* (*A. blazei*) and their antioxidant properties. LWT Food Sci. Techno. 46: 493–499.

25. Mau JL, Tsai SY, Tseng YH, Huang SJ (2005) Antioxidant properties of methanolic extracts from *Ganoderma tsugae*. Food Chem. 93: 641–649.

26. Lee YL, Huang GW, Liang ZC, Mau JL (2007) Antioxidant properties of three extracts from *Pleurotus citrinopileatus*. LWT Food Sci. Technol. 40: 822–833.

27. Reis FS, Martins A, Barros L, Ferreira ICFR (2012) Antioxidant properties and phenolic profile of the most widely appreciated cultivated mushrooms: a comparative study between *in-vivo* and *in-vitro* samples. Food Chem. Toxicol. 50: 1201–1207.

28. Wong KH, Sabaratnam V, Abdullah N, Kuppusamy UR, Naidu M (2009) Effects of cultivation techniques and processing on antimicrobial and antioxidant activities of *Hericium erinaceus* (Bull.:Fr.) Pers. Extracts. Food Technol. Biotechnol. 47: 47–55.

29. Wu GH, Hu T, Li ZY, Huang ZL, Jiang JG (2013) *In-vitro* antioxidant activities of the polysaccharides from *Pleurotus tuber-regium* (Fr.) Sing. Food Chem. 148: 351–356.

30. Du X, Mu H, Zhou S, Zhang Y, Zhu X (2013) Chemical analysis and antioxidant activity of polysaccharides extracted from *Inonotus obliquus* sclerotia. Int. J. Biol. Macromol. 62: 691–696.

31. Geran RI, Greenberg NH, Macdonald MM, Schumacher AM, Abbott BJ (1972) Protocols for screening chemical agents and natural products against animal tumours and other biological systems. Cancer Chemother. Rep. 3: 1–103.

32. Rios JL, Andujar I, Recio MC, Giner RM (2012) Lanostoids from fungi: a group of potential anti-cancer compounds. J. Nat. Prod. 75: 2016–2044.

33. Yang M, Wang X, Guan S, Xia J, Sun J, et al. (2007) Analysis of triterpenoids in *Ganoderma lucidum* using liquid chromatography coupled with electrospray ionisation mass spectrometry. J. Am. Soc. Mass Spectrom. 18: 927–939.

34. Liu YW, Gao JL, Guan J, Qian ZM, Feng K, et al. (2009) Evaluation of anti-proliferative activities and action mechanisms of extracts from two species of *Ganoderma* on tumour cell lines. J. Agric. Food Chem. 57: 3087–3093.

35. Lu W, Gao S, Xiao Y, Zhang L, Li J, et al. (2011) A liquid chromatographic-tandem mass spectrometric method for the quantification of eight components involved in lithospermic acid B biosynthesis pathway in *Salvia miltiorrhiza* hairy root cultures. J. Med. Plant Res. 5: 1664–1672.

36. Ying X, Ma J, Liang Q, Wang Y, Bai G, et al. (2013) Identification and analysis of the constituents in an aqueous extract of *Tricholoma matsutake* by HPLC coupled with diode array detection/electrospray ionisation mass spectrometry. J. Food Sci. 78: 1173–1182.

37. Brudzynski K, Miotto D (2011) Honey melanoidins: analysis of the compositions of the high-molecular-weight melanoidins exhibiting radical-scavenging activity. Food Chem. 127: 1023–1030.

38. Taylor VF, March RE, Longerich HP, Stadey CJ (2005) A mass spectrometric study of glucose, sucrose, and fructose using an inductively coupled plasma and electrospray ionization. Int. J. Mass Spectrom. 243: 71–84.

39. John JA, Shahidi F (2010) Phenolic compounds and antioxidant activity of Brazil nut (*Bertholletia excels*). J. Funct. Foods 2: 196–209.

40. Barros L, Pereira C, Ferreira ICFR (2013) Optimised analysis of organic acids in edible mushrooms from Portugal by ultra-fast liquid chromatography and photodiode array detection. Food Anal. Methods 6: 309–316.

41. Pinto S, Barros L, Sousa MJ, Ferreira ICFR (2013) Chemical characterisation and antioxidant properties of *Lepista nuda* fruiting bodies and mycelia obtained by *in-vitro* culture: effects of collection habitat and culture media. Food Res. Int. 51: 496–502.

42. Von Wright A, Raatikainen O, Taipale H, Karenlampi S, Maki-Paakkanen J (1992) Directly acting geno- and cytotoxic agents from a wild mushroom (*Dermocybe sanguinea*). Mutat. Res. 269: 27–33.

43. Cheung LM, Cheung PCK, Ooi VEC (2003) Antioxidant activity and total phenolics of edible mushroom extracts. Food Chem 81: 249–255.

44. Zhu M, Chang Q, Wong LK, Chong FS, Li RC (1999) Triterpene antioxidants from *Ganoderma lucidum*. Phytother. Res. 13: 529–531.

45. Elias RJ, Kellerby SS, Decker EA (2008) Antioxidant activity of proteins and peptides. Crit. Rev. Food Sci. Nutr. 48, 430–441.

46. Zhang Y, Mills GL, Nair MG (2002) Cyclooxygenase inhibitory and antioxidant compounds from the mycelia of the edible mushroom *Grifola frondosa*. J. Agric. Food Chem. 50: 7581–7585.

47. Ferreira IC, Vaz JA, Vasconcelos MH, Martins A (2010) Compounds from wild mushrooms with anti-tumour potential. Anti-cancer Agents Med. Chem. 10: 424–436.

48. Nomura M, Takakashi T, Uesugi A, Tanaka R, Kobayashi S (2008) Inotodiol, a lanostane triterpenoid, from *Inonotus obliquus* inhibits cell proliferation through caspase-3-dependent apoptosis. Anti-cancer Res. 28: 2691–2696.

49. Chen S, Oh SR, Phung S, Hur G, Ye JJ, et al. (2006) Anti-aromatase activity of phytochemicals in white button mushrooms (*Agaricus bisporus*). Cancer Res. 66: 12026–12034.

50. Harada H, Yamashita U, Kurihara H, Fukushi E, Kawabata J, et al. (2002) Anti-tumour activity of palmitic acid found as a selective cytotoxic substance in a marine red alga. Anti-cancer Res. 22: 2587–2590.

51. Fermor BF, Masters JR, Wood CB, Miller J, Apostolov K, et al. (1992) Fatty acid composition of normal and malignant cells and cytotoxicity of stearic, oleic, and sterculic acids *in vitro*. Eur. J. Cancer 28: 1143–1147.

52. Zhang WX, Zhong JJ (2013) Oxygen limitation improves ganoderic acid biosynthesis in submerged cultivation of *Ganoderma lucidum*. Biotechnol. Bioprocess Eng. 18: 972–980.

5

Purification and Characterization of Flavonoids from the Leaves of *Zanthoxylum bungeanum* and Correlation between Their Structure and Antioxidant Activity

Yujuan Zhang, Dongmei Wang*, Lina Yang, Dan Zhou, Jingfang Zhang

College of Forestry, Northwest A & F University, Yangling, China

Abstract

Nine flavonoids were isolated and characterized from the leaves of *Zanthoxylum bungeanum*. Their structures were elucidated by spectroscopic techniques as quercetin (**1**), afzelin (**2**), quercitrin (**3**), trifolin (**4**), quercetin-3-O-β-D-glucoside (**5**), isorhamnetin 3-O-α-L-rhamnoside (**6**), hyperoside (**7**), vitexin (**8**) and rutin (**9**). All compounds were isolated from the leaves of *Z. bungeanum* for the first time. Five compounds (**2, 4, 5, 6** and **8**) were found for the first time in the genus *Zanthoxylum*. To learn the mechanisms underlying its health benefits, *in vitro* (DPPH, ABTS, FRAP and lipid peroxidation inhibition assays) and *in vivo* (protective effect on *Escherichia coli* under peroxide stress) antioxidant activities of the nine flavonoids were measured. Quercetin and quercetin glycosides (compounds **1, 3, 5, 7, 9**) showed the highest antioxidant activity. Structure-activity relationships indicated that the -OH in 4′ position on the B ring and the -OH in 7 position on the A ring possessed high antioxidant activity; B ring and/or A ring with adjacent -OH groups could greatly increase their antioxidant ability. Also, due to the different structures of various flavonoids, they will certainly exhibit different antioxidant capacity when the reactions occur in solution or in oil-in-water emulsion. These findings suggest that *Z. bungeanum* leaves may have health benefits when consumed. It could become a useful supplement for pharmaceutical products and functional food ingredients in both nutraceutical and food industries as a potential source of natural antioxidants.

Editor: Jamshidkhan Chamani, Islamic Azad University-Mashhad Branch, Mashhad, Iran, Islamic Republic of Iran

Funding: The authors are grateful to the research fund from the Special Fund for Forestry Scientific Research in the Public Interest of China (201304811) and the Fundamental Research Funds for the Central Universities (ZD2013010). The funders had no role in study design, data collection and analysis, decision to publish, or preparation of the manuscript.

Competing Interests: The authors have declared that no competing interests exist.

* Email: dmwli@163.com

Introduction

The genus *Zanthoxylum* (Rutaceae) consists of 250 species in the world, of which there are 45 species and 13 varieties in China. *Zanthoxylum bungeanum*, also called Sichuan pepper, is a common Chinese pepper, growing widely in the area of Sichuan, Shanxi, Shandong and Hebei Provinces of China. Just like other species in this genus, *Z. bungeanum* has a distinctive tingling taste. Due to its fresh aroma and taste, the dried fruits are used as one of the eight cuisine condiments and seasonings in China [1–2]. Apart from its common application as a condiment to make foods more flavoring, each part of *Z. bungeanum* have numerous medicinal virtues. In Traditional Chinese Medicine, the pericarp can be used for gastralgia and dyspepsia; the seed is reported to be antiphlogistic and diuretic; the leaves are considered carminative, stimulant and sudorific; the root can cure epigastric pains and treat bruises, eczema and snake-bites [3–7]. Owing to its high medicinal value, intensive investigation leads to isolate many classes of secondary metabolites such as flavonoids, alkaloids, amides, lignans, coumarins [8–11]. Thirteen flavonoids and ten unsaturated alkylamides were isolated from the pericarps of *Z. bungeanum* [9–12]; six alkaloids were isolated from the root of *Z. bungeanum* [13]. Phytochemicals, such as hydroxy-β-sanshool,

xanthoxylin, hyperoside were found to stimulate the spontaneous beating rate (BR) of myocardial cells in culture [14]. These previous literatures indicated that *Z. bungeanum* can be used as pharmacological active products for improving human health in nutraceutical and pharmaceutical industries.

The young leaves of *Z. bungeanum* have been consumed as foodstuffs, and the mature leaves were used as condiments in traditional Chinese cusine. In some rural areas, local people eat the new leaves in spring seasons [15]. However, only a few people pay attention to the chemical work on the leaves of *Z. bungeanum*. Deng and coworkers found that the leaves of *Z. bungeanum* contained mainly nutritious and nutritional ingredients [15]. Yang and Xu provided preliminary data confirming that the leaves contain abundant flavonoids with good radical scavenging abilities [16–17]. These previous studies revealed that *Z. bungeanum* leaves are rich in flavonoids but studies on the isolation, purification and structure elucidation of individual polyphenols in the leaves of *Z. bungeanum* have not been reported yet. Since flavonols supplements in dietary food may evoke protective effects under peroxide stress, our current study was carried out to isolate the bioactive components present in the leaves of *Z. bungeanum*, in an attempt to gain a deeper understanding of the correlation between diet, health benefits and reduced risk of diseases.

Furthermore, the authors attempted to evaluate the comparative antioxidant activity of the isolated compounds using several *in vitro* (DPPH, ABTS, FRAP and lipid peroxidation inhibition assays) and *in vivo* (protective effect on *Escherichia coli* under peroxide stress) methods. Also, the correlation between the structures of compounds and antioxidant capacity was discussed.

Material and Methods

General

Electrospray ion trap mass spectrometry (ESI-MS) was carried out with a Bruker ESI-TRAP Esquire 6000 plus mass spectrometry instrument. Nuclear magnetic resonance spectra (NMR) were recorded on a Bruker Avance III 500 MHz instrument in DMSO-*d6* using tetramethylsilane (TMS) as the internal standard. Analytical thin-layer chromatography (TLC) was performed with silica gel plates using silica gel 60 GF$_{254}$ (Qingdao Haiyang Chemical Co., Ltd.). Sephadex LH-20 was purchased from GE Healthcare Bio-science AB (Sweden).

Chemicals and reagents

1,1-Diphenyl-2-picrylhydrazyl (DPPH), 2,4,6-Tripyridyl-s-triazine (TPTZ), 2,2-azino-bis (3-ethyl-benzothiazoline-6-sulphonic acid) diammonium salt (ABTS), 6-hydroxy-2,5,7,8- tetramethyl-chroman-2-carboxylic acid (Trolox), (Sigma-Aldrich Co., St. Louis, USA); Thiobarbituric acid (TBA) (Guangdong Guanghua Chemical Factory Co., Ltd. PR China); Yeast extract, tryptone (Oxoid Ltd., Basingstoke, Hampshire, England). Deionized water (18 MΩ cm) was used to prepare aqueous solutions. All the chemicals used, including the solvents, were of analytical grade.

Plant materials

Z. bungeanum leaves were collected from Taibai Mountains of Shaanxi province, in September, 2012, and authenticated by the Herbarium of the Northwest A&F University.

Ethics statement

Specific permissions were not required for the described field sampling studies or for the collection of plant materials. For any locations/activities, no specific permissions were required. All locations where the plants were collected were not privately owned or protected in any way and the field studies did not involve endangered or protected species.

Extraction, fractionation, and isolation

The air-dried and powdered leaves of *Z. bungeanum* (9.40 Kg) were extracted with 95% ethanol (1:5 w/v) at room temperature (6 days×6), and the total filtrate was then concentrated by rotary evaporation under vacuum to obtain the ethanol extracts (1824.4 g). This extracts was further fractioned by column chromatography on silica gel (200–300 mesh, 120*10 cm), successively eluting with petroleum ether, chloroform, ethyl acetate, acetone and methanol. The eluents of the five different polarity solvents were collected separately and concentrated by rotary evaporation under vacuum to obtain five fractions (PEF, 105.9 g; CF, 112.7 g; EAF, 28.0 g; AF, 100.0 g and MF, 624.3 g). The EAF and AF fractions were screened as the most effective fractions. Thus, later purification and isolation were focused on EAF and AF.

The EAF (28.0 g) was initially chromatographed over a silica gel column (80*8 cm) using CHCl$_3$ and MeOH under gradient conditions (9:1→4:1→1:1→1:4→0:1) to yield 7 sub-fractions (E01–E07). Fraction 2 (E02, 8.6 g) was subjected to column chromatography over a silica gel column (60*4 cm) with a solvent

mixture of CHCl$_3$ and MeOH (30:1→15:1→9:1→4:1→1:1) to yield compounds **1** (150.0 mg) and **2** (180.00 mg), respectively. Fraction 3 (E03, 10.6 g) was purified by silica gel with CHCl$_3$:MeOH (20:1→9:1→4:1→1:1) to produce compound **3** (4.0 g). Fraction 4 (E04, 1.0 g) was subjected to column chromatography over a silica gel column (30*2) with a solvent mixture of CHCl$_3$ and MeOH (15:1→9:1→4:1→1:1) and later a Sephadex LH-20 column (150*1 cm) with 100% MeOH to yield compound **8** (20.0 mg).

The AF (100.0 g) was chromatographed over a silica gel column (100*8 cm), using a mixed solvent of CHCl$_3$ and MeOH (9:1→4:1→1:1→1:4→0:1) to afford 5 fractions (A01–A05). Fraction 3 (A03, 5.0 g) was chromatographed over silica gel column (50*3) with a solvent mixture of CHCl$_3$ and MeOH (30:1→15:1→9:1→4:1→1:1) and Sephadex LH-20 column (150*1 cm) with 100% MeOH to yield compounds **5** (38.0 mg) and **6** (40.0 mg). Fraction 5 (A05, 20.0 g) was chromatographed over silica gel column (60*4 cm) with a solvent mixture of CHCl$_3$ and MeOH (9:1→4:1→1:1→1:4) to afford 5 fractions (A0501–A0505). Fraction 2 (A0502, 300.0 mg) was chromatographed over silica gel with CHCl$_3$ and MeOH (4:1), followed by a Sephadex LH20 column (150*1 cm) with 100% MeOH to obtain compound **4** (40.0 mg). Fraction 3 (A0504, 5.8 g) was chromatographed over silica gel with a solvent mixture of CHCl$_3$ and MeOH (15:1→9:1→4:1→1:1) to yield compound **7** (2.0 g) and **9** (45.0 mg).

All the isolated compounds **1–9** were characterized and identified by spectroscopic methods, as well as through comparison with published data. The spectral data were as follows and the structures are shown in Figure 1–3.

Acid Hydrolysis of isolated compounds

Compounds **2–7**, and **9** (5.0 mg) was dissolved in 5.0 mL of 2 N HCl, heated at 90°C for 2 h, and then partitioned between ethyl acetate and water. Aglycon was recovered from the ethyl acetate layer and identified by direct comparison with an authentic sample. The water layers were identified by comparison with an authentic sample by TLC analysis. Sugars liberated from compounds **2**, **3** and **6** were identified as rhamnose; sugars liberated from compounds **4** and **7** were identified as galactose; sugars liberated from compounds **5** were identified as glucose; finally sugars liberated from compounds **9** were identified as glucose and rhamnose [18].

DPPH radical scavenging assay

DPPH radical-scavenging activity was evaluated using the method described by Yen and Chen [19] with the following modifications [20]: a 2 mL volume of the sample solutions (0.003–0.2 mM) were added to 2 mL of DPPH solution (0.1 mM); and the absorbance was measured with a spectrophotometer at 517 nm 30 min later. All the tests and the controls were performed in triplicate. The DPPH free radical-scavenging activity was calculated using the following equation:

$$Scavenging\ Ability\ (\%) = \left(1 - \frac{A_i - A_j}{A_0}\right) \times 100\%$$

Where A_o is the absorbance of ethanol (2 mL) and DPPH· (2 mL), A_i is the absorbance of the tested sample (2 mL sample and 2 mL DPPH·), and A_j is the absorbance of the blank (2 mL sample and 2 mL ethanol). IC$_{50}$ values were the effective concentrations at which DPPH radicals were scavenged by 50%, and were obtained from linear regression analysis.

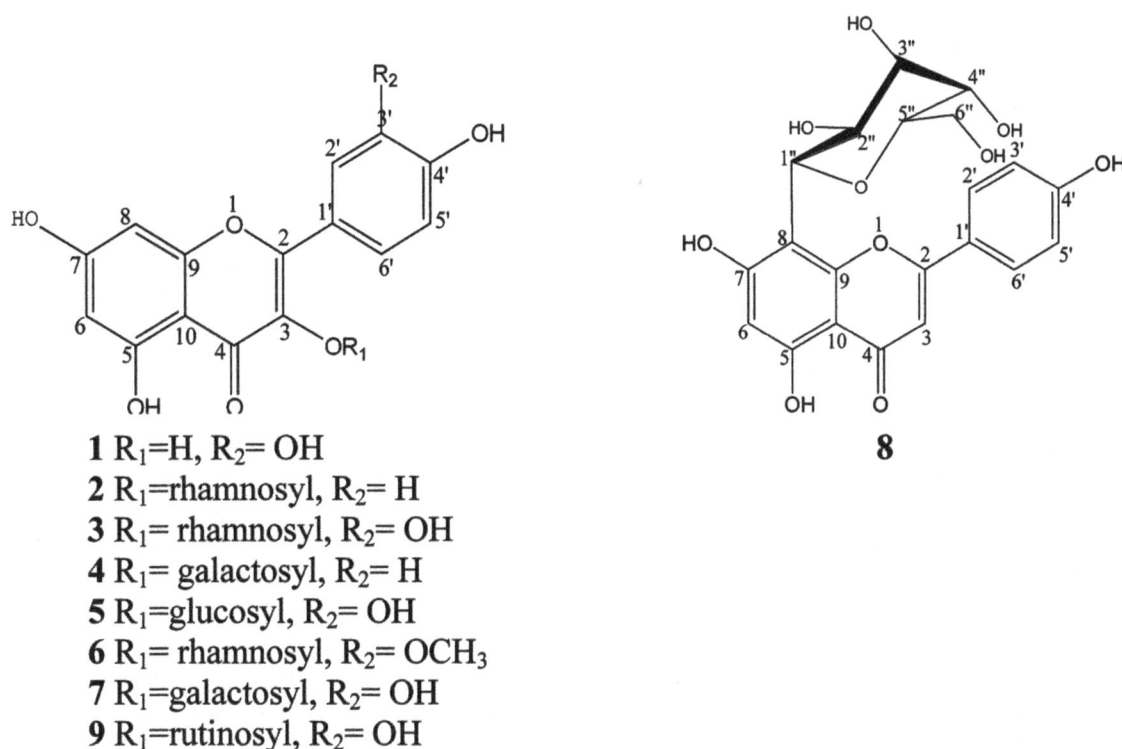

1 R_1=H, R_2= OH
2 R_1=rhamnosyl, R_2= H
3 R_1= rhamnosyl, R_2= OH
4 R_1= galactosyl, R_2= H
5 R_1=glucosyl, R_2= OH
6 R_1= rhamnosyl, R_2= OCH_3
7 R_1=galactosyl, R_2= OH
9 R_1=rutinosyl, R_2= OH

Figure 1. Structures of compounds 1–9.

ABTS·⁺ radical cation decolorization assay

Antioxidant activity was determined using the decolorizing free radical ABTS·⁺ method as described previously [21–23]. An ABTS radical cation (ABTS·⁺) was produced by reacting an ABTS solution (7 mM) with 2.45 mM potassium persulfate, and the mixture was allowed to stand for 12–16 h in the dark at room temperature prior to use. The ABTS solution was diluted with PBS (pH 7.4) to the concentration that provided an absorbance value of 0.70 ($\pm$0.02) at 734 nm. For each analysis, 100 µL of

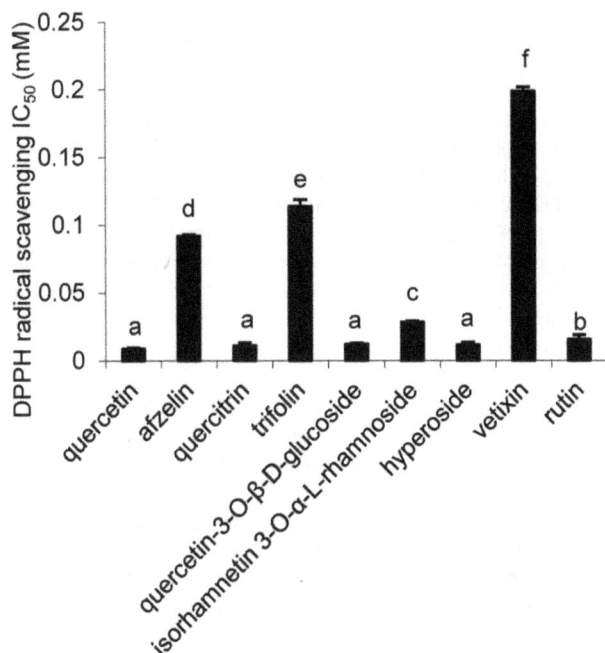

Figure 2. IC$_{50}$ values of isolated compounds for DPPH radical scavenging activity (mean ± SD, n = 3). Bars with no letters in common are significantly different (P<0.05).

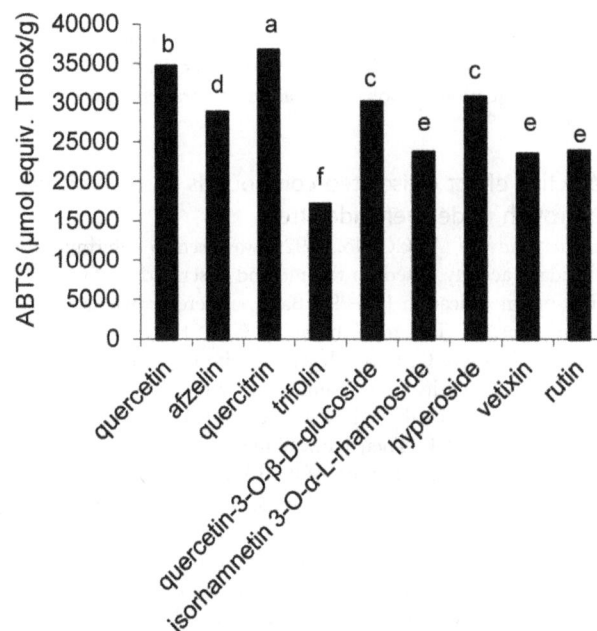

Figure 3. ABTS·⁺ radical scavenging activity of isolated compounds (mean ± SD, N = 3). Bars with no letters in common are significantly different (P<0.05).

samples (0.1 mM) were added to 3.9 mL of the ABTS$\cdot^+$ solution, and the decrease in absorbance at 734 nm was recorded within 6 min. The results were expressed as micromoles of trolox equivalent per gram dry weight of samples (μmol equiv. Trolox/g).

Ferric reducing antioxidant power (FRAP) assay

The FRAP assay was performed with the following modifications [23–24]: the FRAP reagent was prepared daily by mixing 10 mL of a solution of ferric trichloride hexahydrate (20 mM), 10 mL of a solution of TPTZ (10 mM in 40 mM of hydrochloric acid) and 100 mL of 0.3 M acetate buffer (pH 3.6), and incubating them at 37°C. For each analysis, 400 μL of the samples (0.1 mM) were added to 3 mL of the FRAP solution. The increase in absorbance at 593 nm was recorded in 15 s intervals over the course of 30 min at 37°C. The FRAP results were expressed in terms of micromoles trolox equivalent per gram dry weight of samples (μmol equiv. Trolox/g).

Lipid peroxidation inhibition assay

In this assay, a modified thiobarbituric acid reactive species (TBARS) assay was used to measure the lipid peroxide formed using egg yolk homogenates as lipid-rich media [25–27]. Briefly, fresh egg yolk emulsion was diluted to 10% v/v with 1.15% w/v KCl. 50 μL egg yolk emulsion, 50 μL of sample solution in different concentrations (0.003–0.2 mM), 150 μL of 20% (aqueous) trichloroacetic acid and 150 μL of 0.67% w/v thiobarbituric acid were added respectively and mixed thoroughly as the reaction solution. The whole reaction solution was then vortexed thoroughly and followed by incubation at 95°C in water bath for 1 h. After cooling, centrifuged the solution at 3000 rpm for 10 min. Absorbance of the upper layer was measured at 532 nm and percentage inhibition was calculated with the following formula:

$$\text{Inhibition of lipid peroxidation } \% = \left(1 - \frac{t}{c}\right) \times 100\%$$

where c is the absorbance of fully peroxidized control and t is the absorbance of test sample. The IC_{50} value was calculated from the regression equation between sample concentration and rate of inhibition.

Protective effect of isolated compounds on *Escherichia coli* growth under peroxide stress

Escherichia coli (ATCC No. 25922) was used to determine the antioxidant activity based on the method described by Smirnova with some modification [28–29]. Bacteria were grown overnight on Luria-Bertani (LB) medium at 37°C in 150 mL flask with shaking (50 r/min). Cell growth was monitored by measurement of the optical density at 600 nm. The mid-log phase bacteria ($OD_{600} = 0.6$) was diluted by fresh LB medium (final $OD_{600} = 0.25 \pm 0.004$). Then 1 mL of the cell solution were added to test tubes containing 8.9 mL of the LB medium and 100 μL of samples (0.1 mM) and incubated at 37°C with shaking (180 r/min). OD_{600} before cell addition was measured and subtracted for precise determination of growth. In 90 min when OD_{600} reached a value exactly equal to 0.342, bacteria were treated with hydrogen peroxide (6.5 mM) and incubated at 37°C with shaking (180 r/min) for 30 min. The absorbance was measured, and the growth was monitored for another 2–3 h. Specific growth rate was calculated according to the equation:

$$\mu = \ln\left(\frac{N}{N_0}\right)/t$$

where μ is the specific growth rate and N_0 and N are the optical density at time zero and t, respectively. The relative protective effect of compounds in this test was calculated as follows: t = 30 min, the specific growth rate of *E. coli* in medium containing compounds and 6.5 mM H_2O_2/the specific growth rate in medium containing only H_2O_2 (the control).

Statistical analysis

All experiments were performed in triplicate and analyses of all samples were run in triplicate and averaged. Statistical analyses were performed with SPSS 18.0 for windows. The results were expressed as the means ± standard deviation (SD) of triplicate. The data were subjected to one-way analysis of variance (ANOVA) and the significance of difference between samples means was calculated by Duncan's multiple range test (SPSS Inc. Chicago).

Results

Antioxidant activities of the extracts and different fractions from *Z. bungeanum* leaves

Chromatographic fractionation of crude ethanol extracts led to five different polarity fractions (PEF, CF, EAF, AF, and MF fraction). The DPPH, ABTS radical scavenging activities and reducing power of individual fractions were then evaluated. As shown in Table 1, the EAF and AF fractions were screened as the most effective fractions with the highest DPPH radical scavenging abilities. Achieving a DPPH radical-scavenging activity of 50% required an EAF concentration of 13.20 μg/mL (IC_{50}), which was not significantly different from that of AF (18.55 μg/mL), but was 3.1-fold lower than that of ethanol extract (p<0.05). Concerning the ABTS radical scavenging activity, we also found that EAF exhibited the highest activity (2147.83 μmol equiv Trolox/g), followed by AF (2044.58 μmol equiv. Trolox/g, p<0.05). The analysis of the results from the FRAP assay showed the same tendency when compared with the results of the scavenging capacity on DPPH and ABTS radical. It is exhibited that the reducing ability of EAF was 615.88 μmol equiv. Trolox/g, which was not significantly different from that of AF (594.15 μmol equiv. Trolox/g, p<0.05) but was 1.9-fold higher (p<0.05) than that of ethanol extracts. As a result, powerful antioxidants originating from *Z. bungeanum* leaves might mainly exist in two active fractions (EAF and AF). Thus, later purification and isolation were focused on EAF and AF.

Structure elucidation of isolated compounds

Nine compounds were identified from the leaves of *Z. bungeanum*. The individual structures of compounds **1–9** are shown in Figure 1.

Compound **1** was obtained as an amorphous yellow powder with the molecular formula $C_{15}H_{10}O_7$, which was deduced from the ESI-MS *m/z*: 301.29 [M-H]$^-$. The ^{1}H and ^{13}C NMR spectroscopic data indicated that **1** has a flavonol skeleton with 15 carbons, including five aromatic CH; ten quaternary carbons (one carbonyl, five O-bearing, and four aliphatic), suggesting that it is 3,5,7,3′,4′-pentahydroxyflavone, commonly known as quercetin [30–31]. Compound **3** was isolated as an amorphous yellow powder with the molecular formula $C_{21}H_{20}O_{11}$, ESI-MS *m/z*: 447.77 [M-H]$^-$. Further comparison of the ^{1}H NMR and ^{13}C

Table 1. Antioxidant activities of the extracts and different fractions from *Z. bungeanum* leaves.

samples	DPPH IC$_{50}$ (µg/mL)	FRAP (µmol equiv. Trolox/g)	ABTS (µmol equiv. Trolox/g)
ECE	40.75±0.21[b]	317.11±9.71[b]	1122.91±34.62[b]
PEF	377.95±39.39[e]	81.56±7.41[d]	264.20±37.27[e]
CF	169.15±4.60[d]	170.44±10.45[c]	563.86±22.66[d]
EAF	13.20±0.85[a]	615.88±1.86[a]	2147.83±23.08[a]
AF	18.55±0.35[a]	594.15±8.89[a]	2044.58±19.99[a]
MF	85.85±2.19[c]	191.93±2.22[c]	747.69±38.77[c]

ECE: ethanol crude extracts; PEF: petroleum ether fraction; CF: chloroform fraction; EAF: ethyl acetate fraction; AF: acetone fraction; MF: methanol fraction.
Each values represented in tables are means ± SD (N = 3).
Values with different letters (a, b, c, d, e) within same column were significantly different (P<0.05).

NMR spectroscopic data of compounds **3**, **5**, **7** and **9** with those of compound **1** revealed that they all showed a typical flavonol pattern with a quercetin aglycon (Table 2). Compound **3** was determined as quercetin 3-O-α-L-rhamnoside (quercitrin), which was in accordance with the reported data [30,32–33]. Compound **5** gives the molecular formula $C_{21}H_{20}O_{12}$, ESI-MS m/z: 462.98 [M-H]$^{-}$. Compared with the published literature, compound **5** was finally determined as quercetin 3-O-β-D-glucoside [33]. Compound **7** was obtained as an amorphous yellow powder. Its molecular formula was established as $C_{21}H_{20}O_{12}$, ESI-MS m/z: 465.01 [M+H]$^{+}$. Compared with the known literature, the structure of compound **7** was determined as quercetin 3-O-β-D-galactoside (hyperoside) [30,33]. Compound **9** was isolated as an amorphous light yellow powder with the molecular formula $C_{27}H_{30}O_{16}$, ESI-MS m/z: 609.91 [M-H]$^{-}$. Through comparison with reported spectral data, compound **9** was identified as rutin [30–31].

Compound **2** was also obtained as an amorphous light yellow powder with the molecular formula $C_{21}H_{20}O_{10}$, ESI-MS m/z: 431.80 [M-H]$^{-}$. The ^{1}H NMR and ^{13}C NMR (Table 3) spectra of compound **2** and compound **4** were compared with the data of known compounds and showed a typical flavonol pattern with a kaempferol aglycon. Compared with the known literature, the structure of compound **2** was determined as kaempferol 3-O-α-L-rhamnoside (afzelin) [30,34–35]. Compound **4** gives the molecular formula $C_{21}H_{20}O_{11}$, ESI-MS m/z: 447.20 [M-H]$^{-}$. Compared with the known literature, compound **4** was finally determined as kaempferol 3-O-β-D-galactoside (trifolin) [30].

Compound **6** was obtained as an amorphous yellow powder with the molecular formula, $C_{22}H_{22}O_{11}$, ESI-MS at m/z: 462.79 [M-H]$^{-}$. ^{13}C NMR spectrum showed signals for one -OCH$_3$ (Table 3). The ^{1}H NMR and ^{13}C NMR (Table 3) spectra of compound **6** were compared with the data of known compounds and showed a typical flavonol pattern with an isorhamnetin aglycon. Acid hydrolysis of compound **6** gave L-rhamnose, identified by comparision with an authentic sample. The sugar portion was examined by TLC analysis. The ^{1}H-NMR spectrum of compound **6** displayed two doublets at δ 5.26 (1H, d, J = 7.0 Hz) for the anomeric protons. Based on the coupling constant J = 1.15 Hz lesser than 4.0 Hz, the sugar configurations could be identified as α-L-rhamnose, which correlated with signals at 102.29 in the HSQC spectrum. In the HMBC (Figure 2), the anomeric proton signal of the rhamnose at δ 5.26 (H-1″) correlated with the carbon signal at δ 134.67 (C-3). These indicated that the rhamnose was attached to C-3 of the aglycone. Based on the above evidences and detailed analyses of the NMR spectra, the structure

of compound **6** was determined as isorhamnetin 3-O-α-L-rhamnoside [30,36].

Compound **8** was isolated as an amorphous light yellow powder with the molecular formula $C_{21}H_{20}O_{10}$, ESI-MS m/z: 431.28 [M-H]$^{-}$. The ^{1}H NMR and ^{13}C NMR (Table 3) spectroscopic data of compound **8** revealed that compound **8** bears similar flavonoid skeleton as compound **2** and the same sugar moieties as compound **5**. Yet a remarkable upfield shift (Δδ 11.05) for the C-8 of compound **8** (δ 105.26) compared to compound **2** (δ 94.21), and a remarkable downfield shift (Δδ 31.76) for the C-3 of compound **8** (δ 102.94) compared to compound **2** (δ 134.70), indicated that the sugar moiety was attached to C-8 instead of C-3, which was in accordance with the reported NMR spectroscopic data of vitexin [30,37–38]. Unambiguous assignments of the ^{1}H and ^{13}C NMR data and the relative configuration were deduced from the COSY and HMBC experiments.

To the best of the author's knowledge, this is the first report of flavonoids **1–9** isolated from the leaves of *Z. bungeanum*. Compounds **2**, **4**, **5**, **6** and **8** were found for the first time in the genus *Zanthoxylum* [3,8–11]. Furthermore, the 2D-NMR data of the five known compounds **2**, **3**, **6**, **7**, **8** were given in this paper.

DPPH radical scavenging activity of isolated compounds

quercetin and quercetin glycosides (quercetin, quercitrin, quercetin-3-O-β-D-glucoside, hyperoside and rutin) ranked with the most potent antioxidant activity. In particular, quercetin, quercitrin, quercetin-3-O-β-D-glucoside and hyperoside exerted predominant DPPH· radical scavenging activity with respective IC$_{50}$ values of 0.009±0.001, 0.011±0.001, 0.012±0.001 and 0.011±0.001 mM, followed by rutin and isorhamnetin 3-O-α-L-rhamnoside with IC$_{50}$ values of 0.016±0.001 and 0.028±0.001 mM, respectively (p<0.05). Kaempferol glycosides (afzelin and trifolin) were also found to possess significant DPPH· radical scavenging activity, while the C-glycoside flavonol (vitexin) exhibited poor DPPH· radical scavenging activity (Figure 2).

ABTS·$^{+}$ radical cation decolorization of isolated compounds

All the isolated compounds exhibited potent ABTS·$^{+}$ scavenging activity (>17000 µmol equiv. Trolox/g), though quercetin (**1**), quercitrin (**3**), quercetin-3-O-β-D-glucoside (**5**) and hyperoside (**7**) ranked with the most potent ones (>30000 µmol equiv. Trolox/g) (Figure 3).

Table 2. ^{1}H, ^{13}C, HMBC and COSY NMR spectroscopic data (500 MHz, DMSO-d_6) of compounds 1, 3, 5, 7 and 9.

Position	Compound 1		Compound 3				Compound 5		Compound 7				Compound 9	
	δ^1H	$\delta^{13}C$	δ^1H	$\delta^{13}C$	HMBC	COSY	δ^1H	$\delta^{13}C$	δ^1H	$\delta^{13}C$	HMBC	COSY	δ^1H	$\delta^{13}C$
2		147.28		156.93				156.78		156.79				156.91
3		136.14		134.60				133.76		133.99				133.81
4		176.27		178.23				177.90		177.97				178.01
5	12.47 br.s (6.9)	161.10	12.66 s	161.78	C5, C6, C10		12.65 s	161.70	12.63 s	161.71	C5, C6, C10		12.59 s	161.70
6	6.19 d (2.1)	98.65	6.21 br s	99.17	C8, C10, C5, C7	H8	6.21 d (2.0)	99.11	6.21 d (1.9)	99.16	C8, C10, C5, C7	H8	6.19 d (2.0)	99.17
7	10.93 s	164.30		164.69			10.88 s	164.59		164.66				164.69
8	6.42 d (2.1)	93.85	6.40 br s	94.10	C6, C10, C9, C7	H6	6.41 d (2.0)	93.96	6.41 d (1.9)	93.98	C6, C10, C9, C7	H6	6.38 d (2.0)	94.10
9		156.60		157.76				156.62		156.79				157.05
10		103.45		104.56				104.43		104.39				104.42
1'		122.40		121.23				121.62		121.59				121.66
2'	7.66 d (2.1)	115.47	7.30 d (1.5)	115.94	C6', C4', C2		7.53 d (J=2.0)	115.66	7.53 d (2.0)	115.67	C6', C4', C2		7.53 d (2.1)	115.70
3'	9.35 s	145.49		145.68			9.23 s	145.27		145.30				145.23
4'	9.69 s	148.13		148.91			9.73 s	148.92		148.95				148.91
5'	6.88 d (8.4)	116.05	6.87 d (8.4)	116.15	C1', C3'	H6'	6.87 d (9.0)	116.66	6.81 d (8.5)	116.44	C1', C3'	H6'	6.84 d (9)	116.75
6'	7.53 dd (8.4, 2.1)	120.51	7.25 d (8.4)	121.59	C2', C4', C2	H5'	7.59 dd (9.0, 2.0)	122.06	7.67 dd (8.5, 2.0)	122.46	C2', C4'	H5'	7.55 d (9, 2.1)	122.06
1''			5.27 s	102.33	C3, C2''	H2''	5.48 d (7.2)	101.30	5.38 d (7.8)	102.33	C3	H2''	5.35 d (7.4)	101.21
2''			3.99 dd	70.54		H1'', H3''		74.55	3.57 dd	71.70				74.57
3''			3.53 dd	71.05		H2'', H4''		76.95	3.39 dd	73.69				76.98
4''			3.17 dd	71.69		H3''		70.39	3.71 dd	68.41				70.51
5''			3.23 dd	70.86		H6''		78.04	3.33 dd	76.32				76.41
6''			0.82 d (6.0)	17.96		H5''		61.43	3.51 dd	60.62				67.47
1'''													5.12 d (1.9)	101.71
2'''														70.85
3'''														71.07
4'''														72.36
5'''														68.70
6'''													1.00 d (6.1)	18.18

Assignments were done by 1D (^1H, ^{13}C, DEPT) and 2D (COSY, HSQC, HMBC) NMR experiments.

Table 3. ^1H, ^{13}C, HMBC and COSY NMR spectroscopic data (500 MHz, DMSO-d_6) of compounds 2, 4, 6 and 8.

Position	Compound 2 δ^1H	δ^{13}C	HMBC	COSY	Compound 4 δ^1H	δ^{13}C	Compound 6 δ^1H	δ^{13}C	Compound 8 δ^1H	δ^{13}C	COSY
2		156.97				156.83		156.90		164.41	
3		134.70				133.71		134.67	6.78 s	102.94	
4		178.20				178.01		178.21		182.54	
5	12.63 s	161.77	C5, C6, C10		12.62 s	161.69	12.66 s	161.76	13.17 s	156.48	
6	6.21 d (1.6)	99.19	C8, C10, C5,C7	H8	6.21 d (1.9)	99.17	6.21 br s	99.14	6.27 s	98.67	
7	10.86 s	164.68				164.63		164.66	10.83 s	162.91	
8	6.42 d (1.6)	94.21	C6, C10, C9,C7	H6	6.44 d (1.9)	94.14	6.40 br s	94.08	6.78 s	105.26	
9		157.70				156.85		157.76		160.89	
10		104.62				104.43		104.54		104.55	
1'		121.00				121.34		121.18		122.11	
2'	7.77 dd (8.58)	131.06	C6', C2, C4'	H3'	8.07 d (8.8)	131.45	7.30 d (1.5)	115.91	8.02 d (8.7)	129.39	H3'
3'	6.93 dd (8.58)	115.87	C5', C1', C4'	H2'	6.87 d (8.8)	115.53		148.90	6.91 d (8.7)	116.32	H2'
4'	10.2 s	160.46				160.43		145.67	10.35 s	161.61	
5'	6.93 dd (5.58)	115.87	C3', C1', C4'	H6'	6.87 d (8.8)	115.53	6.88 d (8.4)	116.10	6.91 d (8.7)	116.32	H6'
6'	7.77 dd (8.58)	131.06	C2', C2, C4'	H5'	8.07 d (8.8)	131.45	7.25 d (8.4)	121.57	8.02 d (8.7)	129.39	H5'
3'- OCH$_3$							3.19 s	49.06			
1"	5.31 d (1.8)	102.27	C3, C2"	H2"	5.40 d (7.5)	102.14	5.26 s	102.29	4.70 d (9.8)	73.88	H2"
2"	3.99 dd (3.6, 1.5)	70.55		H1", H3"		71.68	3.98 dd	70.51	3.85 dd	71.39	
3"	3.48 dd	71.08		H2", H4"		73.57	3.50 dd	71.05	3.29 dd	79.18	
4"	3.15 dd	71.61		H3"		68.35	3.17 dd	71.63	3.52 dd	71.09	
5"	3.17 dd	70.83		H6"		76.25	3.23 dd	70.80	3.26 dd	82.29	
6"	0.81 d (6.09)	17.93		H5"		60.67	0.82 d (6.0)	17.96	3.76 dd	61.82	

Assignments were done by 1D (^1H, ^{13}C, DEPT) and 2D (COSY, HSQC, HMBC) NMR experiments.

Ferric reducing antioxidant power (FRAP assay) of isolated compounds

In the FRAP assay, quercetin and quercetin glycosides (quercetin, quercitrin, quercetin-3-O-β-D-glucoside, hyperoside and rutin) showed the highest (p<0.05) ferric reducing ability (> 5300 μmol equiv Trolox/g) compared with the isorhanmetin glycoside (**6**), kaempferol glycosides (**2** and **4**), or C-glycoside flavonol (**8**) (<4300 μmol equiv. Trolox/g) (Figure 4).

Lipid peroxidation inhibition of isolated compounds

Vitexin (**8**) and quercitrin (**3**) was observed to have the highest lipid peroxidation inhibitory capacity with the lowest IC_{50} values of 0.014 ± 0.001 and 0.013 ± 0.005 mM, respectively (p<0.05), compared with kaempferol glycosides (0.065 ± 0.003 mM for afzelin and 0.040 ± 0.001 mM for trifolin, p<0.05). Quercetin and other quercetin glycosides, as well as isorhamnetin glycosides, were also found to possess potent lipid peroxidation inhibitory activity (Figure 5).

Protective effect of isolated compounds on E. coli under peroxide stress

In our current research, nine flavonoids were tested on *E. coli* growth under peroxide stress. Addition of 6.5 mM H_2O_2 into growing *E. coli* cells ($OD_{600} = 0.342$) resulted in a 40-min growth arrest (Figure 6). A 90-min pretreatment of *E. coli* by flavonoids **1–9** did not cause considerable effects on duration of the H_2O_2-triggered growth arrest. Addition of 6.5 mM H_2O_2 into the cells pretreated with flavonoids **1–9** inhibited growth considerably but not completely. In these cultures the earlier recovery of rapid growth was observed. The values of optical density statistically different compared to control were reached in 30–40 min after H_2O_2 addition in cultures pretreated with compounds **1–9** (Figure 6). All the isolated compounds exerted a protective effect against the bacterio static action of H_2O_2, increasing cell growth

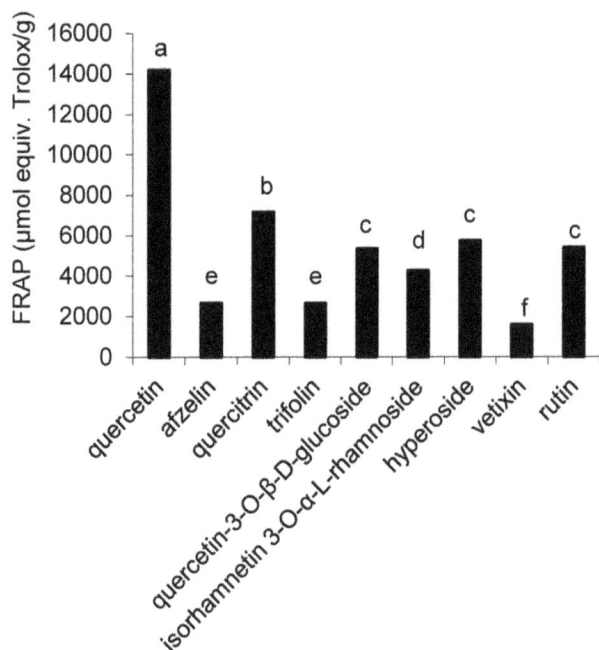

Figure 5. IC_{50} values of isolated compounds for lipid peroxidation inhibition activity (mean ± SD, N = 3). Bars with no letters in common are significantly different (P<0.05).

rate under peroxide stress to 1.88–5.76 fold compared with that of untreated cells (the control) in 30 min (p<0.05). In particular, the highest activity was revealed in quercetin and quercetin glycosides (quercetin, quercitrin, hyperoside, quercetin-3-O-β-D-glucoside, isorhamnetin 3-O-α-L-rhamnoside and rutin), and their growth rate were 5.76-, 5.45-, 4.91-, 4.27-, 3.84-, and 2.96-fold,

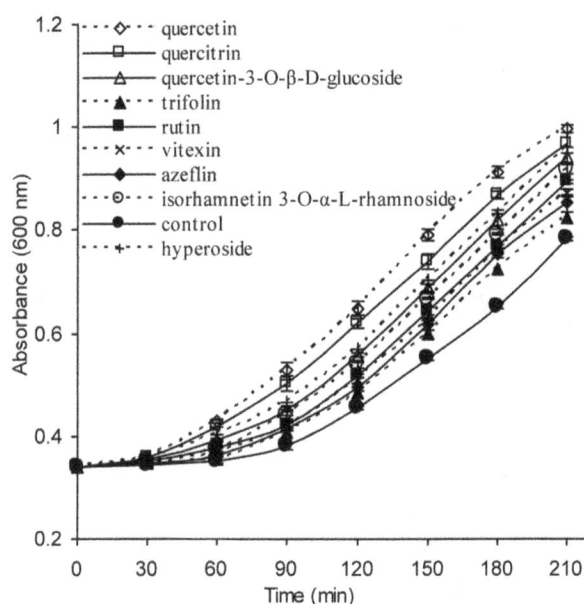

Figure 4. Reducing power of isolated compounds (mean ± SD, N = 3). Bars with no letters in common are significantly different (P<0.05).

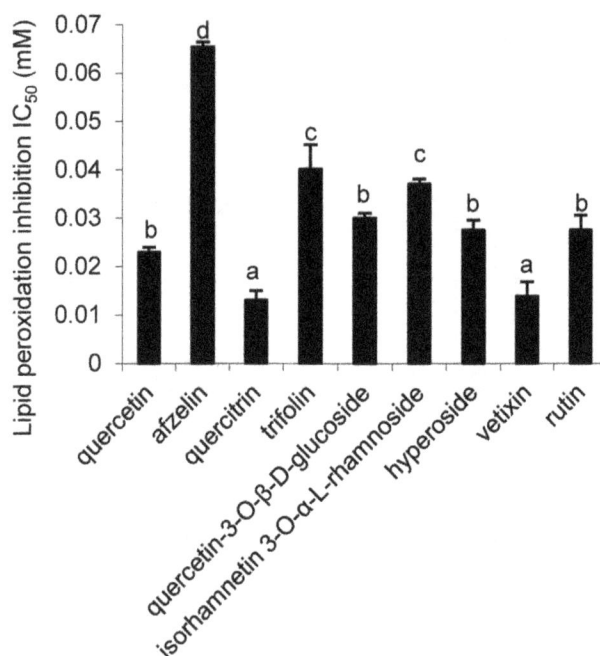

Figure 6. Influence of pretreatment with different isolated compounds on growth in *E. coli* under peroxide stress (mean ± SD, N = 3).

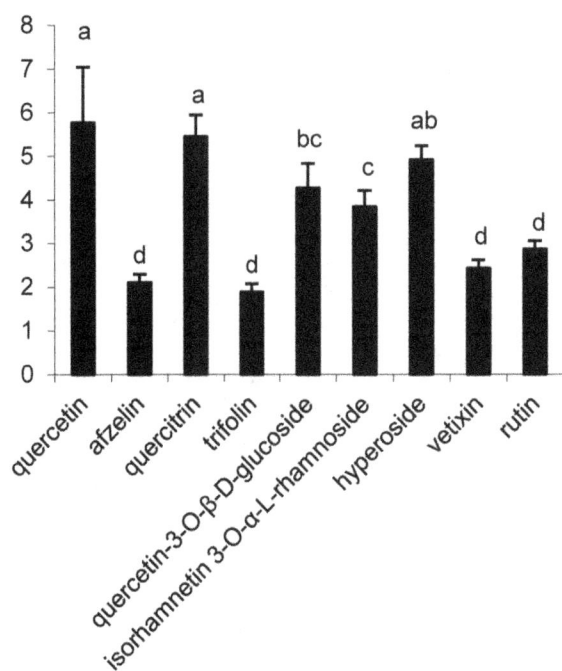

Figure 7. The relative protective effect of isolated compounds on *E. coli* under peroxide stress: t = 30 min, the specific growth rate of *E. coli* in medium containing compounds and 6.5 mM H_2O_2/the specific growth rate in medium containing only 6.5 mM H_2O_2 (the control) (mean ± SD, N = 3). Bars with no letters in common are significantly different ($P < 0.05$).

respectively ($p < 0.05$), compared with that of untreated cells (the control) 30 min later (Figure 7).

Our results indicated that the antioxidant capacity of the nine compounds varied. Taking into account such parameters as ability of DPPH, ABTS radical scavenging, ferric reducing power, lipid peroxidation inhibition and protection of *E. coli* cells under peroxide stress, the highest antioxidant activities were observed in quercetin and quercetin glycosides.

Discussion

Owing to the high medicinal values of *Z. bungeanum*, intensive investigations lead to isolation of many classes of secondary metabolites such as flavonoids, alkaloids, amides, lignans, coumarins [8–11]. Till now, thirteen flavonoids (two new flavonol glucoside, quercetin 3',4'-dimethyl ether 7-glucoside and tamarixetin 3,7-bis-glucoside, together with hyperoside, quercetin, quercitrin, foeniculin, isorhamnetin 7-glucoside, rutin, 3,5,6-trihydroxy-7,4'-dimethoxy flavone, arbutin, sitosterol β-glucoside, L-sesamin and palmitic acid) were isolated from the pericarps of *Z. bungeanum* [9]. Studies on individual polyphenols in the leaves of *Z. bungeanum* have not been reported yet, especially the chemical structure of each polyphenol. In this study, nine compounds were isolated and identified from the leaves of *Zanthoxylum bungeanum*, namely quercetin (**1**), afzelin (**2**), quercitrin (**3**), trifolin (**4**), quercetin-3-O-β-D-glucoside (**5**), isorhamnetin 3-O-α-L-rhamnoside (**6**), hyperoside (**7**), vitexin (**8**) and rutin (**9**). The nine isolated compounds are mainly quercetin and kaempferol glycosides, apart from one isorhamnetin glycoside and one c-glycoside flavanol. They are all part of the flavonol glycosides, a class of flavonoids. All the compounds were isolated for the first time in the leaves of *Z. bungeanum*. This is the first report of nine flavonoids isolated

from *Z. bungeanum* leaves. Compounds **2, 4, 5, 6** and **8** were found for the first time in the genus *Zanthoxylum*. In literatures, rutin, hyperoside and quercetin were reported to have antioxidant, chelation, anti-carcinogenic, cardio protective, bacterio static, and secretory properties [40,45]. Afzelin (kaempferol 3-O-α-L-rhamnoside) is a competitive inhibitor of intestinal SGLT1 cotransporter, and it is very effective in reduction of glucose intestinal absorption [35]. Quercitrin (quercetin 3-O-α-L-rhamnoside) is a well-known flavonoid with anti-diarrhoeic activity, sedative activity, anti-inflammatory effect and antifungal activity [32,46]. Vitexin, a bioactive C-8 glycosylated flavonoid, has been reported to have antiviral, antimicrobial, antioxidant and radioprotection activities [37–38]. The various biological activities of these compounds indicated that the leaves of *Z. bungeanum* have health benefit when consumed. These also could explain its frequent addition to the Chinese diet for promoting human health and for disease prevention.

With more than 5000 species known, flavonoids are among the most potent natural antioxidants in plants, and many of them are stronger reducing agents on a molar basis than ascorbic acid [39–40]. Their chemical structures consist of a backbone with two benzene rings linked by a pyran chain (C_6-C_3-C_6). The number and position of hydroxyl groups on the aromatic rings determine the polyphenol's antioxidant capacity. Matsuda and others (2002) found that flavones and their derivatives were found to be the most effective antioxidants during a comparative study among different flavonoid structures [41]. In the present study, due to the results of several *in vitro* and *in vivo* antioxidant activity tests, all the isolated flavonoids exhibited potential antioxidant activity. As shown in Figure 2–7, the antioxidant capacities of the nine compounds varied. Taking into account such parameters as ability of DPPH, ABTS radical scavenging, ferric reducing power, lipid peroxidation inhibition and protection effect on *E. coli* cells under peroxide stress, the highest antioxidant activity among the isolated compounds were found in quercetin and quercetin glycosides. Furthermore, quercetin (**1**) had the highest antioxidant activity of all. When the -OH at 3 position on the C ring was glycosylated, its antioxidant activity decreased. The effects of quercetin were compared to those of Trolox. The latter compounds exert its antioxidant effects *in vivo* indirectly by free iron chelating and decreasing production of hydroxyl radicals in the Fenton reaction [28].

The five methods used for determine the antioxidant capacity of the nine flavonoids have complementary effects. All the methods are well designed for determination of the antioxidant activity of pure, isolated compounds, though the results varied in the different antioxidant capacity tests. Dubeau (2010) and Bourassa (2013) pointed out the discrepancies concerning the effect of milk on antioxidant capacity of tea polyphenols might be due to the different methods, more specifically on the phase in which the redox reaction occur [42–43]. Due to the different structures of various flavonoids, they will certainly exhibit different antioxidant capacity when the reactions occur in solution or in oil-in-water emulsion. For example, quercitrin (**3**) and vitexin (**8**) displayed higher efficiency to retard lipid peroxidation compared with quercetin (**1**), owing to their chemical structure facilitating their partitioning at the lipid-water interface of the egg yolk micelles [42]. Efficiency of an antioxidant is determined not only by its susceptibility to donate an electron or a hydrogen atom to an oxidant [40] but also by its accessibility to the oxidant [44]. However, quercetin (**1**) displayed higher DPPH, ABTS radical scavenging abilities and better reducing power than quercetin glycosides. Rutin (**9**) displayed less antioxidant activity compared with other quercetin glycosides. This could be attributed to the fact that those larger polyphenols diffuse less readily in aqueous media reducing the overall free radical scavenging ability [42]. Freitas and coworkers proved that the binging affinity of polyphenols increases with the molecular weight and the number of hydrophilic hydroxyl groups. This binding can affect the electron donation capacity of the flavonoids by reducing the number of hydroxyl groups available for oxidation in the media [45]. In *in vivo* test, quercetin and flavonol quercetin, capable of increasing katG expression and catalase activity in *E. coli* and chelating intracellular redox-active iron, showed the most effective protection of bacterial growth after H_2O_2 treatment [46-52]. Structure-activity relationships revealed that several structural requirements for antioxidants were proposed; the -OH in 4' position on the B ring and the -OH in 7 position on the A ring possessed high antioxidant activity. The free 7-OH group on the A ring is the main structural feature for the effectiveness of free radical scavenging and protective effect on cells under oxidation. In addition, 2-phenyl substitution due to its aromatic and lipophilic nature, as well as its specific spatial conformation, was found to be effective for the scavenging of free radicals. Finally, the 4'- OH group seems to play an important role for exhibiting the antioxidant activity of these compounds [53-56]. These explained the whole powerful *in vitro* and *in vivo* antioxidant activity of compounds isolated from *Z. bungeanum* leaves. B ring and/or A ring with adjacent OH groups could greatly increase their antioxidant ability. The rhamnosyl moiety is present in kaempferol 3-O-α-L-rhamnoside (2), isorhamnetin 3-O-α-L-rhamnoside (6) and quercetin 3-O-α-L-rhamnoside (3), but compound 3 has higher antioxidant activity, indicating that the hexose is not a determinant for their biological activity. The presence of an -OH (3) instead of a -H (2) or -OCH_3 (6) in the 3' position of the B ring is a determinant for the potent antioxidant activity. The same tendency of antioxidant activity was also found in kaempferol 3-O-galactoside (trifolin, 4) and quercetin 3-O-β-D-galactoside (hyperoside, 7) in which a galactosyl moiety is present in both 4 and

7, proving that the -OH in the 3' position of the B ring is a determinant for the potent antioxidant activity. These also explain that quercetin and quercetin glycosides (compounds 3, 5, 7, 9) have better antioxidant activity compared with kaempferol glycosides or vitexin.

Conclusions

In our current study, nine flavonoids were isolated for the first time from the leaves of *Z. bungeanum*. Five compounds (2, 4, 5, 6 and 8) were found for the first time in the genus *Zanthoxylum*. To learn the mechanisms underlying its health benefits, we further investigated the antioxidant activities of nine isolated compounds using several *in vitro* (DPPH, ABTS, FRAP and lipid peroxidation inhibition assays) and *in vivo* (protective effect on *Escherichia coli* under peroxide stress) methods. Among them, quercetin and quercetin glycosides showed the highest antioxidant activity. Structure-activity relationships indicated that, the -OH in 4' position on the B ring and the -OH in 7 position on the A ring possessed high antioxidant activity; B ring and/or A ring with adjacent -OH groups could greatly increase their antioxidant ability. Also, due to the different structures of various flavonoids, they will certainly exhibit different antioxidant capacity when the reactions occur in solution or in oil-in-water emulsion. These findings suggest that *Z. bungeanum* leaves may have health benefits when consumed. It could become useful supplements for pharmaceutical products and functional food ingredients in both nutraceutical and food industries as a potential source of natural antioxidants. This study offers a theoretical basis for the further study on the bioactive compounds from *Z. bungeanum* leaves.

Author Contributions

Conceived and designed the experiments: DMW JFZ. Performed the experiments: YJZ LY DZ. Analyzed the data: YJZ DMW. Contributed to the writing of the manuscript: YJZ.

References

1. Wang S, Xie JC, Yang W, Sun BG (2011) Preparative separation and purification of alkylamides from *Zanthoxylum bungeanum* Maxim by high-speed counter-current chromatography. J Liq Chromatogr R T 34: 2640-2652.
2. Yang X (2008) Aroma constituents and alkylamides of red and green huajiao (*Zanthoxylum bungeanum* and *Zanthoxylum schinifolium*). J Agric Food Chem 56: 1689-1696.
3. Lim TK (2012) Zanthoxylum simulans. Edible medicinal and non-medicinal plants: Springer, 904-911.
4. Chang ZQ, Liu F, Wang SL, Zhao TZ, Wang MT (1981) Studies on the chemical constituents of *Zanthoxylum simulans* Hance. Acta Pharm Sinica 16: 394-396.
5. Duke JA, Ayensu ES (1985) Medicinal plants of China. Reference Publications, Inc, Algonac, 705.
6. Yeung HC (1985) Handbook of Chinese herbs and formulas. Institute of Chinese Medicine, Los Angeles.
7. Bauer R, Xiao PG (2011) Pericarpium *Zanthoxyli* Huajiao. *Chromatographic fingerprint analysis of herbal medicines*: Springer 191-202.
8. Mizutani K, Fukunaga Y, Tanaka O, Takasugi N, Saruwatari YI, et al. (1988) Amides from Huajiao, pericarps of *Zanthoxylum bungeanum* Maxim. Chem Pharm Bull 36: 2362-2365.
9. Xiong Q, Dawen S, Mizuno M (1995) Flavonol glucosides in pericarps of *Zanthoxylum bungeanum*. Phytochem 39: 723-725.
10. Tirillini B, Stoppini AM (1994) Volatile constituents of the fruit secretory glands of *Zanthoxylum bungeanum* Maxim. J Essent Oil Res 6: 249-252.
11. Li Y, Zeng J, Liu L, Jin X (2001) GC-MS analysis of supercritical carbon dioxide extraction products from pericarp of *Zanthoxylum bungeanum*. Journal of Chinese medicinal materials 24: 572-573.
12. Xiong Q, Dawen S, Yamamoto H, Mizuno M (1997) Alkylamides from pericarps of *Zanthoxylum bungeanum*. Phytochem 46: 1123-1126.
13. Ren L, Xie F (1981) Alkaloids from the root of *Zanthoxylum bungeanum* Maxim. Acta Pharm Sinica 16: 672-677.
14. Huang XL, Kakiuchi N, Che QM, Huang SL, Sheng L, et al. (1993) Effects of extracts of *Zanthoxylum* fruit and their constituents on spontaneous beating rate of myocardial cell sheets in culture. Phytother Res 7: 41-48.
15. Deng ZY, Sun BY, Kang KG, Dong YG (2005) Analysis of the main nutritional labeling in the tender bud of *Zanthoxylum bungeanum*. Journal of Northwest Forestry University 20: 179-180.
16. Xu HD, Fan JH (2010) Macroporous adsorption resin purification of total flavonoids from the crude extract of *Zanthoxylum bungeanum* leaves and their reducing power and free radical scavenging activity. Food Sci 31: 111-115.
17. Yang LC, Li R, Tan J, Jiang ZT (2013) Polyphenolics composition of the leaves of *Zanthoxylum bungeanum* Maxim. grown in Hebei, China, and their radical scavenging activities. J Agric Food Chem 61: 1772-1778.
18. Oshima R, Yamauchi Y, Kumanotani J (1982) Resolution of the enantiomers of aldoses by liquid chromatography of diastereoisomeric 1-(N-acetyl-[α]-methylbenzylamino)-1- deoxyalditol acetates. Carbohydr Res 2: 169-176.
19. Yen GC, Chen HY (1995) Antioxidant activity of various tea extracts in relation to their antimutagenicity. J Agric Food Chem 43: 27-32.
20. Wang DM, Zhang YJ, Wang SS, Li DW (2013) Antioxidant and antifungal activities of extracts and fractions from *Anemone taipaiensis*, China. Allelopathy J 32: 67-68.
21. Rice-Evans CA, Miller NJ, Paganga G (1996) Structure-antioxidant activity relationships of flavonoids and phenolic acids. Free Radical Bio Med 20: 933-956.
22. Tachakittirungrod S, Okonogi S, Chowwanapoonpohn S (2007) Study on antioxidant activity of certain plants in Thailand: Mechanism of antioxidant action of guava leaf extract. Food Chem 103: 381-388.
23. Sandra G (2012) *Helichrysum monizii* Lowe: Phenolic composition and antioxidant potential. Phytochem Anal 23: 72-83.
24. Benzie IFF, Strain J (1996) The ferric reducing ability of plasma (FRAP) as a measure of "antioxidant power": the FRAP assay. Anal Biochem 239: 70-76.
25. Dorman HJD, Deans SG, Noble RC, Surai P (1995) Evaluation in vitro of plant essential oils as natural antioxidants. J Essent Oil Res 7: 645-651.
26. Miguel G, Simões M, Figueiredo AC, Barroso JG, Pedro LG, et al. (2004) Composition and antioxidant activities of the essential oils of *Thymus caespititius*, *Thymus camphoratus* and *Thymus mastichina*. Food Chem 86: 183-188.
27. Dhar P, Bajpai PK. Tayade AB, Chaurasia OP, Srivastava RB, et al. (2013) Chemical composition and antioxidant capacities of phytococktail extracts from trans-Himalayan cold desert. BMC Complem Altern Med 13: 259.

28. Smirnova GV, Samoylova ZY, Muzyka NG, Oktyabrsky ON (2009) Influence of polyphenols on *Escherichia coli* resistance to peroxide stress. Free Radical Bio Med 46: 759–768.

29. Smirnova GV, Vysochina GI, Muzyka NG, Samoylova ZY, Kukushkina TA, et al. (2010) Evaluation of antioxidant properties of medical plants using microbial test systems. World J Microbiol Biotechnol 26: 2269–2276.

30. Azimova SS, Vinogradova VI (2013) Physicochemical and Pharmacological Properties of Flavonoids. *Natural compounds–flavonoids*, Springer Science+ Business Media New York.

31. Kimura YM, Kubo T, Tani S, Arichi S, Okuda H (1981) Study on *Scutellaria* radix IV. Effects on lipid peroxidation in rat liver. Chem Pharm Bull 29: 2610–2617.

32. Ma XF, Tian WX, Wu LH, Cao XL, Ito Y (2005) Isolation of quercetin-3-O-l-rhamnoside from Acer truncatum Bunge by high-speed counter-current chromatography. J Chromatogr A 1070: 211–214.

33. Jung HA, Islam MD, Kwon YS, Jin SE, Son YK, et al. (2011) Extraction and identification of three major aldose reductase inhibitors from *Artemisia montana*. Food ChemToxicol 49: 376–384.

34. Ibrahim LF, Kawashty SA, El-Hagrassy AM, Nassar MI, Mabry TJ (2002) An acylated kaempferol glycoside from flowers of *Foeniculum vulgare* and *F. Dulce*. Molecules 7: 245–251.

35. Rodríguez P, González-Mujica F, Bermúdez J, Hasegawa M (2010) Inhibition of glucose intestinal absorption by kaempferol 3-O-α-L-rhamnoside purified from *Bauhinia megalandra* leaves. Fitoterapia 81: 1220–1223.

36. Nassar MI, Aboutabl EA, Makled YA, El-Khrisy EA, Osman AF (2010) Secondary metabolites and pharmacology of *Foeniculum vulgare* Mill. Subsp. Piperitum. Revista latinoamericana de química 38: 103–112.

37. Zhou X, Peng JY, Fan GY, Wu YT (2005) Isolation and purification of flavonoid glycosides from *Trollius ledebouri* using high-speed counter-current chromatography by stepwise increasing the flow-rate of the mobile phase. J Chromatogr A 1092: 216–221.

38. Li H, Cao DD, Yi JY, Cao JK, Jiang WB (2012) Identification of the flavonoids in mungbean (*Phaseolus radiatus* L.) soup and their antioxidant activities. Food Chem 135: 2942–2946.

39. Lee KW, Kim YJ, Kim DO, Lee HJ, Lee CY (2003) Major phenolics in apple and their contribution to the total antioxidant capacity. J Agric Food Chem 51: 6516–6520.

40. Rice-Evans CA, Miller NJ, Pagangu G (1997) Antioxidant properties of phenolic compounds. Trends Plant Sci 2: 152–159.

41. Matsuda H, Morikawa T, Toguchida I, Yoshikawa M (2002) Structural requirements of flavonoids and related compounds for aldose reductase inhibitory activity. Chem Pharm Bull 50: 788–795.

42. Dubeau S, Samson G, Tajmir-Riahi HA (2010) Dual effect of milk on the antioxidant capacity of green, Darjeeling and English breakfast teas. Food Chem 122: 539–545.

43. Bourassa P, Côté R, Hutchandani S, Samson G, Tajmir-Riahi H (2013) The effect of milk alpha-casein on the antioxidant activity of tea polyphenols. J Photoch Photobio B 128: 43–49.

44. Decker EA, Warner K, Richards MP, Shahidi F (2005) Measuring antioxidant effectiveness in food. J Agric Food Chem 53: 4303–4310.

45. De Freitas N, Mateus S (2001) Structural features of procyanidin interactions with salivary proteins. J Agric Food Chem 4: 940–945.

46. Aherne SA, O'Brien NM (2000) Mechanism of protection by the flavonoids, quercetin and rutin, against tert-butylhydroperoxide- and menadione-induced DNA single strand breaks in Caco-2 cells. Free Radic Biol Med 29: 507–514.

47. Wätjen W, Michels G, Steffan B, Niering P, Chovolou Y, et al. (2005) Low concentrations of flavonoids are protective in rat H4IIE cells whereas high concentrations cause DNA damage and apoptosis. J Nutr 135: 525–531.

48. Melidou M, Riganakos K, Galaris D (2005) Protection against nuclear DNA damage offered by flavonoids in cells exposed to hydrogen peroxide: The role of iron chelation. Free Radic Biol Med 39: 1591–1600.

49. Ferrali M, Signorini C, Caciotti B, Sugherini L, Ciccoli L, et al. (1997) Protection against peroxide damage of erythrocyte membrane by the flavonoids quercetin and its relation to iron chelating activity. FEBS Lett 416: 123–129.

50. Williams RJ, Spencer JP, Rice-Evans C (2004) Flavonoids: antioxidants or signaling molecules? Free Radic Biol Med 36: 838–849.

51. Harwood M, Danielewska-Nikiel B, Borzelleca J, Flamm G, Williams G, et al. (2007) A critical review of the data related to the safety of quercetin and lack of evidence of in vivo toxicity, including lack of genotoxic/carcinogenic properties. Food Chem Toxicol 45: 2179–2205.

52. Galvez J, Crespo ME, Jimenez J, Suárez A, Zarzuelo A (1993) Antidiarrhoeic activity of quercitrin in mice and rats. J Pharm Pharmacol 45: 157–159.

53. Nijveldt RJ, van Nood E, van Hoorn DEC, Boelens PG, van Norren K, et al. (2001) Flavonoids: a review of probable mechanisms of action and potential applications. Am J Clin Nutr 74: 418–425.

54. Costantino L, Rastelli G, Cignarella G, Vianello P, Barlocco D (1997) New aldose reductase inhibitors as potential agents for the prevention of long-term diabetic complications. Exp Opin Ther Patents 7: 843–858.

55. Rastelli G, Antolini L, Benvenuti S, Constantino L (2000) Structural bases for the inhibition of aldose reductase by phenolic compounds. Bioorg Med Chem 8: 1151–1158.

56. Urzumtzev A, Tete-Favier F, Mitscier A, Barbanton J, Barth P, et al. (1997) A 'specificity' pocket inferred from the crystal structures of the complexes of aldose reductase with the pharmaceutically important inhibitors tolrestat and sorbinil. Structure 5: 601–612.

Ameliorative Effect of Fisetin on Cisplatin-Induced Nephrotoxicity in Rats via Modulation of NF-κB Activation and Antioxidant Defence

Bidya Dhar Sahu[1], Anil Kumar Kalvala[2], Meghana Koneru[1], Jerald Mahesh Kumar[3], Madhusudana Kuncha[1], Shyam Sunder Rachamalla[4], Ramakrishna Sistla[1]*

1 Medicinal Chemistry and Pharmacology Division, CSIR-Indian Institute of Chemical Technology (IICT), Hyderabad, Andhra Pradesh, India, 2 Department of Pharmacology and Toxicology, National Institute of Pharmaceutical Education and Research (NIPER), Hyderabad, Andhra Pradesh, India, 3 Animal House, CSIR-Centre for Cellular and Molecular Biology (CCMB), Hyderabad, Andhra Pradesh, India, 4 Faculty of Pharmacy, Osmania University, Hyderabad, Andhra Pradesh, India

Abstract

Nephrotoxicity is a dose-dependent side effect of cisplatin limiting its clinical usage in the field of cancer chemotherapy. Fisetin is a bioactive flavonoid with recognized antioxidant and anti-inflammatory properties. In the present study, we investigated the potential renoprotective effect and underlying mechanism of fisetin using rat model of cisplatin-induced nephrotoxicity. The elevation in serum biomarkers of renal damage (blood urea nitrogen and creatinine); degree of histopathological alterations and oxidative stress were significantly restored towards normal in fisetin treated, cisplatin challenged animals. Fisetin treatment also significantly attenuated the cisplatin-induced IκBα degradation and phosphorylation and blocked the NF-κB (p65) nuclear translocation, with subsequent elevation of pro-inflammatory cytokine, TNF-α, protein expression of iNOS and myeloperoxidase activities. Furthermore, fisetin markedly attenuated the translocation of cytochrome c protein from the mitochondria to the cytosol; decreased the expression of pro-apoptotic proteins including Bax, cleaved caspase-3, cleaved caspase-9 and p53; and prevented the decline of anti-apoptotic protein, Bcl-2. The cisplatin-induced mRNA expression of NOX2/gp91phox and NOX4/RENOX and the NADPH oxidase enzyme activity were also significantly lowered by fisetin treatment. Moreover, the evaluated mitochondrial respiratory enzyme activities and mitochondrial antioxidants were restored by fisetin treatment. Estimation of platinum concentration in kidney tissues revealed that fisetin treatment along with cisplatin did not alter the cisplatin uptake in kidney tissues. In conclusion, these findings suggest that fisetin may be used as a promising adjunct candidate for cisplatin use.

Editor: Partha Mukhopadhyay, National Institutes of Health, United States of America

Funding: This work was supported by research grant from CSIR project 'SMiLE' (CSC 0111). Senior Research Fellowship to BDS from CSIR, India, is gratefully acknowledged. The funders had no role in study design, data collection and analysis, decision to publish, or preparation of the manuscript.

Competing Interests: The authors have declared that no competing interests exist.

* Email: sistla@iict.res.in

Introduction

Though recent investigations find the new generation of platinum-based cytotoxic agents, cisplatin (cis-Diamminedichloroplatinum II, CDDP) remains a highly effective and widely used anti-neoplastic drug against various solid tumors, including endometrial, testicular, ovarian, breast, bladder, head, neck and lung cancer [1]. Despite being a potent anticancer drug, cisplatin elicits dose and duration dependent nephrotoxicity limiting its clinical utility in 25–35% of hospitalized patients undergoing chemotherapy [2]. Owing to its potent and wide range of therapeutic benefit against various malignancies, establishment of a new adjunct therapeutic strategy which ameliorates the severity of cisplatin elicited toxicity in the field of cancer research is therefore warranted.

Recent studies in molecular mechanism of chemotherapy induced toxicity have revealed that cisplatin-induced nephrotoxicity is multifactorial and numerous signalling pathways are involved in. Studies demonstrated that accumulation of cisplatin in renal tubular cells is five times more in comparison with other

tissues and is considered as one of the prime reason for cisplatin-induced nephrotoxicity [1]. Many studies, including ours, have demonstrated that oxidative stress due to impaired antioxidant status and/or excess generation of free radicals, in particular superoxide, due to cisplatin-induced renal NADPH oxidase NOX4 (RENOX) and phagocyte NADPH oxidase (NOX2/gp91phox) over expression is involved in such deleterious effects [2,3]. Evidences from *in vitro* and *in vivo* studies also have demonstrated that cisplatin induces apoptosis and necrosis of renal tubular cells through activation of both intrinsic and extrinsic mitochondrial pathways [4,5]. In addition, studies also document involvement and activation of p53 mediated pro-apoptotic molecules in cisplatin-induced nephrotoxicity [6,7]. Moreover, activation of pro-inflammatory pathways (TNF-α, NF-κB) and infiltration of inflammatory cells are the other crucial mechanisms involved in cisplatin-induced nephrotoxicity [8]. A growing body of evidence also suggests the mitochondrial dysfunction, generation of mitochondrial reactive oxygen species (ROS) and impairment of mitochondrial antioxidant activities trigger the deleterious cascade of renal tissue injury in cisplatin administered rats [9,10].

Fisetin (3, 7, 3′, 4′-tetrahydroxyflavone) is a bioactive polyphenolic flavonoid, commonly found in many fruits and vegetables such as strawberries, apples, persimmons, onions and cucumbers [11]. In a recent review, fisetin was found to be one of the potent antioxidant flavonoid among the evaluated flavonoids [12]. Fisetin has been shown to posses both direct intrinsic antioxidant as well as indirect antioxidant effects by increasing levels of reduced glutathione in *in vitro* neuronal cells [13]. It exerts multiple beneficial pharmacological activities such as anti-inflammatory, anticancer, hypolipidemic and in rheumatoid arthritis [14]. Recent investigations suggests that fisetin attenuates migration and invasion of cervical cancer cells [15]; inhibits allergic airway inflammation [16], prevents adipocyte differentiation [17], prevents hepatosteatosis [18], attenuates complications of diabetes [19], counters osteoporosis [14] and exerts neuroprotection in cerebral ischemic condition [13]. In addition, fisetin is a potent natural anticancer agent. A study conducted by Tripathi et al. [20] demonstrated that the fisetin treatment along with cisplatin increased the cytotoxic effect of cisplatin by four-fold. However, the effect of fisetin on cisplatin-induced nephrotoxicity has not been evaluated. Based on the aforementioned facts and in continuation with the study conducted by Tripathi et al. [20], the present investigation was designed to evaluate the effect of fisetin on kidney tissues of cisplatin treated rats. We report that fisetin pre-treatment significantly ameliorates cisplatin-induced renal impairments, histopathological alterations and restores antioxidant and mitochondrial respiratory enzyme activities in kidney tissues. Furthermore, the results of the present study at molecular and cellular level revealed that fisetin significantly attenuates cisplatin-induced renal NOX4/RENOX and NOX2/gp91phox expression, apoptosis related protein expressions, modulates NF-κB activation and subsequent inflammation in kidney tissues.

Materials and Methods

Ethics statement and experimental animals

This study was carried out in strict accordance with recommendations on use of experiment animals according to Committee for the Purpose of Control and Supervision of Experiments on Animals (CPCSEA), Govt. of India for safe use and care of experimental animals. The protocol was approved by the committee called "Institutional Animal Ethics Committee (IAEC)" of Indian Institute of Chemical technology (IICT), Hyderabad, India (Approval No: IICT/PHARM/SRK/26/08/2013/09). All efforts were made to minimize sufferings. Male Sprague-Dawley rats weighing between 180 and 200 g were procured from National Institute Nutrition (NIN), Hyderabad, India and were housed under controlled environmental conditions (12 h light: 12 h dark cycle, 22±2°C temperature and 55±15% relative humidity). Animals were acclimatized for 7 days and allowed free access of food and water at all times.

Drugs and chemicals

Cisplatin, fisetin (purity: ≥98%), superoxide dismutase (SOD) assay kit, cytochrome c oxidase assay kit, o-dianisidine, MTT [3-(4, 5-dimethylthiazol-2-yl)-2, 5-diphenyltetrazolium bromide, β-nicotinamide adenine dinucleotide 3-phosphate reduced form (NADPH), lucigenin, β-nicotinamide adenine dinucleotide hydrate (NADH), succinic acid, 2, 6-dichlorophenolindophenol (DCIP), reduced glutathione (GSH), catalase (CAT), oxidized glutathione (GSSG), 5, 5-dithio-bis (2-nitrobenzoic acid) (DTNB), Bradford

reagent, 2-thiobarbituric acid (TBA) and cytochrome c were purchased from Sigma-Aldrich Co, St Louis, MO, USA. Bicinchoninic acid (BCA) protein assay kit was purchased from Pierce Biotechnology, Rockford, IL, USA. NF-κB (p65) transcription factor assay kit was obtained from Cayman Chemical Company, Ann Arbor, MI. Rat TNF-α (BD OptEIA) and IL-6 (BD OptEIA) ELISA kits were obtained from BD Bioscience, San Diego, CA, USA. All other chemicals were of analytical grade.

Experimental design

Prior to initiation of main experiment, a pilot study was conducted to confirm the safety and effect of fisetin alone on renal tissues. Based on the previous literature [20], fisetin at two different doses i.e. 1.25 and 2.5 mg/kg was selected and administered intraperitoneally for 7 consecutive days. On 8th day, blood samples were collected through retro-orbital plexus and serum was separated for the estimation of serum specific renal injury (BUN and creatinine) biomarkers. Kidneys were dissected and processed for histopathological evaluation. From this pilot study, it was observed that serum levels of BUN and creatinine were not showing statistically significant (p>0.05) differences between vehicle control, fisetin at 1.25 mg/kg and fisetin at 2.5 mg/kg administered group of rats (Figure 1). Histopathological findings of kidneys from the rats treated with vehicle (Control) and fisetin at a dose of 1.25 mg/kg (Fis-1.25) revealed normal kidney histo-morphology. To our surprise, a moderate infiltration of inflammatory cells was observed in kidney tissues of rats administered with 2.5 mg/kg fisetin (Fis-2.5) (Figure 1). Based on this finding, we further evaluated the effect of fisetin alone on heart and liver tissues. Cardiac injury (serum levels of CK-MB and LDH, and histopathology of heart, Figure S1) and hepatic injury (serum levels of SGOT and SGPT, and histopathology of liver, Figure S2) markers revealed that fisetin administration did not produce any significant alterations in the heart and liver tissue compared to vehicle control rats. Based on the above findings, fisetin at two different doses i.e. 0.625 and 1.25 mg/kg was selected for the main study. Fisetin, being poorly soluble in aqueous medium, was dissolved in PEG200/DMSO (7:3; v:v) and was administered intraperitoneally to the rats once in a day for 7 consecutive days [21]. Same vehicle system was administered to control rats.

In the main study, forty animals were randomly divided into 5 groups containing 8 rats in each. Group I (Vehicle control, Control), animals were administered with 40 μl of PEG200/DMSO (7:3) intraperitoneally once daily for 7 consecutive days and a single intraperitoneal injection of normal saline on 3rd day. In group II (Fisetin control, Fis), animals were intraperitoneally administered with fisetin [1.25 mg/kg dissolved in PEG200/DMSO (7:3)] once daily for 7 consecutive days and a single intraperitoneal injection of normal saline on 3rd day. In group III (Cisplatin control, Cis), animals were intraperitoneally administered with 40 μl of PEG200/DMSO (7:3) once daily for 7 consecutive days and a single intraperitoneal injection of cisplatin (5 mg/kg dissolved in normal saline) on 3rd day. In group IV (Cis+Fis-0.625 mg), animals were intraperitoneally administered with 40 μl of 0.625 mg/kg of fisetin [dissolved in PEG200/DMSO (7:3)] once daily for 7 consecutive days and a single intraperitoneal injection of cisplatin (5 mg/kg dissolved in normal saline) on 3rd day. In group V (Cis+Fis-1.25 mg), animals were intraperitoneally administered with 40 μl of 1.25 mg/kg of fisetin [dissolved in PEG200/DMSO (7:3)] once daily for 7 consecutive days and a single intraperitoneal injection of cisplatin (5 mg/kg dissolved in normal saline) on 3rd day. At the end of the experiment (i.e. on 8th day), body weight of all animals was recorded. Blood samples were

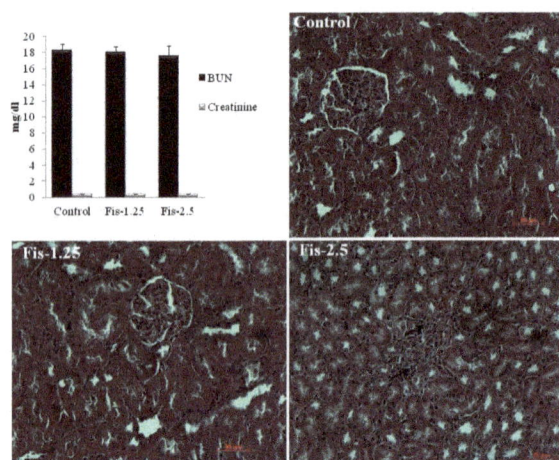

Figure 1. Effect of fisetin itself on serum renal function parameters and kidney histology. Intraperitoneal administration of fisetin at two different doses i.e. 1.25 mg/kg and 2.5 mg/kg body weight for 7 consecutive days showing normal BUN (blood urea nitrogen) and serum creatinine levels compared to vehicle control rats. Histopathological examination of kidney tissues (X200 magnification, scale bar: 50 μm) from vehicle control (Control) and fisetin at 1.25 mg/kg (Fis-1.25) treated rats showing apparently normal histo-morphology. Histopathological findings from fisetin at 2.5 mg/kg treated rats (Fis-2.5) showing moderate infiltration of inflammatory cells (black arrow) in kidney tissues.

collected from all the experimental animals and serum was separated. The animals were euthanized with CO_2 asphyxiation; kidney tissues were isolated, relative weights of kidneys (i.e kidney to body weight ratio normalized to 100 g body weight of animals) were determined and then stored at $-80°C$ for further study.

Estimation of serum blood urea nitrogen (BUN) and creatinine

To assess the renal failure, serum levels of BUN and creatinine were estimated using auto blood analyzer (Siemens, Dimension Xpand[plus], USA) by employing BUN and creatinine assay kits (Siemens, India).

Histopathology of kidney tissues

For histopathological evaluation of kidney tissues, tissue samples were fixed in 10% neutral buffered formalin and were processed for embedding in paraffin wax. Thin sections (5 μm) of the kidney tissues were cut using a microtome (Leica, Bensheim, Germany) and were stained with hematoxylin and eosin (H & E). The sections were examined for histopathological changes under light microscopy using Zeiss microscope (Axioplan 2 Imaging, Axiovision software).

Tissue preparation and biochemical estimations

A 10% kidney tissues homogenate was prepared with phosphate buffer saline (50 mM, pH 7.4) containing 1% protease inhibitor cocktail (Sigma-Aldrich Co, St Louis, MO, USA) using Teflon homogenizer. Homogenates were centrifuged at 14000 g for 45 min and supernatant obtained was used for estimation of various biochemical parameters. The kidney tissues content of reduced glutathione (GSH) [22], glutathione reductase (GR) [23], glutathione S-transferase (GST) [24], catalase (CAT) [25], superoxide dismutase (SOD) (SOD assay kit, Sigma–Aldrich

Co., St. Louis, MO, USA), NAD (P) H: quinine oxidoreductase 1 (NQO1) [26], vitamin C [27] and thiobarbituric acid reactive substances (TBARS) as an index of lipid peroxidation [28] were estimated as described in earlier literatures. Total protein content in kidney tissues homogenate was estimated using Bradford reagent (Sigma-Aldrich Co, St Louis, MO, USA) and bovine serum albumin (BSA) as standard.

Isolation of mitochondrial and cytosolic fraction from kidney tissues

The mitochondrial and cytosolic fractions of kidney tissues were isolated as described by Wei et al. [29]. Briefly, a portion of the fresh kidney tissue was homogenized in ice-cold isolation buffer [270 mM Sucrose, 1 mM EGTA and 5 mM Tris (pH 7.4)]. The homogenates were centrifuged at 600 g for 10 min at 4°C to remove cell debris and nuclei. The obtained supernatant was once again centrifuged at 10000 g for 10 min at 4°C to collect the mitochondrial fraction in the pellets. The supernatant fractions were centrifuged again at 100000 g for 1 h (Thermo Scientific Sorvall Discovery M150 SE Ultra centrifuge) to collect the cytosolic fraction. Protein concentration in mitochondrial and cytosolic fraction was determined using Bradford reagent and bovine serum albumin (BSA) as standard.

Estimation of mitochondrial respiratory enzyme activities

The mitochondrial respiratory enzyme activities such as NADH dehydrogenase [30], succinate dehydrogenase [31] and cytochrome c oxidase (cytochrome c oxidase assay kit, Sigma-Aldrich Co, St Louis, MO, USA) were estimated as reported in earlier standard procedure. In addition, mitochondrial redox activity was assessed by incubating the isolated mitochondria with MTT solution (5 mg MTT/ml of 50 mM phosphate buffer saline, pH 7.4) for 3 h at 37°C [32]. Depending upon the mitochondrial intactness, formazan crystals thus formed were solubilised in dimethyl sulfoxide and optical density was recorded at 580 nm.

Estimation of mitochondrial antioxidant and lipid peroxidation parameters

The mitochondrial non-enzymatic antioxidant, GSH [22] and enzymatic antioxidants i.e. GR [23], GST [24], CAT [25] and SOD (SOD assay kit, Sigma-Aldrich Co, St Louis, MO, USA) activities were analysed as described in earlier methods. Kidney tissue mitochondrial TBARS content was determined, as an index of lipid peroxidation, as described in earlier literature [28].

Estimation of TNF-α and IL-6 in kidney tissues

For estimation of pro-inflammatory cytokines in kidney tissues, a 10% tissue homogenate was prepared with phosphate buffer saline (50 mM, pH 7.4) containing 1% protease inhibitor cocktail (Sigma-Aldrich Co, St Louis, MO, USA). Then, the homogenates were centrifuged at 4000 g for 20 min and the supernatant obtained were used for the estimation of TNF-α and IL-6 using rat TNF-α (BD OptEIA) and IL-6 (BD OptEIA) ELISA kits (BD Bioscience, San Diego, CA, USA) respectively. The concentrations of TNF-α and IL-6 in kidney tissues were expressed as pg/mg protein.

Estimation of myeloperoxidase (MPO) activity in kidney tissues

Myeloperoxidase activity was estimated as described in previously published procedure [33]. Briefly, a 10% kidney tissues homogenate was prepared using ice-cold 50 mM potassium phosphate buffer (pH 6) containing 0.5% hexadecyltrimethylammonium bromide (HTAB) and 10 mM EDTA. Homogenates were subjected to one cycle of freeze and thaw, followed by brief sonication. Then the homogenates were centrifuged at 13100 g for 20 min. The supernatant obtained was used for estimation of MPO activity using 0.167 mg/ml of o-dianisidine hydrochloride and 0.0005% hydrogen peroxide at 460 nm. The MPO activity was expressed as U/g of tissue.

Estimation of NADPH oxidase activity in kidney tissues

The activity of NADPH oxidase was estimated according to Oh et al [34]. Briefly, a 10% kidney tissues homogenate were prepared using ice-cold phosphate buffer saline (50 mM, pH 7.4). The homogenates were centrifuged at 4000 g for 15 min at 4°C to remove any cellular debris. The protein concentration in homogenates was estimated using Bradford reagent and BSA as standard. Around 50 μg protein of kidney homogenates was incubated with a reaction mixture contained 1 mM ethylene glycol tetra acetic acid and 5 μM lucigenin in 50 mM phosphate buffer, pH 7.0. The reaction was started by the addition of 50 μM NADPH to the reaction mixture and the activity was estimated by recording the luminescence for 5 min at every 31 s interval. No activity could be measured in the absence of NADPH. Relative luminescence unit (RLU) was calculated and was expressed as fold increase over control.

RNA isolation and quantitative Real-time PCR

Total RNA was isolated from frozen kidney tissues using TRI reagent (Sigma-Aldrich Co, St Louis, MO, USA) according to manufacturer's protocol. RNA concentration was quantified using NanoDrop 2000/2000c (Thermo Fisher Scientific, Wilmington, DE 19810 U.S.A). cDNA were synthesized using Thermo Scientific Verso cDNA kit according to manufacturers' instructions. NADPH oxidase subunits of mRNA (NOX2/gp91phox and NOX4/RENOX) expression were measured by quantitative real-time polymerase chain reaction (PCR) analysis using the StepOnePLUS (Applied Biosystems, USA) real-time PCR detection system. The reaction mixture of real-time PCR contained 100 ng 5' and 3' primer and 2 μl of cDNA product in Thermo Scientific DyNAmo ColorFlash SYBR Green qPCR Kit. All primers were synthesized by Geno Bioscience Pvt Ltd (India). The primers used for rat NOX2/gp91phox were (forward) 5'-CCCTTTGGTACAGCCAGTGAAGAT-3' and (reverse) 5'-CAATCCCAGCTCCCACTAACATCA-3', rat NOX4/RE-NOX (forward) 5'-GGATCACAGAAGGTCCCTAGCAG-3' and (reverse) 5'-GCAGCTACATGCACACCTGAGAA-3', and those used for rat GAPDH were (forward) 5'-TCAA-GAAGGTGGTGAAGCAG-3' and (reverse) 5'-AGGTGGAA-GAATGGGAGTTG-3'. The real-time PCR results for the mRNA levels of each gene were normalized to GAPDH levels. Relative mRNA expression was quantified using the ΔΔCt method. Results were expressed as fold change.

Nuclear, cytoplasmic and total protein extraction

Nuclear, cytoplasmic and total protein extracts from kidney tissues of different experimental groups were prepared as described in our previously published literature [35]. Protein concentrations in all extracts were determined using Bicinchoninic acid (BCA) assay kit (Pierce Biotechnology, Rockford, IL, USA) against bovine serum albumin (BSA) as standard and stored them at −80°C till further analysis.

Antibodies and immunoblot analysis

Antibodies against cleaved caspase 3, cleaved caspase 9, Bax, Bcl-2, cytochrome c, p53, NF-κB (p65), IκBα, phospho-IκBα, β-actin, lamin B and HRP-conjugated secondary anti-rabbit and anti-mouse antibodies were purchased from Cell Signaling Technology, Boston, MA. Antibody against iNOS was obtained from Sigma-Aldrich Co, St Louis, MO, USA. For detection of protein expression in renal cortex, an equal amount of protein (40 μg/lane) samples were resolved in 10% SDS-PAGE and electrophoretically transferred onto polyvinylidine difluoride membranes (PVDF, Pierce Biotechnology, Rockford, IL, USA) for immunoblot analysis. The following dilutions were made for primary antibody preparation for detection of total protein; monoclonal rabbit cleaved caspase 3 (1:1000), monoclonal mouse cleaved caspase 9 (1:1000), monoclonal rabbit Bax (1:1000), monoclonal rabbit Bcl-2 (1:1000), monoclonal mouse p53 (1:1000), monoclonal rabbit iNOS (1:500); for detection of nuclear protein, monoclonal rabbit NF-κB (p65) (1:500); for detection of cytoplasmic protein, monoclonal mouse IκBα (1:500) and mono-clonal rabbit phospho-IκBα (1:500) antibodies; HRP-conjugated anti-rabbit and anti-mouse secondary antibodies at a dilution of 1:8000 were used. Similarly, mitochondrial and cytosolic fractions were used for detection of cytosolic translocation of cytochrome c (rabbit monoclonal, 1:1000). β-actin (1:1000) was used for equal loading of total, cytoplasmic and cytosolic proteins and Lamin B (1:1000) was used for equal loading of nuclear proteins. Coomassie blue stain was used for equal loading of mitochondrial protein. The membranes were visualized with SuperSignal West Pico Chemiluminescent Substrate (Pierce Biotechnology, Rockford, IL, USA) and developed to X-ray film (Pierce Biotechnology, Rockford, IL, USA). Band intensities were quantified by using Image J software (NIH).

NF-κB (p65) transcription assay

As per the manufacturer instructions, NF-κB (p65) transcription factor ELISA assay kit (Cayman Chemical Company, Ann Arbor, MI) was used to evaluate the NF-κB DNA-binding activity of nuclear samples extracted from kidney tissues of different experimental groups.

Estimation of platinum in kidney tissues

To analyze the platinum content in kidney tissues, as described in earlier method, known quantity of kidney tissues were digested in concentrated nitric acid at a temperature of 40°C for 3 h under constant agitation [36]. The platinum concentration of digested samples were assessed by Inductively Coupled Plasma Mass Spectrometry (ICP-MS) method and expressed as μg/g of tissue.

Statistical analysis

Experimental data was represented as Mean ± S.E.M. Graph Pad Prism software (version 5.0) was used for statistical analysis of the study. Significance difference was evaluated by performing one-way analysis of variance (ANOVA) with Dunnett's multiple comparison procedure; p values of less than 0.05 were regarded as significant.

Results

Effect of fisetin on cisplatin-induced renal injury parameters

Cisplatin administration significantly ($p < 0.001$) increased the BUN (Figure 2A) (from 18.25 ± 0.53 mg/dl to 108.33 ± 7.64 mg/dl) and creatinine (Figure 2B) (from 0.34 ± 0.03 mg/dl to

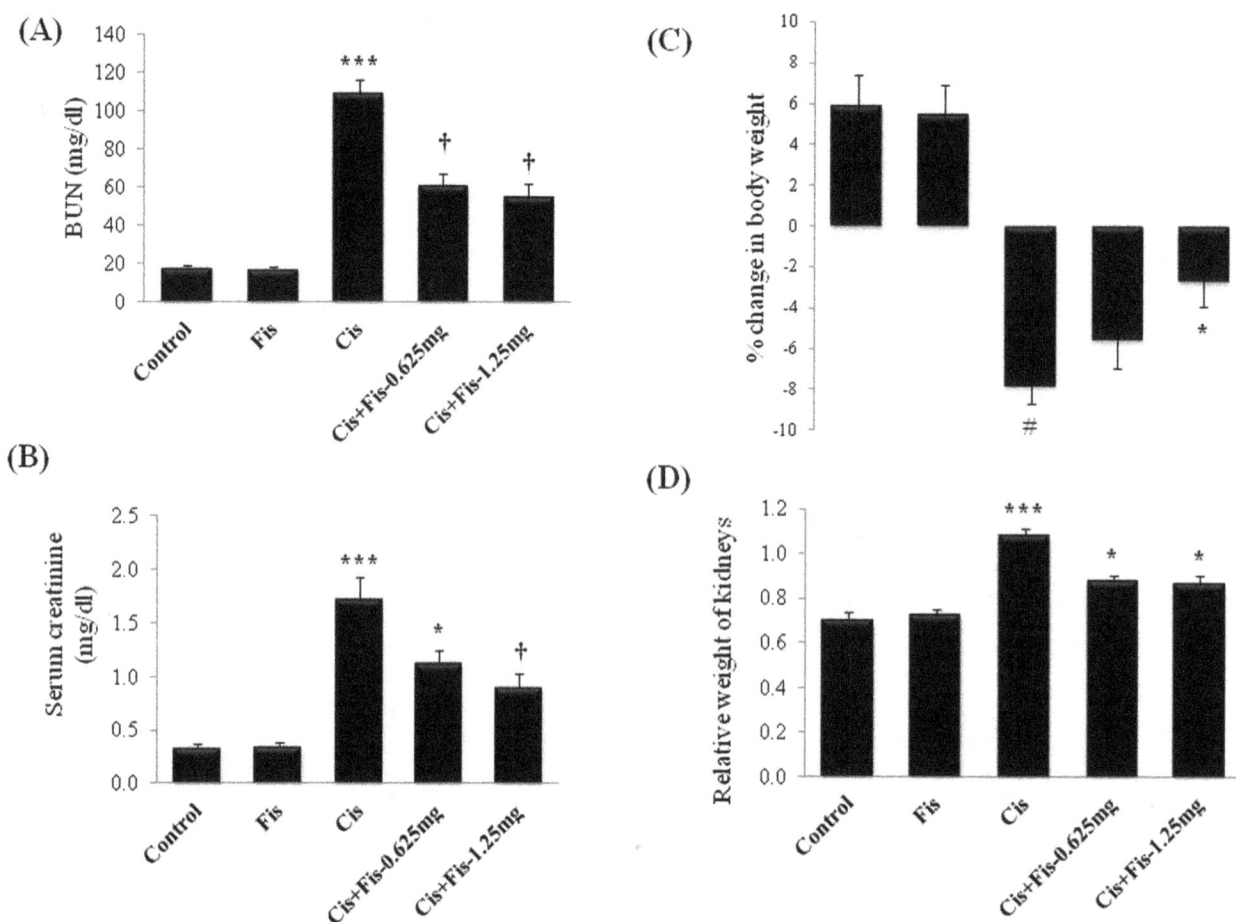

Figure 2. Effect of fisetin on cisplatin-induced changes in serum renal function parameters. (**A**) Blood urea nitrogen (BUN), (**B**) Creatinine, (**C**) Percentage change in body weight and (**D**) Relative weight of kidneys in different experimental groups. Fisetin was administered intraperitoneally at two different doses i.e. 0.625 mg/kg and 1.25 mg/kg for 7 consecutive days and a single dose of cisplatin (5 mg/kg, i.p) on 3rd day. On 8th day, serum levels of BUN and creatinine, percentage change in body weight and relative weight of kidneys were recorded. The results were expressed as mean ± S.E.M of 8 animals in each group. Where, Control, vehicle control; Fis, fisetin control; Cis, cisplatin control; Cis+Fis-0.625 mg, cisplatin plus fisetin (0.625 mg/kg) treated groups; Cis+Fis-1.25 mg, cisplatin plus fisetin (1.25 mg/kg) treated groups. *$p<0.05$ and †$p<0.001$ vs cisplatin control group; #$p<0.05$ and ***$p<0.001$ vs vehicle control group.

1.72 ± 0.21 mg/dl) levels when compared to vehicle control groups. Fisetin treatment at both the doses (0.625 and 1.25 mg/kg) along with cisplatin significantly attenuated the increase in BUN and creatinine levels when compared to cisplatin alone treated group. No mortality was observed in animals treated with cisplatin and/or fisetin during the study period. For calculation of percentage of body weight gain/loss, the day of cisplatin administration is considered as day 0 and average body weight of all animals was taken as 0%. After 5 days of cisplatin administration, body weight of all animals from each group was recorded and percentage of weight gain or loss was calculated. In vehicle control and fisetin (1.25 mg/kg) alone treated group, we observed a 5.9% and 5.5% body weight gain respectively, when compared to day 0 body weight of respective group animals. In cisplatin alone administered group of rats, we observed a 7.8% body weight loss when compared to day 0 body weight of same animals. Fisetin (1.25 mg/kg) treatment along with cisplatin significantly ($p<0.05$) attenuated the body weight loss when compared to cisplatin alone treated rats (Figure 2C). Similarly, relative weight of kidneys in cisplatin alone treated rats were significantly ($p<0.001$) increased compared to vehicle control

group. Fisetin treatment at both doses (0.625 and 1.25 mg/kg) significantly ($p<0.05$) attenuated the increase in relative weight of kidneys compared to cisplatin alone treated rats (Figure 2D). Histopathological findings (Figure 3) of kidney tissues from vehicle control rats (Control) showed intact histo-morphology, whereas kidney tissues from cisplatin alone administered rats (Cis) revealed severe tubular degeneration and necrosis in the tubular epithelium (white arrow), infiltration of inflammatory cells (black arrow) and accumulation of homogenous eosinophilic casts (yellow arrow) in the lumen of the tubules. Kidney tissue sections from fisetin at low dose (0.625 mg/kg) with cisplatin treated group (Cis+Fis-0.625) showed decrease in tubular degeneration (white arrow), infiltration of inflammatory cells (black arrow) and accumulation of eosino-philic casts (yellow arrow) in tubular lumen when compared to cisplatin alone treated group of rats. However, animals treated with fisetin at 1.25 mg/kg along with cisplatin showed predom-inantly normal renal histology with occasional degenerative changes (white arrow) when compared to cisplatin alone treated rats (Cis+Fis-1.25).

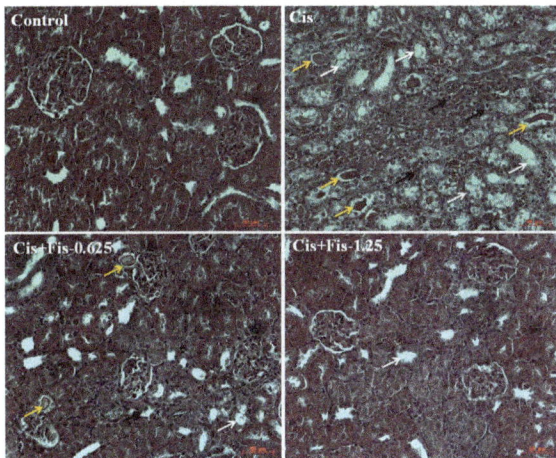

Figure 3. Histopathological observations of kidney tissues from rats treated with fisetin and cisplatin. Representative photomicrographs of kidney sections stained with Hematoxylin-eosin (H and E). Light microscopic examination of kidney sections from vehicle control rats (Control) showing normal glomeruli and tubules (X200 magnification, scale bar: 50 μm). Kidney tissue sections from cisplatin alone treated rats (Cis) showing severe tubular degeneration and necrosis (white arrow), infiltration of inflammatory cells (black arrow) and accumulation of homogenous eosinophilic casts (yellow arrow) in the lumen of the tubules (X200 magnification, scale bar: 50 μm). Kidney tissue sections from low dose of fisetin (0.625 mg/kg) plus cisplatin treated rats (Cis+Fis-0.625) showing recovery of tubular degeneration (white arrow), mild infiltration of inflammatory cells (black arrow) and decreased accumulation of eosinophilic casts in tubular lumen (yellow arrow) when compared to cisplatin alone treated rats (X200 magnification, scale bar: 50 μm). Kidney tissue sections from high dose of fisetin (1.25 mg/kg) plus cisplatin treated rats (Cis+Fis-1.25) showing predominantly normal renal histology with occasional degenerative changes (white arrow) when compared to cisplatin alone treated rats (X200 magnification, scale bar: 50 μm).

Effect of fisetin on cisplatin-induced changes in kidney tissues antioxidant and lipid peroxidation parameters

Table 1 depicts the activities and/or levels of various enzymatic (GR, GST, CAT, SOD and NQO1) and non-enzymatic (GSH and Vit C) antioxidants in the kidney tissues of all experimental groups. The activities of enzymatic and levels of non-enzymatic antioxidants were significantly (p<0.05: GSH, GR, NQO1 and Vit C; p<0.01: GST, CAT and SOD) decreased in kidney tissues of cisplatin alone administered rats. Fisetin administration at high dose (1.25 mg/kg) along with cisplatin significantly (p<0.05: GSH; p<0.01: GR, CAT and Vit C; p<0.001: GST, SOD and NQO1) restored the activities of these antioxidants to nearly normal. Though fisetin treatment at low dose (0.625 mg/kg) along with cisplatin was able to significantly (p<0.05: GSH; p<0.01: GST, SOD and NQO1) restore the activities and/or levels of GSH, GST, SOD and NQO1, the activities of GR, CAT and levels of vitamin C remains unaltered. Similarly, the level of TBARS, which measures the extent of lipid peroxidation was significantly (p<0.05) increased in cisplatin treated group of rats when compared to vehicle control rats. Treatment with fisetin at both the doses (0.625 and 1.25 mg/kg) significantly (p<0.05) decreased the level of TBARS when compared to cisplatin alone treated group of rats.

Effect of fisetin on NADPH oxidase activity and mRNA expression of NOX2 and NOX4

As shown in Figure 4, cisplatin administration significantly (2.4 fold) increased the activity of NADPH oxidase, whereas fisetin (1.25 mg/kg) treatment significantly (p<0.05) attenuated the cisplatin-induced increase in NADPH oxidase activity in kidney tissues. qPCR data revealed that the expressions of NOX2 and NOX4 were significantly (NOX2, 2.6 fold; NOX4, 3.4 fold) increased in cisplatin alone treated rats compared to vehicle control rats. Fisetin treatment at the high dose (1.25 mg/kg) significantly (NOX2, p<0.05; NOX4, p<0.01) attenuated the cisplatin-induced expressions of NOX2 and NOX4 in kidney tissues when compared to cisplatin alone treated rats. Rats treated with fisetin at low dose (0.625 mg/kg) along with cisplatin did not

Table 1. Effect of fisetin on cisplatin-induced changes in renal antioxidants.

	Control	Fis	Cis	Cis+Fis-0.625 mg	Cis+Fis-1.25 mg
GSH levels	0.59±0.034	0.63±0.02	0.44±0.03$^\$$	0.67±0.05*	0.70±0.05*
GR activity	7.36±0.56	7.26±0.83	5.22±0.48$^\$$	7.10±0.31	8.49±0.95**
GST activity	15.98±0.79	16.46±1.33	7.42±1.47$^\#$	14.78±0.61**	18.69±2.04***
CAT activity	12.26±0.50	13.24±0.51	6.58±0.81$^\#$	7.87±0.38	11.10±0.93**
SOD activity	100.00±0.85	99.47±0.88	84.41±1.80$^\#$	96.4±1.02**	99.69±1.18***
NQO1 activity	127.81±5.97	132.38±4.20	89.62±18.22$^\$$	137.3±8.51**	152.86±4.49***
Vit C levels	0.82±0.05	1.37±0.13	0.38±0.08$^\$$	0.69±0.17	1.06±0.11**
TBARS levels	11.47±0.74	11.60±2.98	22.82±4.96$^\$$	10.79±2.46*	10.51±2.00*

All data were expressed as mean ± S.E.M. Where, Control, vehicle control (N=7); Fis, fisetin alone treated group (N=8); Cis, cisplatin alone treated group (N=8); Cis+Fis-0.625 mg, fisetin (0.625 mg/kg) treated cisplatin-induced rats (N=8); Cis+Fis-1.25 mg, fisetin (1.25 mg/kg) treated cisplatin-induced rats (N=8); GSH, reduced glutathione (mg/g of tissue); GR, glutathione reductase (U/mg protein); GST, glutathione S-transferase (nmoles of CDNB conjugated/min/ml); CAT, catalase (U/mg protein); SOD, superoxide dismutase (% of control); NQO1, NAD (P) H: quinine oxidoreductase 1 (nmoles of DCIP reduced/min/mg protein); Vit C, vitamin C (mg/g of tissue); TBARS, thiobarbituric acid reactive substances (nmoles/g of tissue); CDNB, 1-chloro-2, 4-dinitrobenzene; DCIP, 2,6-dichlorophenolindophenol.
$^\$$p<0.05,
$^\#$p<0.01 vs control,
*p<0.05,
**p<0.01,
***p<0.001 vs Cis control.

(A)

(B)

(C)

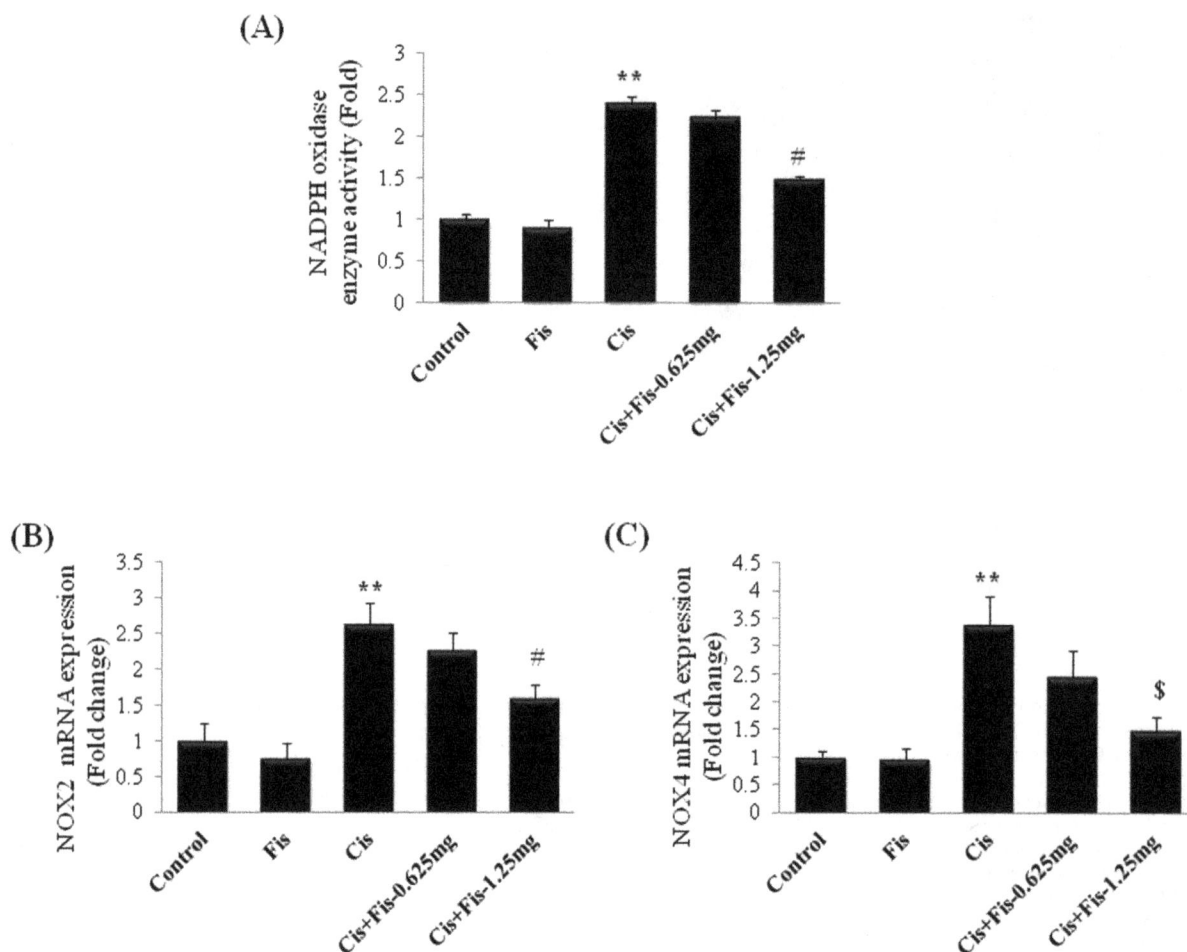

Figure 4. Effect of fisetin on cisplatin-induced changes in (A) NADPH oxidase enzyme activity and mRNA expressions of (B) NOX2/ gp91phox and (C) NOX4/RENOX. The results were expressed as mean ± S.E.M of 4 animals in each group. Where, Control, vehicle control; Fis, fisetin control; Cis, cisplatin control; Cis+Fis-0.625 mg, cisplatin plus fisetin (0.625 mg/kg) treated groups; Cis+Fis-1.25 mg, cisplatin plus fisetin (1.25 mg/kg) treated groups. **$p < 0.05$ vs vehicle control group; #$p < 0.05$, $p < 0.01$ vs cisplatin control group.

produce any significant ($p > 0.05$) change in activity or mRNA expression of NOX compared to those of cisplatin alone treated rats.

Effect of fisetin on cisplatin-induced changes in mitochondrial respiratory enzyme activities in kidney tissues

In order to investigate the effect of fisetin on mitochondrial function, mitochondrial respiratory enzymes such as NADH dehydrogenase, succinate dehydrogenase, cytochrome c oxidase and mitochondrial redox activity were evaluated in isolated mitochondrial fraction of kidney tissues of all experimental rats. As shown in Figure 5, the activities of NADH dehydrogenase (NDH) (Figure 5A), succinate dehydrogenase (SDH) (Figure 5B), cytochrome c oxidase (COX) (Figure 5C) and mitochondrial redox activities (Figure 5D) were significantly ($p < 0.01$) decreased in cisplatin alone treated rats. Rats treated with fisetin at low dose (0.625 mg/kg) along with cisplatin did not produce any significant ($p > 0.05$) change in these enzymes compared to those of cisplatin control group of rats. In contrast, the activities of these mitochondrial respiratory enzymes and mitochondrial redox activity were significantly ($p < 0.05$ for NDH, SDH and redox

activity; $p < 0.001$ for COX) restored in high dose (1.25 mg/kg) fisetin treatment group when compared to cisplatin alone treated group of rats.

Effect of fisetin on cisplatin-induced changes in mitochondrial antioxidant and lipid peroxidation parameters in kidney tissues

Table 2 represents the effect of fisetin on the cisplatin-induced changes in the mitochondrial antioxidants in the kidney tissues of all experimental rats. Cisplatin administration caused a significant ($p < 0.05$: GR, GST and CAT; $p < 0.01$: GSH and SOD) reduction in the levels of mitochondrial antioxidants such as GSH, GR, GST, SOD and CAT in kidney tissues. Fisetin treatment at high dose (1.25 mg/kg) along with cisplatin significantly ($p < 0.05$: GR, GST and CAT; $p < 0.01$: GSH and SOD) restored the level of these enzymes when compared to cisplatin control rats. Though fisetin treatment at low dose (0.625 mg/kg) restored the levels of mitochondrial activities of SOD and CAT, the levels of GSH, GR and GST remain unaltered when compared to those of cisplatin alone treated rats. In addition, to assess the mitochondrial lipid peroxidation status, TBARS content was estimated. TBARS content in cisplatin alone treated rats was significantly ($p < 0.01$) increased when compared to those of vehicle control rats. Fisetin

(A)

(B)

(C)

(D)

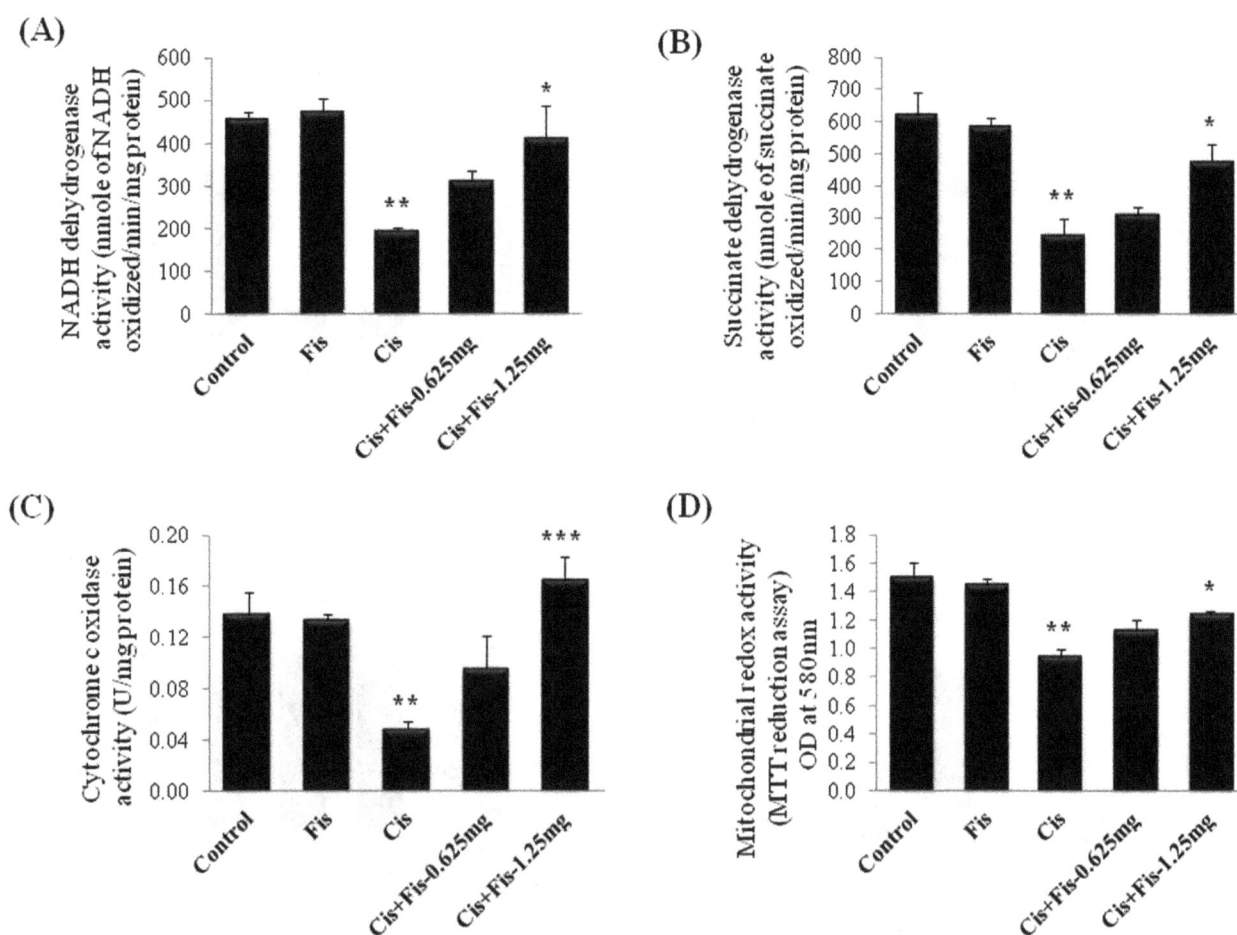

Figure 5. Effect of fisetin on cisplatin-induced changes in mitochondrial respiratory enzyme activities in kidney tissues. (A) NADH dehydrogenase, **(B)** Succinate dehydrogenase, **(C)** Cytochrome c oxidase and **(D)** Mitochondrial redox activities. The results were expressed as mean ± S.E.M of 6 animals in each group. Where, Control, vehicle control; Fis, fisetin control; Cis, cisplatin control; Cis+Fis-0.625 mg, cisplatin plus fisetin (0.625 mg/kg) treated groups; Cis+Fis-1.25 mg, cisplatin plus fisetin (1.25 mg/kg) treated groups. *p<0.05 and ***p<0.001vs cisplatin control group; **p<0.01 vs vehicle control group.

Table 2. Effect of fisetin on cisplatin-induced changes in renal mitochondrial antioxidants.

	Control	Fis	Cis	Cis+Fis-0.625 mg	Cis+Fis-1.25 mg
GSH levels	45.72±0.98	45.33±1.99	34.50±1.15[#]	38.88±1.22	45.10±1.66**
GR activity	3.20±0.17	3.32±0.23	2.34±0.06[$]	2.70±0.21	3.48±0.28*
GST activity	126.72±8.56	127.40±5.48	105.92±3.45[$]	123.67±5.36	142.3±3.84*
CAT activity	3.59±0.19	3.74±0.30	2.55±0.19[$]	3.40±0.25*	3.51±0.14*
SOD activity	100.00±0.38	99.86±1.18	88.61±0.86[#]	95.88±1.32*	98.19±0.52**
TBARS levels	5.92±0.54	5.95±0.32	10.68±0.87[#]	7.04±0.69*	5.57±0.26**

All data were expressed as mean± S.E.M. Where, Control, vehicle control (N=7); Fis, fisetin alone treated group (N=8); Cis, cisplatin alone treated group (N=8); Cis+Fis-0.625 mg, fisetin (0.625 mg/kg) treated cisplatin-induced rats (N=8); Cis+Fis-1.25 mg, fisetin (1.25 mg/kg) treated cisplatin-induced rats (N=8); GSH, reduced glutathione (μg/g of tissue); GR, glutathione reductase (U/mg protein); GST, glutathione S-transferase (nmoles of CDNB conjugated/min/ml); CAT, catalase (U/mg protein); SOD, superoxide dismutase (% of control); TBARS, thiobarbituric acid reactive substances (nmoles/g of tissue); CDNB, 1-chloro-2, 4-dinitrobenzene.
$p<0.05,
#p<0.01 vs control,
*p<0.05,
**p<0.01 vs Cis control.

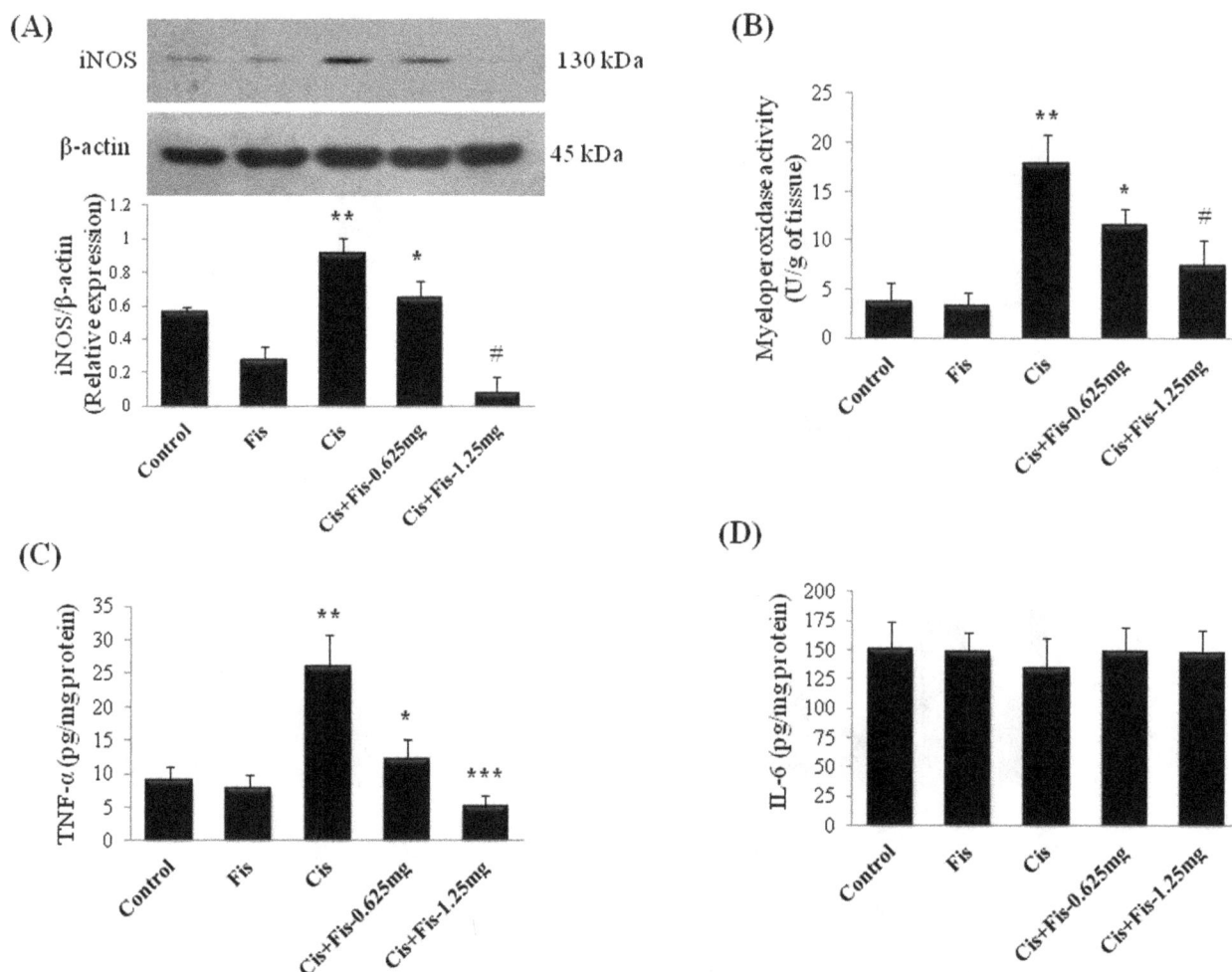

Figure 6. Effect of fisetin on cisplatin-induced renal inflammation. (A) Effect of fisetin on cisplatin-induced changes in protein expression of inducible nitric oxide (iNOS) in kidney tissues of control and treated groups. Renal tissue total protein extract was prepared and analyzed for expression level of iNOS by western blotting using specific antibodies. β-actin was used as loading control for equal loading of proteins. Protein expression was studied for three independent experiments and representative blots are shown. Quantitative densitometry was performed for each blot. Fisetin at both the doses (0.625 and 1.25 mg/kg) significantly decreased the protein expression of iNOS compared to cisplatin control. Effect of fisetin on cisplatin-induced changes in **(B)** Myeloperoxidase activity, as an index for neutrophil infiltration; changes in proinflammatory cytokine levels **(C)** Tumor necrosis factor-α (TNF-α) and **(D)** Interleukin-6 (IL-6). The results were expressed as mean ± S.E.M of 6 animals in each group. Where, Control, vehicle control; Fis, fisetin control; Cis, cisplatin control; Cis+Fis-0.625 mg, cisplatin plus fisetin (0.625 mg/kg) treated groups; Cis+Fis-1.25 mg, cisplatin plus fisetin (1.25 mg/kg) treated groups. $^*p<0.05$, $^\#p<0.01$ and $^{***}p<0.001$vs cisplatin control group; $^{**}p<0.01$ vs vehicle control group.

treatment at both the doses (0.625 and 1.25 mg/kg) significantly (p<0.05: for 0.625 mg/kg fisetin dose; p<0.01: for 1.25 mg/kg fisetin dose) decreased the TBARS level when compared to cisplatin alone treated group of rats.

Effect of fisetin on cisplatin-induced changes in inflammatory markers

In order to evaluate whether fisetin was able to attenuate the cisplatin evoked renal inflammation, we analyzed the pro-inflammatory cytokine levels i.e. TNF-α, IL-6, myeloperoxidase activity and renal protein expression of inducible nitric oxide synthase (iNOS) in kidney tissues. As shown in Figure 6, the protein expression of iNOS (Figure 6A); the myeloperoxidase (MPO) activity (Figure 6B), a marker for leukocytes/macrophage infiltration; TNF-α (Figure 6C), a pro-inflammatory cytokine were significantly (p<0.01) increased in kidney tissues of cisplatin alone treated rats when compared to those of vehicle control rats. Fisetin

administration significantly attenuated the cisplatin-induced increase in the renal protein expression of iNOS, the activities of MPO and pro-inflammatory cytokines, TNF-α when compared to cisplatin alone treated rats. The IL-6 levels remain unchanged in fisetin plus cisplatin treated rats when compared to vehicle as well as cisplatin control group of rats (Figure 6D). Fisetin alone did not produce any significant change in these inflammatory markers when compared to vehicle control group of rats

To further investigate the mechanisms underlying the beneficial effects of fisetin on cisplatin induced renal inflammation, we analyzed the protein expression and activation of NF-κB (p65) in kidney tissues. Results from immunoblot analysis showed the amount of NF-κB (p65) in the nuclear protein fraction of kidney tissues from cisplatin alone administered rats were significantly (p<0.05) increased when compared to vehicle control rats (Figure 7A). To further clarify the involvement of NF-κB, NF-κB (p65) transcription assay was performed. As expected the DNA

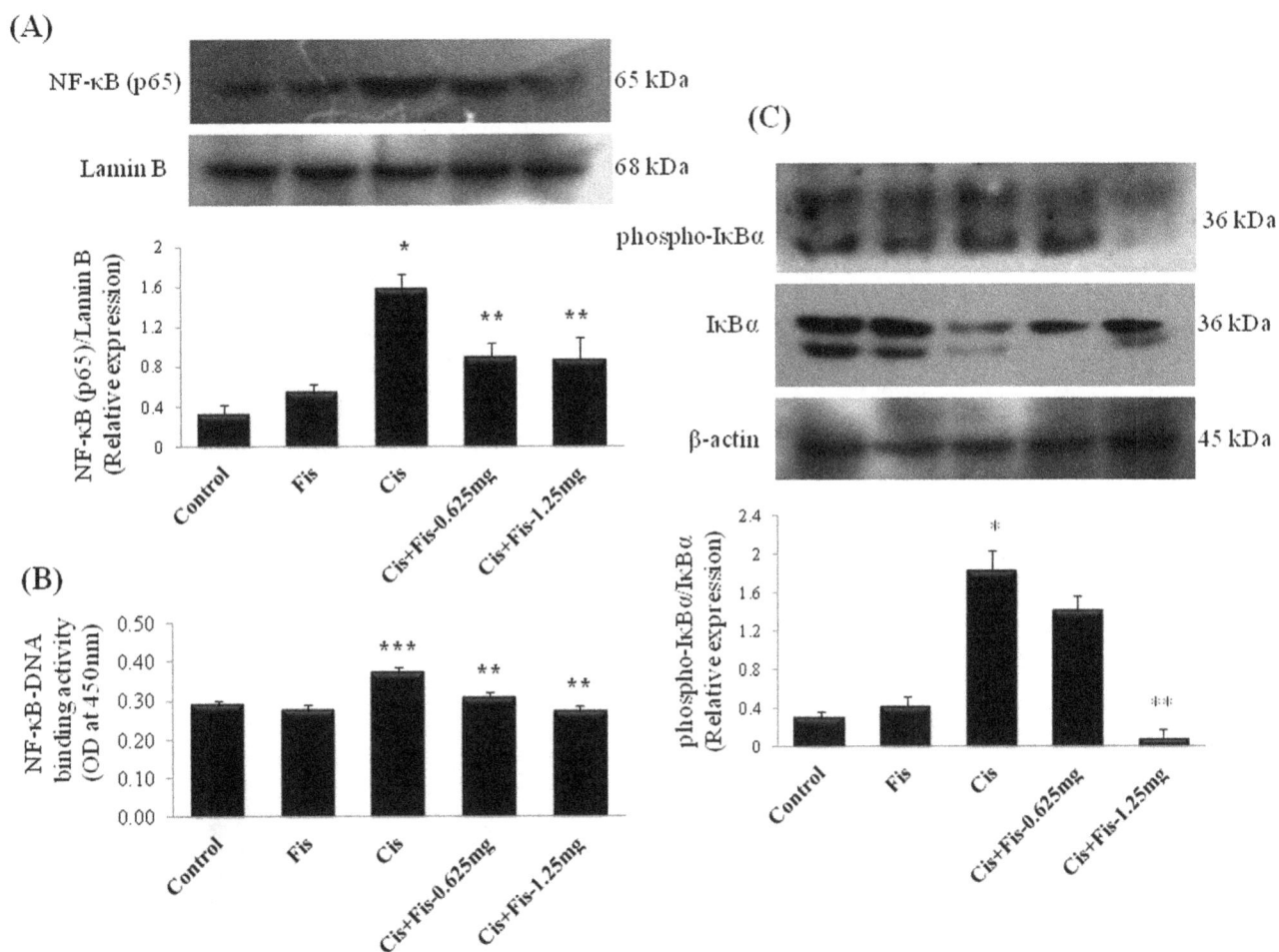

Figure 7. Effect of fisetin on cisplatin-induced NF-κB (p65) protein activation and IκBα phosphorylation. (A) Effect of fisetin on cisplatin-induced changes in kidney tissue nuclear protein expression of NF-κB (p65) and (B) DNA binding activity of NF-κB (p65). Kidney tissue nuclear protein extract was used for estimation of NF-κB (p65) - DNA binding activity using NF-κB (p^{65}) transcription factor ELISA assay kit (Cayman Chemical Company, Ann Arbor, MI). The results were expressed as mean ± S.E.M of 6 animals in each group. (C) Kidney tissue cytoplasmic protein expression of phospho-IκBα and IκBα. β-actin and Lamin-B were used as loading control for equal loading of cytoplasmic and nuclear proteins respectively. Protein expressions were studied for three independent experiments and representative blots are shown. Quantitative densitometry was performed for each blot. Where, Control, vehicle control; Fis, fisetin control; Cis, cisplatin control; Cis+Fis-0.625 mg, cisplatin plus fisetin (0.625 mg/kg) treated groups; Cis+Fis-1.25 mg, cisplatin plus fisetin (1.25 mg/kg) treated groups. *p<0.05, ***p<0.001vs vehicle control group; **p<0.01 vs cisplatin control group.

binding activity of NF-κB (p65) was significantly (p<0.001) increased in cisplatin alone treated rats (Figure 7B). Fisetin administration at both the doses (0.625 and 1.25 mg/kg) along with cisplatin significantly (p<0.01) decreased the amount of NF-κB (p65) and NF-κB (p65)-DNA binding activity when compared to cisplatin alone treated rats. Additionally, the amount of IκBα protein was down regulated with concomitant increase in phospho-IκBα in cisplatin alone administered rats (Figure 7C). Fisetin administration at higher dose (1.25 mg/kg) decreased the phosphorylation of IκBα protein and enhanced the IκBα protein level in kidney tissues when compared to cisplatin alone treated rats.

Effect of fisetin on cisplatin-induced changes in renal apoptosis related proteins

In order to elucidate the effect of fisetin on cisplatin-induced renal cell apoptosis, we analysed the various apoptosis related proteins such as cytochrome c, Bax, Bcl-2, cleaved caspase-3, cleaved caspase-9 and p53 in kidney tissues. Cisplatin significantly

(p<0.05) enhanced the cytosolic translocation of cytochrome c from mitochondrial fraction when compared to vehicle control rats (Figure 8). Furthermore, the protein expressions of Bax, p53, cleaved caspase-3 and cleaved caspase-9 were significantly (p< 0.01) increased and the protein expression of Bcl-2 was significantly (p<0.05) decreased in cisplatin alone administered rats when compared to those of vehicle control rats (Figure 9). Fisetin treatment at higher dose (1.25 mg/kg) along with cisplatin significantly attenuated the cytosolic translocation of cytochrome c, protein expression of Bax, p53, cleaved caspase-3 and cleaved caspase-9 and increased the expression of Bcl-2 when compared to those of cisplatin alone treated rats.

Effect of fisetin on platinum accumulation in kidney tissues

To assess the effect of fisetin on platinum accumulation in renal tissue, we next analyzed the platinum concentration in kidney tissues of all experimental groups. As shown in Figure 10, we have not observed any significant change in platinum concentration between cisplatin alone and cisplatin plus fisetin treated (0.625 and 1.25 mg/kg) group of rats.

Discussion

Nephrotoxicity is a frequent devastating adverse effect of cisplatin chemotherapy. The unique pharmacological profile of fisetin has attracted considerable attention in the field of cancer research. In a previous study, it has been demonstrated that combination of fisetin along with cisplatin exhibited four times more anticancer potential than individual treatment group [20]. In the present study, we demonstrated that fisetin treatment along with cisplatin largely reduced the nephrotoxicity, a clinical-utility limiting side effect of the cisplatin chemotherapy, by employing rat model. The findings of the present study revealed that fisetin treatment reduced the cisplatin-induced renal and mitochondrial oxidative stress, restored mitochondrial respiratory enzyme activities and attenuated expressions of apoptosis and inflammation related proteins, thus forming the molecular basis for

protective mechanism of fisetin against cisplatin-induced nephrotoxicity.

In agreement with previous reports, we found that a single intraperitoneal injection of cisplatin (5 mg/kg) induced marked elevation of BUN and creatinine (renal function biomarkers) in serum (Figure 2) and also demonstrated histopathological damage with tubular degeneration, tubular necrosis and infiltration of inflammatory cells in kidneys (Figure 3). Although the precise molecular mechanism of cisplatin-induced nephrotoxicity is complex and remains uncertain, induction of oxidative damage, apoptosis/necrosis of renal tubular cells and activation of inflammatory pathways have been demonstrated in kidneys of cisplatin treated animals [37]. Most of the literatures, including our earlier studies, have revealed that cisplatin induces free radicals and produces oxidative damage and lipid peroxidation in kidney tissues [38,39]. Cisplatin generates highly reactive free radicals such as superoxide and hydroxyl radicals which can directly interact and modify many subcellular components including DNA, proteins, lipids and other macromolecules and eventually causes cell death [40]. Cisplatin induced ROS

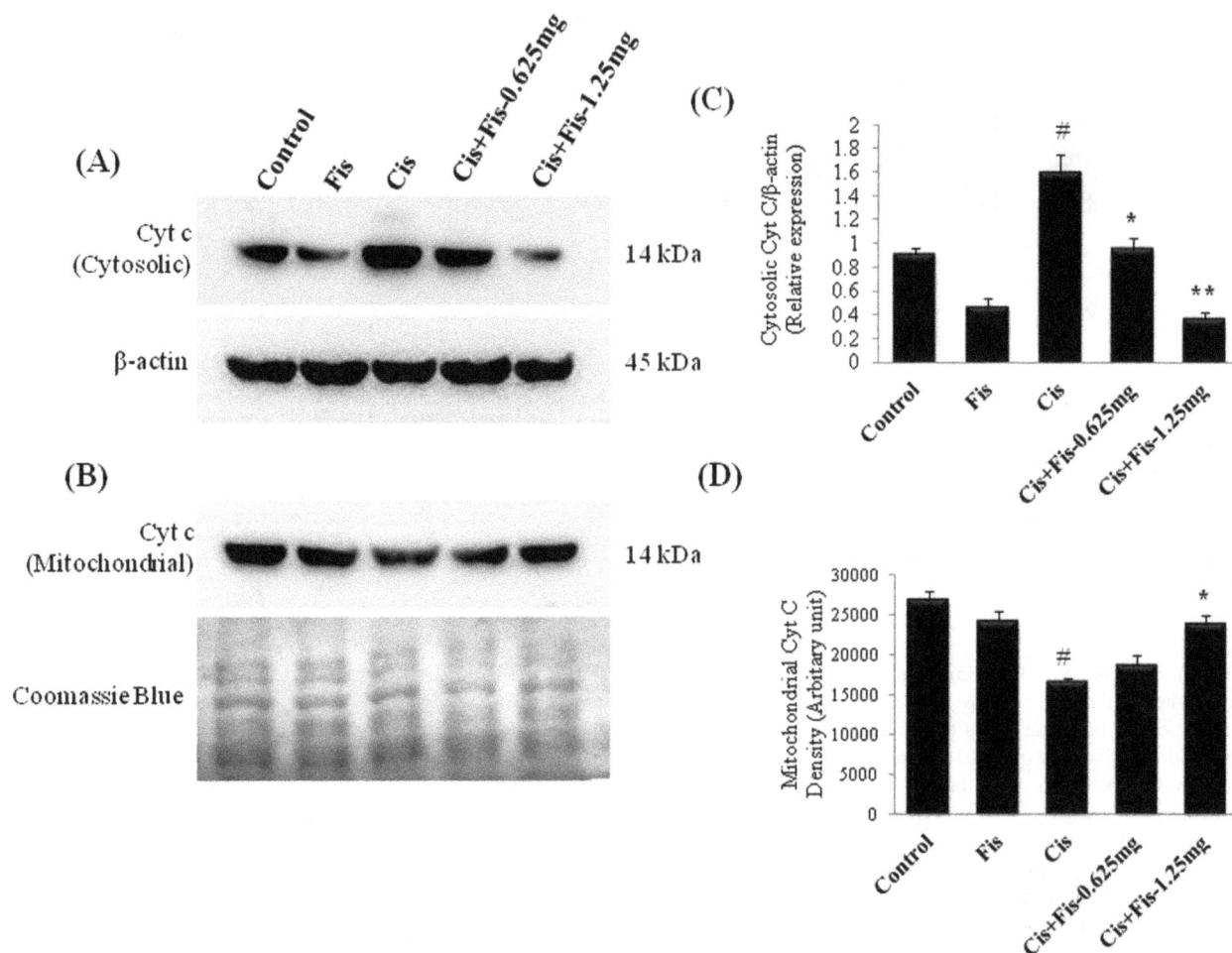

Figure 8. Effect of fisetin on cisplatin-induced cytosolic translocation of cytochrome c. Fisetin at both the doses (0.625 and 1.25 mg/kg) significantly attenuated the cytosolic translocation of cytochrome c protein (**A**) from the mitochondria (**B**) compared to cisplatin alone treated group. β-actin and Coomassie blue were used as loading control for equal loading of cytosolic and mitochondrial proteins respectively. Protein expressions were studied for three independent experiments and representative blots are shown. Quantitative densitometry was performed for each blot. (**C**) Densitometric analysis of cytosolic cytochrome c. (**D**) Densitometric analysis of mitochondrial cytochrome c. Where, Control, vehicle control; Fis, fisetin control; Cis, cisplatin control; Cis+Fis-0.625 mg, cisplatin plus fisetin (0.625 mg/kg) treated groups; Cis+Fis-1.25 mg, cisplatin plus fisetin (1.25 mg/kg) treated groups. #p<0.05 vs vehicle control group; *p<0.05 and **p<0.01 vs cisplatin control group.

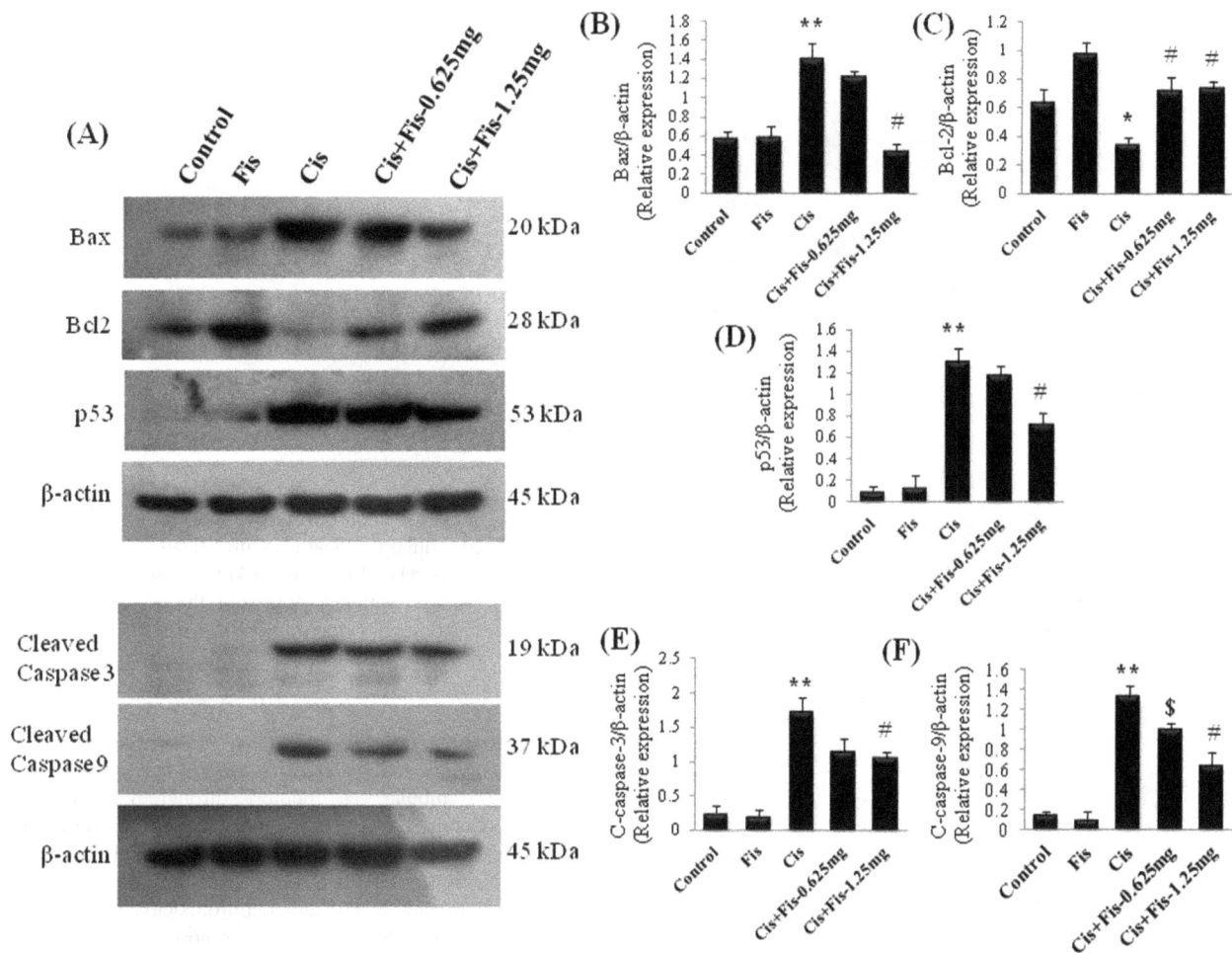

Figure 9. Western blot analyses of apoptosis related proteins in kidney tissues. Kidney tissue total protein extract was prepared and aliquots of 40 μg of protein extracts from different experimental groups were separated by SDS-PAGE and transferred to PVDF membrane. Western blot analyses were performed for Bax, Bcl2, p53, cleaved caspase-3 and cleaved caspase-9 using specific antibodies. β-actin was used as loading control for equal loading of proteins. Quantitative densitometry was performed for each blot. Protein expressions were studied for three independent experiments and representative blots are shown. Where, Control, vehicle control; Fis, fisetin control; Cis, cisplatin control; Cis+Fis-0.625 mg, cisplatin plus fisetin (0.625 mg/kg) treated groups; Cis+Fis-1.25 mg, cisplatin plus fisetin (1.25 mg/kg) treated groups. *p<0.05 and **p<0.01 vs vehicle control group; $p<0.05 and #p<0.01 vs cisplatin control group.

generation in renal tubular cells activates NF-κB and thus is responsible for augmentation of iNOS and pro-inflammatory cytokine, chemokines and adhesion molecules expressions along with infiltration of inflammatory cells in renal tubules [3,10,41]. The importance of NOX2/gp91phox and NOX4/RENOX as a source of ROS generation, in particular superoxide, in the kidneys under cisplatin insult and other pathological conditions is also documented [3,10]. The present study also revealed oxidative damage in kidney tissues of cisplatin treated group of rats. Fisetin has been reported to possess anti-inflammatory properties as well as good antioxidant property with a trolox equivalent antioxidant capacity (TEAC) value of ~3. The NF-E2-related factor 2 (Nrf2) dependent inductions of phase II detoxifying enzymes of fisetin in HT22 cells, retinal ganglion cells, and primary cortical neurons has also been reported [12]. In the present study, cisplatin-induced increase in renal NADPH oxidase activities, expressions of NOX2 and NOX4 and thereby cellular ROS production was attenuated by fisetin treatment (Figure 4). Fisetin treatment also restored the renal antioxidants including the level and/or activity of GSH, CAT, GR, GST, NQO1 and SOD (Table 1). These findings

strengthen the hypothesis that renoprotective effect of fisetin could be attributed to its free radical scavenging and strong antioxidant properties.

A growing body of evidence suggests the role of inflammation and iNOS-mediated increased nitrosative stress in cisplatin-induced kidney toxicity [3,42]. A significant elevation of nuclear translocation (Figure 7A) and DNA binding activity of NF-κB (p65) (Figure 7B), increased iNOS expression (Figure 6A), myeloperoxidase activity (Figure 6B) and concentration of TNF-α (Figure 6C) in the kidneys of cisplatin administered rats was observed in the present study. Our results revealed that fisetin treatment effectively scavenged the cisplatin-induced ROS and suppressed NF-κB activation and subsequent NF-κB mediated inflammatory protein expression in kidney tissues. The findings of the present study also corroborates with earlier study in which fisetin attenuated NF-κB (p65) nuclear translocation and DNA-binding activity and attenuates allergic airway inflammation in mice [16]. Under physiological condition, NF-κB is sequestered in cytoplasm by IκBα subunit, however, on exposure to ROS, sequestration complex breaks down and IκBα is phosphorylated at

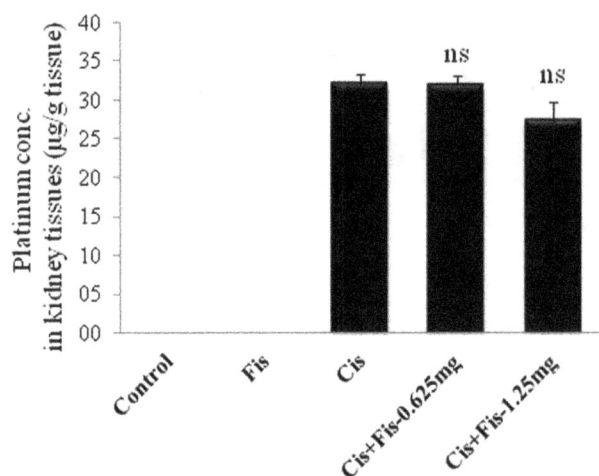

Figure 10. Effect of fisetin on platinum accumulation in kidney tissues. Platinum concentration in kidney tissues was estimated by using Inductively Coupled Plasma Mass Spectrometry (ICP-MS) and was expressed as µg/g of kidney tissues. Data revealed that fisetin treatment did not alter the cisplatin uptake in kidney tissues. The results were expressed as mean ± S.E.M of 6 animals in each group. Where, Control, vehicle control; Fis, fisetin control; Cis, cisplatin control; Cis+Fis-0.625 mg, cisplatin plus fisetin (0.625 mg/kg) treated groups; Cis+Fis-1.25 mg, cisplatin plus fisetin (1.25 mg/kg) treated groups. ns: non-significant.

serine residues by IKKs, allows NF-κB to translocate into the nucleus to promote transcription of inflammatory genes [43]. In the present study, we observed a significant increase in phosphorylated IκBα protein and decrease in intact IκBα protein in cisplatin treated rats when compared to those of vehicle control rats (Figure 7C). Fisetin treatment along with cisplatin preserved IκBα degradation, attenuated IκBα phosphorylation and subsequent NF- κB nuclear translocation. A number of studies reported that cisplatin administration caused a significant elevation of IL-6 levels in kidneys [44]. Contrary to this, we did not find any significant change in IL-6 levels in kidney tissues of cisplatin treated rats (Figure 6D). Studies have revealed dual role of IL-6 in terms of pro-inflammatory and anti-inflammatory response. Interleukin-6 is known to alleviates reactive oxygen species generation through heme oxygenase-1 (HO-1) induction and protecting renal tissues from cisplatin-induced toxicity [5]. At the same time, IL-6 also acts as downstream mediator in TNF-α/NF-κB signalling pathway. However, further studies are advocated to confirm this.

In physiological conditions, mitochondria continuously generate small quantity of superoxide free radicals by converting 1–2% of consumed oxygen and act as important source of ROS. Oxidative damage of mitochondria alters the mitochondrial redox function and respiratory chain enzymes, leading to over production of free radicals and cellular dysfunction [45]. Generation of free radicals and mitochondrial oxidative stress-induced dysfunction have been implicated as early events in the pathogenesis of cisplatin-induced nephrotoxicity [9,10,40,46]. It has been reported that endogenous free radical scavengers such as vitamin C and E, glutathione, ubiquinol, superoxide dismutase and glutathione peroxidase protect mitochondria from cisplatin-induced oxidative damage [45]. Recently, Mukhopadhyay et al. [10] demonstrated that a single systemic dose of mitochondrial targeted antioxidants,

MitoQ and Mito-CP, that deliver superoxide dismutase mimetics preferentially in to mitochondria, significantly prevented cisplatin-induced renal dysfunction in mice. We found that fisetin treatment (1.25 mg/kg) significantly restored the cisplatin-induced decrease in mitochondrial respiratory chain enzyme activities such as NADH dehydrogenase (Figure 5A), succinate dehydrogenase (Figure 5B) and cytochrome c oxidase (Figure 5C). The decrease in mitochondrial redox activity which is a measure of functional intact mitochondria was also significantly increased in fisetin (1.25 mg/kg) treated cisplatin challenged rats (Figure 5D). Additionally, mitochondrial content of GSH, GST, GR, SOD and CAT activities were significantly restored and mitochondrial content of TBARS was significantly decreased in fisetin (1.25 mg/kg) treated cisplatin challenged group of rats (Table 2). These data indicates fisetin ameliorated mitochondrial oxidative damage and associated dysfunction and exhibited renoprotective effect against cisplatin-induced nephrotoxicity.

Apoptosis is a major cause of cisplatin-induced nephrotoxicity [6,47]. To elucidate the nephroprotective mechanism of fisetin on cisplatin-induced tubular cell death, we investigated the expression of various apoptosis related proteins in kidney tissues. Our result revealed that fisetin treatment decreased the cisplatin-induced cytosolic translocation of cytochrome c (Figure 8) and protein expression of Bax and significantly increased the anti-apoptotic protein, Bcl-2 in kidney tissues (Figure 9). Additionally, the amount of cleaved caspase-9 and cleaved caspase-3 were significantly decreased in fisetin treated cisplatin-induced rats. These findings in our study indicate that fisetin may attenuate the cisplatin-induced tubular cell apoptosis through modulation of intrinsic mitochondrial apoptosis pathway. In addition to this, the p53 protein expression was also significantly increased in cisplatin control rats (Figure 9). It has been reported that pro-apoptotic role of p53 is predominant in cisplatin nephrotoxicity, inhibition of which significantly reduces tubular cell apoptosis [6]. Reports also suggest that the cisplatin-induced renal oxidative stress may activate cisplatin-induced p53 activation. Thus, in the present study, we believe that fisetin effectively scavenged cisplatin-induced ROS generation directly and/or indirectly through antioxidant mechanism and attenuated renal p53 expression and subsequent renal tubular cell death. Additionally, ICP-MS data revealed that fisetin did not alter the renal uptake of cisplatin (Figure 10). Hence, we suggest, fisetin by virtue of its free radical scavenging properties attenuates the cisplatin-induced nephrotoxicity in rats.

In conclusion, the results of the present study clearly demonstrated the renoprotective effect of fisetin against cisplatin-induced acute renal injury in rats. The mechanisms underlying the renoprotective effect of fisetin could be by reducing oxidative stress, restoring mitochondrial respiratory enzyme activities and suppressing apoptosis in renal tissues. In addition, the mechanism of this renoprotective effect may also involve inhibition of NF-κB activation and attenuation of subsequent pro-inflammatory mediators release in kidney tissues. However, further studies are needed to explore the additional mechanisms responsible for renoprotective effect of fisetin and to establish its feasible use in clinical setup as an adjunct candidate to cisplatin therapy.

Supporting Information

Figure S1 Effect of fisetin itself on serum cardiac injury biomarkers and heart histology. Intraperitoneal administration of fisetin at two different doses i.e. 1.25 mg/kg and 2.5 mg/kg body weight for 7 consecutive days showing normal serum CK-MB (creatine kinase-MB isoenzyme) and LDH (lactate dehydrogenase) levels compared to vehicle control rats. Histopathological examination of heart tissue (X200 magnification, scale bar: 50 μm) from vehicle control (Control), fisetin at 1.25 mg/kg (Fis-1.25) and fisetin at 2.5 mg/kg (Fis-2.5) treated rats showing apparently normal histo-morphology.

Figure S2 Effect of fisetin itself on serum liver injury biomarkers and liver histology. Intraperitoneal administration of fisetin at two different doses i.e. 1.25 mg/kg and 2.5 mg/kg body weight for 7 consecutive days showing normal serum SGOT (serum glutamic oxaloacetic transaminase) and SGPT (serum glutamic pyruvic transaminase) levels compared to vehicle control rats. Histopathological examination of liver tissue (X200 magnification, scale bar: 50 μm) from vehicle control (Control), fisetin at

1.25 mg/kg (Fis-1.25) and fisetin at 2.5 mg/kg (Fis-2.5) treated rats showing apparently normal histo-morphology.

Acknowledgments

We gratefully acknowledge the Director, CSIR-IICT for providing necessary facilities. We thank Dr Anil Gopala, Application Scientist, Perkin-Elmer, Hyderabad, India for his kind help in ICP-MS.

Author Contributions

Conceived and designed the experiments: BDS RS. Performed the experiments: BDS AKK M. Koneru JMK M. Kuncha. Analyzed the data: BDS RS. Contributed reagents/materials/analysis tools: JMK SSR. Wrote the paper: BDS RS.

References

1. Angelen AAV, Glaudemans B, van der Kemp AWCM, Hoenderop JGJ, Bindels RJM (2013) Cisplatin-induced injury of the renal distal convoluted tubule is associated with hypomagnesaemia in mice. Nephrol Dial Transplant 28: 879–889.

2. Sahu BD, Kuncha M, Sindhura GJ, Sistla R (2013) Hesperidin attenuates cisplatin-induced acute renal injury by decreasing oxidative stress, inflammation and DNA damage. Phytomedicine 20: 453–460.

3. Mukhopadhyay P, Pan H, Rajesh M, Batkai S, Patel V, et al. (2010) CB1 cannabinoid receptors promote oxidative stress/nitrosative stress, inflammation and cell death in a murine nephropathy model. Br J Pharmacol 160, 657–668.

4. Pabla N, Dong Z (2008) Cisplatin nephrotoxicity: mechanisms and renoprotective strategies. Kidney Int 73: 994–1007.

5. Mitazaki S, Hashimoto M, Matsuhashi Y, Honma S, Suto M, et al. (2013) Interleukin-6 modulates oxidative stress produced during the development of cisplatin nephrotoxicity. Life Sci 92: 694–700.

6. Jiang M, Dong Z (2008) Regulation and Pathological Role of p53 in Cisplatin Nephrotoxicity. J Phamacol Exp Therap 327: 300–307.

7. Jaiman S, Sharma AK, Singh K, Khanna D (2013) Signaling mechanisms involved in renal pathological changes during cisplatin-induced nephropathy. Eur J Clin Pharmacol 69: 1863–1874.

8. Luo J, Tsuji T, Yasuda H, Sun Y, Fujigaki Y, et al. (2008) The molecular mechanisms of the attenuation of cisplatin-induced acute renal failure by N-acetylcysteine in rats. Nephrol Dial Transplant 23: 2198–2205.

9. Santos NAG, Catao CS, Martins NM, Curti C, Bianchi MLP, et al. (2007) Cisplatin-induced nephrotoxicity is associated with oxidative stress, redox state unbalance, impairment of energetic metabolism and apoptosis in rat kidney mitochondria. Arch Toxicol 81: 495–504.

10. Mukhopadhyay P, Horváth B, Zsengellér Z, Zielonka J, Tanchian G, et al. (2012) Mitochondrial-targeted antioxidants represent a promising approach for prevention of cisplatin-induced nephropathy. Free Rad Biol Med 52, 497–506.

11. Kim HJ, Kim SH, Yun J-M (2012) Fisetin Inhibits Hyperglycemia-Induced Proinflammatory Cytokine Production by Epigenetic Mechanisms. Evid Based Complement Alternat Med doi:10.1155/2012/639469. In Press.

12. Maher P (2009) Modulation of multiple pathways involved in the maintenance of neuronal function during aging by fisetin. Genes Nutr 4: 297–307.

13. Gelderblom M, Leypoldt F, Lewerenz J, Birkenmayer G, Orozco D, et al. (2012) The flavonoid fisetin attenuates postischemic immune cell infiltration, activation and infarct size after transient cerebral middle artery occlusion in mice. J Cereb Blood Flow Metab 32: 835–843.

14. Leotoing L, Wauquier F, Guicheux J, Miot-Noirault E, Wittrant Y, et al. (2013) The polyphenol fisetin protects bone by repressing NF-κB and MKP-1-dependent signaling pathways in osteoclasts. PLoS ONE 8: e68388.

15. Chou R-H, Hsieh S-C, Yu Y-L, Huang M-H, Huang Y-C, et al. (2013) Fisetin inhibits migration and invasion of human cervical cancer cells by down-regulating urokinase plasminogen activator expression through suppressing the p38 MAPK-dependent NF-κB signaling pathway. PLoS ONE 8: e71983.

16. Goh FY, Upton N, Guan S, Cheng C, Shanmugam MK, et al. (2012) Fisetin, a bioactive flavonol, attenuates allergic airway inflammation through negative regulation of NF-κB. Eur J Pharmacol 679: 109–116.

17. Lee Y, Bae EJ (2013) Inhibition of mitotic clonal expansion mediates fisetin-exerted prevention of adipocyte differentiation in 3T3-L1 cells. Arch Pharm Res 36: 1377–1384.

18. Jeon T-I, Park JW, Ahn J, Jung CH, Ha YU (2013) Fisetin protects against hepatosteatosis in mice by inhibiting miR-378. Mol Nutr Food Res 57: 1931–1937.

19. Maher P, Dargusch R, Ehren JL, Okada S, Sharma K, et al. (2011) Fisetin Lowers Methylglyoxal Dependent Protein Glycation and Limits the Complications of Diabetes. PLoS ONE 6: e21226.

20. Tripathi R, Samadder T, Gupta S, Surolia A, Shaha C (2011) Anticancer Activity of a Combination of Cisplatin and Fisetin in Embryonal Carcinoma Cells and Xenograft Tumors. Mol Cancer Ther 10: 255–268.

21. Touil YS, Auzeil N, Boulinguez F, Saighi H, Regazzetti A, et al. (2011) Fisetin disposition and metabolism in mice: Identification of geraldol as an active metabolite. Biochem Pharmacol 82: 1731–9.

22. Ellman GL (1959) Tissue sulfhydryl group. Arch Biochem Biophys 82: 70–77.

23. Carlberg I, Mannervik B (1975) Glutathione reductase levels in rat brain. J Biol Chem 250: 5475–5480.

24. Habig WH, Pabst MJ, Jakob WB (1974) Glutathione s-transferases. The first enzymatic step in mercapturic acid formation. J Biol Chem 249: 7130.

25. Aebi H (1974) Catalase. In: Bergmeyer HU, editor. Methods of enzymatic analysis. Academic Press: New York and London, pp. 673–677.

26. Zhu H, Itoh K, Yamamoto M, Zweier JL, Li Y (2005) Role of Nrf2 signaling in regulation of antioxidants and phase 2 enzymes in cardiac fibroblasts: Protection against reactive oxygen and nitrogen species-induced cell injury. FEBS Lett 579: 3029–3036.

27. Omaye ST, Turbull TP, Sauberchich HC (1979) Selected methods for determination of ascorbic acid in cells, tissues and fluids. Meth Enzymol 6: 3–11.

28. Ohkawa H, Ohishi N, Yagi K (1979) Assay for lipid peroxides in animal tissues by thiobarbituric acid reaction. Anal Biochem 95: 351–358.

29. Wei Q, Dong G, Franklin J, Dong Z (2007) The pathological role of Bax in cisplatin nephrotoxicity. Kidney Int 72, 53–62.

30. King TE, Howard RL (1967) Preparations and properties of soluble NADH dehydrogenases from cardiac muscle. Meth Enzymol 10: 275–294.

31. King TE (1967) Preparation of succinate dehydrogenase and reconstitution of succinate oxidase. Meth Enzymol 10: 322–331.

32. Liu H, Bowes RC, Van de Water B, Sillence C, Nagelkerke JF, et al. (1997) Endoplasmic reticulum chaperones GRP78 and calreticulin prevent oxidative stress, Ca2+ disturbances, and cell death in renal epithelial cells. J Biol Chem 272: 21751–21759.

33. Thippeswamy BS, Mahendran S, Biradar MI, Raj P, Srivastava K, et al. (2011) Protective effect of embelin against acetic acid induced ulcerative colitis in rats. Eur J Pharmacol 654: 100–105.

34. Oh G, Kim H, Choi J, Shen A, Choe S, et al. (2014) Pharmacological activation of NQO1 increases NAD+ levels and attenuates cisplatin-mediated acute kidney injury in mice. Kidney Int 85, 547–560.

35. Sahu BD, Tatireddy S, Koneru M, Borkar RM, Jerald MK, et al. (2014) Naringin ameliorates gentamicin-induced nephrotoxicity and associated mitochondrial dysfunction, apoptosis and inflammation in rats: possible mechanism of nephroprotection. Tox Appl Pharmacol 277, 8–20.

36. Rubera I, Duranton C, Melis N, Cougnon M, Mograbi B, et al. (2013) Role of CFTR in oxidative stress and suicidal death of renal cells during cisplatin-induced nephrotoxicity. Cell Death Disease 4: e817.

37. EI-Naga RN (2014) Pre-treatment with cardamonin protects against cisplatin-induced nephrotoxicity in rats: Impact on NOX-1, inflammation and apoptosis. Toxicol Appl Pharmacol 274: 87–95.

38. Sahu BD, Rentam KKR, Putcha UK, Kuncha M, Vegi GMN, et al. (2011) Carnosic acid attenuates renal injury in an experimental model of rat cisplatin-induced nephrotoxicity. Food Chem Toxicol 49: 3090–3097.

39. Park H-M, Cho J-M, Lee H-R, Shim G-S, Kwak M-K (2008) Renal protection by 3H-1, 2-dithiole-3-thione against cisplatin through the Nrf2-antioxidant pathway. Biochem Pharmacol 76: 597–607.

40. Satoh M, Kashihara N, Fujimoto S, Horike H, Tokura T, et al. (2003) A novel free radical scavenger, edaravone, protects against cisplatin-induced acute renal damage in vitro and in vivo. J Pharmacol Exp Therap 305: 1183–1190.

41. Sung MJ, Kim DH, Jung YJ, Kang KP, Lee AS, et al. (2008) Genistein protects the kidney from cisplatin-induced injury. Kidney Int 74:1538–1547.

42. Santos NAGD, Rodrigues MAC, Martins NM, Santos ACD (2012) Cisplatin-induced nephrotoxicity and targets of nephroprotection: an update. Arch Toxicol 86:1233–1250.

43. Lee I-C, Kim S-H, Baek H-S, Moon C, Kang S-S, et al. (2014) The involvement of Nrf2 in the protective effects of diallyl disulfide on carbon tetrachloride-induced hepatic oxidative damage and inflammatory response in rats. Food Chem Tox 63: 174–185.

44. Miller RP, Tadagavadi RK, Ramesh G, Reeves WB (2010) Mechanisms of cisplatin nephrotoxicity. Toxins 2: 2490–2518.

45. Kruidering M, Water BVD, Heer ED, Mulder GJ, Nagelkerke JF (1997) Cisplatin-induced nephrotoxicity in porcine proximal tubular cells: mitochondrial dysfunction by inhibition of complexes I to IV of the respiratory chain. J Pharmacol Exp Therap 280: 638–649.

46. Marullo R, Werner E, Degtyareva N, Moore B, Altavilla G, et al. (2013) Cisplatin induces a mitochondrial-ROS response that contributes to cytotoxicity depending on mitochondrial redox status and bioenergetic functions. PLoS ONE 8: e81162.

47. Kang KP, Park SK, Kim DH, Sung MJ, Jung YJ, et al. (2011) Luteolin ameliorates cisplatin-induced acute kidney injury in mice by regulation of p53-dependent renal tubular apoptosis. Nephrol Dial Transplant 26: 814–822.

7

Role of Nrf2/ARE Pathway in Protective Effect of Electroacupuncture against Endotoxic Shock-Induced Acute Lung Injury in Rabbits

Jian-bo Yu*¶, Jia Shi¶, Li-rong Gong¶, Shu-an Dong, Yan Xu, Yuan Zhang, Xin-shun Cao, Li-li Wu

Department of Anesthesiology, Tianjin Nankai Hospital, Tianjin Medical University, Tianjin, China

Abstract

NF-E2 related factor 2 (Nrf2) is a major transcription factor and acts as a key regulator of antioxidant genes to exogenous stimulations. The aim of current study was to determine whether Nrf2/ARE pathway is involved in the protective effect of electroacupuncture on the injured lung in a rabbit model of endotoxic shock. A dose of lipopolysaccharide (LPS) 5 mg/kg was administered intravenously to replicate the model of acute lung injury induced by endotoxic shock. Electroacupuncture pretreatment was handled bilaterally at Zusanli and Feishu acupoints for five consecutive days while sham electroacupuncture punctured at non-acupoints. Fourty anesthetized New England male rabbits were randomized into normal control group (group C), LPS group (group L), electroacupuncture + LPS group (group EL) and sham electroacupuncture + LPS (group SEL). At 6 h after LPS administration, the animals were sacrificed and the blood samples were collected for biochemical measurements. The lungs were removed for calculation of wet-to-dry weight ratios (W/D), histopathologic examination, determination of heme oxygenase (HO)-1 protein and mRNA, Nrf2 total and nucleoprotein, as well as Nrf2 mRNA expression, and evaluation of the intracellular distribution of Nrf2 nucleoprotein. LPS caused extensive morphologic lung damage, which was lessened by electroacupuncture treatment. Besides, lung W/D ratios were significantly decreased, the level of malondialdehyde was inhibited, plasma levels of TNF-α and interleukin-6 were decreased, while the activities of superoxide dismutase, glutathione peroxidase and catalase were enhanced in the electroacupucnture treated animals. In addition, electroacupuncture stimulation distinctly increased the expressions of HO-1 and Nrf2 protein including Nrf2 total protein and nucleoprotein as well as mRNA in lung tissue, while these effects were blunted in the sham electroacupuncture group. We concluded that electroacupuncture treatment at ST36 and BL13 effectively attenuates lung injury in a rabbit model of endotoxic shock through activation of Nrf2/ARE pathway and following up-regulation of HO-1 expression.

Editor: John Calvert, Emory University, United States of America

Funding: This research was supported by grants No. 81372096 from the National Natural Science Foundation of China, Beijing, China, and grant 12ZCZDSY03300 from the Key Project of Tianjin Science and Technology Support, Tianjin, China. The funders had no role in study design, data collection and analysis, decision to publish,or preparation of the manuscript.

Competing Interests: The authors have declared that no competing interests exist.

* Email: yujianbo11@126.com

¶ JbY, JS and LrG are co-first authors on this work.

Introduction

Pulmonary dysfunction is documented as a hallmark of sepsis, and diffuse lung injury resulting in acute respiratory distress syndrome has been put forward as the major characters of pulmonary dysfunction after endotoxin administration [1]. Intravenous infusion of LPS induced acute lung injury and caused alterations in lung physiologic processes, which was similar to those in humans [2]. However, the precise mechanism involved in the formation of acute lung injury (ALI) or acute respiratory distress syndrome (ARDS) induced by endotoxin is not, as yet, fully understood. Our previous study showed that the increased oxidative stress may be a major cause of organ failure and high mortality during endotoxic shock [3]. Oxidative stress is defined as a condition of imbalance between reactive oxygen species (ROS) formation and cellular antioxidant capacity owing to overproduction of ROS or dysfunction of the antioxidant system [4].Thus,

repairing the imbalance status by scavenging ROS or enhancing cellular antioxidant capacity may have implication for a wide array of pathology and disease models.

Acupuncture as an integral part of a traditional Chinese medical system for more than 2500 years. It is worthy of applying acupuncture to specific acupoints to achieve favorable regional or systemic effects [5]. It was proven that electroacupuncture pretreatment significantly inhibited systemic inflammatory responses and improved survival rate in rats with lethal endotoxemia [6]. Feishu (BL13) acupoint is considered of choice to treat lung diseases and regulate pulmonary functions, Pan et al. reported that treatment with electroacupuncture on BL13 showed beneficial effects on hypoxia-induced pulmonary hypertension in rats [7]. Traditionally, acupuncture at Zusanli (ST36) acupoints was known as the modulation of immune functions and is often used in clinical disorders of the immune system [8].

Nuclear factor erythroid-2 related factor-2 (Nrf2) as a Cap "n" Collar basic leucine zipper transcription factor plays a crucial role in regulating antioxidant and cytoprotective genes in response to oxidative stress [9]. An accumulation of ROS or electrophilic compounds give rise to the disruption of Nrf2/Kelch-like ECH-associated protein-1 (Keap1) complex and result in the translocation of Nrf2 from the cytoplasm into the nucleus where it dimerizes with antioxidant response element (ARE) DNA sequence ultimately activates the expression of ARE-dependent genes [10]. Downstream targets of Nrf2 include direct antioxidant proteins such as catalase (CAT) and glutathione peroxidase (GPx), stress-response proteins such as HO-1 and phase II metabolizing enzymes such as glutathione S-transferase (GST), and others [11]. Among these, CAT and GPx directly neutralize ROS and are regarded as very important antioxidant enzymes [4]. Heme oxygenases (HO-1), and with the productions of heme catabolism including biliverdin, bilirubin, carbon monoxide (CO) andiron, shows anti-inflammatory and antioxidant properties [12]. Our preliminary studies have confirmed that up-regulation of HO-1 protein followed by CO increasing could lessen the mortality in septic shock rats [13]. Another previous study elucidated the protective effects of electroacupuncture stimulation at ST36 and BL13 acupoints against acute lung injury evoked by endotoxic shock in rabbits were dependent on up-regulated HO-1 expression [14].

However, it is not known whether electroacupuncture stimulation play a protective role in impaired lung by activating Nrf2/ARE pathway during endotoxic shock. Based on these previous data, we hypothesized that electroacupuncture treatment at ST36 and BL13 acupoints protect against endotoxic shock-induced lung injury via modulating Nrf2/ARE pathway.

Materials and Methods

Animals

The current study was conducted in accordance with the Institutional Animal Use Guidelines and approved by the Animal Care Committee of Tianjin Nankai Hospital (NKH-20120818, Tianjin, China). Two-month-old male New England white rabbits (1.5~2.0 kg) was provided by Laboratory Animal Center of Nankai Clinical Institution of Tianjin Medical University. The animals were housed at 18~22°C on a 12-h light-dark cycle. Besides, food and water were supplied *ad libitum* for a 5-day period prior to the experiment protocols.

Prior to the induction of anesthesia, the rabbits were fasted for 12 h but allowed free access to water. All the rabbits were anesthetized with 20% urethane (5 ml/kg) via the marginal ear vein and anesthesia was maintained with intravenous infusion of ketamine at 3 mg/Kg/h throughout the experiment. Then, a 3.5 mm non-cuffed endotracheal tube was inserted and tied in place through the tracheotomy. Mean arterial pressure was continuously monitored with a PE-50 catheter inserted through the right carotid artery by using Hellige monitor instruments (Germany), while a 24-g catheter was inserted into internal jugular vein for intravenous injections. As Nishina et al. described [15], the lungs of animals were mechanically ventilated with an infant ventilator (IV100B, Sechrist, Anaheim, CA) at an inspired oxygen concentration of 40%. Tidal volume was set to 10 ml/kg (peak inspiratory pressure was 11–13 cm H_2O), as measured by pneumotachograph, and 2 cm H_2O of peak expiratory pressure was added [16]. Respiratory rate was controlled to produce initial $PaCO_2$ of 35–40 mmHg while the inspiratory/expiratory time ratio was set at 1:2 [17]. The rabbits were placed on a heating pad under a radiant heat lamp so as to keep the body temperature at 37.7–40.3°C. Lactated Ringer's solution was administered intravenously at a rate of 8 ml/kg/h.

Electroacupuncture protocols

All animals were lightly immobilized using hands to minimize stress during acupuncture stimulation which was initiated for a five-consecutive-day before the experiment. Besides, electroacupunture stimulation was performed throughout the operating steps for 6 h during the experimental day [18]. The selected acupoints in this study were Zusanli (ST36), located between the tibia and fibular approximately 5 mm lateral to the anterior tubercle of the tibial, and Feishu (BL13), located between T3 and T4 of the spine approximately 1.5 cm lateral to the midline. A set of non-acupoints located on 5 mm lateral to the ST36 or BL13 original location as controls. Two pairs of stainless steel needles (diameter, 0.3 mm) were inserted bilaterally to a depth of 5 mm into the acupoints and kept in place. The parameter of electroacupuncture was applied for 15 minutes with a disperse-dense wave (ie, alternating frequencies of 2 Hz and 15 Hz) once a day by an electrical stimulation device (HANS G6805-1A, Huayi Co, Shanghai, China) [19], and the intensity was adjusted to induce moderate muscle contract of the hindlimb (≤1 mA) [20]. Acupuncture points were identified by an experienced acupuncturist (ZY).

Experimental design

Forty Rabbits were randomized into four different groups (n = 10/group): group C, group L, group EL and group SEL. Rabbits in group L, EL and SEL were treated with intravenous injection of 0.5 ml (5 mg/kg) LPS (L2630, sigma, USA) to replicate the experimental model of acute lung injury induced by endotoxic shock, while group C received 0.5 ml normal saline intravenously as a control. Electroacupuncture treatment at ST36 and BL13 bilaterally was conducted in group EL from the preparation of the model until the end of the experiment for 6 h. Meanwhile, group SEL was acupunctured at non-acupoints with the same frequency and intensity described as above. There was no electroacupuncture stimulation in group C. Mean arterial blood pressure (MAP) was monitored continuously and recorded at 30, 60, 90 and 120 min after the start of administration of LPS. MAP did not decrease within 2 h or rabbits died within 6 h after LPS administration were regarded as the exclusion criteria of the study.

Preparation and analysis of samples

The whole blood was withdrawn from the right carotid artery at 6 h after LPS or normal saline administration. Approximately 1 ml of the arterial blood samples were analyzed for the calculation of oxygenation indexes by a blood gas analyzer (Gem premier 3000, USA) before death. Meanwhile, the blood specimen remained were centrifuged at 4°C (3000 rpm for 15 min). The plasma was removed and the aliquots of the supernatant were separately frozen at −80°C for subsequent analysis. At the end of the experiment, the rabbits were sacrificed by exsanguination, and the lungs were removed and quickly flushed with phosphate-buffered saline (PBS) to remove the blood. Ultimately, sections of the left lung tissues were put in 10% formaldehyde for histopathological analysis, and the remaining tissues were snap frozen in liquid nitrogen and stored at −80°C for subsequent analysis.

Biochemical measurements

The tissue homogenate was prepared for biochemical assays by the upper lobe of the right lung. The levels of superoxide dismutase (SOD) activities and malondialdehyde (MDA) contents in the lung tissues were determined by spectrophotometry [21] and measured by means of Loewenberg [22], respectively. And both were expressed as per unit of protein determined by the Lowry method [23]. Moreover, the serum level of glutathione peroxidase (GPx) and catalase (CAT) activities were measured using commercial assay kits supplied by the Nanjing Jiancheng Bioengineering Institute (Nanjing, China). Plasma tumor necrosis factor-alpha (TNF-α) and interleukin-6 (IL-6) were assayed with a commercial enzyme-linked immunosorbent assay kit (R&D systems, USA). All the procedures were performed according to manufacturer protocols.

Lung wet-to-dry (W/D) weight ratios

The W/D weight ratio of the lung was calculated to evaluate the severity of pulmonary edema. The harvested tissue of left upper lobe was rinsed with normal sodium to scour off the superfluous water. After that, the lung tissue was weighed as wet weight. Dry weight was recorded after the specimen was dried to a constant weight at 70°C for 24 h in an electric air blast drier. The W/D weight ratio was then calculated [14].

Preparation of BALF and Measurements

The Bronchoaleveolar lavage fuild (BALF) analysis was measured to quantify the magnitude of the pulmonary edema. 40 ml saline with ethylendiamine tetraacetic acid (EDTA)-2Na at 4°C was slowly infused for 5 times through the right mainstem bronchus and withdrawn. BALF was analyzed for cell differentiation and cell counts by the Bürker-Türk method [2]. Lavage samples were centrifuged at 250 g at 4°C for 10 min to remove the cells. The cell-free supernatant was analyzed for albumin determined by immune-nephelometry.

Histopathological examination

Immediately after the rabbits were killed (<5 min), the middle lobe of right lung was fixed in 10% formaldehyde for 24 h, and then dehydrated with graded alcohol followed by embedding in paraffin at 60°C. A battery of microsections (4 μm) stained with hematoxylin and eosin were examined under a light microscope (×400) and ten different visual fields were observed for each slice. The lung pathologic change was assessed by alveolar edema, airway congestion, widening of the interstitium or hyaline membrane formation and reactive cell infiltration or aggregation [24]. Each item was graded according to a 5-point scale described as follows [25]: 0 = minimal damage, 1+ = mild damage, 2+ = moderate damage, 3+ = severe damage, and 4+ = maximal damage. The individual scores were added together to rack up a final score ranging from 0 to 16. The lung injury assessment was quantified by a blinded expert pathologist.

Western blot analysis

The expression of HO-1, Nrf2 total protein and Nrf2 nucleoprotein of the lung samples were analyzed by Western blot technique. The tissues stored at −80°C were homogenized in 13.2 mmol/L Tris-HCl, 5.5%glycerol, 0.44%SDS and 10% β-mercaptoethanol. The proteins were extracted according to the instructions of the total protein and nuclear protein extraction kit (Thermo, USA), while the protein concentration was detected on the basis of the BCA protein assay kit (Thermo, USA). Equal amounts of soluble protein were fractionated by 12% SDS-PAGE and were transferred to a PVDF membrane (Bio-Rad, USA). Blots were washed triple for 5 min in TBS and then were incubated overnight at 4°C with polyclonal rabbit antibodies against HO-1 (1:800, Abcam, UK) or Nrf2 (1:300, Biorbyt, UK). Primary antibodies were diluted in blocking solution containing 1% nonfat milk plus 0.5% BSA in TBS-0.05% Tween 20. After three washes with TBS-0.05% Tween 20, blots was incubated at 37°C for 1 h with the horseradish peroxidase (HRP)-conjugated goat anti-rabbit IgG (1:3000 dilation, Biorbyt, UK). The blots were visualized with the enhanced chemiluminesence (Bio-Rad, USA) according to the manufacturer's instruction [26], and the relative density of bands was quantified by densitometry (Molecular Analyst Image-analysis Software, Bio-Rad, USA).

RNA isolation and Real-time PCR

Total RNA was extracted from lung tissue using a Total Quick RNA kit (TA200TQR, Talent, Italy). Tissue was lysed in the provided buffer and RNA was eluted by RNAse-free water. Total amount of RNA was determined by absorbance at 260 nm, while the purity of RNA were measured by 260/280 nm absorbance ratio, respectively. 1 μg total RNA was reversely transcribed into cDNA with random hexamers using a Revert Aid™ First Strand cDNA Synthesis kit (MetaBiosInc, Canada). The final PCR reaction of volume of 20 μl was established with SYBR Green master mix. Predegeneration of the PCR mix was at 95°C for 10 min, and the thermal cycle profile was denaturing for 20 s at 94°C, annealing for 20 s at 59°C and extension for 20 s at 59°C. A total of 40 PCR cycles were used. The primers were as followed: β-actin sense, 5′-CGCGACATCAAGGAGAAGCTG-3′ and β-actin antisense, 5′-ATTGCCAATGGGTGATACCTG-3′, 128 bp; HO-1sense, 5′-TGCCGAGGGTTTTAAGCTGGT-3′ and HO-1 antisense, 5′-AGAAGGCCATGTCCAGCTCCA-3′, 158 bp; Nrf2 sense, 5′-CCCACAAGTTCGGCATCCAC-3′ and Nrf2 antisense, 5′-TGGCGATTCCTCTGGCGTCT-3′, 182 bp. The comparative C_t (threshold cycle) method was applied for quantitiation of target gene expression as described by Schmittgen et al [27]. The relative gene expression of HO-1 and Nrf2 mRNA were normalized to that of β-actin.

Immunoflurescence assay

The intracellular distribution of Nrf2 was displayed by immunofluorescence technique. Paraffin-embedded tissue sections (4 μm) were dewaxed in xylene and rehydrated in graded ethanol solutions. The antigen retrieval method with citric acid solution was proceeded for 5 min at 95°C, and then was washed in PBS. Follwing that, the tissues were permeabilized with 0.5%Triton X-100 for 15 min and blocked in normal goat serum at room temperature for 30 min [28]. After incubation with the polyclonal Nrf2 primary antibody conjugated to FITC (1:150, Biorbyt, UK) at 4°C overnight, the sections was washed triple with PBS. At last, the nuclei were counterstained with DAPI (Roche, Shanghai, China) and visualized using a fluorescent microscope (Olympus U-25ND25, Tokyo, Japan). The green staining was showed in cytoplasm and nucleus, while blue staining was in nucleus. The overlay color was considered to be positive. Ultimately, the results were evaluated by semi-quantitative analysis based on the proportion of the Nrf2 nucleoprotein to the number of nuclei in five fields of each slices at a 400 multiple signal magnification [29].

Statistical analysis

Values were expressed in mean±SD or median (range), except for lung injury scores, for which the Kruskal-Wallis rank test was used. Parametric data were analyzed by one-way analysis of variance (ANOVA) and variations of different groups were compared using the Tukey-Kramer post hoc test. Repeated-measures data (eg. PaO_2/FiO_2) were determined by repeated-measures ANOVA. The W/D ratios were analyzed by two-way ANOVA followed by Bonferroni correction. Data of real-time PCR were tested by a t test with two-tailed hypothesis testing. SPSS19.0 statistical software was used for data analysis and $P<0.05$ was deemed as statistically significant.

Results

Death

The death rate in group EL (1 rabbits) was lower than that in group L (4 rabbits), and was higher than that in group C (zero rabbits). There was no significant difference in death rate between group SEL (3 rabbits) and group L. And, the animals were supplemented according to the randomized crossover principles.

Hemodynamic and oxygenation indexes

MAP in rabbits were remained stable throughout the experiment and the baseline MAP of each group were similar (105~109 mmHg) (table 1). Sixty minutes after LPS injection, MAP in group EL was distinctly lower than that in group C, and higher than that in group L ($P<0.05$). There was no significant difference between group L and SEL ($P>0.05$). Oxygenation indexes were decreased to less than 300 mmHg in group L, EL and SEL at the end of LPS administration. However, electro-acupuncture treatment, rather than sham electroacupuncture stimulation could attenuate the reduction, which revealed that oxygenation indexes in group EL was higher than group L ($P<0.05$).

Lung W/D weight ratio

The lung wet-to-dry weight ratio was calculated as an indicator of pulmonary edema. W/D ratio was increased in the rabbits received LPS (Group L, EL and SEL) compared to group C ($P<0.05$) (Table 2). Electroacupuncture treatment attenuated the increase of W/D weight ratio (attenuation 50.5%, $P<0.05$), while group SEL did not show the protective effect ($P>0.05$).

MDA contents and SOD activities in the lung tissue

The comparisons of MDA contents and SOD activities were showed in Table 2. Intravenous administration of LPS showed an apparent increase of MDA contents and decrease of SOD activities compared to group C ($P<0.05$). However, electroacupuncture reduced the contents of MDA by 53.5% and enhanced activities of SOD by 34.3% to counteract the effects induced by LPS ($P<0.05$). No significant influence in above parameters were discovered when compared group SEL with group L ($P>0.05$).

GPx and CAT activities in serum

Our data revealed that rabbits from group L, EL plus group SEL possessed lower GPx and CAT activities than control group ($P<0.05$) (Table 2). Moreover, the activities of GPx and CAT, which were known as the ROS direct scavengers were enhanced significantly in group EL (augment 30.6% for GPx and 50.4% for CAT) compared with group L or group SEL ($P<0.05$). There were no significant differences between group SEL and group L ($P>0.05$).

Plasma levels of TNF-α and IL-6

As shown in Table 3, plasma levels of TNF-α and IL-6 in group L, EL and SEL were significantly higher than in group C. However, the EL group showed lower levels of TNF-α and IL-6

Table 1. Changes in MAP and Oxygenation indexes among four groups.

Groups	Baseline MAP (mmHg)	MAP after electro-acupuncture for 30 min (mmHg)	Time after the start of LPS or saline				Oxygenation Indexes (mmHg)
			30 min	60 min	90 min	120 min	
C	105±18	103±15	106±17	102±19	103±16	104±13	436±42
L	109±26	105±24	95±29*	83±27*	75±21*	62±28*	195±83*
EL	106±23	123±16*+	109±22+	97±24*+	88±28*+	79±19*+	248±59*+
SEL	107±29	110±19	92±27*	84±22*	71±24*	65±22*	189±94*

Abbreviations: MAP, mean arterial pressure; LPS, lipopolysaccharide. Values as means ± SD (n = 10).
*$P<0.05$ versus control group;
+$P<0.05$ versus LPS group.

Table 2. Comparisons of W/D ratio, MDA contents and SOD activities in the lung tissue, as well as CAT and GPx activities in serum among four groups.

Groups	W/D Ratio	MDA (nmol/mg protein)	SOD (U/mg protein)	CAT (U/ml serum)	GPx (U)
C	3.76±0.21	1.96±0.48	88.93±23.76	67.85±15.25	381.37±74.94
L	5.68±0.25*	4.37±0.72*	47.24±12.28*	36.83±12.19*	213.53±61.94*
EL	4.71±0.17*+	3.25±0.52*+	63.48±18.76*+	50.74±19.13*+	303.46±86.29*+
SEL	5.65±0.35*	4.33±0.83*	46.37±13.34*	31.26±11.99*	219.88±70.54*

Abbreviations: W/D, wet to dry weight ratio; *MDA,* malondialdehyde; *SOD,* superoxide dismutase; *CAT,* catalase; *GPx,* glutathione peroxidase. Values as means ± SD (n = 10).
*$P<0.05$ versus control group;
+$P<0.05$ versus LPS group.

than the L group (TNF-α, 19.62±4.89 and 26.79±7.65, $P<0.05$; IL-6, 87.53±16.23 and 112.32±25.76, $P<0.05$). We did not find a significant difference in plasma levels of TNF-α and IL-6 between group SEL and group L ($P>0.05$).

Analysis of BALF

The recovery percentages of BALF among the four groups were 83%~88%, indicating no differences in the groups. Compared with group C, the number of leukocytes and albumin concentrations in the supernatant of BALF were obviously higher in group L, EL and SEL (Table 3). However, electroacupuncture treatment mitigated the increase in leukocyte counts and albumin concentrations in the BALF in rabbits receiving LPS. In contrast, there was no significant difference in BALF concentrations between group SEL and group L.

Histopathological grading

Figure 1. illustrated the histopathological changes following LPS and the effects of electroacupuncture treatment at acupoints or non-acupoints. Administration of LPS gave rise to diffuse edema in alveolar spaces, infiltration and exudation of inflammatory cells into alveolar space, hemorrhage and thickened alveolar septum under light microscopy. In comparison, the morphological changes were far less pronounced with pretreatment of electroacupuncture. The scores of acute lung injury were summarized in Table 4. The levels of the lung injury scores were decreased in group EL compared with group L ($P<0.05$). However, in rabbits treated with sham electroacupuncture, the lung injury scores were similar with group L ($P>0.05$).

The expressions of HO-1 mRNA, Nrf2 mRNA, HO-1 protein, Nrf2 total protein and Nrf2 nucleoprotein in lung tissue

The results of the experiment determined by Real-time PCR and Western blot were shown in Figure 2 and Figure 3. Exposure to LPS notably increased the mRNA expression of HO-1 and Nrf2 as well as the protein expression of HO-1 and Nrf2 containing nucleoprotein and total protein compared with group C ($P<0.05$). In addition, the expression of HO-1 m RNA and Nrf2 mRNA plus the levels of HO-1 protein and Nrf2 total and nucleoprotein were markedly up-regulated in group EL in contrast to group L and group SEL ($P<0.05$). Nevertheless, there were no significant differences between group SEL and group L in terms of the above mentioned mRNA or protein expressions ($P>0.05$).

The distribution ratio of Nrf2 nucleoprotein expression

Immunofluorescence analysis of Nrf2 expression was represented in Figure 4. Group C showed a negligible Nrf2 nucleoprotein expression, while an enhanced expression with concomitant increase in Nrf2 positive protein was apparent in group L ($P<0.05$). Meanwhile, electroacupuncture stimulation at acupoints of ST36 and BL13 resulted in a significant increase in the number of Nrf2 nucleoprotein comparison with group L (augment 70.2%, $P<0.05$). However, sham electroacupuncture treatment exhibited the similar expression of Nrf2 to that of group L ($P>0.05$).

Discussion

Data from the current study demonstrated that electrostimulation at ST36 and BL13 acupoints dramatically mitigated LPS-induced ALI in endotoxic shock rabbits. Furthermore, the

Table 3. Comparisons of plasma TNF-α and IL-6 levels and analysis of Bronchoalveolar lavage fluid.

Groups	TNF-α (pg/ml)	IL-6 (pg/ml)	Leukocytes (cells/mm³)	BALF Albumin (mg/dl)
C	12.17±2.93	26.53±4.45	197±19	1.2±0.2
L	26.79±7.65*	112.32±25.76*	481±34*	10.3±1.4*
EL	19.62±4.89*+	87.53±16.23*+	375±26*+	6.4±0.7*+
SEL	27.18±8.11*	106.58±20.87*	455±31*	9.7±1.1*

Abbreviations: BALF, Bronchoalveolar lavage fluid; *TNF-α,* tumor necrosis factor-alpha; *IL-6,* interleukin-6. Values as means ± SD (n = 10).
*$P<0.05$ versus control group;
+$P<0.05$ versus LPS group.

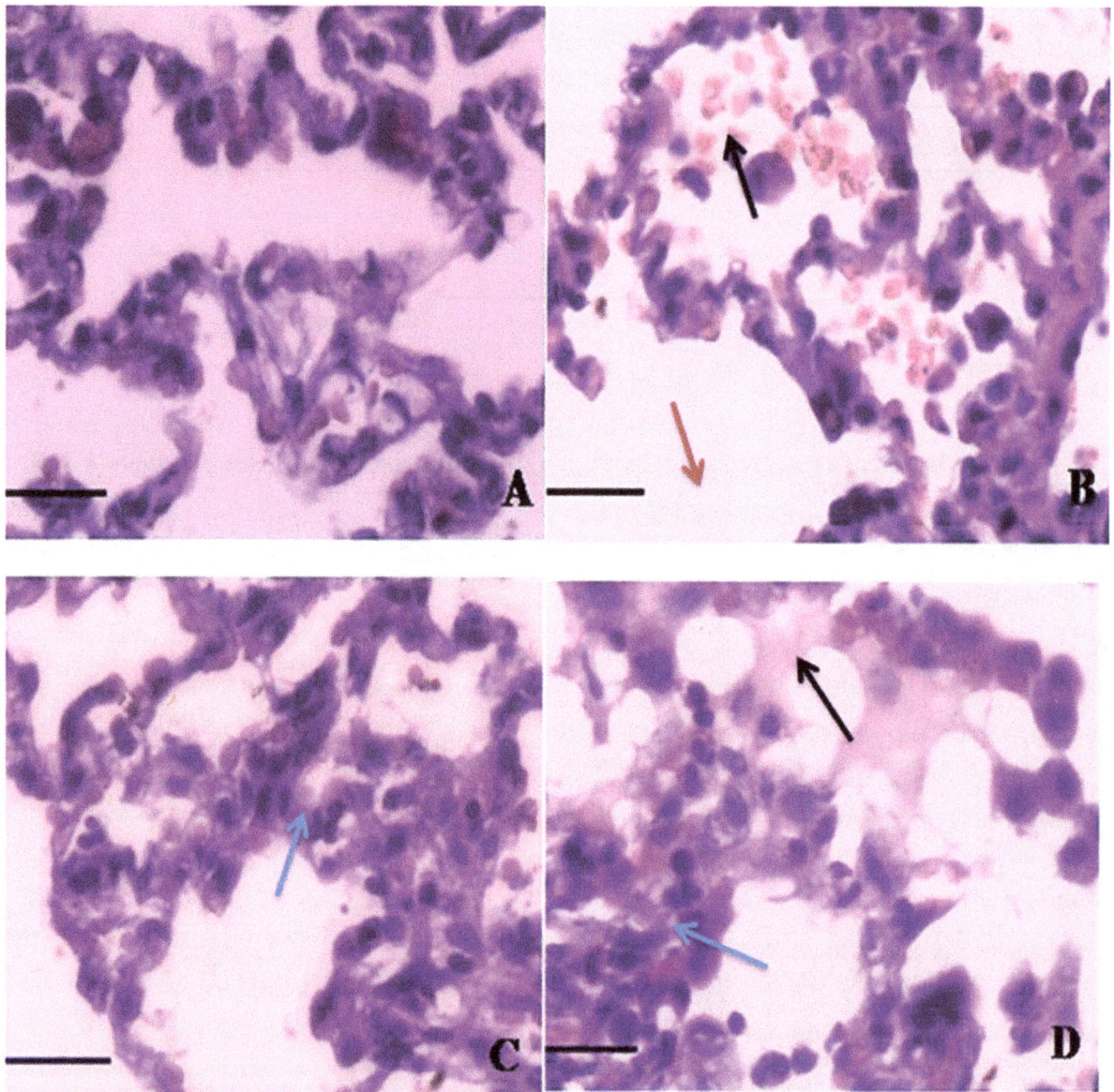

Figure 1. Microphotographs of representative lung section stained with hematoxylin and eosin (original magnification ×400). A. The normal structure of lung from the control group (Group C); B. Severe alveolar edema, hemorrhage, the infiltration of leukocytes and thickened alveolar septum were observed in LPS group (Group L); C. Slight attenuation of the lung injury were displayed in treatment with electroacupuncture (Group EL); D. No improvement of the lung pathology were reflected in sham electroacupuncture stimulation (Group SEL). Black arrows: hemorrhage and infiltration of leukocytes in alveolar space; Red arrows: fracture of alveolar septum; Blue arrows: thickened alveolar septum. Scale bars: 50 μm.

Table 4. Analysis of the histological assessment among four groups.

Groups	Acute lung injury scores
C	1(0–2)
L	10.5(8–14)*
EL	7.5(5–10)*+
SEL	11.5(7–15)*

Values were expressed as medians (range). The lung injury was scored by a 5-point scale according to combined assessments of alveolar congestion, hemorrhage and edema, infiltration or aggregation of neutrophils in the airspace or vessel wall and thickness of alveolar wall/hyaline membrane formation: Score of 0 = minimal (little) damage; 1+ = mild damage; 2+ = moderate damage; 3+ = severe damage; and 4+ = maximal damage. Minimum and maximum possible lung injury scores are 0 and 16, respectively.
*$P<0.05$ versus Group C;
+$P<0.05$ versus Group L.

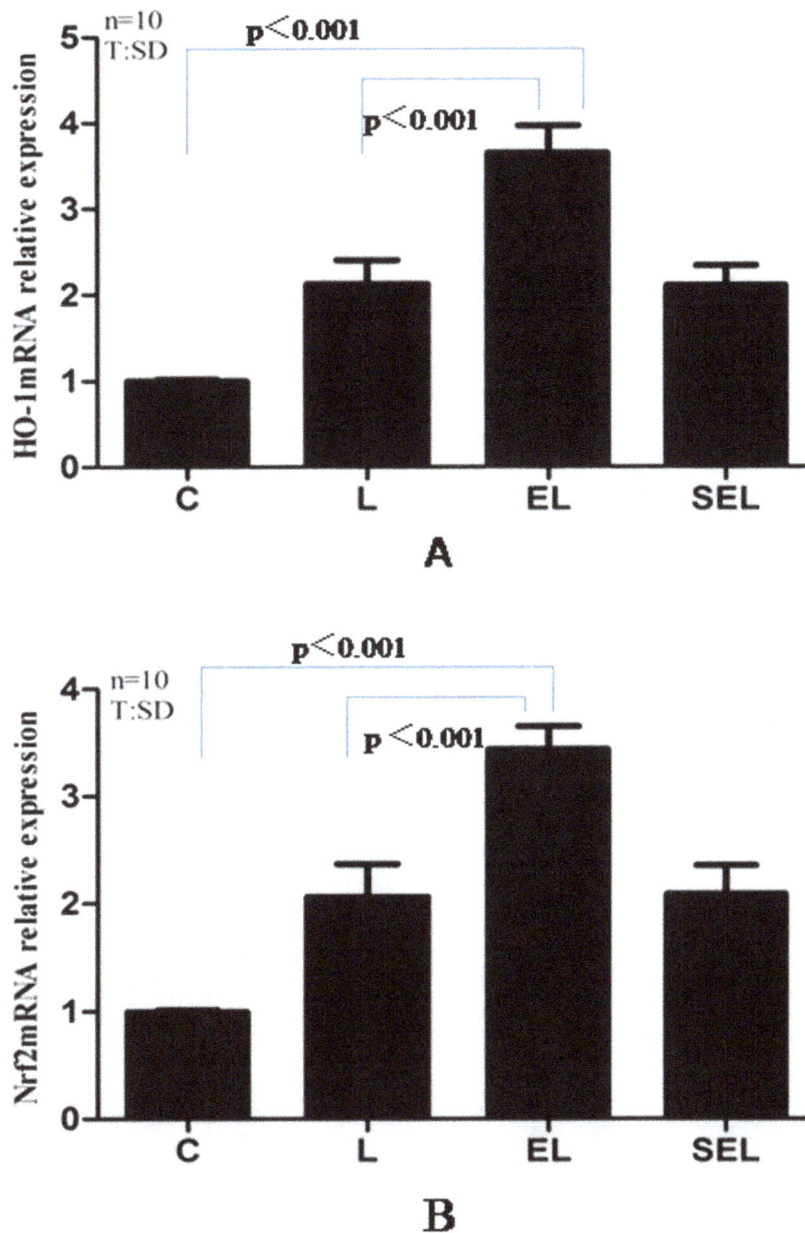

Figure 2. Real-time PCR analysis of HO-1 mRNA (A) and Nrf2 mRNA (B) expressions in the lung tissue of four groups. Data were presented as mean±SD. "C" presents group C, "L" presents group LPS, "EL" presents group electroacupuncture + LPS and "SEL" presents group sham electroacupuncture + LPS. The relative expressions of HO-1 mRNA and Nrf2 mRNA in group EL were higher than that in group C and group L ($P<0.05$), while no significant differences were found between group SEL and group L in terms of the above mentioned mRNA expressions($P>0.05$). Ten control, LPS, EL and SEL experiments were performed for each group.

production of HO-1 mRNA and HO-1 protein in group EL were notably higher than group L, which were consistent with the expression of Nrf2 mRNA, Nrf2 total protein and nucleoprotein. In addition, electroacupuncture treatment enhanced the activities of SOD, GPx and CAT with the increase of Nrf2 and following HO-1 expression. In brief, the present study for the first time confirmed that electroacupuncture at bilaterally ST36 and BL13 produced powerful protection against lung injury through activation of the Nrf2/ARE pathway and induction of the following antioxidant enzymes.

Systemic LPS exposure was used for establishing the standard model of endotoxic shock, which was invariably associated with

ALI or ARDS and even multiorgan dysfunction [30]. ALI induced by endotoxin was manifested with hypoxemia and pulmonary edema depended on severe leukocytes infiltration, increased microvascular permeability and endothelial barrier disruption. The excess production of ROS by polymorphonuclear leucocytes exceeded the antioxidant defense capacity of cells and extracellular fluids, which leaded to oxidative damage in multiple organs [31]. MDA was a reliable marker of oxidative stress mediated lipid peroxidation [32]. Therefore, we applied it to reflect the degree of cell damage caused by reactive oxygen metabolites. The first line of defense against ROS mediated oxidative stress injury involved endogenous antioxidant enzymes such as SOD, GPx and CAT

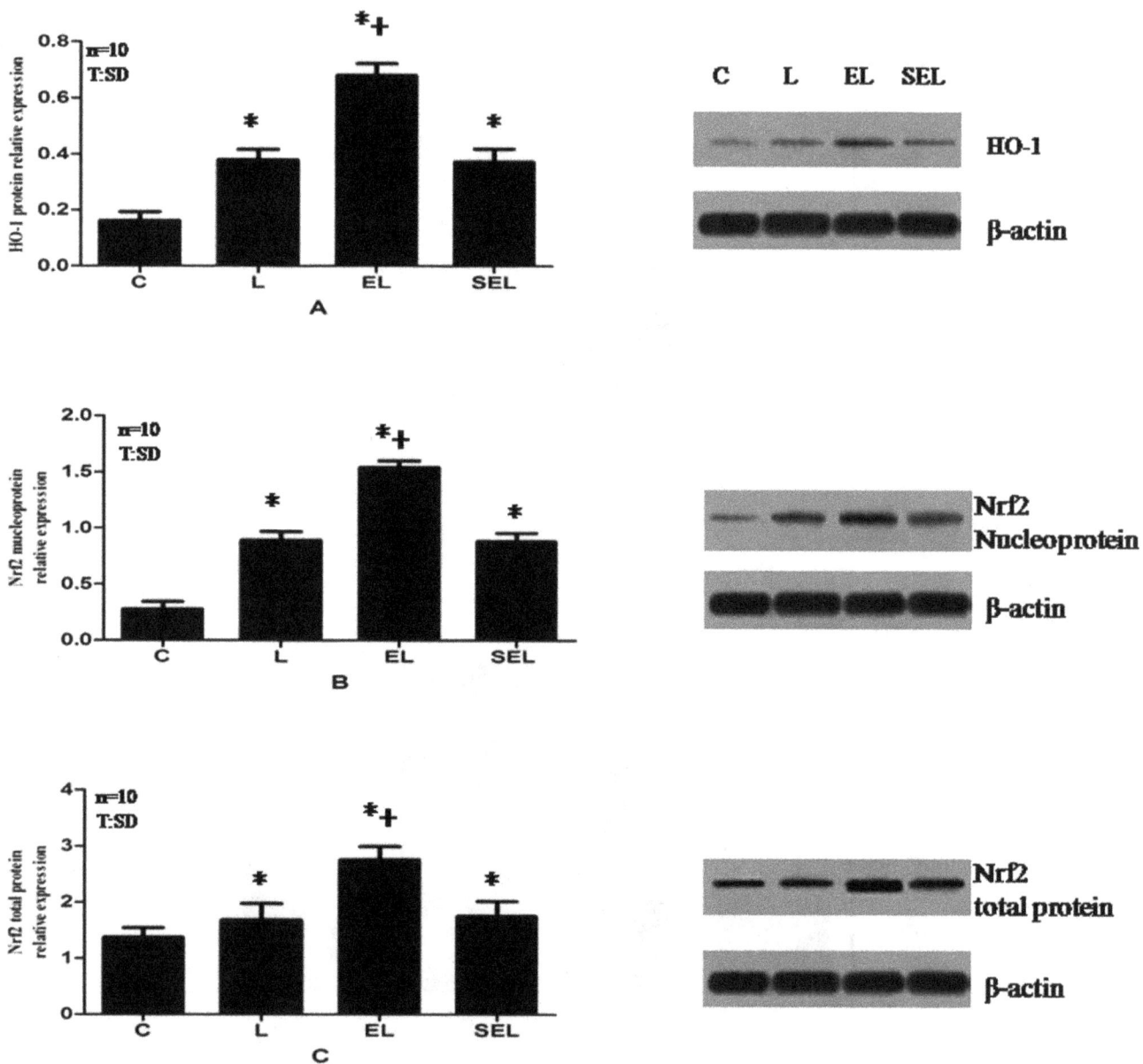

Figure 3. Western blot analysis of HO-1 protein (A), Nrf2 nucleoprotein (B) and Nrf2 total protein (C) relative expressions in lung tissue of four groups. Data were presented as mean±SD. "C" presents group C, "L" presents group LPS, "EL" presents group electroacupuncture + LPS and "SEL" presents group sham electroacupuncture + LPS. The expressions of HO-1 protein, Nrf2 nucleoprotein and total protein in group EL were higher compared with group C and group L ($P<0.05$). While sham electroacupuncture treatment exhibited the similar expressions of the above mentioned proteins to that of group L ($P>0.05$). Ten control, LPS, EL and SEL experiments were performed for each group. *$P<0.05$ versus control group; $^+P<0.05$ versus LPS group.

[4,33]. Moreover, plasma levels of TNF-α and IL-6 were measured as indicators of systemic inflammatory responses [34,35]. In our study, the activities of SOD, GPx and CAT were significantly decreased in LPS induced groups, accompanied with the increased MDA contents as well as the higher levels of TNF-α and IL-6. Concordant with previous studies [13], the rabbit model of injured lung induced by endotoxic shock in this research was defined by the lowered MAP<75% of the baseline values and oxygenation index (PaO$_2$/FiO$_2$) $\leq$300 mmHg.

HO-1 is highly inducible under conditions of ischemia/reperfusion injury or inflammatory cytokines and serves as one of the most prominent lines of defense of the cell against oxidative stress [36]. Previous research showed that hemin pretreatment

with ulinastatin in endotoxin treated rats resulted in an improved response by upregulating HO-1 protein followed by increasing CO and restraining increased oxidative stress [37]. Furthermore, Takaki et al. indicated [12], oxidative stress was closely related to HO-1 expression, and the expression of HO-1 protein was increased in critically ill patients, especially those with severe sepsis or septic shock. To our knowledge, the parameters of electro-acupuncture treatment including the frequency of EA were critical for producing prophylactic effects [38]. Therefore, acupuncture was performed with a disperse-dense wave with 2 Hz/15 Hz lasted 15 min for 5 days consecutively before the experiment in the current study [8,19]. Data from our research revealed that electroacupuncture stimulation attenuated ALI induced by LPS in

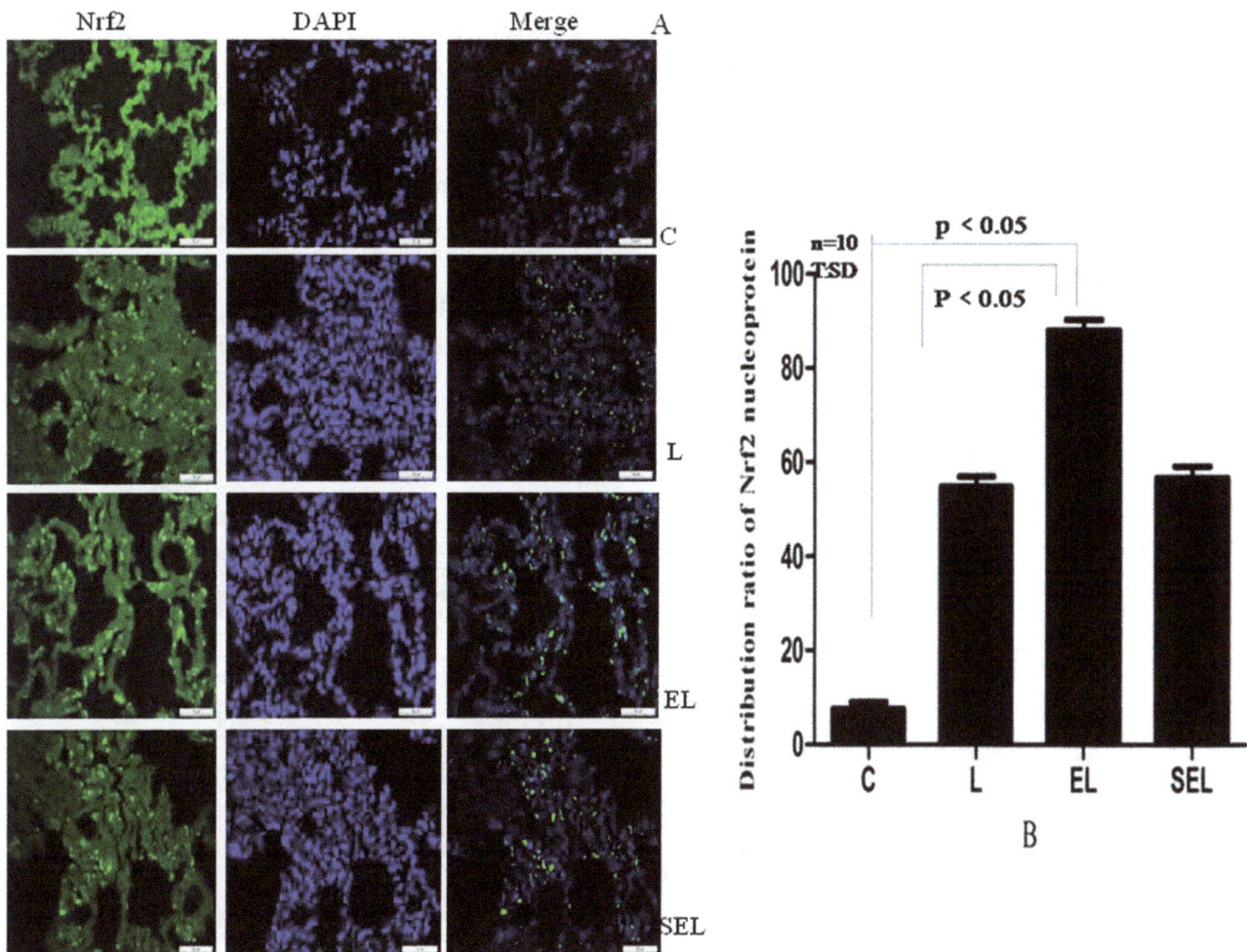

Figure 4. Immunofluorescence assays of nuclear localization of Nrf2 protein using fluorescence microscope (original magnification×400). "A" showed the pictures of immunofluorescence staining, while "B" presented the distribution ratios of Nrf2 nucleoprotein to the number of nuclei in unit area of five fields among four groups. Green standed for Nrf2-FITC stained sections, while blue standed for images of DAPI stained nuclei. It was confirmed that Nrf2 increasingly translocated from cytoplasm into the nucleus by electroacupuncture protocols ($P<0.05$) rather than sham electroacupuncture stimulation ($P>0.05$). Data were representative of three independent experiments. Values were mean ± SD, and ten control, LPS, EL and SEL experiments were performed for each group.

rabbits through upregulation of HO-1 and reduction of MDA content, W/D weight ratio and lung injury scores as well as augment of SOD activities, which was compatible with our prior study [14].

Nrf2/ARE signaling pathway is essential for upregulating the expression of numerous antioxidant genes in response to a wide array of stimuli, and also protecting the cell against oxidative stress and inflammation [39]. Under physiological conditions, Keap1 promoted cytosolic Nrf2 degradation via the Cul3-dependent ubiquitin proteasome pathway. When exposure to redox modulators, the reactive cysteine residues of Keap1 was modified, leading to Nrf2 translocation and accumulation in the nucleus. Subsequently, Nrf2 dimerized with small Maf or Jun proteins that binded to the ARE sequence in the promoter regions of phase II detoxification enzymes and antioxidant proteins, which were activated ultimately to protect cells from ROS generation [40]. As mentioned, SOD, GPx and CAT are directly involved in ROS scavenge, thus, are deemed as very important antioxidant enzymes. The SOD decomposes superoxide radicals and produces H_2O_2. And, H_2O_2 is subsequent removed to water by CAT in the peroxisomes or by GPx oxidizing GSH in the cytosol [10].

Consequently, the activities of above antioxidant enzymes are proportional to the Nrf2 expression. Moreover, Nrf2 had been clarified in Tsai PS et al. research to mediate upregulation of HO-1 by LPS in human monocytic cells [41]. Judging from the present study, upregulation of HO-1mRNA and protein by electroacupuncture in endotoxic shock rabbits was displayed in the lung tissue, which showed the same tendency with expression of Nrf2 mRNA as well as Nrf2 total and nucleoprotein, followed the increase of SOD, GPx and CAT activities. Consistent with Western blot and real-time PCR analyses, immunofluorescence staining identified increments in Nrf2 protein accumulation in the nucleus of cells subsequent to electroacupuncture stimulation.

Collectively, our research has several limitations. First of all, the experimental model of injured lung induced by endotoxic shock was established by intravenous LPS injection, which was extracted from the cell wall of Gram-negative bacteria. However, the infection of pathogenic bacteria was not the common cause in clinical patients with endotoxic shock. As a result, it is not easily for us to extrapolate our conclusions to the clinical setting. Secondly, lung hyper-permeability causing pulmonary edema was deemed as the main mechanism of ALI/ARDS [42]. Therefore, the determination of albumin content and leukocyte count in bronchoalveolar lavage fluid should be added in further exploration to more fully evaluate the effect of electroacupuncture on the impaired lung. Finally, the expression levels of Nrf2 including total protein and nucleoprotein in our present study were both increased significantly. The findings were in coincidence with the research of Chen et al. [43], which still need further exploration for a suitable explanation.

In summary, treatment with electroacupuncture at ST36 and BL13 acupoints alleviated ALI induced by LPS in rabbits, which was counted on activation of Nrf2/ARE pathway and up-regulation of HO-1 expression. In addition, several kinases and signaling pathways have been implicated in the activation of Nrf2/ARE pathway. It has been reported that the MAPK cascade, PI3K/AKT and PKC signaling pathways facilitate the dissociation of Nrf2 from Keap1 and thereby influence the Nrf2/ARE pathway [44]. Nevertheless, the accurate mechanism by which electroacupuncture activates Nrf2 pathway has not been fully understood and requires further investigation. Above all, the present findings provide the scientific foundation for the development of electroacupuncture as a prophylactic treatment for acute lung injury induced by endotoxic shock in clinic.

Acknowledgments

We thank Dr. Jian-bo Wang (Washington University School of Medicine) for his technical support and valuable advice.

Author Contributions

Conceived and designed the experiments: JbY LrG. Performed the experiments: JS SaD YX YZ XsC LlW. Analyzed the data: JS LrG SaD. Contributed reagents/materials/analysis tools: JbY SaD. Contributed to the writing of the manuscript: JbY JS.

References

1. Ghosh S, Latimer RD, Gray BM, Harwood RJ, Oduro A (1993) Endotoxin-induced organ injury. Crit Care Med 21: S19–24.
2. Mikawa K, Maekawa N, Nishina K, Takao Y, Yaku H, et al. (1994) Effect of lidocaine pretreatment on endotoxin-induced lung injury in rabbits. Anesthesiology 81: 689–699.
3. Yu JB, Zhou F, Yao SL, Tang ZH, Wang M, et al. (2009) Effect of heme oxygenase-1 on the kidney during septic shock in rats. Transel Res 153: 283–287.
4. Jung KA, Kwak MK (2010) The Nrf2 system as a potential target for the development of indirect antioxidants. Molecules 15: 7266–7291.
5. Chernyak GV, Sessler DI (2005) Perioperative acupuncture and related techniques. Anesthesiology 102: 1031–1049.
6. Song JG, Li HH, Cao YF, Lv X, Zhang P, et al. (2012) Electroacupuncture improves survival in rats with lethal endotoxemia via the autonomic nervous system. Anethesiology 116: 406–414.
7. Pan P, Zhang X, Qian H, Shi W, Wang J, et al. (2010) Effects of electro-acupuncture on endothelium-derived endothelin-1 and endothelial nitric oxide synthase of rats with hypoxia-induced pulmonary hypertension. Exp Bio Med (Maywood) 235: 642–648.
8. Ferreira Ade S, Lima JG, Ferreira TP, Lopes CM, Meyer R (2009) Prophylactic effects of short-term acupuncture on Zusanli (ST36) in Wistar rats with lipopolysaccharide-induced acute lung injury. Zhong Xi Yi Jie He Xue Bao 7: 969–975.
9. Sarkar S, Payne CK, Kemp ML (2013) Conditioned media downregulates nuclear expression of Nrf2. Cell Mol Bioeng 6: 130–137.
10. Yu M, Xu M, Liu Y, Yang W, Rong Y, et al. (2013) Nrf2/ARE is the potential pathway to protect Sprague-Dawley rats against oxidative stress induced by quinocetone. Regul Toxicol Pharmacol 66: 279–285.
11. Francis RC, Vaporidi K, Bloch KD, Ichinose F, Zapol WM (2011) Protective and Detrimental Effects of Sodium Sulfide and Hydrogen Sulfide in Murine Ventilator-induced Lung Injury. Anesthesiology 115: 1012–1021.
12. Takaki S, Takeyama N, Kajita Y, Yabuki T, Noguchi H, et al. (2010) Beneficial effects of the heme oxygenase-1/carbon monoxide system in patients with severe sepsis/septic shock. Intensive Care Med 38: 42–48.
13. Yu JB, Yao SL (2009) Effect of hemeoxygenase-endogenous carbon monoxide on mortality during septic shock in rats. Ir J Med Sci 178: 491–496.
14. Yu JB, Dong SA, Luo XQ, Gong LR, Zhang Y, et al.(2013) Role of HO-1 in protective effect of electro-acupuncture against endotoxin shock-induced acute lung injury in rabbits. Exp Biol Med (Maywood) 238: 705–712.
15. Nishina K, Mikawa K, Takao Y, Maekawa N, Shiga M, et al. (1997) ONO-5046, an elastase inhibitor, attenuates endotoxin-induced acute lung injury in rabbits. Anesth Analg 84: 1097–1103.
16. Takao Y, Mikawa K, Nishina K, Obara H (2005) Attenuation of acute lung injury with propofol in endotoxemia. Anesth Anaig 100: 810–816.
17. Nishina K, Mikawa K, Maekawa N, Takao Y, Obara H (1995) Does early posttreatment with lidocaine attenuate endotoxin-induced acute lung injury in rabbits? Anesthesiology 83: 169–177.
18. Wang W, Wu H, Wang G, Li M, Zhang Z, et al. (2009) The effects of electroacupuncture on TH1/TH2 cytokine mRNA expression and mitogen-activated protein kinase signaling pathways in the splenic T cells of traumatized rats. Anesth Analg 109: 1666–1673.
19. Zhang H, Liu L, Huang G, Zhou L, Wu W, et al. (2009) Protective effect of electroacupuncture at the Neiguan point in a rabbit model of myocardial ischemia-reperfusion injury. Can J Cardiol 25: 359–363.
20. Zhang Z, Wang C, Gu G, Li H, Zhao H, et al. (2012) The effects of electroacupuncture at the ST36 (Zusanli) acupoint on cancer pain and transient receptor potential vanilloid subfamily 1 expression in Walker 256 tumor-bearing rats. Anesth Analg 114: 879–885.
21. Misra HP, Fridovich I (1971) The generation of superoixide radical during the autoxidation of ferredoxins. J BiolChem 246: 6886–6890.
22. Ohkawa H, Ohishi N, Yagi K (1979) Assay for lipid peroxides in animal tissues by thiobarbituric acid reaction. Anal Biochem 95: 351–358.
23. Loewenberg JR (1967) Cyanide and the determination of protein with the Folin phenol reagent. Anal Biochem 19: 95–97.
24. Chen HI, Hsieh NK, Kao SJ, Su CF (2008) Protective effects of propofol on acute lung injury induced by oleic acid in conscious rats. Crit Care Med 36: 1214–1221.
25. MikawaK, Nishina K, Tamada M, Takao Y, Maekawa N, et al. (1998) Aminoguanidine attenuates endotoxin-induced acute lung injury in rabbits. Crit Care Med 26: 905–911.
26. Balogun E, Hogue M, Gong P, Killeen E, Green CJ, et al. (2003) Curcumin activates the haem oxygenase-1 gene via regulation of Nrf2 and the antioxidant-responsive element. Biochem J 371: 887–895.
27. Schmittgen TD, Livak KJ (2008) Analyzing real-time PCR data by the comparative C(T) method. Nat Protoc 3: 1101–1108.
28. Fer ND, Shoemaker RH, Monks A (2010) Adaphostin toxicity in a sensitive. non-small cell lung cancer model is mediated through Nrf2 signaling and heme oxygenase 1. J Exp Clin Cancer Res 29: 91.
29. Zhao HD, Zhang F, Shen G, Li YB, Li YH, et al. (2010) Sulforaphane protects liver injury induced by intestinal ischemia reperfusion through Nrf2-ARE pathway. World J Gastroenterol 16: 3002–3010.
30. Mikawa K, Nishina K, Takao Y, Obara H (2003) ONO-1714, a nitric oxide synthase inhibitor, attenuates endotoxin-induced acute lung injury in rabbits. Anesth Analg 97: 1751–1755.
31. Goraca A, Józefowicz-Okonkwo G (2007) Protective effects of early treatment with lipoic acid in LPS-induced lung injury in rats. J Physiol Pharmacol 58: 541–549.
32. Wang C, Wang HY, Liu ZW, Fu Y, Zhao B (2011) Effect of endogenous hydrogen sulfide on oxidative stress in oleic acid-induced acute lung injury in rats. Chin Med J (Engl) 124: 3476–3480.
33. Goraca A, Piechota A, Huk-Kolega H (2009) Effect of alpha-lipoic acid on LPS-induced oxidative stress in the heart. J Physiol Pharmacol 60: 61–68.
34. Di Filippo A, Ciapetti M, Prencipe D, Tini L, Casucci A, et al. (2006) Experimentally-induced acute lung injury: the protective effect of hydroxyethyl starch. Ann Clin Lab Sci 36: 345–352.
35. Xia J, Zhang H, Sun B, Yang R, He H, et al. (2014) Spontaneous Breathing with Biphasic Positive Airway Pressure Attenuates Lung Injury in Hydrochloric Acid-induced Acute Respiratory Distress Syndrome. Anesthesiology 120: 1441–9.
36. Bauer M, Huse K, Settmacher U, Claus RA (2008) The hemeoxygenase-carbon monoxide system: regulation and role in stress response and organ failure. Intensive Care Med 34: 640–648.
37. Yu JB, Yao SL (2008) Protective effects of hemin pretreatment combined with ulinastatin on septic shock in rats. Chin Med J (Engl) 121: 49–55.
38. Yu JB, Zhao C, Luo X (2013) The effects of electroacupuncture on the extracellular signal-regulated kinase 1/2/P2x3 signal pathway in the spinal cord of rats with chronic constriction injury. Anesth Analg 116: 239–246.
39. Kilic U, Kilic E, Tuzcu Z, Tuzcu M, Ozercan IH, et al. (2013) Melatonin suppresses cisplatin-induced nephrotoxicity via activation of Nrf-2/HO-1 pathway. Nutr Metab (Lond) 10: 7.
40. Kim JH, Choi YK, Lee KS, Cho DH, Baek YY, et al. (2012) Functional dissection of Nrf2-dependent phase II genes in vascular inflammation and endotoxic injury using Keap1 siRNA. Free Radic Biol Med 53: 629–640.
41. Tsai PS, Chen CC, Tsai PS, Yang LC, Huang WY, et al. (2006) Hemeoxygenase 1, nuclear factor E2-related factor 2, and nuclear factor kappa B are involved in hemin inhibition of type 2 cationic amino acid transporter

expression and L-Arginine transport in stimulated macrophages. Anesthesiology 105: 1201–1210.

42. Voigtsberger S, Lachmann RA, Leutert AC, Schläpfer M, Booy C, et al. (2009) Sevoflurane ameliorates gas exchange and attenuates lung damage in experimental lipopolysaccharide-induced lung injury. Anesthesiology 111: 1238–1248.

43. Chen XJ, Zhang B, Hou SJ, Shi Y, Xu DQ, et al. (2013) Osthole improves acute. lung injury in mice by up-regulating Nrf-2/thioredoxin 1. Respir Physiol Neurobiol 188: 214–222.

44. Chen HH, Chen YT, Huang YW, Tsai HJ, Kuo CC (2012) 4-Ketopinoresinol, a novel naturally occurring ARE activator, induces the Nrf2/HO-1 axis and protects against oxidative stress-induced cell injury via activation of PI3K/AKT signaling. Free Radic Biol Med 52: 1054–1066.

8

The Importance of Intraoperative Selenium Blood Levels on Organ Dysfunction in Patients Undergoing Off-Pump Cardiac Surgery: A Randomised Controlled Trial

Ana Stevanovic[1], Mark Coburn[1], Ares Menon[2], Rolf Rossaint[1], Daren Heyland[3], Gereon Schälte[1], Thilo Werker[1], Willibald Wonisch[4,5], Michael Kiehntopf[6], Andreas Goetzenich[2], Steffen Rex[7¶]*, Christian Stoppe[1,2,8¶]*

1 Department of Anaesthesiology, University Hospital of the RWTH Aachen, Aachen, Germany, 2 Department of Thoracic, Cardiac and Vascular Surgery, University Hospital, RWTH Aachen, Aachen, Germany, 3 Kingston General Hospital, Kingston, Ontario, Canada, 4 Institute of Physiological Chemistry, Centre for Physiological Medicine, Medical University of Graz, Graz, Austria, 5 Clinical Institute of Medical and Chemical Laboratory Diagnostics, Medical University of Graz, Graz, Austria, 6 Institute of Clinical Chemistry, Friedrich-Schiller University, Jena, Germany, 7 Department of Anaesthesiology and Department of Cardiovascular Sciences, University Hospitals Leuven, KU Leuven, Belgium, 8 Institute of Biochemistry and Molecular Cell Biology, RWTH Aachen University, Aachen, Germany

Abstract

Introduction: Cardiac surgery is accompanied by an increase of oxidative stress, a significantly reduced antioxidant (AOX) capacity, postoperative inflammation, all of which may promote the development of organ dysfunction and an increase in mortality. Selenium is an essential co-factor of various antioxidant enzymes. We hypothesized a less pronounced decrease of circulating selenium levels in patients undergoing off-pump coronary artery bypass (OPCAB) surgery due to less intraoperative oxidative stress.

Methods: In this prospective randomised, interventional trial, 40 patients scheduled for elective coronary artery bypass grafting were randomly assigned to undergo either on-pump or OPCAB-surgery, if both techniques were feasible for the single patient. Clinical data, myocardial damage assessed by myocard specific creatine kinase isoenzyme (CK-MB), circulating whole blood levels of selenium, oxidative stress assessed by asymmetric dimethylarginine (ADMA) levels, antioxidant capacity determined by glutathionperoxidase (GPx) levels and perioperative inflammation represented by interleukin-6 (IL-6) levels were measured at predefined perioperative time points.

Results: At end of surgery, both groups showed a comparable decrease of circulating selenium concentrations. Likewise, levels of oxidative stress and IL-6 were comparable in both groups. Selenium levels correlated with antioxidant capacity (GPx: $r = 0.720$; $p < 0.001$) and showed a negative correlation to myocardial damage (CK-MB: $r = -0.571$, $p < 0.001$). Low postoperative selenium levels had a high predictive value for the occurrence of any postoperative complication.

Conclusions: OPCAB surgery is not associated with less oxidative stress and a better preservation of the circulating selenium pool than on-pump surgery. Low postoperative selenium levels are predictive for the development of complications.

Trial registration: ClinicalTrials.gov NCT01409057

Editor: Giovanni Landoni, San Raffaele Scientific Institute, Italy

Funding: D.H. has received less than $5000 for travel support honorarium for lectures. W.W. has received 1050€ for special lab work from biosyn Arzneimittel GmbH, http://www.biosyn.de/. S.R. and C.S. have received in each case 10000€, to perform clinical studies and special lab work, from biosyn Arzneimittel GmbH, http://www.biosyn.de/. The funders had no role in study design, data collection and analysis, decision to publish, or preparation of the manuscript. All the funding declared in the FD statement is specific to a previous study [Stoppe C, Spillner J, Rossaint R, Coburn M, Scha¨lte G, Wildenhues A, Marx G, Rex S]. Selenium blood concentrations in patients undergoing elective cardiac surgery and receiving perioperative sodium selenite. Nutrition 2013;29:158-65] and the present study.

Competing Interests: D.H. has received less than $5000 for travel support honorarium for lectures. W.W. has received 1050€ for special lab work from biosyn Arzneimittel GmbH, http://www.biosyn.de/. S.R. and C.S have received in each case 10000€, to perform clinical studies and special lab work, from biosyn Arzneimittel GmbH, http://www.biosyn.de/. There are no patents, products in development or marketed products to declare.

* Email: steffen.rex@uzleuven.be (SR); Christian.stoppe@gmail.com (CS)

¶ SR and CS contributed equally to this work and are last co-authors on this work.

Introduction

In cardiac surgery, the use of cardioplegic cardiac arrest and cardiopulmonary bypass (CPB) is known to trigger a significant release of reactive oxygen and nitrogen species (ROS, RNOS) [1]. Termination of cardioplegic arrest by reperfusion leads to oxidative stress, which is a major contributor to the complex pathophysiology of ischaemia-reperfusion injury (I/R) [2,3].

Moreover, extracorporeal circulation (ECC) [4] by itself is known to stimulate the production of ROS in neutrophils and monocytes [5]. When exceeding the endogenous antioxidant (AOX) capacity [6], oxidative stress results in the oxidation of proteins, membrane lipids and deoxyribonucleic acids [3,7]. Although often remaining sub-clinical and resolving promptly, postoperative inflammation and oxidative stress can contribute to the development of the systemic inflammatory response syndrome, which is frequently observed after cardiac surgery and may further progress to multiple organ dysfunctions (MOD) and eventual death of patients [2,8]. Amongst the various endogenous AOX defence lines, selenium plays a unique role as essential co-factor for different AOX-enzymes that are involved in the detoxification of both ROS and RNOS [9]. We recently demonstrated that patients undergoing cardiac surgery with cardiopulmonary bypass (CPB) and cardioplegic arrest showed a significant intraoperative decrease in circulating selenium levels, which was independently associated with the postoperative development of MOD [10]. The underlying mechanisms for the decrease in selenium levels have not been comprehensively elaborated yet. It is known from other antioxidants that their circulating concentrations are depleted when scavenging reactive oxygen species during/after CPB [11]. Furthermore, selenium might be trans located into the interstitial compartment during inflammation [12,13] and/or might be adsorbed by an extracorporeal circuit [14]. Although off-pump coronary artery bypass grafting (OPCAB) has become increasingly popular in selected patients, the effects of OPCAB-surgery on perioperative circulating selenium levels have not been thoroughly studied yet. Due to the abstinence of cardioplegic arrest and preservation of normothermia, OPCAB-surgery should theoretically be associated with less oxidative stress [6,15]. In contrast, various studies repeatedly demonstrated for OPCAB-surgery a systemic inflammation that is comparable to on-pump cardiac surgery [16,17]. Comparing on-pump with OPCAB patients should therefore allow to distinguish the effects of oxidative stress from those of inflammation on circulating selenium levels.

Therefore we analysed perioperative selenium levels, the overall inflammatory response and oxidative stress in patients undergoing OPCAB-surgery in comparison to patients that underwent on-pump coronary artery bypass cardiac surgery (CABG). We hypothesized a less pronounced decrease of circulating selenium levels in the off-pump group, owing to less intraoperative oxidative stress than in the on-pump group. This is an additional analysis of data collected in cardiac surgical patients in which the effects of on- and off-pump surgery on the release of macrophage migration inhibitory factor have been previously been reported [18].

Methods

Study design and patients

This mono-centre study was designed as a randomised, interventional clinical trial at the University Hospital of the RWTH Aachen, Germany. It was registered at ClinicalTrials.gov (NCT01409057). The protocol for this trial and supporting CONSORT checklist are available as supporting information; see Checklist S1 and Protocol S1. Data on the perioperative release of Macrophage Migration Inhibitory Factor (MIF) obtained from the same patients have been recently published elsewhere [18].

After approval of the institutional review board (*Ethics commission RWTH Aachen, EK 086/10*), written informed consent was obtained. We initially screened for enrolling a total of 60 patients scheduled for elective isolated CABG. From 50 randomised patients, a total of 40 were followed until final

analysis. We included only patients ≥ 18 years. Exclusion criteria were severe hepatic and renal failure (serum creatinine $>$ 200 µmol l^{-1}) and patients for whom either on- or off-pump techniques were not considered feasible. Furthermore patients with a severe ischemic cardiomyopathia, a recent (<7 days) myocardial infarction and an emergency operation were excluded. Preoperatively, the attending surgeon assessed the eligibility of the potential study participants for off-pump coronary artery bypass grafting. After obtainment of written informed consent, eligible patients were then randomised by a closed envelope technique into either the on-pump or into the OPCAB-group (Fig. 1). Of note, the randomisation list has been created prior to the start of study. The investigators who assessed postoperative outcome remained blinded throughout the whole study.

Anaesthesia

As usual in our institution, anaesthesia was induced by etomidate (0.1–0.2 mg·kg^{-1}) and sufentanil (0.5–1 µg·kg^{-1}). The patient was endotracheally intubated after application of rocuronium (1 mg·kg^{-1}). General anaesthesia was maintained with sevoflurane (0.6–1.0%) and sufentanil (0.5–1 µg·kg^{-1}·h^{-1}). Balanced crystalloid solutions 1 ml·kg^{-1}·h^{-1} were used to manage the fluid balance. Upon the end of the surgery all patients were admitted to the ICU.

Surgical procedure

All patients underwent a midline sternotomy. For performing the bypasses, the internal mammaria artery was harvested as a pedicle. Moreover, venous conduits were used. Heparin was administered to achieve an activated clotting time (ACT) of $>$ 400 s (on-pump) and 250–300 s (OPCAB) and antagonized in proportion of 1:1 with protamine at end of surgery. The patients' temperatures were either maintained normothermic (OPCAB) or had a minimum of 32°C (on-pump) during CPB.

On-pump CABG. CPB was performed on a conventional CPB circuit. Cardiac arrest was generated by the antegrade infusion of cold crystalloid cardioplegic solution. A nonpulsatile pump flow of 2.2 L min^{-1}·m^{-2} was maintained throughout CPB.

OPCAB. Patients were operated in a right rotated Trendelenburg position. To prevent myocardial ischaemia during placement of the distal anastomoses, an intracoronary shunt was inserted. To facilitate the performance of distal anastomosis, commercially available mechanical stabilizers were used.

Hemodynamic management

The intraoperative hemodynamic management was according to our clinical routine. Hypovolaemia was treated with colloid solutions (hydroxyethylstarch 130/0.4, Voluven, Fresenius Kabi, Bad Homburg, Germany). The threshold value for transfusion of packed red blood cells (PRBC) was a haemoglobin content $<$ 7.5 g·dl^{-1}. If further haemodynamic stabilization was necessary, norepinephrine was administered. If required, epinephrine was applied for inotropic support.

Data collection

Preoperatively, we documented relevant medical data and baseline characteristics. According to the ACCP/SCCM consensus conference criteria [19], we recorded during the observation period the duration of mechanical ventilation, the ICU- and hospital length of stay and the incidence of systemic inflammatory response syndrome (SIRS), sepsis, severe sepsis and septic shock. Furthermore, the incidence of any organ dysfunction was evaluated by established organ failure variables [19]. In addition,

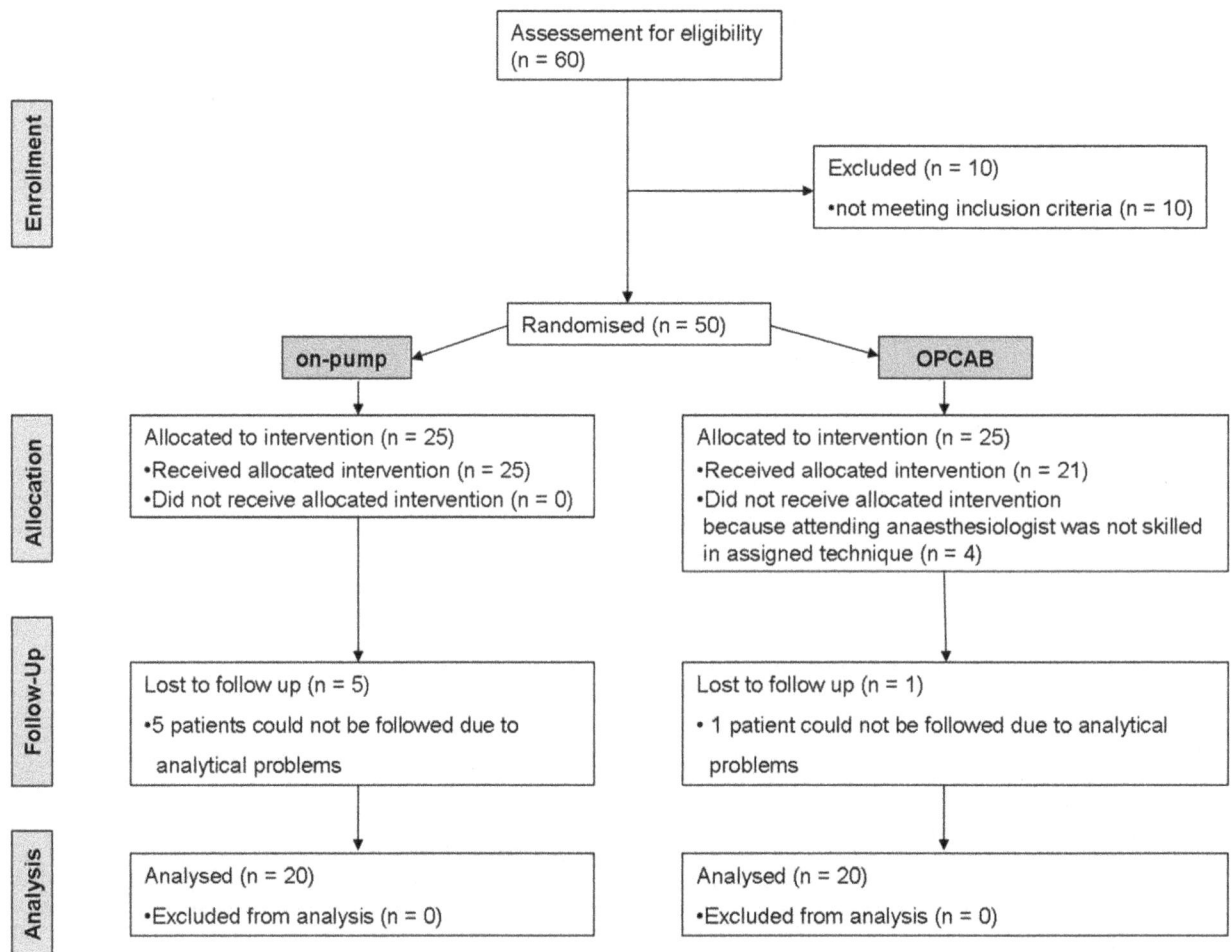

Figure 1. Flowchart. According to the *CONSORT-statement* for randomised clinical trials. From the initially screened 60 patients, 46 patients received the allocated intervention. 6 patients had to be excluded from further analysis.

organ dysfunction was assessed on the 1st postoperative day (1.POD) by the means of the simplified acute physiology score (SAPS II) [20] and the sequential organ failure assessment (SOFA) [21].

Laboratory assessment

Serum and whole blood probes for the measurement of selenium, GPX and markers of oxidative stress were drawn from the central venous catheter after induction of anaesthesia and at ICU-admission. Whole blood samples were stored at room temperature and serum samples were immediately stored at $-80°C$ until final analysis.

Electrothermal atomic absorption spectroscopy (ASS) (5100 PC, Perkin-Elmer, Paris France) was used to determine whole blood selenium-concentrations [22].

We measured serum levels of interleukin-6 (IL-6), to assess the inflammatory response [23] by a commercially available enzyme-linked immunosorbent assay (ELISA) kit (IL-6, R&D Systems, Minneapolis, MN, USA).

The myocard specific creatine kinase isoenzyme (CK-MB) was analyzed by a centrifugal analyzer (Cobas 8000, Roche, Switzerland) to evaluate the extent of myocardial damage [24,25].

The GPx-activity was assessed in serum. Reduction of oxidized glutathione (GSH) was coupled with a peroxidase reaction and one unit of GPx-activity leads to oxidation of 1 mol NADPH min^{-1}

[26]. Selenium and GPx-activities were determined in the laboratories of biosyn Arzneimittel GmbH, Fellbach, Germany and GPx-activities in addition in the Institute of Clinical Chemistry, Friedrich-Schiller Universität Jena, Germany.

Serum levels of ADMA were measured using an ADMA ELISA Kit [27].

Statistical analysis

Statistical analysis was performed with a commercially available software package (SPSS 21.0 (IBM Corporation, Armonk, NY, USA).

Originally, we had planned to include 100 patients in this study as a pre-study power-analysis was not possible due to a complete lack of available data in the literature. Unfortunately, the cooperating surgeon unexpectedly left our institution before the enrolment of patients was terminated. At that time 46 patients had been enrolled. All participating investigators discussed whether it is possible to continue the study under these circumstances. Since we could no longer assure comparable conditions for all patients we decided to stop the recruitment of patients and close the study. Due to analytical problems and missing outcome data, only 40 patients of the 46 randomised patients were analysed per modified ITT-analysis according to the CONSORT-recommendations. A power analysis on the basis of the so far obtained data, using nQuery Advisor Version 7.0 (Statistical Solutions, Saugus,

Table 1. Patient baseline characteristics and data on surgery in the two groups.

		All patients (n = 40)	[95% CI]	Groups		OPCAB (n = 20)	[95% CI]	p-value
				on-pump (n = 20)	[95% CI]			
Demographic Data								
Age	years	67±10	[64–70]	67±12	[61–72]	67±9	[63–71]	0.787
Sex, male	n (%)	32 (80)		16(80)		16 (80)		1.000
Height	cm	172±10	[169–175]	171±10	[166–175]	173±9	]168–177]	0.524
Weight	kg	82±14	[77–86]	82±13	[75–88]	82±15	[75–89]	0.896
euroSCORE		5±3	[5–6]	5±2	[4–7]	5±3	[4–7]	0.857
Prior or pre-existing disease								
Hypertension	n (%)	31 (78)		17 (85)		14 (70)		0.451
Chronic pulmonary disease	n (%)	10 (25)		7 (35)		3 (15)		0.144
Extra cardiac arteriopathy	n (%)	14 (35)		7 (35)		7 (35)		1.000
Cerebral dysfunction	n (%)	3 (7.5)		2 (10)		1 (5)		1.000
Unstable angina	n (%)	11 (28)		5 (13)		6 (15)		0.723
Recent myocardial infarction (<90d)	n (%)	14 (35)		7 (35)		7 (35)		1.000
Chronic kidney disease	n (%)	7 (18)		3 (15)		4 (20)		1.000
Liver disease	n (%)	1 (2.5)		0 (0)		1 (5)		1.000
Diabetes	n (%)	6 (15)		3 (15)		3 (15)		1.000
LVEF > 50%	n (%)	31 (78)		17 (85)		14 (70)		0.451
LVEF 30 - 50%	n (%)	7 (18)		3 (15)		4 (20)		1.000
LVEF < 30%	n (%)	2 (5)		0 (0)		2 (10)		0.487
Intraoperative data								
Intraoperative fluid balance	ml	2413±1146	[2007–2819]	2890±1077	[2337–3444]	1906±1016*	[1365–2447]	0.011
PRBC	n	0 (0–5)		0 (0–5)		0 (0–4)		0.416
Fluid balance within first 24h	ml	2673±1313	[2248–3099]	2810±1340	[2183–3437]	2529±1305	[1900–3158]	0.511
Haemoglobin at admission	g/dl	10±1	[10–10]	10±1	[9–10]	10±1	[9–10]	0.094
Duration of surgery	min	214±54	[196–231]	230±51	[206–254]	198±54	[172–223]	0.058
Ischaemia Time	min			56±20	[46–66]		n.a.	
Time of recirculation	min			32±11	[26–37]		n.a.	
CPB Time	min			100±28	[87–113]		n.a.	

Data are presented as median (range) (not normally distributed data), as mean ± SD (normally distributed data) or as absolute numbers (with the percentage (%) of the whole). * p<0.05

[95% CI] = 95% Confidence interval on the mean

CABG = coronary artery bypass grafting; CPB = cardiopulmonary bypass, MI = myocardial infarction, PRBC = packed red blood cells.

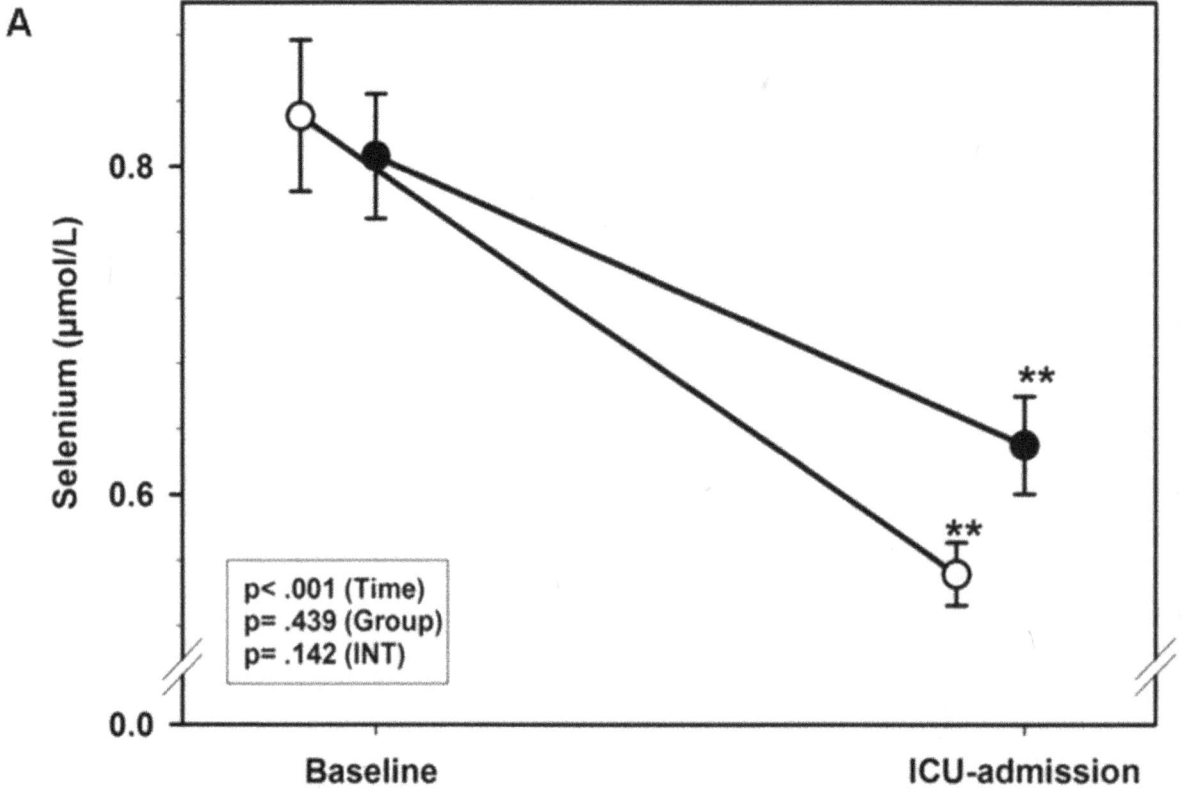

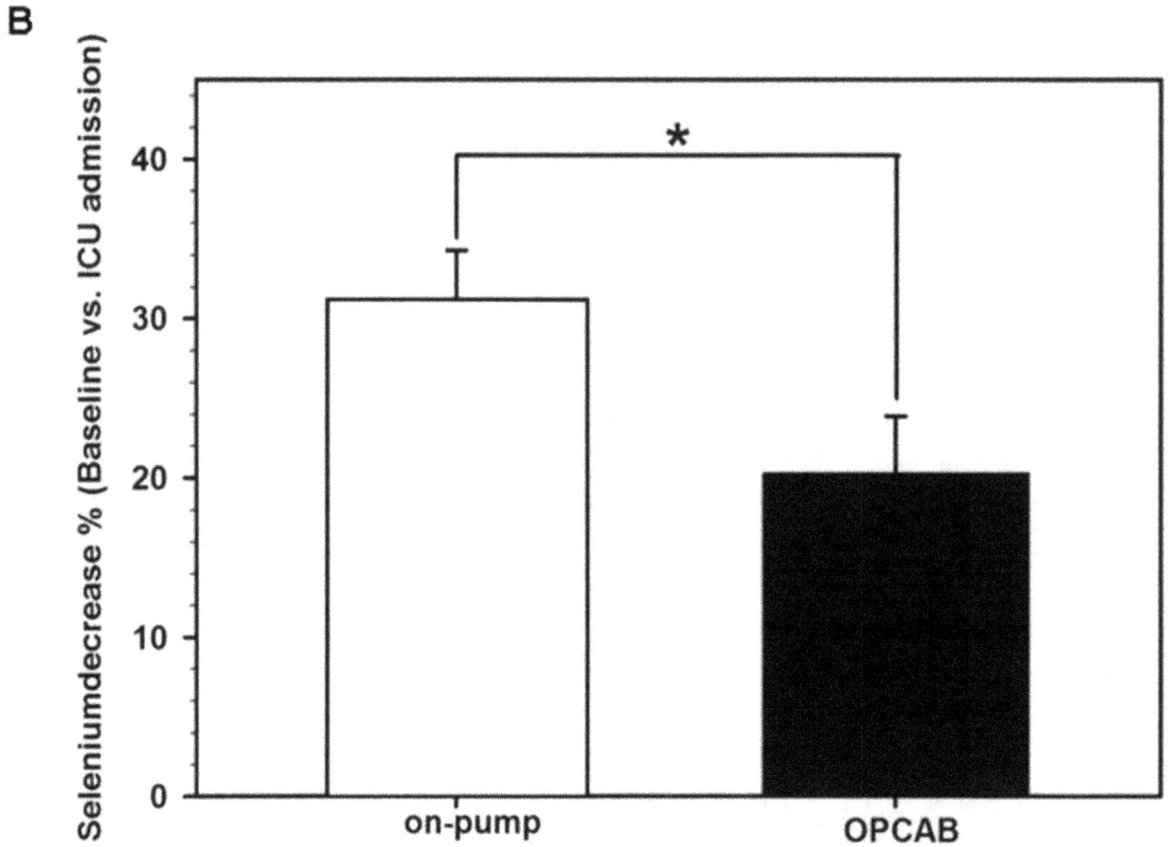

Figure 2. Perioperative selenium-levels. A) Comparison of whole blood selenium levels between the on-pump group (open circles) and the OPCAB-group (closed circles) at baseline (preoperative) and at ICU admission (postoperative). Data are presented as mean ± standard deviation. *$p<$ 0.05, **$p<0.01$ versus baseline, analyzed with 2-way ANOVA. **B)** Comparison of the intraoperative percentual decrease of whole blood selenium between the on-pump group (white bar) and the OPCAB-group (black bar). *$p<0.05$, **$p<0.01$ between the two groups, analyzed with the Mann–Whitney U test.

Massachusetts, USA), was performed to get an impression of how large the probability for a type II error is. This analysis revealed that the observed difference from pre- to postoperative selenium concentrations in the OPCAB and on-pump group had a statistical power of greater than 80% with a significance level of 0.05. The decision to close the study was not influenced by the results of the power analysis.

Our primary endpoint was the difference in selenium decrease during two different techniques of coronary artery bypass grafting.

As secondary endpoints, we investigated the association between circulating selenium levels and the extent of oxidative stress as reflected by ADMA and GPx. The degree of perioperative inflammation was assessed by serum levels of IL-6. Furthermore, we evaluated clinically relevant outcome parameters involving

SAPS II and SOFA score, duration of mechanical ventilation, the hospital- and ICU-length of stay. The degree of perioperative myocardial damage was assessed by CK-MB. The occurrence of postoperative complications was assessed separately with respect to the single organs and as composite outcome, which evaluated the occurrence of organ dysfunction and death.

Normal distribution was tested by the Shapiro-Wilk W-test. We compared single measurements, if with the Students t-test. A two-way ANOVA was used to compare the results of repeated measurements to take into account the correlated observations within the groups. We included as fixed effects the grouping factor treatment (OPCAB vs. on-pump) and the within-factor time. We used the Mann–Whitney U test for nonparametric data. Significant results were post hoc tested with the Bonferroni adjustment

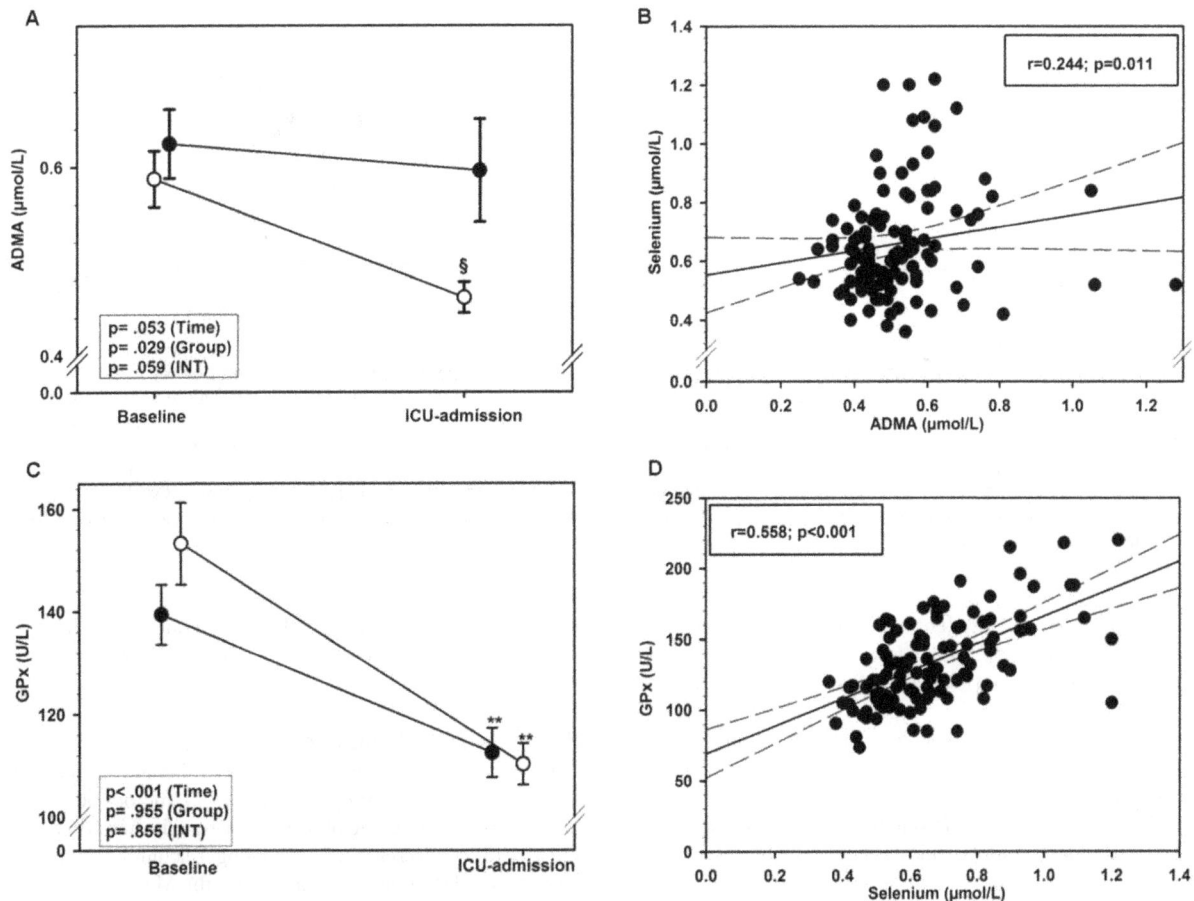

Figure 3. Perioperative time course of markers of oxidative stress and antioxidant capacity. A) Comparison of the intraoperative decrease of ADMA levels in serum between the on-pump group (open circles) and the OPCAB-group (closed circles) at baseline (preoperative) and at ICU admission (postoperative). Data are presented as mean ± standard deviation. $^\S p<0.05$, $^{\S\S}p<0.01$ versus OPCAB group, analyzed with 2-way ANOVA. **B)** Correlation of whole blood selenium levels and ADMA in serum between the two groups. Data are depicted as linear regression (black line) with 95% confidence intervals (long dashed line). **C)** Comparison of GPx levels between the on-pump group (open circles) and the OPCAB-group (closed circles) at baseline (preoperative) and at ICU admission (postoperative). Data are presented as mean ± standard deviation. *$p<0.05$, **$p<0.01$ versus baseline, analyzed with 2-way ANOVA. **D)** Correlation of whole blood selenium and GPx content in serum between the two groups. Data are depicted as linear regression (black line) with 95% confidence intervals (long dashed line).

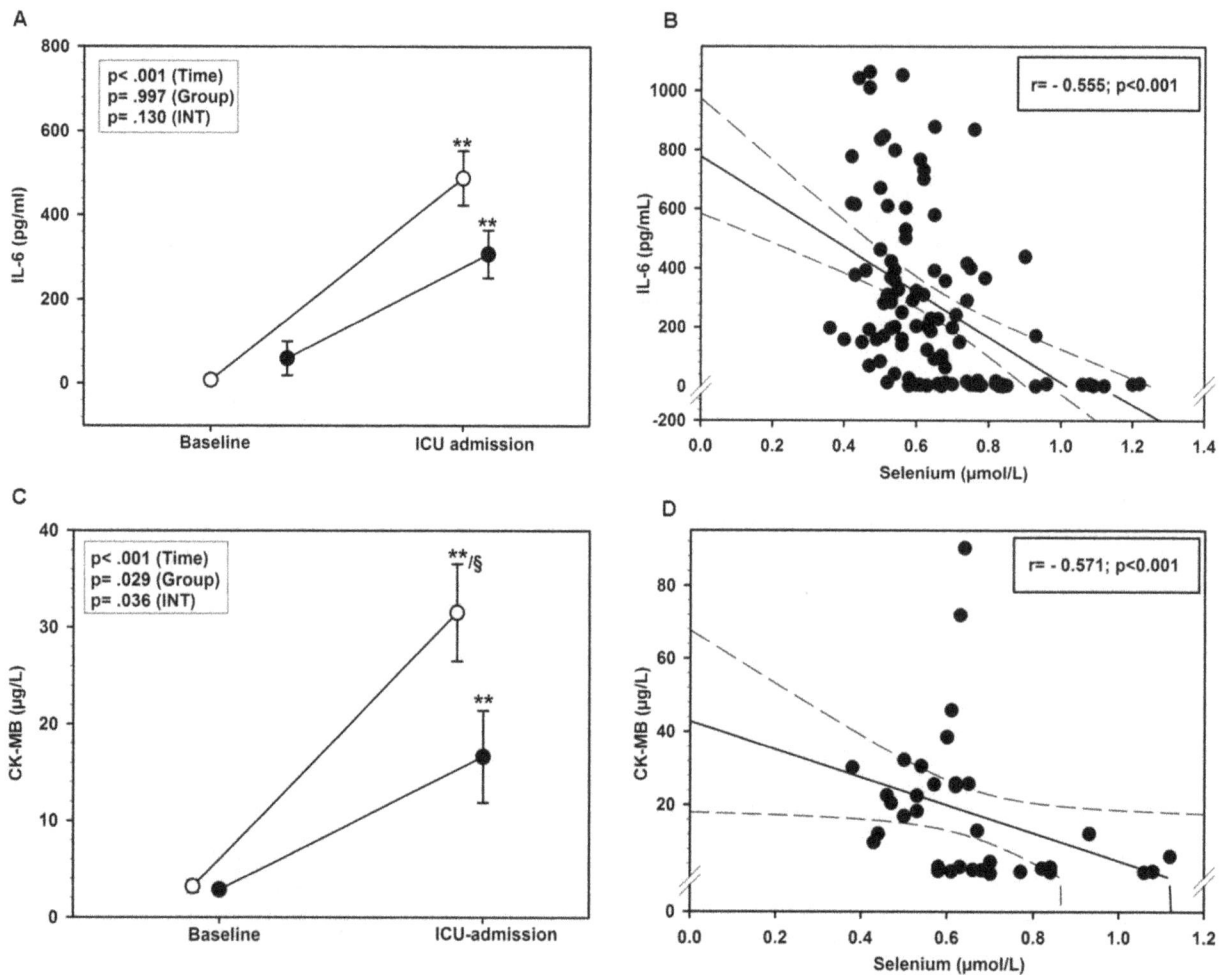

Figure 4. Perioperative inflammatory response and myocardial damage. A) Comparison of serum IL-6 levels between the on-pump group (open circles) and the OPCAB-group (closed circles) at baseline (preoperative) and at ICU admission (postoperative). Data are presented as mean ± standard deviation. *$p<0.05$, **$p<0.01$ versus baseline, analyzed with 2-way ANOVA. **B)** Correlation of whole blood selenium levels and IL-6 levels in serum, between the two groups. Data are depicted as linear regression (black line) with 95% confidence intervals (long dashed line). **C)** Comparison of serum CK-MB levels between the on-pump group (open circles) and the OPCAB-group (closed circles) at baseline (preoperative) and at ICU admission (postoperative). Data are presented as mean ± standard deviation. *$p<0.05$, **$p<0.01$ versus baseline, analyzed with 2-way ANOVA. **D)** Correlation of whole blood selenium levels and CK-MB in serum between the two groups. Data are depicted as linear regression (black line) with 95% confidence intervals (long dashed line).

for multiple measurements (not normally distributed data), respectively. The Fisher's exact test was used to compare proportions of data with an incidence of <5, the Chi-square test was used for incidences >5.

The predictive value of selenium concentrations for the occurrence of organ dysfunction was determined by calculating the area under the curve (AUC) of the receiver-operating characteristic curves (ROC). *P*-values <0.05 were considered statistically significant in all statistical analyses.

Results

Enrolled patients

From sixty screened patients scheduled for CABG-surgery, fifty patients fulfilled all inclusion criteria and were randomised between June 2010 and December 2012 (Fig.1). We performed a modified ITT-analysis of 40 patients. The enrolled patients were also part of the previously published trial of our group, which analysed the significance of perioperative release of macrophage

inhibitory factor (MIF) [18]. Preoperative baseline patient characteristics did not show any significant differences between the two groups (table 1).

Perioperative selenium-levels

Baseline selenium values were comparable in both study groups and lower than the European reference value of 100–140 µg l^{-1} [28,29]. Both groups demonstrated a significant and comparable intraoperative decrease of circulating selenium levels (Fig. 2A). The extent of decrease, measured with the Mann–Whitney *U* test, was more pronounced in the on-pump group (31.2±13.6 (mean ± SD) % vs. 20.2±16.3%; p = 0.040) (Fig. 2B).

Perioperative time course of markers of oxidative stress and antioxidant capacity

Time course of ADMA-levels was comparable in both groups. Intraoperatively, there was a decrease (although statistically not significant) only in the on-pump group (Fig. 3A). Within the OPCAB group, ADMA levels remained unchanged throughout

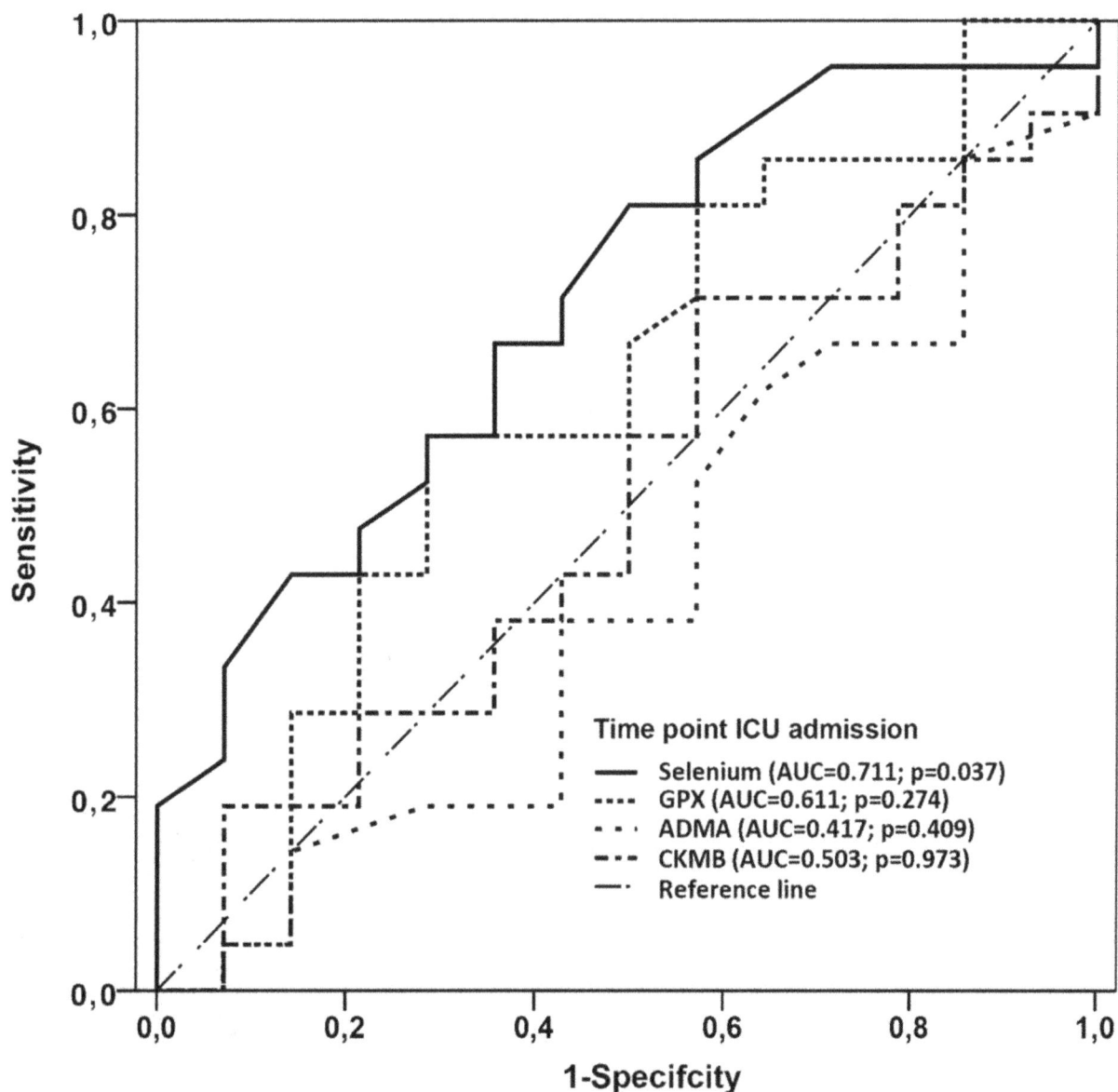

Figure 5. Receiver operating characteristic curve (all patients). Receiver operating characteristic curve for the significance of postoperative (admission to ICU) selenium, GPx, ADMA and CK-MB concentrations in all patients to predict the development of organ dysfunction in the postoperative period. AUC, area under the receiver operating curve.

surgery (Fig. 3A). Circulating selenium levels were directly correlated with ADMA levels (Fig. 3B).

GPx-activity was significantly reduced in both groups after termination of surgery (Fig. 3C). The extent of intraoperative decrease showed a trend towards a significant higher reduction in the on-pump group when compared to the OPCAB (26.1±11.6% (mean ± SD) vs. 19.1±15.6%; p = 0.121). Circulating selenium levels and GPx were significantly correlated (Fig. 3D).

Perioperative inflammatory response and myocardial damage

The time course of perioperative IL-6 levels showed a comparable increase in both groups (Fig. 4A). Circulating selenium levels demonstrated a negative correlation to the IL-6 values within the entire observation period (Fig. 4B).

CK-MB showed a significant intraoperative increase in both groups (Fig. 4C) with a significantly higher percentual increase in the on-pump group (1195.9±1230.3% (mean ± SD) vs. 478.8±582.1; p = 0.044). We found an inverse correlation between circulating selenium- and CK-MB-levels (Fig. 4D).

Biomarkers and postoperative organ dysfunction

Postoperative complications and organ dysfunction are shown in table 2. No significant differences were detected between the groups. In comparison with GPX, ADMA and CKMB, only postoperative measured selenium levels had predictive accuracy for the development of postoperative organ dysfunction and death in the later time course (Fig. 5, 6).

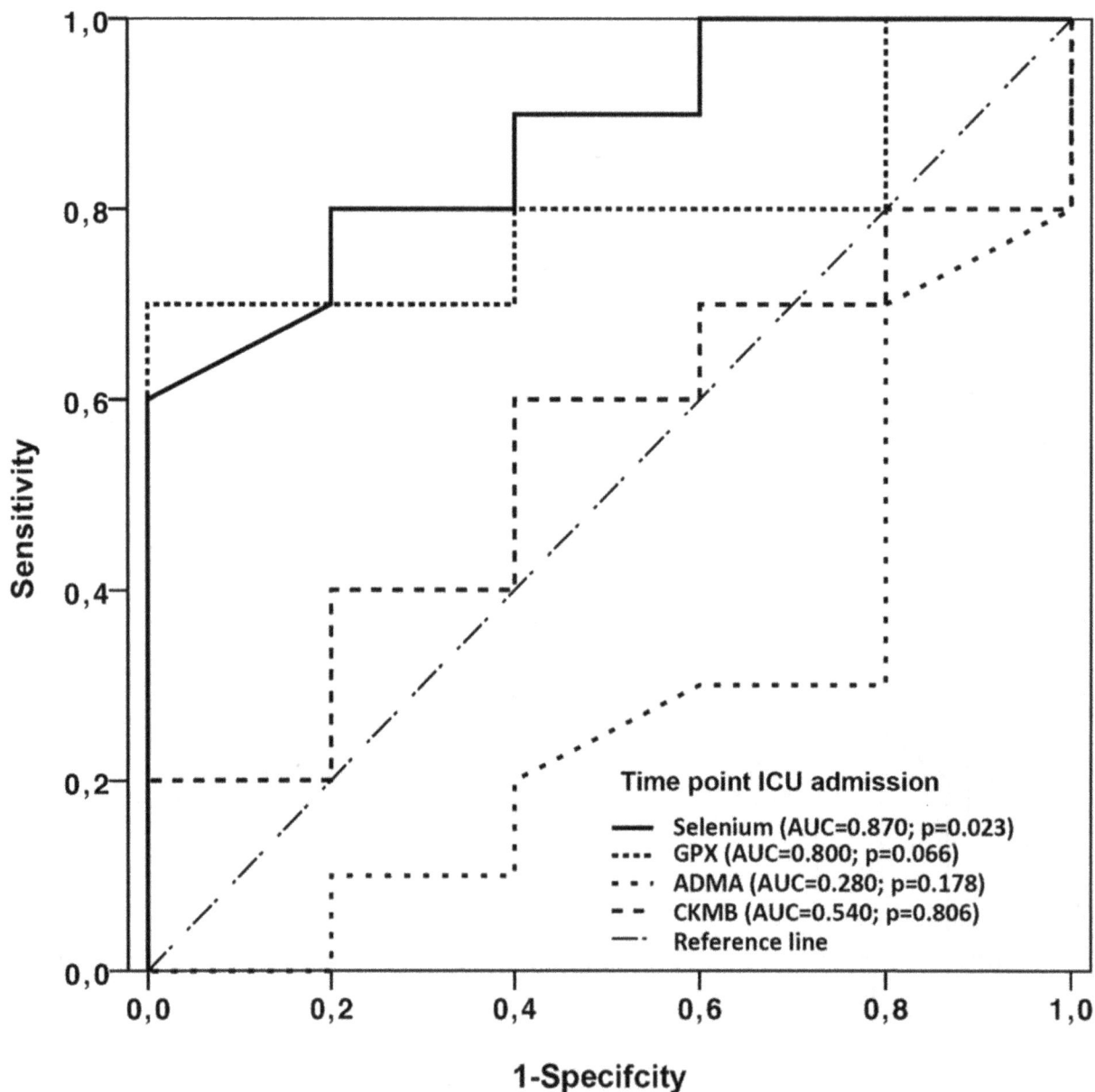

Figure 6. Receiver operating characteristic curve (OPCAB group). Receiver operating characteristic curve for the significance of postoperative (admission to ICU) selenium, GPx, ADMA and CK-MB concentrations in the OPCAB group to predict the development of organ dysfunction in the postoperative period. AUC, area under the receiver operating curve.

Discussion

In the present trial, we observed an intraoperative decrease of circulating selenium levels independent from the type of cardiac surgery. The extent of oxidative stress and inflammatory response was comparable in both groups. Of note, postoperative circulating selenium levels showed a significant predictive accuracy for the occurrence of postoperative organ dysfunction.

In accordance with our previous observations [10,30], the majority of the enrolled patients exhibited significantly reduced selenium levels already *prior* to surgery [28]. These preoperative low selenium-levels were significantly decreased in both groups during surgery, but the extent of this decrease was significantly higher in the on-pump group. We showed recently that the extent of perioperative selenium decrease after on-pump surgery was

independently associated with postoperative occurrence of organ dysfunction, indicating a pivotal role of selenium within the antioxidant and anti-inflammatory defence mechanisms [10,30]. In the present trial, we could translate these findings to patients that underwent OBCAB surgery. As low postoperative selenium levels were predictive for the occurrence of any organ dysfunction, this might indicate a key role of circulating selenium in cardiac surgical patients *per se*. Our data might also demonstrate the potential usefulness of routinely measuring the preoperative and postoperative selenium status. Due to the small sample size of our trial we cannot prove that high selenium levels may protect from adverse events. This can only be tested in a large randomised controlled trial in which the efficacy of an intraoperative optimization of selenium status in cardiac surgical patients should be studied. Of note, in a most recent meta-analysis, high-dose

Table 2. Outcome characteristics of the two groups.

		All patients (n = 40)	[95% CI]	Groups on-pump (n = 20)	[95% CI]	OPCAB (n = 20)	[95% CI]	p-value
Postoperative organ failure/complication								
Atrial fibrillation	n (%)	6 (15)		1 (5)		5 (25)		0.182
Stroke	n (%)	1 (3)		0 (0)		1 (5)		1.000
Delir	n (%)	6 (15)		2 (10)		4 (20)		0.661
Acute Kidney Injury	n (%)	3 (8)		0 (0)		3 (15)		0.231
Pneumonia	n (%)	6 (15)		4 (20)		2 (10)		0.661
Cardiogenic shock	n (%)	1 (3)		0 (0)		1 (5)		1.000
Wound infections	n (%)	1 (3)		1 (5)		0 (0)		1.000
SOFA score 1.POD	n	6 (3–11)		6 (3–9)		6 (3–11)		0.719
SAPS II 1.POD	n	29 (18–39)		29 (18–38)		29 (21–39)		0.951
Incidence of SIRS/Sepsis								
SIRS	n (%)	14 (35)		6 (30)		8 (40)		0.507
Severe SIRS and Sepsis	n (%)	1 (3)		1 (5)		0 (0)		1.000
Septic shock	n (%)	3 (8)		1 (5)		2 (10)		1.000
Longterm - Outcome								
Sedation	hours	10±8	[7–12]	9±5	[7–11]	11±10	[6–15]	0.539
Duration of mechanical ventilation	hours	12±9	[9–15]	11±5	[9–14]	13±12	[7–19]	0.586
ICU stay	hours	80±108	[45–115]	67±90	[25–109]	93±126	[33–154]	0.459
Hospital length of stay	days	15±10	[12–18]	14±11	[9–20]	16±8	[12–20]	0.635
Mortality	n (%)	1 (3)		0 (0)		1 (5)		1.000

Data are presented as median (range) (not normally distributed data), as mean ± SD (normally distributed data) or as absolute numbers (with the percentage (%) of the whole). CABG = coronary artery bypass grafting; CPB = cardiopulmonary bypass, MI = myocardial infarction.

supplementation of selenium has been shown to reduce mortality in patients with severe sepsis [31].

Various studies repeatedly indicated ROS to represent one of the major factors contributing to myocardial I/R injury during cardiac surgery [32,33]. Normally, a sophisticated endogenous defence system including the AOX-enzyme GPx protects tissues from oxidative stress. Assessment of GPx activity was performed to determinate the antioxidant capacity (AOX), hereby indirectly reflecting oxidative stress [7]. The activity of GPx is known to be critically dependent upon circulating selenium levels [9] which is confirmed by our findings of a strong correlation between GPX and selenium in both groups. Interestingly, it has recently been reported that cardiac surgical patients who received a perioperative selenium supplementation showed a reduced extent of myocardial damage [34]. Also in our patients, we could observe an inverse correlation between the postoperatively measured selenium and CK-MB levels. It is tempting to speculate that this inverse correlation may indicate antioxidant and cardioprotective properties of selenium.

ADMA has previously been shown to be crucially involved in the regulation of vascular tone via endothelial nitric oxide synthase (e-NOS) and inducible nitric oxide synthase (i-NOS) [4]. The increased consumption of glutathione during oxidative stress results in increased consumption of homocysteine (Hcy) [35], a potent inhibitor of the ADMA metabolising enzyme dimethylarginine-dimethylaminohydrolase (DDAH) [36]. Thus, a postoperative decrease of Hcy as a consequence of oxidative stress leads to an increased metabolism of ADMA by DDAH. Therefore, ADMA levels decrease with increasing oxidative stress. As we found only an insignificant postoperative reduction in ADMA levels in the on-pump group, we conclude that there was a comparable oxidative stress in both study groups.

The underlying mechanisms of the observed perioperative selenium decrease still have to be elucidated. It has been speculated that selenium levels decrease due to intraoperative blood losses, dilution by resuscitation fluids, extravasation due to systemic inflammation, and depletion owing to the scavenging of reactive oxygen species during/after CPB [12,37,38]. We observed a comparable inflammatory response in both groups, as reflected by a comparable perioperative time course of IL-6. This is according to previous studies which demonstrated the surgical trauma itself to represent the main reason for the pro-inflammatory response after both kinds of cardiac surgery [16,17,39]. Previous findings in on-pump cardiac surgical patients [4,40,41] [42] indicate a higher level of oxidative stress when compared to patients undergoing OPCAB-surgery. In fact, the inevitable use of cardioplegic arrest during conventional on-pump surgery exposes patients to a significant longer duration of myocardial ischaemia and hence more pronounced reperfusion injury when compared to OPCAB-patients where myocardial I/R is minimized by the use of intracoronary shunts during performance of the distal anastomoses. Furthermore, activation of immune cells (e.g. neutrophils and monocytes) after contact with the artificial surfaces of the extracorporeal circuit [5] and the use of mild hypothermia [43] might also contribute to the increased oxidative stress after CPB. Our findings of a comparable level of oxidative stress in both groups do however not support these

considerations and do not allow to distinguish the effects of inflammation from those of oxidative stress.

Interestingly, haemodilution and blood loss represent further possible causes for a decrease of circulating selenium levels. Of note, the intraoperative fluid balance differed significantly between our two study groups, most likely due to priming of the extracorporeal circuit with 1500 ml crystalloid fluid. However, postoperative haemoglobin concentration and the transfusion of packed red blood cells (PBRC) were comparable in both groups, suggesting that blood loss and haemodilution contributed only marginally to the observed selenium decreases.

We acknowledge that the present trial suffers from several limitations, including a small sample size, which only allows an adequately analysis of the primary outcome parameter (differences in selenium decrease during two different techniques of coronary artery bypass grafting) with sufficient statistical power. Analyses of the various secondary outcome parameters have to be considered to be purely explorative and hypothesis-generating. A lack of statistical power might also explain why we could not observe differences in the incidence of postoperative organ dysfunction in the OPCAB-group despite a better preservation of the intraoperative selenium status. Furthermore we were unable to clarify whether the observed intraoperative selenium decrease in both groups was truly causative or only "indicative" for increased oxidative stress and the development of organ dysfunction. This question can only be resolved by a large-scale clinical trial in which the efficacy of an intraoperative selenium supplementation strategy has to be tested.

Conclusion

Cardiac surgery (irrespective of the use of CPB) is associated with an intraoperative decrease of circulating selenium levels. It still remains to be elucidated which mechanisms precisely underlie the intraoperative decrease of selenium levels as both the extent of oxidative stress and inflammatory response were comparable in both groups. Of note, postoperative circulating selenium levels showed a significant predictive accuracy for the occurrence of postoperative organ dysfunction also in OPCAB-patients.

Acknowledgments

We thank Dr. Horst Dawczynski (biosyn Arzneimittel GmbH) for his excellent scientific and technical assistance.

Author Contributions

Conceived and designed the experiments: AS AG SR CS. Performed the experiments: AS AM GS TW SR CS. Analyzed the data: AS MC RR DH WW MK AG SR CS. Contributed reagents/materials/analysis tools: AS MC RR DH WW MK AG SR CS. Wrote the paper: AS MC AM RR DH WW MK AG SR CS.

References

1. McDonald CI, Fraser JF, Coombes JS, Fung YL (2014) Oxidative stress during extracorporeal circulation. Eur J Cardiothorac Surg.
2. Raja SG, Berg GA (2007) Impact of off-pump coronary artery bypass surgery on systemic inflammation: current best available evidence. J Card Surg 22: 445–455.
3. Gerritsen WB, van Boven WJ, Boss DS, Haas FJ, van Dongen EP, et al. (2006) Malondialdehyde in plasma, a biomarker of global oxidative stress during mini-CABG compared to on- and off-pump CABG surgery: a pilot study. Interact Cardiovasc Thorac Surg 5: 27–31.

4. Karu I, Taal G, Zilmer K, Pruunsild C, Starkopf J, et al. (2010) Inflammatory/oxidative stress during the first week after different types of cardiac surgery. Scand Cardiovasc J 44: 119–124.

5. Sohn N, Marcoux J, Mycyk T, Krahn J, Meng Q (2009) The impact of different biocompatible coated cardiopulmonary bypass circuits on inflammatory response and oxidative stress. Perfusion 24: 231–237.

6. Chandrasena LG, Peiris H, Waikar HD (2009) Biochemical changes associated with reperfusion after off-pump and on-pump coronary artery bypass graft surgery. Ann Clin Lab Sci 39: 372–377.

7. Blankenberg S, Rupprecht HJ, Bickel C, Torzewski M, Hafner G, et al. (2003) Glutathione peroxidase 1 activity and cardiovascular events in patients with coronary artery disease. N Engl J Med 349: 1605–1613.

8. Mei YQ, Ji Q, Liu H, Wang X, Feng J, et al. (2007) Study on the relationship of APACHE III and levels of cytokines in patients with systemic inflammatory response syndrome after coronary artery bypass grafting. Biol Pharm Bull 30: 410–414.

9. Fairweather-Tait SJ, Bao Y, Broadley MR, Collings R, Ford D, et al. (2011) Selenium in human health and disease. Antioxid Redox Signal 14: 1337–1383.

10. Stoppe C, Schalte G, Rossaint R, Coburn M, Graf B, et al. (2011) The intraoperative decrease of selenium is associated with the postoperative development of multiorgan dysfunction in cardiac surgical patients. Crit Care Med 39: 1879–1885.

11. Frass OM, Buhling F, Tager M, Frass H, Ansorge S, et al. (2001) Antioxidant and antiprotease status in peripheral blood and BAL fluid after cardiopulmonary bypass. Chest 120: 1599–1608.

12. Forceville X, Mostert V, Pierantoni A, Vitoux D, Le Toumelin P, et al. (2009) Selenoprotein P, rather than glutathione peroxidase, as a potential marker of septic shock and related syndromes. Eur Surg Res 43: 338–347.

13. Oster O, Schmiedel G, Prellwitz W (1988) The organ distribution of selenium in German adults. Biol Trace Elem Res 15: 23–45.

14. McDonald CI, Fung YL, Fraser JF (2012) Antioxidant trace element reduction in an in vitro cardiopulmonary bypass circuit. ASAIO J 58: 217–222.

15. Cavalca V, Sisillo E, Veglia F, Tremoli E, Cighetti G, et al. (2006) Isoprostanes and oxidative stress in off-pump and on-pump coronary bypass surgery. Ann Thorac Surg 81: 562–567.

16. Diegeler A, Doll N, Rauch T, Haberer D, Walther T, et al. (2000) Humoral immune response during coronary artery bypass grafting: A comparison of limited approach, "off-pump" technique, and conventional cardiopulmonary bypass. Circulation 102: III95–100.

17. Biglioli P, Cannata A, Alamanni F, Naliato M, Porqueddu M, et al. (2003) Biological effects of off-pump vs. on-pump coronary artery surgery: focus on inflammation, hemostasis and oxidative stress. Eur J Cardiothorac Surg 24: 260–269.

18. Stoppe C, Werker T, Rossaint R, Dollo F, Lue H, et al. (2013) What is the significance of perioperative release of macrophage migration inhibitory factor in cardiac surgery? Antioxid Redox Signal 19: 231–239.

19. Levy MM, Fink MP, Marshall JC, Abraham E, Angus D, et al. (2003) 2001 SCCM/ESICM/ACCP/ATS/SIS International Sepsis Definitions Conference. Crit Care Med 31: 1250–1256.

20. Le Gall JR, Lemeshow S, Saulnier F (1993) A new Simplified Acute Physiology Score (SAPS II) based on a European/North American multicenter study. JAMA 270: 2957–2963.

21. Vincent JL, de Mendonca A, Cantraine F, Moreno R, Takala J, et al. (1998) Use of the SOFA score to assess the incidence of organ dysfunction/failure in intensive care units: results of a multicenter, prospective study. Working group on "sepsis-related problems" of the European Society of Intensive Care Medicine. Crit Care Med 26: 1793–1800.

22. Tiran B, Tiran A, Rossipal E, Lorenz O (1993) Simple decomposition procedure for determination of selenium in whole blood, serum and urine by hydride generation atomic absorption spectroscopy. J Trace Elem Electrolytes Health Dis 7: 211–216.

23. Levy JH, Tanaka KA (2003) Inflammatory response to cardiopulmonary bypass. Ann Thorac Surg 75: S715–S720.

24. Mediratta N, Chalmers J, Pullan M, McShane J, Shaw M, et al. (2013) In-hospital mortality and long-term survival after coronary artery bypass surgery in young patients. Eur J Cardiothorac Surg 43: 1014–1021.

25. Lamy A, Devereaux PJ, Prabhakaran D, Taggart DP, Hu S, et al. (2012) Off-pump or on-pump coronary-artery bypass grafting at 30 days. N Engl J Med 366: 1489–1497.

26. Paglia DE, Valentine WN (1967) Studies on the quantitative and qualitative characterization of erythrocyte glutathione peroxidase. J Lab Clin Med 70: 158–169.

27. Schulze F, Wesemann R, Schwedhelm E, Sydow K, Albsmeier J, et al. (2004) Determination of asymmetric dimethylarginine (ADMA) using a novel ELISA assay. Clin Chem Lab Med 42: 1377–1383.

28. Rayman MP (2000) The importance of selenium to human health. Lancet 356: 233–241.

29. Winnefeld K, Streck S, Treff E, Jutte H, Kroll E, et al.(1999) [Reference ranges of antioxidant parameters in whole blood (erythrocytes) in a Thuringen region]. Med Klin (Munich) 94 Suppl 3: 101–102.

30. Stoppe C, Spillner J, Rossaint R, Coburn M, Schalte G, et al.(2013) Selenium blood concentrations in patients undergoing elective cardiac surgery and receiving perioperative sodium selenite. Nutrition 29: 158–165.

31. Huang TS, Shyu YC, Chen HY, Lin LM, Lo CY, et al. (2013) Effect of parenteral selenium supplementation in critically ill patients: a systematic review and meta-analysis. PLoS One 8: e54431.

32. Barta E, Pechan I, Cornak V, Luknarova O, Rendekova V, et al. (1991) Protective effect of alpha-tocopherol and L-ascorbic acid against the ischemic-reperfusion injury in patients during open-heart surgery. Bratisl Lek Listy 92: 174–183.

33. Kharazmi A, Andersen LW, Baek L, Valerius NH, Laub M, et al.(1989) Endotoxemia and enhanced generation of oxygen radicals by neutrophils from patients undergoing cardiopulmonary bypass. J Thorac Cardiovasc Surg 98: 381–385.

34. Leong JY, van der Merwe J, Pepe S, Bailey M, Perkins A, et al. (2010) Perioperative metabolic therapy improves redox status and outcomes in cardiac surgery patients: a randomised trial. Heart Lung Circ 19: 584–591.

35. Storti S, Cerillo AG, Rizza A, Giannelli I, Fontani G, et al. (2004) Coronary artery bypass grafting surgery is associated with a marked reduction in serum homocysteine and folate levels in the early postoperative period. Eur J Cardiothorac Surg 26: 682–686.

36. Stuhlinger MC, Tsao PS, Her JH, Kimoto M, Balint RF, et al. (2001) Homocysteine impairs the nitric oxide synthase pathway: role of asymmetric dimethylarginine. Circulation 104: 2569–2575.

37. Manzanares W, Biestro A, Galusso F, Torre MH, Manay N, et al. (2009) Serum selenium and glutathione peroxidase-3 activity: biomarkers of systemic inflammation in the critically ill? Intensive Care Med 35: 882–889.

38. Frass OM, Buhling F, Tager M, Frass H, Ansorge S, et al. (2001) Antioxidant and antiprotease status in peripheral blood and BAL fluid after cardiopulmonary bypass. Chest 120: 1599–1608.

39. Franke A, Lante W, Fackeldey V, Becker HP, Kurig E, et al. (2005) Pro-inflammatory cytokines after different kinds of cardio-thoracic surgical procedures: is what we see what we know? Eur J Cardiothorac Surg 28: 569–575.

40. Karu I, Zilmer K, Starkopf J, Zilmer M (2006) Changes of plasma asymmetric dimethylarginine levels after coronary artery bypass grafting. Scand Cardiovasc J 40: 363–367.

41. Loukanov T, Arnold R, Gross J, Sebening C, Klimpel H, et al. (2008) Endothelin-1 and asymmetric dimethylarginine in children with left-to-right shunt after intracardiac repair. Clin Res Cardiol 97: 383–388.

42. Bellinger FP, Raman AV, Reeves MA, Berry MJ (2009) Regulation and function of selenoproteins in human disease. Biochem J 422: 11–22.

43. Caputo M, Bays S, Rogers CA, Pawade A, Parry AJ, et al. (2005) Randomized comparison between normothermic and hypothermic cardiopulmonary bypass in pediatric open-heart surgery. Ann Thorac Surg 80: 982–988.

9

Antidiabetic Property of *Symplocos cochinchinensis* Is Mediated by Inhibition of Alpha Glucosidase and Enhanced Insulin Sensitivity

Kalathookunnel Antony Antu[1], Mariam Philip Riya[1], Arvind Mishra[2], Karunakaran S. Anilkumar[3], Chandrasekharan K. Chandrakanth[1], Akhilesh K. Tamrakar[4], Arvind K. Srivastava[2], K. Gopalan Raghu[1]*

1 Agroprocessing and Natural Products Division, Council of Scientific and Industrial Research-National Institute for Interdisciplinary Science and Technology (CSIR-NIIST), Thiruvananthapuram, Kerala, India, 2 Division of Biochemistry, Council of Scientific and Industrial Research-Central Drug Research Institute (CSIR-CDRI), Lucknow, Uttar Pradesh, India, 3 Medicinal Chemistry Division, CSIR-CDRI, Lucknow, Uttar Pradesh, India, 4 Division of Pharmacology, CSIR-CDRI, Lucknow, Uttar Pradesh, India

Abstract

The study is designed to find out the biochemical basis of antidiabetic property of *Symplocos cochinchinensis* (SC), the main ingredient of '*Nisakathakadi*' an *Ayurvedic* decoction for diabetes. Since diabetes is a multifactorial disease, ethanolic extract of the bark (SCE) and its fractions (hexane, dichloromethane, ethyl acetate and 90% ethanol) were evaluated by *in vitro* methods against multiple targets relevant to diabetes such as the alpha glucosidase inhibition, glucose uptake, adipogenic potential, oxidative stress, pancreatic beta cell proliferation, inhibition of protein glycation, protein tyrosine phosphatase-1B (PTP-1B) and dipeptidyl peptidase-IV (DPP-IV). Among the extracts, SCE exhibited comparatively better activity like alpha glucosidase inhibition (IC_{50} value-82.07$\pm$2.10 μg/mL), insulin dependent glucose uptake (3 fold increase) in L6 myotubes, pancreatic beta cell regeneration in RIN-m5F (3.5 fold increase) and reduced triglyceride accumulation (22% decrease) in 3T3L1 cells, protection from hyperglycemia induced generation of reactive oxygen species in HepG2 cells (59.57% decrease) with moderate antiglycation and PTP-1B inhibition. Chemical characterization by HPLC revealed the superiority of SCE over other extracts due to presence and quantity of bioactives (beta-sitosterol, phloretin 2'glucoside, oleanolic acid) in addition to minerals like magnesium, calcium, potassium, sodium, zinc and manganese. So SCE has been subjected to oral sucrose tolerance test to evaluate its antihyperglycemic property in mild diabetic and diabetic animal models. SCE showed significant antihyperglycemic activity in *in vivo* diabetic models. We conclude that SC mediates the antidiabetic activity mainly via alpha glucosidase inhibition, improved insulin sensitivity, with moderate antiglycation and antioxidant activity.

Editor: Hitoshi Ashida, Kobe University, Japan

Funding: The funder for this work was the Indian Council of Medical Research, Govt. of India, as Research Fellowship (3/1/3/JRF-2009/MPD-34 (21404)) of Antu KA. The funder had no role in study design, data collection and analysis, decision to publish, or preparation of the manuscript.

Competing Interests: The authors have declared that no competing interests exist.

* Email: raghukgopal2009@rediffmail.com

Introduction

Diabetes mellitus is a global health threat associated with increased morbidity, mortality and poor quality of life which is characterized by chronic hyperglycemia [1]. Hyperglycemia leads to vascular complications via glucose toxicity and oxidative stress [2] and its proper control is an important therapeutic strategy to prevent diabetic complications [3]. Major determinants of postprandial hyperglycemic variations include gut digestion and absorption rate, available insulin response and tissue insulin sensitivity [4]. A medication that can address these abnormalities along with oxidative stress may be quite beneficial to diabetes. Current therapies include insulin and various oral agents such as sulfonylureas, biguanides, alpha-glucosidase inhibitors and glip-tins, which are used as monotherapy or in combination to achieve better glycemic regulation [3]. These medications have some undesirable effects [5] and managing diabetes without side effects is still being a challenge. Hence the search for more effective and safer therapeutic agents of natural origin has been found to be valuable.

Traditional medicines are frequently used in urban settings as an alternative in daily healthcare and it recommends complex herbal mixtures and multi-compound extracts [6]. Synergistic properties of herbal medicines due to the presence of variety of components within a single herbal extract are beneficial to multifactorial diseases like diabetes [7]. Herbal medicines have played an important role in treating diabetes in various parts of the world for centuries. *Ayurveda*, a system of traditional medicine native to Indian subcontinent always plays major role in primary health care of both rural and urban populations of India [8]. *Symplocos cochinchinensis* (Lour.) S. Moore. (SC) from the family Symplocaceae, is a medicinal plant with anti-inflammatory, antitumor, antimicrobial and antidiabetic properties [9,10]. The bark of SC is one of the key ingredients of *Nisakathakadi Kashayam* (decoction); a very effective *Ayurvedic* preparation for diabetes mentioned in the ancient script '*Sahasrayogam*' [11]. For wider acceptability of the health benefits of SC, a detailed scientific investigation on its mode of action on various biochemical targets relevant to diabetes is mandatory. But any thorough study illustrating the mechanism of action of SC or its biochemical targets relevant to diabetes is not available in literature. Here, attempts are made to see the main bioactives responsible for its antidiabetic property and to elucidate the mode of action of SC using selected biochemical targets relevant to diabetes.

Materials and Methods

Chemicals and Reagents

Streptozotocin (≥98%), 2,2 diphenyl-1-1-picryl hydrazyl (DPPH), 4-nitro phenyl alpha-D- glucopyranoside, yeast alpha-glucosidase, acarbose, gallic acid, tannic acid, quercetin, trolox, diprotin A, suramin, beta-sitosterol, phloretin 2′glucoside, olea-nolic acid, rosiglitazone, metformin, cytochalasin B, 2-deoxyglu-cose, 3-isobutyl-1-methylxanthine (IBMX), dexamethasone, insu-lin, dimethyl sulphoxide (DMSO) and all other chemicals and biochemicals unless otherwise noted were from Sigma (St. Louis, MO, USA). 2-deoxy-d-[^{3}H]-glucose (2-DG) was from GE Healthcare, UK. All the positive controls used were of HPLC grade.

Ethics statement

No specific permission was required for the collection of the plant material. This plant is plenty available in this specific area (Palode, Thiruvanathapuram) and there is no restriction for the collection of the plant. It is not an endangered or protected species. The location is not privately-owned or protected in any way. According to the guidelines of the Committee for the Purpose of Control and Supervision of Experiments on Animals (CPCSEA) formed by the Government of India in 1964, proper sanction had been obtained for animal experiments from CSIR-CDRI institu-tional animal ethics committee (Ethics Committee Approval Reference No. IAEC/2008/63/Renewal 04 dated 16.05.2012). Approval was obtained specifically for the animal experiments of this study from CSIR-CDRI institutional animal ethics committee. Animals were sacrificed by cervical dislocation under light ether anaesthesia as per ethics committee guidelines.

Plant material

The bark of SC was collected from Palode, Thiruvanantha-puram (8°29′N, 76°59′E) during July 2011 and authenticated by Dr. Biju Haridas, Taxonomist, Jawaharlal Nehru Tropical Botanic Garden and Research Institute (JNTBGRI), Thiruvanantha-puram, Kerala. A voucher specimen (No. 66498) was stored at the herbarium of JNTBGRI. 2 kg dry powder was extracted by maceration at 35–37°C; five times for 18 to 20 hrs with 70% ethanol [12]. Then it was filtered under vacuum and dried using rotary evaporator (Heidolph, Schwabach, Germany) at 35–40°C. This *Symplocos cochinchinensis* hydroethanol extract was desig-nated as SCE. SCE was fractionated using 4 different solvents based on polarity; n-hexane (SCH), dichloromethane (SCD), ethyl acetate (SCEC) & 90% ethyl alcohol (SCEL). The SCE and its fractions were stored at 4°C, protected from light and humidity.

HPLC analysis

The HPLC analysis was carried out as described previously with slight modifications [13] on LC-20AD HPLC system (Shimadzu, Tokyo, Japan) equipped with the PDA detector, SPD-M20A and LC solutions software. The chromatographic separations were performed using Phenomenex Luna C-18 Column (150 mm×4.6 mm I. D, 5 µm), with a flow rate of 0.5 mL/min

and a sample injection volume of 20 µL. The mobile phase used was acetonitrile (A) and water (B) with an isocratic elution ratio of 85:15 (A:B (v/v)) in 20 min. The sample was monitored with UV detection at 210 nm at 40°C.

Atomic Absorption Spectrophotometer (AAS) analysis

SCE (25 mg/mL) was digested in dilute HCl (7:3). The concentration of minerals was quantified (mg/g of sample) by atomic absorption spectrophotometer (Perkin Elmer Inc. USA).

Quantification of Total Phenolic Content (TPC), Total Tannin Content (TTC) and Total Flavonoid Content (TFC)

TPC was determined as described previously [14], and were expressed as milligram gallic acid equivalents per gram of extract (mg GAE/g). Tannin estimation was done by the indirect method [15]. TTC was expressed as milligram tannic acid equivalents per gram of extract (mg TAE/g). TFC estimation was done as described previously [16] and expressed as milligram quercetin equivalents per gram of extract (mg QE/g).

In vitro alpha glucosidase (AG), dipeptidyl peptidase-IV (DPP-IV) & protein tyrosine phosphatase-1B (PTP-1B) inhibition assay

Yeast and rat intestinal AG (EC 3.2.1.20) inhibitory property of the extracts were determined as described previously [17] using acarbose as standard. All the extracts were checked for DPP-IV (EC 3.4.14.5) inhibition using the kit from Cayman chemicals (Ann Arbor, MI, USA). Diprotin A was used as the standard. PTP-1B (EC 3.3.3.48) inhibitory property of extracts was evaluated using the kit from Calbiochem (Darmstadt, Germany). Percentage inhibition values were plotted against the corresponding concen-trations of the sample to obtain IC$_{50}$ value.

Determination of antioxidant potential and metal chelation activity

The antioxidant activity of extracts was assessed by DPPH method [18] with gallic acid as standard. 2,2′-azino-bis(3-ethylbenzothiazoline-6-sulphonic acid) (ABTS) radical scavenging activity was determined using assay kit (Zen-Bio Inc., NC, USA) and trolox was the standard. The hydroxyl radical scavenging activity was measured by the deoxyribose method [19] with catechin as standard. The chelation of ferrous ions by the extracts was estimated using ferrozine method [20] and EDTA was used as the standard. IC$_{50}$ values were calculated and compared with the respective standards.

Determination of antiglycation activity

Advanced glycation end products (AGEs) derived from bovine serum albumin (BSA) were quantified using the previous method [21]. BSA in the presence of ribose in phosphate buffered saline was served as control. AGE fluorescence (λ_{ex}370 nm; λ_{em} 440 nm) was measured in terms of relative fluorescence unit (RFU) after 24 h and 7 days of incubation. Investigations after 24 h and 7 days incubation are designated as day1 and day7 experiments respectively for future references. The data was compared with the reference compound quercetin (100 µM). AGEs formed were also processed for complexity analysis to check whether test material has capacity to block the formation of glycated products [21] using scanning electron microscope (SEM; Carl Zeiss, Munich, Germany).

Cell culture

HepG2 and L6 cell lines were obtained from National Centre for Cell Science, Pune, India. The HepG2 cells were maintained in low glucose (5.5 mM) DMEM supplemented with 10% FBS and 1% antibiotic/antimycotic solution (10,000 U/mL penicillin G, 10 mg/mL streptomycin, 25 μg/mL amphotericin B), with 5% CO_2 at 37°C. L6 skeletal muscle cells were maintained in alpha-MEM supplemented with 10% FBS and 1% antibiotic/antimycotic solution at 5% CO_2 at 37°C. Differentiation was induced by switching confluent cells to medium supplemented with 2% FBS. Experiments were performed in differentiated myotubes. RIN-m5F cells (ATCC, USA) were cultured in RPMI-1640 supplemented with 10% FBS and 1% antibiotic/antimycotic solution at 5% CO_2 at 37°C. MIN-6 cells (ATCC, USA) were maintained in DMEM supplemented with 10% FBS, 1% antibiotic/antimycotic solution, 100 μg/mL L-glutamine, 10 μL/L beta - mercaptoethanol at 5% CO_2 at 37°C. 3T3-L1 murine preadipocytes (ATCC, USA) were cultured in DMEM supplemented with 10% FBS and antibiotics. Differentiation was induced by switching to DMEM with 500 μM 3-isobutyl-1-methylxanthine (IBMX), 10 μM dexamethasone and 500 nM insulin (MDI) for 48 h. Differentiation was then maintained in DMEM containing 10% FBS and 500 nM insulin for 8 days.

Determination of cell viability

The extracts were dissolved in DMSO for application to cell cultures and final concentration of DMSO was fixed at 0.1% for all cell based assays. The cytotoxicity was checked by MTT assay kit (Cayman chemicals, Ann Arbor, MI, USA). HepG2, RIN-m5F, MIN-6, L6 and 3T3-L1 cells were seeded at a density of 4×10^4 cells/well in 24 well plate and incubated for 24 h. Cells were treated with various concentrations of extract and incubated for 24 h. Then, cell viability in HepG2, L6 and 3T3-L1 or proliferation in RIN-m5F and MIN-6 was evaluated.

Hyperglycemia induced oxidative stress

The cells were maintained in low glucose medium (5.5 mM) for the initial 24 h, then switched over to high glucose (25 mM) medium with or without the extracts or quercetin (positive control) to check whether test material prevent the generation of oxidative stress. The intracellular reactive oxygen species (ROS) production was monitored with the fluorescent probe CM-H₂DCFDA [22].

Glucose uptake

The 2-deoxy glucose uptake in L6 myotubes was performed as described previously [23]. Glucose uptake measured in triplicate and normalized to total protein, was expressed as fold induction with respect to unstimulated cells. Rosiglitazone and metformin were the standards.

Adipocyte differentiation

The adipogenic potential of all the extracts (30 μg/mL) was assessed in 3T3-L1 preadipocyte over untreated cells by quantifying the accumulation of triglycerides using oil red O staining on day 8 [24,25]. Rosiglitazone was used as standard. The cell lysate from all experimental groups was prepared according to the previous method [21] and assayed for GPDH (EC 1.1.1.8) activity using a Takara GPDH Assay Kit (Takara Bio Inc, Otsu, Japan). The membrane fraction for DGAT1 assay was collected as described previously [26]. DGAT1 activity was measured using the kit from MyBioSource.com (San Diego, CA, USA). Total cellular TG was extracted as reported previously [27]. TG content was assayed using a TG assay kit (Cayman Chemicals). The protein content was measured and normalized for GPDH, DGAT 1 and TG assays using a bicinchoninic acid kit (Pierce, Rockford, IL USA). The adiponectin level in the residual media was measured using adiponectin assay kit (Cayman Chemicals).

Animals

Male albino rats of Sprague Dawley (SD) strain (7–8 weeks old, 160 ± 20 g), bred at animal facility of CSIR-CDRI, Lucknow were selected for this study. Rats were housed in polypropylene cages (5 rats per cage) under an ambient temperature of $23 \pm 2°C$; 50–60% relative humidity; light 300 lux at floor level with regular 12 h light/dark cycle. Animals were maintained on a standard pellet diet and water *ad libitum*.

Oral Sucrose Tolerance Test (OSTT) in normal rats

For this normal SD rats were fasted for 16 h. Animals showing fasting blood glucose level (BGL) between 70 to 90 mg/dL were divided into 6 groups containing 6 animals each. Animals of experimental groups were orally administered SCE (100, 250 and 500 mg/kg body weight (bw)), metformin (100 mg/kg bw) or acarbose (50 mg/kg bw) dissolved in 1.0% gum acacia. The dose of SCE was selected on the basis of dosage of 'Nisakathakadi Kashayam' for human use. 10–15 mL of this preparation containing approximately 3.5 g of SC bark including other 7 herbs in equal amount, thrice in a day is generally prescribed for patients [11]. Ethanol extract has been used for *in vivo* study due to the yield of more bioactive molecules and less toxicity of the solvent [28]. Since its selective nature, 70% ethanol is the most suitable solvent for *in vivo* pharmacological evaluation compared to other solvents; it will dissolve only the required bioactive constituents with minimum amount of the inert materials [28]. Animals of control group were given an equal volume of 1.0% gum acacia. Rats were loaded with sucrose (10 g/kg bw) orally 30 min after administration of test sample or vehicle. BGL was estimated at 30, 60, 90 and 120 min post administration. Food but not water was withheld during the course of experimentation [29].

OSTT in sucrose loaded mild diabetic rat model (SLM)

Animals were made diabetic by injecting streptozotocin (60 mg/kg in 100 mM citrate buffer-pH 4.5) intraperitoneally after overnight fasting. Animals showing fasting BGL<200 mg/dL after 72 h were selected, termed as mild diabetic [30] and divided into 4 groups of 6 animals each. Animals of experimental group were administered SCE (500 mg/kg bw), metformin (100 mg/kg bw) or acarbose (50 mg/kg bw). Mild diabetic control group were given an equal amount of 1.0% gum acacia. A sucrose load (10 g/kg) was given to each animal orally 30 min after test sample or vehicle. BGL was determined at 30, 60, 90 and 120 min post-administration of sucrose [29].

OSTT in sucrose-challenged streptozotocin-diabetic rat model (STZ-S)

Like SLM, rats were made diabetic. Animals of BGL>350 mg/dL after 72 h were selected, termed as diabetic [30], and divided into 4 groups of 6 animals each. Experimental groups were administered SCE, metformin or acarbose like SLM. Diabetic control group received equal amount of 1.0% gum acacia. Rats were loaded with sucrose (3 g/kg bw) orally 30 min after test sample or vehicle. BGL was checked at 30, 60, 90, 120, 180, 240, 300 and 1440 min (24 h), respectively [29]. Acarbose has been selected as one of the positive control as it is the alpha-glucosidase inhibitor which can improve long term glycemic control in patients with diabetes [31]. Metformin is the widely used

Table 1. Dry yield, Total Phenolic Content (TPC), Total tannin Content (TTC) and Total Flavonoid Content (TFC) of test materials.

SI no.	Sample	Dry yield as % weight of dry plant material	Total phenolic content (TPC) in mg GAE/g	Total flavonoid content (TFC) in mg QE/g	Total tannin content (TTC) in in mg TAE/g
1	SCE	12.35	53.72	19.35	10.47
2	SCH	0.50	13.40	8.56	-
3	SCD	0.32	36.27	19.85	-
4	SCEC	0.55	57.28	26.35	17.54
5	SCEL	2.91	54.68	22.85	13.26

antidiabetic to treat the cardinal symptoms of diabetes like polyphagia, polydipsia, polyuria and insulin resistance due to its pleiotropic effect via various targets and it shows wide tolerance and less toxicity compared to other antidiabetics [32]. Due to the wider acceptability of metformin as an antidiabetic drug, we used it as a positive control.

Statistical analysis

Quantitative glucose tolerance of each group was calculated by the area under the curve (AUC) method using GraphPad Prism software version 3 (GraphPad Software Inc., La Jolla, CA, USA). All other results were analyzed using a statistical program SPSS/PC+, version 11.0 (SPSS Inc., Chicago, IL, USA). Data are presented as mean $\pm$ SD, from 3 independent experiments with triplicates. $P \leq 0.05$ was considered to be significant.

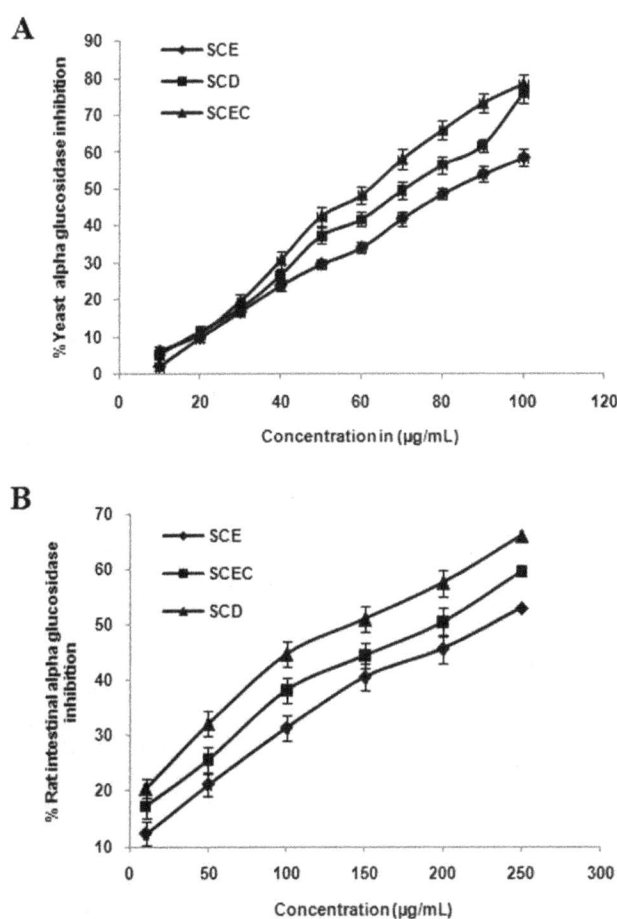

Figure 1. SCE SCD and SCEC exhibited alpha-glucosidase inhibitory property. (A) Yeast alpha glucosidase inhibition. (B) Rat intestinal alpha glucosidase inhibition. Values are means $\pm$ SD; n = 3. SCE, S. cochinchinensis (SC) ethanol extract; SCD, SC dichloromethane fraction and SCEC, SC ethyl acetate fraction.

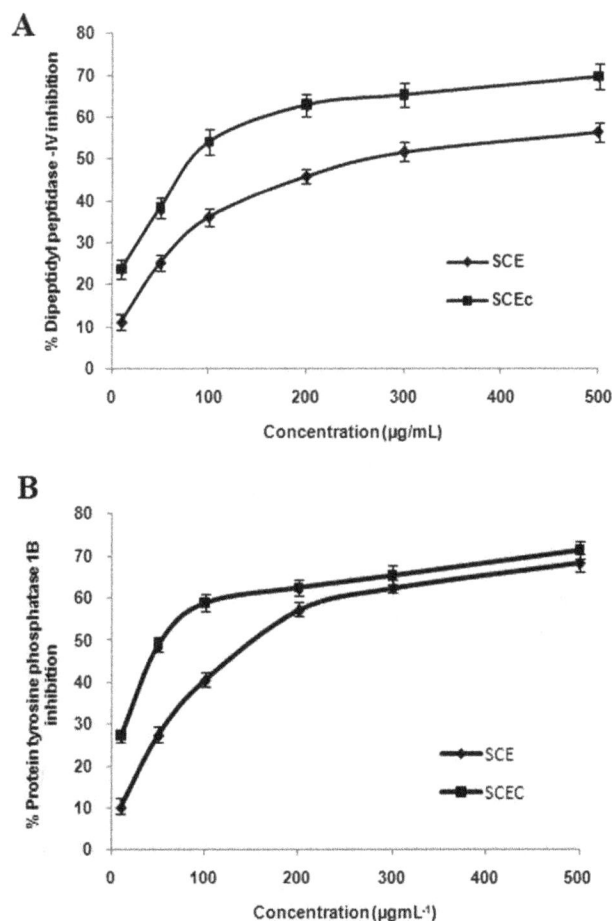

Figure 2. DPP-IV and PTP-1B inhibitory property of SCE and SCEC. (A) DPP-IV inhibition by SCE & SCEC. SCEC IC$_{50}$- 87.63$\pm$1.88 μg/mL and SCE IC$_{50}$- 269.98$\pm$2.95 μg/mL. Values are means $\pm$ SD; n = 3. (B) PTP-1B inhibitory property of SCE & SCEC. SCEC IC$_{50}$- 55.83$\pm$1.24 μg/mL and SCE IC$_{50}$- 159.10$\pm$1.91 μg/mL. Values are means $\pm$ SD; n = 3. SCE, S. cochinchinensis (SC) ethanol extract and SCEC, SC ethyl acetate fraction.

Table 2. IC$_{50}$ values of antioxidant (DPPH, ABTS and hydroxyl radical scavenging) and metal chelation assays.

Sl no.	Samples	IC$_{50}$ values of DPPH radical scavenging assay in µg/mL	IC$_{50}$ values of ABTS radical scavenging assay in µg/mL	IC$_{50}$ values of hydroxyl radical scavenging assay in µg/mL	IC$_{50}$ values of metal chelation activity in µg/mL
1	SCE	133.20±2.45	54.95±1.12	34.74±1.06	89.31±1.82
2	SCH	402.62±3.41	321.12±2.94	364.23±3.17	295.21±4.67
3	SCD	541.65±3.61	96.29±1.90	164.37±2.56	211.38±3.61
4	SCEC	129.43±1.84	35.72±1.02	31.64±0.98	86.49±1.76
5	SCEL	130.04±1.92	36.47±1.21	42.81±1.52	94.38±2.04

Results

Phytochemical characterization

HPLC analysis confirmed the presence of beta-sitosterol (111.62±4.12 mg/g), phloretin 2′glucoside (98.32±4.87 mg/g) and oleanolic acid (63.89±3.03 mg/g) in *Symplocos cochinchinensis* ethanolic extract (SCE); phloretin 2′glucoside (508.46±11.63 mg/g) and oleanolic acid (39.09±1.73 mg/g) in ethyl acetate fraction of SCE (SCEC); beta-sitosterol (145.56±4.63 mg/g) in hexane fraction of SCE (SCH); beta-sitosterol (152.29±6.31 mg/g) and phloretin 2′glucoside (188.97±6.41 mg/g) in dichloromethane fraction of SCE (SCD); phloretin 2′glucoside (273.65±7.63 mg g−1) in ethyl acetate fraction of SCE (SCEL) (Fig. S1 and S2 in Supporting Information S1) [33]. Analysis of minerals by AAS for micro-nutrients revealed presence of various minerals like zinc (0.014±0.0005 mg/g) manganese (0.096±0.0041 mg/g), iron (0.147±0.005 mg/g), sodium (1.387±0.062 mg/g), potassium (2.496±0.11 mg/g), magnesium (4.368±0.203 mg/g) and calcium (46.799±2.15 mg/g). The dry yield, TPC, TTC and TFC of the extracts were shown in Table 1. Since SCE exhibited comparatively better activity with respect to various *in vitro* targets and its high content of bioactives, SCE was taken forward for *in vivo* study. Moreover, in Indian traditional system of medicine (*Ayurveda*) most of the decoctions are hydroalcohol based (eg. *Arishta* and *Kashaya*).

In vitro AG, DPP-IV and PTP1-B inhibitory property

The extracts were evaluated for AG inhibition utilizing rat intestinal and yeast enzymes. SCEC, SCD and SCE showed significant yeast AG inhibition with IC$_{50}$ values of 62.30±1.53, 71.26±1.94 and 82.07±2.10 µg/mL respectively (Fig. 1A). Rat intestinal AG inhibition (IC$_{50}$) of the extracts was found to be 194.93±2.67 (SCEC), 143.02±2.91 (SCD) and 232.05± 3.34 µg/mL (SCE) (Fig. 1B). Acarbose showed an IC$_{50}$ of 45±1.12 for yeast and 49.78±1.45 µg/mL for rat AG enzymes. SCEC fraction showed DPP-IV inhibition with an IC$_{50}$ of 87.63±1.88 µg/mL while IC$_{50}$ of SCE was 269.98±2.95 µg/mL (Fig. 2A). Standard compound diprotin A showed an IC$_{50}$ of 1540±11.2 µg/mL. PTP-1B inhibition was noticed in SCEC fraction with an IC$_{50}$ of 55.83 µg/mL and SCE exhibited an IC$_{50}$ of 159.10 µg/mL (Fig. 2B). Standard was suramin (IC$_{50}$ 14.01 µg/mL (10.8 µM)).

SC fractions exhibited antioxidant and metal chelation potential

SCEC, SCEL and SCE showed better DPPH radical scavenging property compared to SCH and SCD (Table 2). IC$_{50}$ of gallic acid was 6.5±0.73 µg/mL. Similarly SCEC, SCEL and SCE exhibited promising ABTS cation decolorization potential compared to SCH and SCD (Table 2). IC$_{50}$ of the standard trolox was 5±0.51 µg/mL. SCEC, SCEL and SCE showed potent hydroxyl

radical scavenging and metal chelation activity compared to SCH and SCD (Table 2). IC$_{50}$ of catechin was 9±0.86 µg/mL and that of EDTA was 4.67±0.36 µg/mL.

Antiglycation property was observed in SCE, SCEC and SCEL

AGEs derived from BSA was analysed using 2 methods; by RFU measurements and SEM analysis. There were 22 groups under day1 and day7 experiments. In detail, 2 untreated control groups (one each with day1 and day7 experiments), 2 quercetin treated groups (day1 and day7 experiments) and 18 extract treated groups (3 doses- 100, 500 & 1000 µg/mL of SCE, SCEC and SCEL under day1 and day7). Quercetin (100 µM) showed significant (P≤0.05) antiglycation property in RFU measurement and also in SEM analysis (Fig. 3B b). Significant decrease (P≤0.05) in fluorescence in dose dependent manner was observed in day1 and day7 experiments at 500 and 1000 µg/mL doses of three extracts, indicative of antiglycation property (Fig. 3A). SEM analysis of the microstructure of control group of day1 showed highly granular agglomeration with uneven pores and highly complex cross linking (Fig. 3B a). 500 and 1000 µg/mL doses of SCE, SCEC and SCEL reduced highly complex microstructure to simple membranous structure without any cross linking in day1 and day7 experiments (Fig. 3B c–h; day7 SEM data not shown). The SEM results were analyzed based on the previous report [21].

Protection from hyperglycemia induced oxidative stress

High glucose treatment induced the generation of significant amount of ROS in HepG2 cells (64.23%; Fig. 4A and B, b), but co-treatment with SCE or SCEC significantly attenuated ROS in a dose dependent manner (P≤0.05). SCE and SCEC were selected on the basis of their potent *in vitro* antioxidant property. Results showed that 41.94, 51.28 and 59.57% decrease of ROS level with 10, 50 and 100 µg/mL SCE respectively (Fig. 4B, d–f) compared to high glucose control group. Similarly SCEC caused 34.92, 45.79 and 56.72% decrease of ROS level with 10 and 50 and 100 µg/mL dose respectively (Fig. 4B, g–i). Quercetin (25 µM) showed significant (P≤0.05) decrease (60.04%) of ROS (Fig. 4B, c). All extracts were found to be absolutely safe up to 100 µg/mL in all 5 cell lines; HepG2, L6, 3T3L1, RIN-m2F and MIN-6 (data not shown).

Proliferation potential of pancreatic beta cells in RIN-m5F and MIN-6 cell lines

Treatment with SCE (10 µg/mL) exhibited significant cell proliferation rate; 3.5 fold and 0.5 fold respectively compared to control both in RIN-m5F and MIN-6 cells (Fig. 5) and other extracts did not show any positive effect.

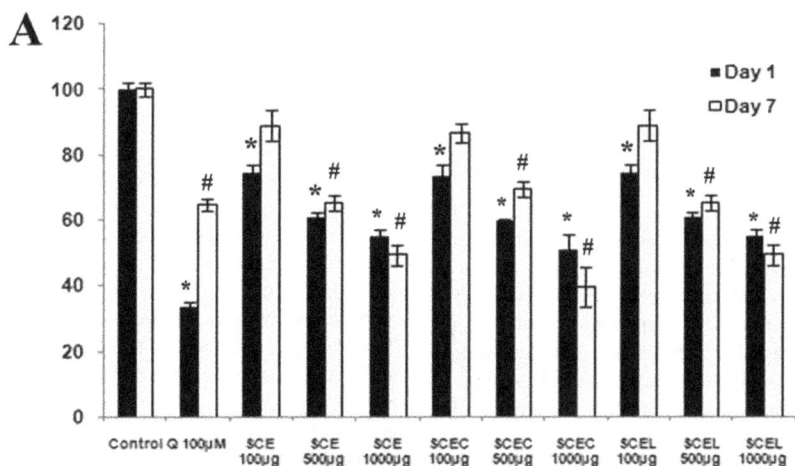

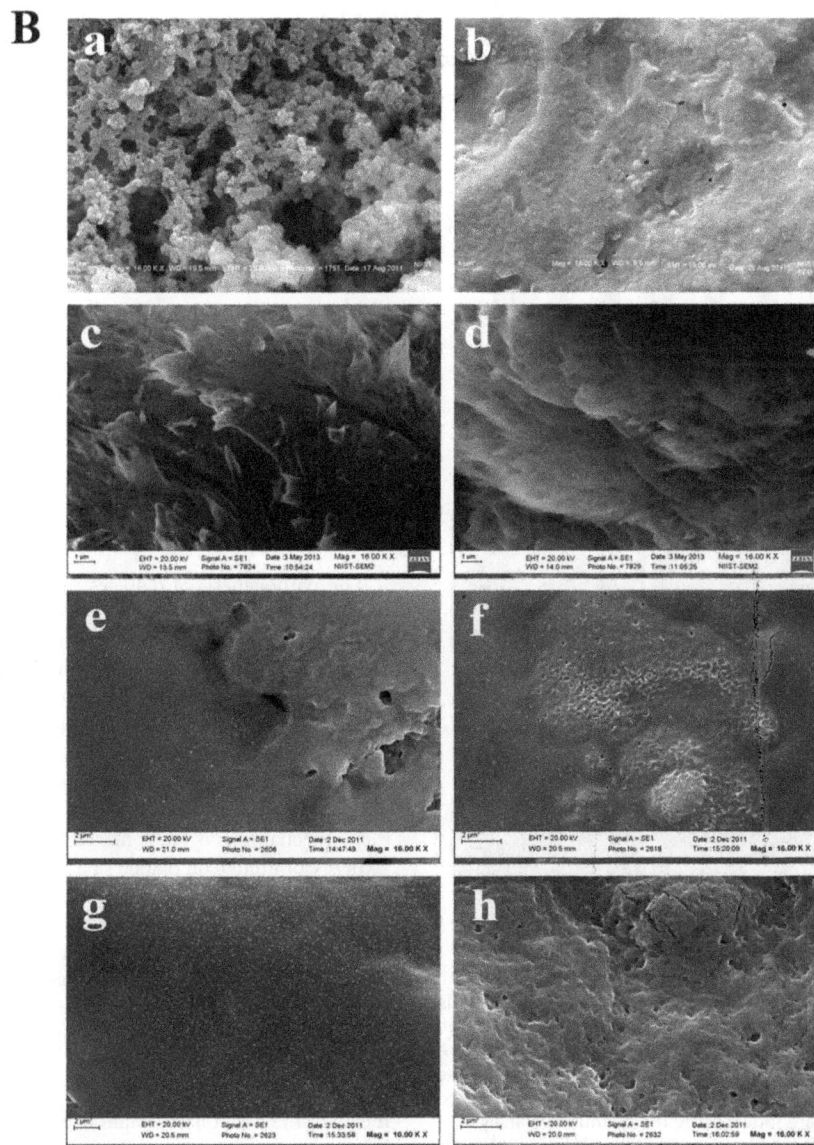

Figure 3. Fluorescence quantification and SEM microstructure analysis of advanced glycation end products revealed the antiglycation property of SCE SCEC and SCEL. (A) Quantification of fluorescence intensity of glycated products in presence of various concentrations of SCE, SCEC and SCEL (100, 500, 1000 µg/mL) under 2 different time intervals (day1 & day7) in terms of relative fluorescence units (RFU). Quercetin (100 µM) was used as reference compound. RFU are normalized to 100. Values are means ± SD; n = 3. *represents groups differ

significantly from day 1 control group (P≤0.05) and ≠represents groups differ significantly from day 7 control group (P≤0.05). (B) Representative SEM microstructures of glycated products formed under various groups of day1 experiments (a–h), (a, control; b, quercetin 100 µM, c & d, SCE 100 µg/mL and SCE 1000 µg/mL; e & f, SCEC 100 µg/mL and SCEC 1000 µg/mL and g & h, SCEL 100 µg/mL and SCEL 1000 µg/mL. All samples were visualized at 16000× magnification.

Enhancement of glucose uptake in L6 myotubes

Pre-treatment of myotubes with SCE and its fractions for 16 h with insulin (100 nM) resulted in increase of glucose uptake in an additive manner (Fig. 6A, P≤0.05). Among various fractions studied, both SCE and SCEL exhibited better activity both in the absence and presence of insulin in a dose dependent manner (Fig. S3 in Supporting Information S1; P≤0.05). Insulin alone showed a significant increase in glucose uptake (1.9 fold of basal, P≤0.05) in L6 myotubes. Metformin and rosiglitazone were standards (Fig. 6A).

Adipogenesis

The treatment with SCE and its fractions (30 µg/mL) induced a moderate level of differentiation of 3T3-L1 preadipocytes to adipocytes, but less than rosiglitazone. This was based on the morphological observation and quantification of triglycerides by oil red O staining (Fig. 6B). SCE at 50 µg/mL dose exhibited a significant decrease in GPDH activity compared to MDI positive group (P<0.05, Fig. 7B), at the same time 25 and 50 µg/mL doses of SCE exhibited a significant decrease in DGAT1 activity and TG content compared to MDI positive group (P<0.05, Fig. 7C and D). However, adiponectin level was significantly increased by SCE treatment (25 and 50 µg/mL) compared to MDI positive group (P<0.05, Fig. 7 E). Rosiglitazone was the reference standard.

Antihyperglycemic effect of SCE in normal and diabetic *in vivo* models

In acute toxicity study, SCE did not show any observable toxic effects in behaviour or physiology of animals up to 2 g/kg bw. In normal and SLM, the rise in BGL at 30 min of oral sucrose load was significantly reduced in SCE treated group compared to control group. SCE treatment at doses of 100, 250 and 500 mg/kg bw exhibited 7.56, 10.23 and 15.53% reduction respectively in plasma glucose in normal sucrose loaded rats and 18.18 and 20.42% by acarbose and metformin treatment (Fig. 8A and B). In SLM, treatment with 500 mg/kg bw of SCE reduced the whole glycemic response by 12.88% while acarbose and metformin caused 15.73 &and 17.12% reduction respectively (Fig. 8C and D). SCE treatment (500 mg/kg bw) in STZ-S caused 23.48% improvement in blood glucose profile after 5 h of treatment and acarbose and metformin showed 30.27 and 33.18% respectively (Fig. 8E and F).

Discussion

The pathogenesis of diabetes mellitus is complex and involves many mechanisms leading to several complications and demands a multiple therapeutic approach. Nowadays, medicinal plants have re-emerged as an effective source for the treatment of diabetes as it hold diverse group of compounds. Metformin exemplifies an efficacious oral glucose lowering agent derived from the research based on medicinal plants [34]. To date many antidiabetic medicinal plants have been reported although only a small number of these have received scientific evaluation to elucidate their mechanism of action. The World Health Organisation Expert Committee on diabetes has stressed the need of research on traditional medicine for future drugs [35]. In this study, phytochemically characterized SC was subjected to investigation on various biochemical targets relevant to diabetes like AG, glycation, DPP-IV, PTP-1B and hyperglycemia induced oxidative stress along with pancreatic beta cell proliferation, insulin dependent glucose uptake and adipogenesis using *in vivo* and *in vitro* models. Cell line based *in vitro* models are very much important in diabetic research as it is helpful to determine the mechanism of action of a plant extract with traditional use and/or human or *in vivo* data to support the antidiabetic effect [36]. In addition, the cell line based model allows the use of less amount of test material with reduced variability in results [36].

Oxidative stress due to hyperglycemia and dyslipidemia is one of the physiological parameter evident in diabetes [37]. Depletion of antioxidant level has been demonstrated in diabetic patients and

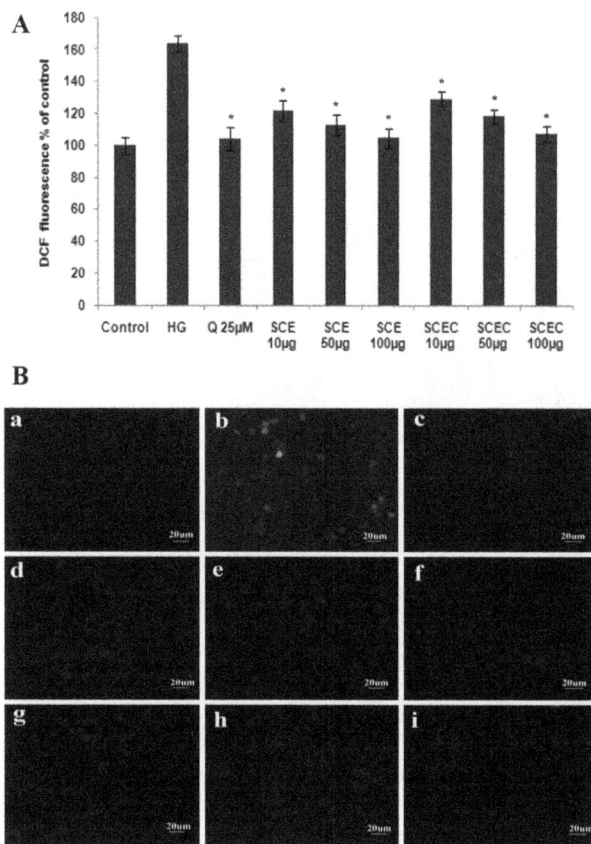

Figure 4. SCE and SCEC fractions protected HepG2 cells against ROS generation during hyperglycemia. Analysis of high glucose induced intracellular ROS levels in HepG2 cells by DCFDA method. Cultured HepG2 cells were treated with SCE or SCEC in the presence of high glucose (HG; 25 mM) for 24 h and then incubated with H_2DCFDA. The results are shown as (A) the quantitative analysis of fluorescence from three independent experiments. Values are means ± SD; n = 3. *represents groups differ significantly from HG group (P≤ 0.05). (B) Representative microscopic scans a–i (a, vehicle control; b, high glucose (HG); c, HG+Quercetin; d–f, HG+10 µg, 50 µg and 100 µg SCE; g–i, HG+10 µg, 50 µg and 100 µg SCEC). All samples were visualized at 20× magnification.

A

B

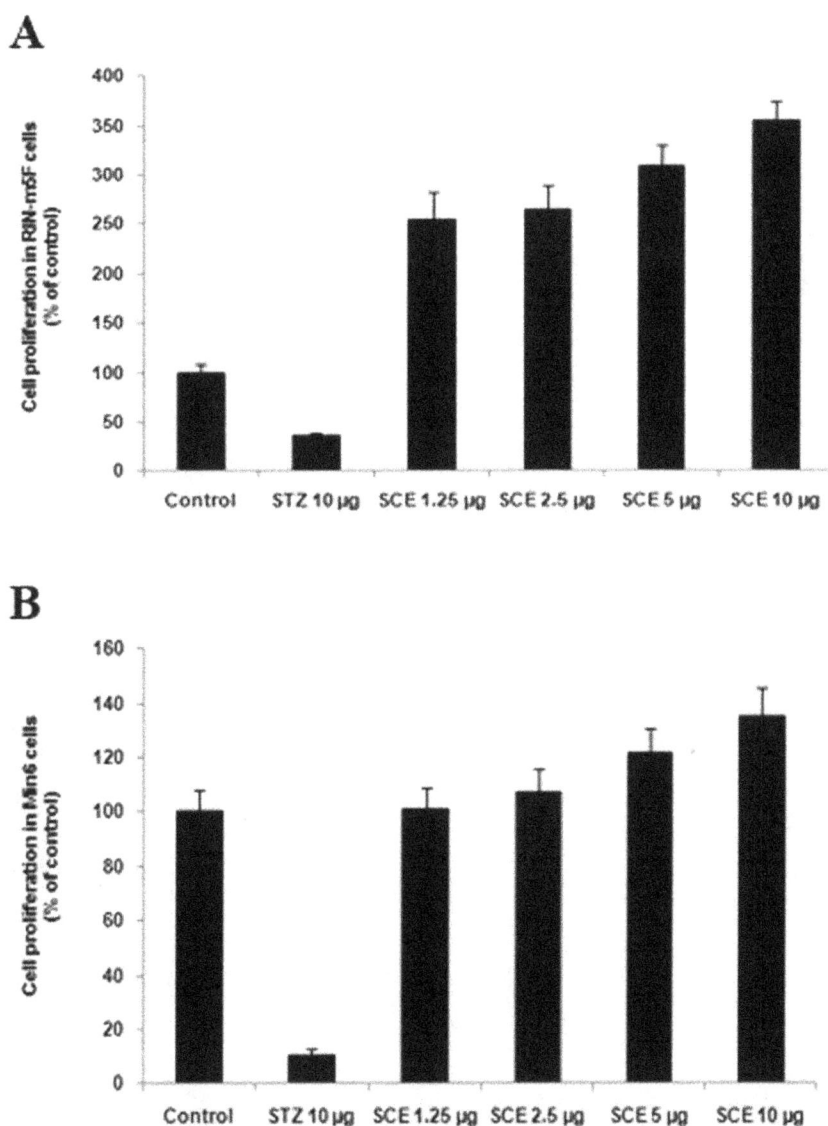

Figure 5. Pancreatic beta cell proliferation potential of SCE in RIN-m5F and MIN-6 cells. (A) Beta cell proliferation potential of SCE in RIN-m5F cells. (B) Beta cell proliferation potential of SCE in MIN-6 cells. Results are normalised to 100 based on control readings. Values are means ± SD; n = 3. *represents groups differ significantly from control group (P≤0.05).

extra administration of antioxidants to compensate the depletion, had helped to prevent diabetes complications [38]. Hyperglycemia induces accelerated hydroxyl radical generation and reactive oxygen species production which could represent the key event in the development of diabetic complications [39,40]. So we had analysed the antioxidant potential and hydroxyl radical scavenging activity of various extracts and the ability of extracts to prevent ROS generation under hyperglycemia. The results revealed significant antioxidant potential of SCE, SCEC and SCEL in an *in vitro* cell free system (Table 2) and protected HepG2 cells from hyperglycemia induced oxidative stress by preventing generation of ROS (Fig. 4A & B, P≤0.05). This result is in line with the reported protective effect of SCE on hepatic oxidative stress markers in STZ diabetic animal model [41]. It has been suggested that during hyperglycemic conditions, a non-enzymatic reaction occur between proteins and monosacharides (glycation) leading to the formation of pathologically significant AGEs [42]. Due to far reaching consequences of AGEs in the body, the estimation of

glycated haemoglobin (% HbA1c) has been advised by clinicians in addition to glucose in diagnosing metabolic syndrome. Biologically AGEs alter enzyme activity, modify protein and are main culprit in diabetes induced cardiomyopathy, retinopathy and neuropathy. Moreover, AGEs induce oxidative stress and vice versa [42]. So there is a tremendous interest in aniglycation agents for diabetes therapy. But as of today no specific drug is available with antiglycation potential. Our study revealed significant antiglycation activity of SCE, SCEC & SCEL (Fig. 5A and B) which could possibly one prominent mechanism of its known antidiabetic property. The two categories of antiglycation agents (AGE inhibitors and AGE breakers) act primarily as chelators by inhibiting metal-catalyzed oxidation reactions that catalyze AGE formation [43]. From the SEM microstructure analysis of AGEs, it is clear that SCE, SCEC and SCEL exhibited antiglycation via its AGE inhibitor property [44]. The *in vitro* method had shown potent metal chelation capacity of SC which may be the mechanism behind the better antiglycation potential of this plant

A

B

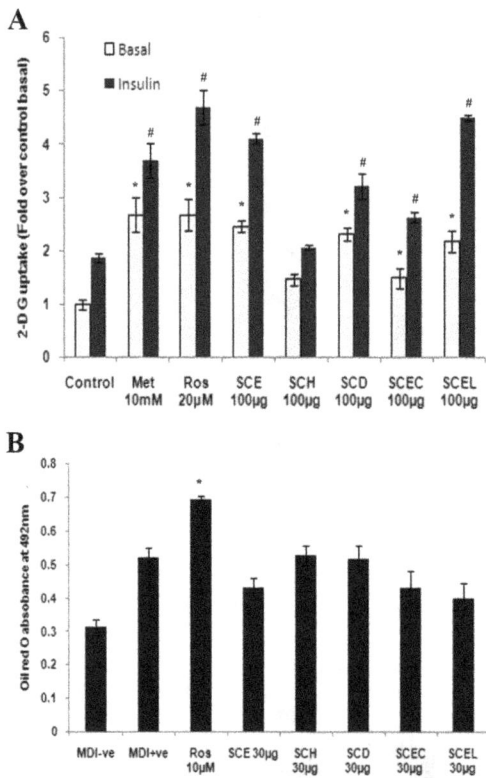

Figure 6. Glucose uptake and adipocyte differentiation studies in all 5 extracts. (A) 2-deoxy glucose uptake in L6 myotubes. Cells were incubated for 16 h with different extracts (100 μg/mL) or standards. After incubation myotubes were left untreated (white bars) or stimulated with 100 nM insulin (black bars) for 20 min, followed by the determination of 2-DG uptake. Results are expressed as fold stimulation over control basal. Metformin (10 mM) & rosiglitazone (20 μM) were the standards. Values are means ± SD; n = 3. *represents groups differ significantly from basal control group (P≤0.05). ≠represents groups differ significantly from insulin control group (P≤0.05). (B) Quantification of triglyceride content in differentiating 3T3-L1 adipocytes treated with different extracts (30 μg/mL) or rosiglitazone (10 μM) for 8 days by oil red O staining. Data are expressed as the means ± SD; n = 3; * represents groups differ significantly from MDI positive group (P≤0.05). MDI−ve, media without 3-isobutyl-1-methylxanthine (IBMX), dexamethasone & insulin; MDI+ve, media with IBMX, dexamethasone & insulin and Ros 10 μM, rosiglitazone 10 μM.

and the reported diminished %HbA1c level in the SCE treated STZ diabetic animal model [41].

PTP-1B is an abundant and widely expressed enzyme localized in endoplasmic reticulum. Theoretically, inhibition of action of PTP-1B that terminates insulin signalling would be expected to increase insulin sensitivity [45]. SCEC and SCE showed better PTP -1B inhibitory property (Fig. 2B). DPP-IV is a serine exopeptidase which regulates the half- life of two key glucoregulatory incretin hormones like glucose dependent insulinotropic polypeptide (GIP) and glucagone like peptide-1 (GLP-1) [46]. Inhibition of DPP-IV prolongs and enhances the activity of endogenous GIP and GLP-1, which serve as important prandial stimulators of insulin secretion in response to glucose and it inhibit glucagon secretion and conserve beta cell mass [46]. SCEC & SCE exhibited moderate DPP-IV inhibitory potential (Fig. 2A).

Significantly enhanced pancreatic beta cell proliferation was noticed in RIN-m5F and MIN-6 cells by SCE treatment. This pancreatic beta cell proliferation potential of test material

represent an extremely useful criteria to evaluate anti-diabetic activity, which could protect the beta cells from degeneration due to gluco-lipotoxicity during type 2 diabetes mellitus or protect from autoimmune mediated destruction as in the case of type 1 diabetes mellitus [47]. We had seen the protective property of SCE against streptozotocin induced toxicity in pancreas [41]. With this result, we strongly believe that this beneficial property contribute significantly to its antidiabetic efficiency.

Since insulin resistance is a major metabolic abnormality of type 2 diabetes, there has been considerable interest in insulin sensitizing agents to counteract insulin resistance for the treatment of this disease [3]. The result of the present study showed significant insulin dependent and independent glucose uptake proving insulin sensitizing property of SCE (Fig. 6A, P≤0.05). Further studies are required to find out the mechanism behind this effect. The peroxisome proliferator activated receptor (PPAR) gamma, the master regulator of adipogenesis is abundantly present in adipocytes which can maintain whole body insulin sensitivity and thiazolidinedione group of drugs (rosiglitazone and pioglitazone) act as PPAR modulators [48,49]. Analysis of the effect of SCE treatment on various markers of adipogenesis such as diminished activity of GPDH and reduced TG content compared to rosiglitazone, the full PPAR gamma agonist allude partial PPAR gamma agonist property of SCE (Fig. 6B and D). Adiponecin, solely secreted from adipocytes acts as a hormone with anti-inflammatory and insulin sensitizing properties [50]. There are reports to suggest the risk of T2DM appeared to decrease monotonically with increasing adiponectin level by several mechanisms [51]. So the potential of SCE to increase adiponectin level in 3T3-L1, suggest a role in its antidiabetic property and this is the first report in this regard (Fig. 6E). But detailed study on transactivation is required to confirm this [52]. Rosiglitazone is effective insulin sensitizer [48], act through its PPAR agonism. It enhances glucose uptake and adipocyte differentiation in a variety of insulin-resistant states [53]. So rosiglitazone has been taken as positive control for both glucose uptake and adipocyte differentiation studies.

Obesity is characterized by the accumulation of triacylglycerol in adipocytes and is an important risk factor for diabetes. Diacylglycerol acyltransferase (DGAT) catalyzes the final reaction of triacylgycerol synthesis and has two isoforms DGAT1 and DGAT2. DGAT1 plays a role in VLDL synthesis; increased plasma VLDL concentrations may promote obesity and thus DGAT1 is considered a potential therapeutic target of obesity and associated complications [54]. Here, a decrease in the DGAT1 activity by the treatment of SCE was observed in the study may attribute to its potential to reduce development of obesity as well hyperglycemia induced dis/hyperlipidemia (Fig. 6C).

The postprandial hyperglycemia (PPH) became a relevant target clinically and scientifically due to the importance in cardiovascular diseases and other complications [55]. The enzyme AG, present in the intestinal brush border cells hydrolyses complex carbohydrates to simple sugars. Inhibition of AG modulate carbohydrate digestion rate and prolong overall carbohydrate digestion time, causing a reduction in the rate of glucose absorption and consequently blunting PPH and insulin levels [56]. Additional therapeutic properties of AG inhibitors include protection against pancreatic beta cell apoptosis, inhibition of attachment of macrophage to vascular endothelium and amelioration of development of atherosclerosis [57]. Initial in vitro screening using yeast AG is required to see whether the study material has some alpha glucosidase inhibitory property [58]. Our in vitro studies showed promising AG inhibitory activity against both yeast derived and rat intestinal enzymes by SCE and its

A

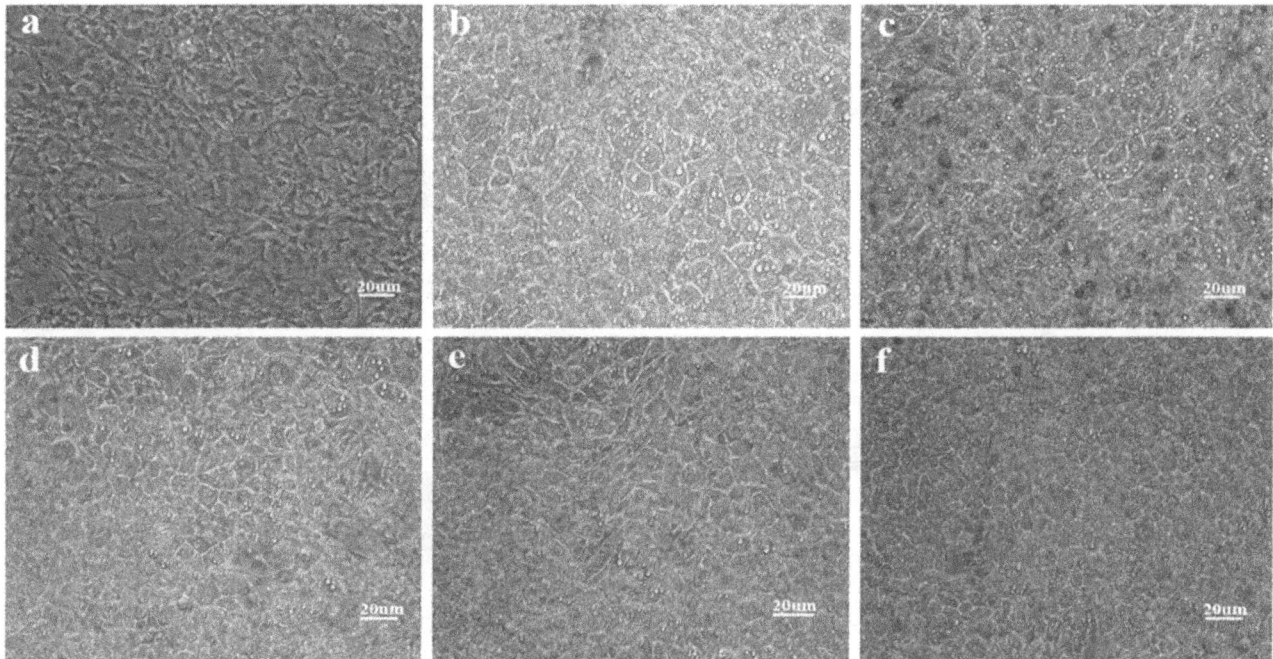

B

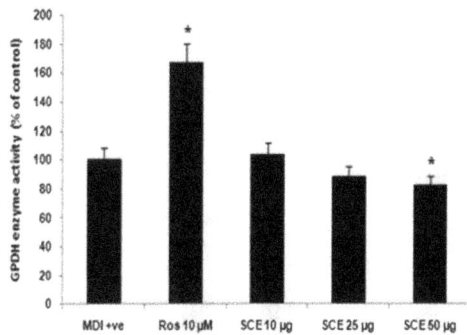

C

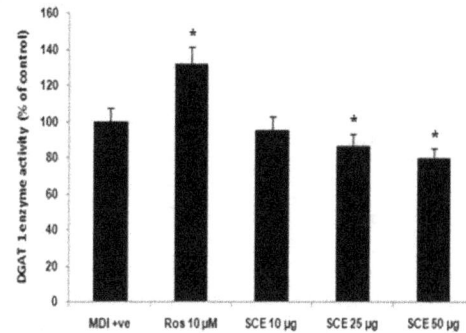

D

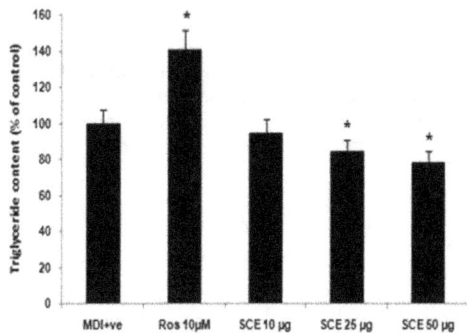

E

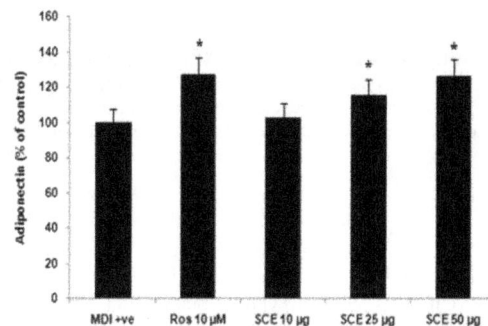

Figure 7. Estimation of adipogenesis in SCE tretment. (A) Cellular morphology. (Panels a–f) Micrographs (×10) showing (a) MDI negative, (b) MDI positive (vehicle control) (c) differentiating 3T3-L1 adipocytes treated for 8 days with rosiglitazone (10 µM), and (d–f) various concentrations of SCE (10, 25 and 50 µg/mL, respectively). DMSO (0.1%, vehicle) in differentiation media served as the vehicle control group i.e MDI positive. (B) Glycerol-3-phosphate dehydrogenase activity in various groups (MDI positive, rosiglitazone at 10 µM) and various concentrations of SCE (10, 25 and 50 µg/mL, respectively). (C) Diacyl glycerol -3 phosphate activity in various groups (MDI positive, rosiglitazone at 10 µM) and various concentrations of SCE (10, 25 and 50 µg/mL, respectively). (D) The triglyceride content in various groups (MDI positive, rosiglitazone at 10 µM) and various concentrations of SCE (10, 25 and 50 µg/mL, respectively). (E) Adiponectin level in various groups (MDI positive, rosiglitazone at 10 µM) and various concentrations of SCE (10, 25 and 50 µg/mL, respectively). Results are normalised to 100 based on control readings. Data are expressed as the means ± SD; n = 3. *Represents groups that differ significantly from the MDI positive (vehicle control) group (P≤0.05).

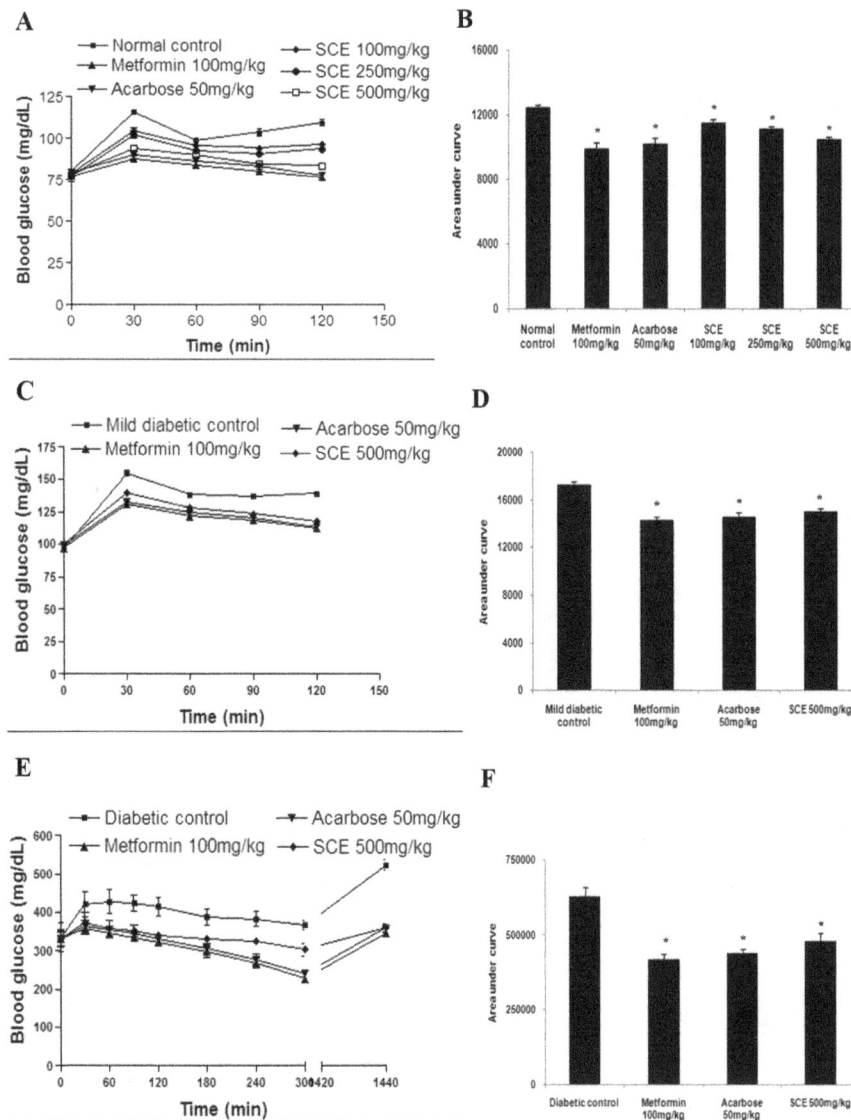

Figure 8. The antihyperglycemic effect of SCE in normal rats, mild diabetic rat model (SLM) and streptozotocin-induced diabetic rat model (STZ-S) after sucrose administration. (A) The glycemic response curve and (B) incremental AUC_{0-120} min in normal rats. (C) The glycemic response curve and (D) incremental AUC_{0-120} min in SLM model. (E) The glycemic response curve and (F) incremental AUC_{0-1440} min in STZ-S model. Data are expressed as the mean $\pm$ SD, n = 6. * represents groups differ significantly from control group (p<0.05). SCE, *S. cochinchinensis* (SC) ethanol extract.

fractions SCD and SCEC (Fig. 1A and B). In sucrose loaded normal and SLM models, SCE prevented acute PPH effectively compared to normal control and mild diabetic control (Fig. 8A–D, P≤0.05). This reveals the efficacy of SCE to control sucrose induced PPH significantly. This antihyperglycemic activity of SCE at 500 mg/kg bw was comparable with the existing drugs like acarbose and metformin. So we selected only 500 mg/kg dose for SLM and STZ-S studies. In streptozotocin models of diabetes, due to the destruction of pancreatic beta-cells, insulin secretion has been impaired and cause blood glucose elevation [55]. SCE treatment in STZ model resulted in attenuation of PPH, whereas diabetic control animals showed elevated blood glucose even after 5 h of sucrose load (Fig. 8E and F, P≤0.05). From this it is clear that SCE negate PPH by inhibiting AG that modify sucrose breakdown rate in small intestine in normal and diabetic rats.

Deficiency of specific vitamins and minerals play important roles in glucose metabolism and insulin signalling contribute to the development of diabetes [59]. In the present investigation, SCE was found to have high amount of calcium, moderate amount of sodium, potassium and magnesium and traces of manganese and zinc. There are also reports to link the role of these minerals in ameliorating complications arising from diabetes [59]. In addition, our TPC and TFC measurement showed the presence of high content of phenolics and flavanoids (Table 1). Accordingly, HPLC analysis revealed the presence of beta-sitosterol, phloretin 2′ glucoside and oleanolic acid. All these compounds are reported to have beneficial role in diabetes as well as to attenuate diabetes induced complications via different ways: beta-sitosterol improves glucose uptake and lipid metabolism [60,61] and alpha glucosidase inhibition [62]; phloretin enhances glucose uptake [63,64] and oleanolic acid improves insulin response [65,66] and possesses alpha glucosidase inhibitory property [67]. The results exhibited by SC in the present study may be due to the synergistic action of these three compounds in addition to other polyphenolic components.

Conclusion

Overall results reveal potent antihyperglycemic activity via inhibition of alpha glucosidase and enhanced insulin sensitivity with moderate antiglycation and antioxidant potential of SC which contribute significantly to its antidiabetic property. The presence of known insulin sensitizers and AG inhibitors like phloretin-2 glucoside, oleanolic acid and beta-sitosterol in SC play an important role in these multifaceted activities of SC with respect to diabetes.

Acknowledgments

We thank Director, CSIR-NIIST, Thiruvananthapuram & Director, CSIR-CDRI, Lucknow for providing necessary laboratory facilities via networking research programme of NaPAHA CSC 0130 of CSIR 12th FYP.

Author Contributions

Conceived and designed the experiments: KGR. Performed the experiments: KAA MPR AM CKC AKT. Analyzed the data: KAA KSA CKC. Contributed reagents/materials/analysis tools: AKT. Contributed to the writing of the manuscript: KAA. Designed the in vivo experiments: AKS. Revised the manuscript critically for important intellectual content: KGR.

References

1. Zimmet P, Alberti KG, Shaw J (2001) Global and societal implications of the diabetes epidemic. Nature 414: 782–787.
2. Fowler MJ (2008) Microvascular and macrovascular complications of diabetes. Clin Diabetes 26: 77–82.
3. Moller DE (2001) New drug targets for type 2 diabetes and the metabolic syndrome. Nature 414: 821–827.
4. Hanfeld M (2008) Alpha glucosidase inhibitors. In: Goldstein BJ, Dirk MW, editors. Type 2 diabetes Principles and Practice. New York: Informa Healthcare. 121.
5. Cheng AY, Fantus IG (2005) Oral antihyperglycemic therapy for type 2 diabetes mellitus. CMAJ 172: 213–226.
6. Leonti M, Casu L (2013) Traditional medicines and globalization: current and future perspectives in ethnopharmacology. Front Pharmacol 4: 92.
7. Graziose R, Lila MA, Raskin I (2010) Merging traditional Chinese medicine with modern drug discovery technologies to find novel drugs and functional foods. Curr Drug Discov Technol 7: 2–12.
8. Meena AK, Bansal P, Kumar S (2009) Plants-herbal wealth as a potential source of ayurvedic drugs. Asian J Tradit Med 4: 152–170.
9. Sunil C, Ignacimuthu S, Agastian P (2011) Antidiabetic effect of Symplocos cochinchinensis (Lour.) S. Moore. in type 2 diabetic rats. J Ethnopharmacol 134: 298–304.
10. Sunil C, Agastian P, Kumarappan C, Ignacimuthu S (2012) In vitro antioxidant, antidiabetic and antilipidemic activities of Symplocos cochinchinensis (Lour.) S. Moore bark. Food Chem Toxicol 50: 1547–1553.
11. Krishnanvaidyan KV, Pillai SG (Eds.) (2000) Sahasrayogam-Sujanapriya Commentary. Alappuzha: Vidyarambham Publishers. 93p.
12. Jones WP, Kinghorn AD (2012) Extraction of plant secondary metabolites. Methods Mol Biol 864: 341–366.
13. Pellati F, Orlandini G, Benvenuti S (2012) Simultaneous metabolite fingerprinting of hydrophilic and lipophilic compounds in Echinacea pallida by high performance liquid chromatography with diode array and electron spray ionization mass spectrometry detection. J Chromatogr A 1242: 43–58.
14. Singleton VL, Rossi JA (1965) Colorimetry of total phenolics with phosphomolybdic-phosphotungstic acid reagents. Am J Enol Vitic 16: 144–158.
15. Makkar HPS, Bluemmel M, Borowy NK, Becker K (1993) Gravimetric determination of tannins and their correlations with chemical and protein precipitation methods. J Sci Food Agric 61: 161–165.
16. Chang C, Yang M, Wen H, Chern J (2002) Estimation of total flavonoid content in propolis by two complementary colorimetric methods. J Food Drug Anal 10: 178–182.
17. Apostolidis E, Kwon YII, Shetty K (2007) Inhibitory potential of herb, fruit, and fungal-enriched cheese against key enzymes linked to type 2 diabetes and hypertension. Innov Food Sci Emerg Technol 8: 46–54.
18. Shimada K, Fujikawa K, Yahara K, Nakamura T (1992) Antioxidative properties of Xantan on the autooxidation of soybean oil in cyclodextrin emulsion. J Agric Food Chem 40: 945–948.
19. Halliwell B, Gutteridge JMC, Aruoma OI (1987) The deoxyribose method: A simple test tube assay for determination of rate constants for reactions of hydroxyl radicals. Anal Biochem 165: 215–219.
20. Stookey LL (1970) Ferrozine - A new spectrophotometric reagent for iron. Anal Chem 42: 779–781.
21. Riya MP, Antu KA, Vinu T, Chandrakanth KC, Anilkumar KS, et al. (2014) An in vitro study reveals nutraceutical properties of Ananas comosus (L.) Merr. var. Mauritius residue beneficial to diabetes. J Sci Food Agric 94: 943–950.
22. Sankar V, Pangayarselvi B, Prathapan A, Raghu KG (2013) Desmodium gangeticum (Linn.) DC exhibits antihypertrophic effect in isoproterenol-induced cardiomyoblasts via amelioration of oxidative stress and mitochondrial alterations. J Cardiovasc Pharmacol 61: 23–34.
23. Tamrakar AK, Schertzer JD, Chiu TT, Foley KP, Bilan PJ, et al. (2010) NOD2 activation induces muscle cell-autonomous innate immune responses and insulin resistance. Endocrinology 151: 5624–5637.
24. Shi C, Wang X, Wu S, Zhu Y, Chung LW, et al. (2008) HRMAS ^{1}H-NMR measured changes of the metabolite profile as mesenchymal stem cells differentiate to targeted fat cells in vitro: implications for non-invasive monitoring of stem cell differentiation in vivo. J Tissue Eng Regen Med 2: 482–490.
25. Nerurkar PV, Lee YK, Nerurkar VR (2010) Momordica charantia (bitter melon) inhibits primary human adipocyte differentiation by modulating adipogenic genes. BMC Complement Altern Med 10: 34.
26. Yu YH, Zhang Y, Oelkers P, Sturley SL, Rader DJ, et al. (2002) Post transcriptional control of the expression and function of diacyl glycerol acyl transferase-1 in mouse adipocytes. J Biol Chem 277: 50876–50884.
27. Zou C, Shen Z (2007) One-step intracellular triglycerides extraction and quantitative measurement in vitro. J Pharmacol Toxicol Methods 56: 63–66.
28. Trease GE, Evans WC (1989) Pharmacognosy I, 12th ed, London: WB Saunders Co. Ltd.
29. Singh AB, Yadav DK, Maurya R, Srivastava AK (2009) Antihyperglycaemic activity of alpha-amyrin acetate in rats and db/db mice. Nat Prod Res 23: 876–882.
30. Thomson M, Al-Amin ZM, Al-Qattan KK, Ali M (2007) Hypoglycemic effects of ginger in mildly and severely diabetic rats. FASEB J 21: 103.4
31. Chiasson JL, Josse RG, Hunt JA, Palmason C, Rodger NW, et al. (1994) The efficacy of acarbose in the treatment of patients with non-insulin-dependent diabetes mellitus. A multicenter controlled clinical trial. Ann Intern Med 121: 928–935.
32. Viollet B, Guigas B, Sanz Garcia N, Leclerc J, Foretz M, et al. (2012) Cellular and molecular mechanisms of metformin: an overview. Clin Sci (Lond) 122: 253–270.
33. Abbasi MA (2004) Bioactive chemical constituents of Symplocos recemosa and Comiphora mukul [Dissertation]. Pakistan: University of Karachi.
34. Bailey CJ, Day C (2004) Metformin: its botanical background. Pract Diab Int 21: 115–117.
35. Bailey CJ, Day C (1989) Traditional plant medicines as treatments for diabetes. Diabetes Care 12: 553–564.
36. Soumyanath A (2006) Traditional medicines for modern times - Antidiabetic plants. Boca Raton, Florida: CRC press, Taylor & Francis Group.
37. Evans JL, Goldfine ID, Maddux BA, Grodsky GM (2002) Oxidative stress and stress activated signaling pathways: a unifying hypothesis of type 2 diabetes. Endocr Rev 23: 599–622.
38. Porasuphatana S, Suddee S, Nartnampong A, Konsil J, Harnwong B, et al. (2012) Glycemic and oxidative status of patients with type 2 diabetes mellitus following oral administration of alpha-lipoic acid: a randomized double-blinded placebo-controlled study. Asia Pac J Clin Nutr 21: 12–21.
39. Winiarska K, Drozak J, Wegrzynowicz M, Fraczyk T, Bryla J (2004) Diabetes-induced changes in glucose synthesis, intracellular glutathione status and hydroxyl free radical generation in rabbit kidney-cortex tubules. Mol Cell Biochem 261: 91–98.

40. Nishikawa T, Edelstein D, Du XL, Yamagishi S, Matsumura T, et al. (2000) Normalizing mitochondrial superoxide production blocks three pathways of hyperglycaemic damage. Nature 404: 787–790.

41. Antu KA, Riya MP, Mishra A, Sharma S, Srivastava AK, et al. (2014) *Symplocos cochinchinensis* attenuates streptozotocin-diabetes induced pathophysiological alterations of liver, kidney, pancreas and eye lens in rats. Exp Toxicol Pathol 66: 281–291.

42. Ahmed N (2005) Advanced glycation end products: role in pathology of diabetic complications. Diabetes Res Clin Pract 67: 3–21.

43. Nagai R, Murray DB, Metz TO, Baynes JW (2012) Chelation: a fundamental mechanism of action of AGE inhibitors, AGE breakers, and other inhibitors of diabetes complications. Diabetes 61: 549–559.

44. Vasan S, Foiles PG, Founds HW (2001) Therapeutic potential of AGE inhibitors and breakers of AGE protein cross-links. Expert Opin Investig Drugs 10: 1977–1987.

45. Shilpa K, Sangeetha KN, Muthusamy VS, Sujatha S, Lakshmi BS (2009) Probing key targets in insulin signaling and adipogenesis using a methanolic extract of Costus pictus and its bioactive molecule, methyl tetracosanoate. Biotechnol Lett 31: 1837–1841.

46. Drucker DJ, Nauck MA (2006) The incretin system: glucagon-like peptide-1 receptor agonists and dipeptidyl peptidase-4 inhibitors in type 2 diabetes. Lancet 368: 1696–1705.

47. Vetere A, Choudhary A, Burns SM, Wagner BK (2014) Targeting the pancreatic β-cell to treat diabetes. Nat Rev Drug Discov 13: 278–289.

48. Lebovitz HE, Dole JF, Patwardhan R, Rappaport EB, Freed M (2001) Rosiglitazone monotherapy is effective in patients with type 2 diabetes. J Clin Endocrinol Metab 86: 280–288.

49. Tontonoz P, Spiegelman BM (2008) Fat and beyond: the diverse biology of PPAR gamma. Annu Rev Biochem 77: 289–312.

50. Kadowaki T, Yamauchi T, Kubota N, Hara K, Ueki K (2006) Adiponectin and adiponectin receptors in insulin resistance, diabetes, and the metabolic syndrome. J Clin Invest 116: 1784–1792.

51. Li S, Shin HJ, Ding EL, van Dam RM (2009) Adiponectin levels and risk of type 2 diabetes: a systematic review and meta-analysis. JAMA 302: 179–188.

52. Atanasov AG, Blunder M, Fakhrudin N, Liu X, Noha SM, et al. (2013) Polyacetylenes from *Notopterygium incisum* - new selective partial agonists of peroxisome proliferator activated receptor-gamma. PLoS One 8: e61755.

53. Nugent C, Prins JB, Whitehead JP, Savage D, Wentworth JM, et al. (2001) Potentiation of glucose uptake in 3T3-L1 adipocytes by PPAR gamma agonists is maintained in cells expressing a PPAR gamma dominant-negative mutant: evidence for selectivity in the downstream responses to PPAR gamma activation. Mol Endocrinol 15: 1729–1738.

54. Yamazaki T, Sasaki E, Kakinuma C, Yano T, Miura S, et al. (2005) Increased very low density lipoprotein secretion and gonadal fat mass in mice over expressing liver DGAT 1. J Biol Chem 280: 21506–21514.

55. Chiasson JL, Rabasa-Lhoret R (2004) Prevention of type 2 diabetes: insulin resistance and beta-cell function. Diabetes 53: S34–S38.

56. Ross SA, Gulve EA, Wang M (2004) Chemistry and biochemistry of type 2 diabetes. Chem Rev 104: 1255–1282.

57. Osonoi T, Saito M, Mochizuki K, Fukaya N, Muramatsu T, et al. (2010) The alpha-glucosidase inhibitor miglitol decreases glucose fluctuations and inflammatory cytokine gene expression in peripheral leukocytes of Japanese patients with type 2 diabetes mellitus. Metabolism 59: 1816–1822.

58. Brindis F, González-Trujano ME, González-Andrade M, Aguirre-Hernández E, Villalobos-Molina R (2013) Aqueous extract of *Annona macroprophyllata*: a potential α-glucosidase inhibitor. Biomed Res Int 2013: 591313.

59. Martini LA, Catania AS, Ferreira SR (2010) Role of vitamins and minerals in prevention and management of type 2 diabetes mellitus. Nutr Rev 68: 341–354.

60. Gupta R, Sharma AK, Dobhal MP, Sharma MC, Gupta RS (2011) Antidiabetic and antioxidant potential of β-sitosterol in streptozotocin-induced experimental hyperglycemia. J Diabetes 3: 29–37.

61. Chai JW, Lim SL, Kanthimathi MS, Kuppusamy UR (2011) Gene regulation in β-sitosterol-mediated stimulation of adipogenesis, glucose uptake, and lipid mobilization in rat primary adipocytes. Genes Nutr 6: 181–188.

62. Tiabou Tchinda A, Nahar Khan S, Fuendjiep V, Ngandeu F, Ngono Ngane A, et al. (2007) Alpha-glucosidase inhibitors from *Millettia conraui*. Chem Pharm Bull (Tokyo) 55: 1402–1403.

63. Masumoto S, Akimoto Y, Oike H, Kobori M (2009) Dietary phloridzin reduces blood glucose levels and reverses SGLT1 expression in the small intestine in streptozotocin - induced diabetic mice. J Agric Food Chem 57: 4651–4656.

64. Najafian M, Jahromi MZ, Nowroznejhad MJ, Khajeaian P, Kargar MM, et al. (2012) Phloridzin reduces blood glucose levels and improves lipids metabolism in streptozotocin-induced diabetic rats. Mol Biol Rep 39: 5299–5306.

65. Wang X, Liu R, Zhang W, Zhang X, Liao N, et al. (2013) Oleanolic acid improves hepatic insulin resistance via antioxidant, hypolipidemic and anti-inflammatory effects. Mol Cell Endocrinol 376: 70–80.

66. Castellano JM, Guinda A, Delgado T, Rada M, Cayuela JA (2013) Biochemical basis of the antidiabetic activity of oleanolic acid and related pentacyclic triterpenes. Diabetes 62: 1791–1799.

67. Ali MS, Jahangir M, Hussan SS, Choudhary MI (2002) Inhibition of alpha-glucosidase by oleanolic acid and its synthetic derivatives. Phytochemistry 60: 295–299.

10

Properties and Antioxidant Action of Actives Cassava Starch Films Incorporated with Green Tea and Palm Oil Extracts

Kátya Karine Nery Carneiro Lins Perazzo[1]*, Anderson Carlos de Vasconcelos Conceição[1], Juliana Caribé Pires dos Santos[1], Denilson de Jesus Assis[2], Carolina Oliveira Souza[1], Janice Izabel Druzian[1]

1 Federal University of Bahia, College of Pharmacy, Department Food Science, Ondina, Salvador, BA, Brazil, **2** Federal University of Bahia, Department of Chemical Engineering, Federação, Salvador, BA, Brazil

Abstract

There is an interest in the development of an antioxidant packaging fully biodegradable to increase the shelf life of food products. An active film from cassava starch bio-based, incorporated with aqueous green tea extract and oil palm colorant was developed packaging. The effects of additives on the film properties were determined by measuring mechanical, barrier and thermal properties using a response surface methodology design experiment. The bio-based films were used to pack butter (maintained for 45 days) under accelerated oxidation conditions. The antioxidant action of the active films was evaluated by analyzing the peroxide index, total carotenoids, and total polyphenol. The same analysis also evaluated unpacked butter, packed in films without additives and butter packed in LDPE films, as controls. The results suggested that incorporation of the antioxidants extracts tensile strength and water vapor barrier properties (15 times lower) compared to control without additives. A lower peroxide index (231.57%), which was significantly different from that of the control ($p < 0.05$), was detected in products packed in film formulations containing average concentration of green tea extracts and high concentration of colorant. However, it was found that the high content of polyphenols in green tea extract can be acted as a pro-oxidant agent, which suggests that the use of high concentration should be avoided as additives for films. These results support the applicability of a green tea extract and oil palm carotenoics colorant in starch films totally biodegradable and the use of these materials in active packaging of the fatty products.

Editor: Vipul Bansal, RMIT University, Australia

Funding: Funding was provided by the Brazilian Committee for Postgraduate Courses in Higher Education (FAPESB) for graduate scholarships for KP for this study. The National Council for Scientific and Technological Development funded this technology research - CNPq 505831/2008-2, with a financial incentive to acquisition of laboratory supplies and equipment for the laboratory, and also scholarships for scientific initiation and post-graduate levels MA and PhD, for other researchers working on the same study. The funders had the role in study design, data collection and analysis, decision to publish, or preparation of the manuscript.

Competing Interests: Cargill Agrícola SA and Chr Hassen SA donated ingredients used in this study. There are no further patents, products in development or marketed products to declare.

* Email: katyanery@gmail.com

Introduction

The interest in biodegradable films produced from natural sources has increased in recent years due to the concerns about the environment and the consumer demand for the improvement of overall product characteristics (quality and appearance) [1], [2]. The basic materials used to produce biodegradable films are polysaccharides, proteins and lipids compounds [3]. Regarding polysaccharides, starch can produce biodegradable films at low cost and on a large scale [4]. Furthermore, starch-based materials may contribute to utilization of nonrenewable resources and the environmental impact caused by synthetic plastics [4]. Global production of cassava has nearly doubled over the past 30 years to about 260 million tons in 2012, making it an abundant and attractive starch source for researchers. Over half is grown in Africa, with a third in Asia and 14% in Latin America. Nigeria and Brazil are the largest producers, growing about 50 and 25 million tons in 2012, respectively [5].

A number of recent studies have focused on extending the functional properties of biodegradable films by adding different natural compounds to yield a biodegradable totally bioactive packaging material [2], [3], [6]. Active packaging films with antioxidant properties, developed by incorporating active functional ingredients into packaging systems, can offer protection against chemical and biological contamination [7], [8], and can delay oxidative changes in packaged products containing fatty components [9].

Oxidation is one of the most common mechanisms of degradation in foodstuffs and can limit the shelf life of food [10]. This process can decreased nutritional quality, increased toxicity, development of off-odor, and altered texture and color. Consequently, the shelf life and sales this products decrease. The direct addition of antioxidants in products, especially in foods, in one large initial dose is limited by the potential for rapid depletion of the antioxidants, in addition to very high initial concentrations [11]. Producers and packaging companies can inhibit the food oxidation process by adding antioxidants compounds at pack.

Several packing formulations using synthetic compounds, such as butylated hydroxytoluene and butylated hydroxyanisole, have been developed [12], [13]. However, modern consumer trends show increasing concern with the use of synthetic chemicals and the belief that natural antioxidants are safer and of greater nutritional benefit. Therefore, a need exists in the food industry to develop polymer packaging which can deliver natural antioxidants in a controlled manner throughout the product shelf life [11], [14], [15].

Research has focused on natural edible antioxidants, such as phenolic compounds, flavonoids [6], [12], [17], [18], [19] and carotenoids [8], [16], which are commonly found in natural sources, such as the green tea and palm oil and kernels.

Green tea is an excellent source of polyphenols, which are natural antioxidants that can be used as alternatives to synthetic antioxidants [17], [18], [19]. Polyphenols are trends to substitute them with naturally available antioxidants and can inhibit oxidation [20], [21]. Tea catechins can act as antioxidants by donating hydrogen atoms, by accepting free radicals, by interrupting chain oxidation reactions, or by chelating metals [22]. Wanasundara and Shahidi [23] suggested that the annexation of hydroxide groups to catechin molecules is the main factor that causes the strong antioxidant proprieties found in green tea extracts, reducing the formation of peroxides more effectively than BHT, BHA and a-tocopherol. It must be noted that some studies have suggested pro-oxidative proprieties of some polyphenols [24]. The activity of polyphenols depends on many factors, for example, the reductive potential, the chelating ability of the metals, the pH of the medium, solubility, bioavailability and stability in tissues and cells [21].

Carotenoids, such as α and β-carotene provide antioxidant protection because of their capacity to scavenge free radicals [25], [26], and palm oil reaching a world production of 50 million tons in 2012 [5] is a major sources [27]. The antioxidant activity of carotenoids in organic solutions is related to oxygen concentration, the chemical structure of the carotenoids, and the presence of other antioxidants [26], [27], [28].

The commercial use of edible films has been limited because these materials have poor mechanical and barrier properties as compared to synthetic polymers [29]. The successful use of natural compounds, such as phenolic and carotenoids, in packaging films is greatly dependent on the final characteristics of the films. The most important properties to be evaluated in biodegradable films are microbiological stability, adhesion, cohesion, wettability, solubility, transparency, mechanical properties, sensory and permeability to water vapor and gases. Once these properties are known, the composition and behavior of the material can be predicted and optimized [9].

Thus, studies to develop an innovative active food packaging that inhibits oxidation and behaves as a scavenger of oxygen radicals are of great interest [17]. The objective of this study was to develop films totally biodegradable from cassava starch containing green tea and palm oil carotenoids extracts, as actives natural compounds to be used as packaging fatty products, adding value to different agro-industrials chains.

Experimental

2.1 Materials

Cassava starch (amylose −23.5% and amylopectin −64,2%) was donated by Cargill Agrícola S.A. (Porto Ferreira, SP, Brazil). Glycerol, analytical grade, was purchased from Synth S.A (Diadema, SP, Brazil). Green Tea (Camellia sinensis) was purchased from Mãe Terra Ltda (Osasco, SP, Brazil). Commercial colorant VEGEX NC 3c WSP mct extracted of the palm oil (Elaeis guineensis), containing 35% α-Carotene and 65% β-Carotene, was provided by Chr. Hansen (Hørsholm, Denmark). Commercial butter without antioxidant was obtained from Imperial (BA, Brazil). Low-density polyethylene (LDPE) film (0.020 mm thickness and 15.86×10^{-8} g_{H2O}.mm/m^2.h.kPa water vapor permeability) was purchased from local markets (Salvador, BA, Brazil).

2.2. Film Preparation

Preliminary experiments were conducted to evaluate the maximum concentrations of additives that could be incorporated to the films, in order to obtain homogeneous materials, flexible and easy to handle. Therefore, different concentrations of colorant (0.01, 0.05 and 1.00%) and green tea (2.5, 5.0 and 7.5%) were alternately tested. At the end of this stage, the maximum concentrations were fixed in 0.05% for colorant and 5.0% for the green tea. The other concentrations used did not show desirable characteristics in the films obtained.

For films production, film-forming dispersions was prepared by with an aqueous green tea extract obtained from green tea powder (0–5.0% of dry leaves, g/100 g) by method of percolation with 2 L hot deionized water (80°C) for optimal extraction and preservation of antioxidant compounds [30]. The extract was cooled to room temperature and then filtered through Whatman No.1 filter paper.

Then, the aqueous green tea extract was added into cassava starch (4.0%, g/100 g) of previously dried (40°C, 6 h), glycerol (1.0%, g/100 g) and colorant powder (0–0.05%, g/100 g) to form the starch-plasticizer dispersions with approximately 90 wt %(w/v) solid concentration. The colorant and green tea extract were added according to a 2^2 central composite design, which was used to investigate the influence of two independent variables, namely, the concentrations of the colorants and the green tea extract. Film forming solutions were heated to 70°C, and the films were prepared by a casting technique, in which 66–67 g of the film-forming suspension was dehydrated on 150 mm diameter polycarbonate petri dishes kept at 30°C under renewable circulated air (Nova Etica, 400ND, SP, Brazil). All the dried starch film were preserved in a humidity chambers (25°C, RH = 75%) for further testing.

2.3. Film Characterization

The film was characterized by the thicknesses, total solid content, mechanical (tensile strength and elongation at break), barrier (water vapor permeability) and thermal properties (TGA).

2.3.1. Film Thickness. The average film thicknesses of the preconditioned samples (75% RH, 25°C) were measured using a flat parallel surface external digital micrometer (Digimess, Ip40 0–25 mm, São Paulo, Brazil) with 0.001 mm resolution. Five replications were conducted for each sample treatment. Five measurements were taken at random positions around the film sample and the mean values were calculated.

2.3.2. Mechanical Properties. Test filmstrips (8×2.5 cm) cut from preconditioned samples (25°C; 75% RH) were characterized for tensile strength resistance and elongation at break percentage by Universal Testing Instrument, electromechanical and microprocessor (EMIC, model DL-200MF, Instron, Paraná, Brazil). The tests were conducted according to the ASTM D882-00 method [31], [32]. Ten specimens were tested for each formulation.

2.3.3. Water Vapor Permeability (WVP). The samples were analyzed using the ASTM E96-80 method [31], modified by Gontard and others [33]. The relative humidity outside of the cell was fixed at 100% (pure water) and at 0% within the cell (dry silica). Four cells were prepared for each analysis and were weighed daily until a 4% weight gain of the silica was attained. Two control cells were prepared without the film and conditioned similarly. The weight gain of each cell was measured with time, and the water vapor permeability was calculated by equation 1 [34].

$$WVP = \frac{w}{t} \frac{e}{A.ps(RH_1 - RH_2)} \qquad (1)$$

where w/t is calculated from the linear regression of the weight gain over time, A is the film area, e is the film thickness, ps is the vapor saturation pressure (kPa), RH_1 is the relative humidity inside the chamber and RH_2 is the relative humidity inside the cells.

2.3.4. Thermogravimetric analyze (TGA). To investigate the thermal stability of the films, curves were generated with a thermogravimetric analyzer (TGA) from Pyris 1 TGA (Perkin Elmer, Pyris, Shelton, USA). Samples of approximately 5–6 mg were tested in an atmosphere of nitrogen (20 mL min-1), and the temperature was increased at a rate of $20°C$ min^{-1} from room temperature to $600°C$. The temperatures at which the rate of decomposition of the sample was at a maximum (Td) were obtained from thermogravimetric derivative curves (DTG).

2.4. Bio-Based Film used to Pack a Product

Butter was packed in cassava starch-plasticizer materials containing both actives green tea and carotenoids extracts colorant. Square-shaped films (5×2 cm–10 cm^2) of 0.164 and 0.212 mm in thickness were molded (Sealer Sulpack SM 400 TE, Brazil) with the open top. Butter homogenized and congealed in blocks with 3×2 cm (10.00 g ±0.54) were involved in films, bubbles of oxygen were removed, and the film was sealed on top.

The antioxidant capacity and the stability of packaged butter during extended storage at 0, 7, 15, 30, and 45 days under storage conditions of accelerate oxidation (64% relative humidity at $30±2°C$). The film storage with butter and analyses were carried out in a dark room to avoid the effects of light interference.

Unpackaged butter (C1), butter packaged in cassava starch-plasticizers without antioxidant additive (C2) and butter packaged in LDPE (C3) were used as controls.

2.5. Packaged Product Oxidative Stability

The oxidative stability from a packaged product was evaluated through the peroxide index, conjugated diene and total carotenoids of the butter, and these parameters were analyzed at 0, 7, 15, 30 and 45 days of storage.

2.5.1. Peroxide Index (PI). The peroxide index (PI) was determined by the titration method described by the Association of Official Analytical Chemists [35].

2.5.2. Total Carotenoids Content (TC). To measure the total carotenoids content (TC), 1.0 g of the packaged butter was dissolved in petroleum ether. The TC was determined spectrophotometrically at 440 nm (UV/Vis Spectrometer Lambda 20, Perkin-Elmer, Norwalk, Connecticut, USA) and was calculated according to equation 2 [36].

$$TC(\mu g.g^{-1}) = \frac{A.V.10^4}{(A1\%.1\,cm).W} \qquad (2)$$

Where TC is the total carotenoids, A is the absorbance at 435 nm, V is dilution volume (mL), $A1\%.1\,cm$ is the absorptivity coefficient value (2592) and W is the sample weight (g).

2.6. Actives Films Stability

The actives bio-based films stability used to pack butter was evaluated by analyzing the total carotenoids, total polyphenols and total flavonoids at 0, 7, 15, 30, 45 days of storage.

2.6.1. Total Carotenoids Content (TC). For total carotenoids (TC) values, films samples (1.00 g) were prepared according to Silva and Mercadante [37] and analyzed spectrophotometrically at 440 nm (UV/Vis Spectrometer Lambda 20, Perkin-Elmer, Norwalk, Connecticut, USA). The TC concentration was determined according to equation 2 at 440 nm.

2.6.2. Total Phenolic Content (TP). The total phenolic content (TP) of the film samples (100 mg) was extracted with water after centrifugation (4400 rpm/5°C/3 min; Eppendorf, 5702R, Hamburg, Germany), being evaluated at 0 and 45 days of storage. The TP in the supernatant was spectrophotometrically determined at 760 nm (UV/Vis Spectrometer Lambda 20, Perkin-Elmer, Norwalk, Connecticut, USA) using Folin-Ciocalteu reagent, and the results were expressed as gallic acid equivalents [8].

2.6.3. Total Flavonoid Content (TF). The total flavonoid concentration was measured using the same supernatant final sample of the total phenolic content. The final sample (1 mL) was added to a 10 mL volumetric flask containing 4 mL of distilled water. Then, 0.3 mL of 5% sodium nitrite solution was added to the volumetric flask, and 0.3 mL of 10% aluminum chloride was added after 5 min. One minute later, 2 mL of 1 M sodium hydroxide was also added. The reaction flask was then filled with distilled water and mixed. The absorbance was measured at 510 nm. Total flavonoid compounds were calculated using a standard curve prepared with dilutions of an epicatechin standard, [38].

2.7. Statistical Analysis

A complete factorial experimental design, 2^2 with 3 center points for a total of 11 experiments (Table 1), was applied to enable to evaluate of the influence of different concentrations of the antioxidant additives incorporated into the bio-based films. Palm oil carotenoid colorant (0.00 to 0.05%; X_1) and green tea extract (0.00 to 5.00%; X_2) were chosen as independent variables. The PI and TC from the packaged product (butter) and the TP, TF, TC, physical, barrier, mechanical and thermic properties from the films were used as dependent variables (Y). The data were subjected to variance analysis and Tukey's test for comparison of means at a 5% significance level using Statistica 7.0 software (Minneapolis, USA).

A quadratic regression model was employed to predict each response:

$$Y = b_0 + b_1X_1 + b_2X_2 + b_{11}X_1^2 + b_{22}X_2^2 + b_{12}X_1X_2 \qquad (3)$$

where Y are the predicted responses, X_1 and X_2 are the independent variables, b_0 is the offset term, b_1 and b_2 are the linear effects, b_{11} and b_{22} are the squared effects and b_{12} is the interaction term.

The goodness of fit of the models was evaluated by the determination coefficient (R^2), an analysis of variance (ANOVA) and Fischer's t test.

Table 1. Coded and real values of green tea extracts and colorants added to cassava starch bio based films according to a (2^2) second order experimental design with 3 central points.

Formulation Films	Coded Values		Real Values (% w/w)	
	Colorant (X_1)	Green tea extract (X_2)	Colorant	Green tea extract
F1	−1.00	−1.00	0.01	0.70
F2	−1.00	1.00	0.01	4.30
F3	1.00	−1.00	0.04	0.70
F4	1.00	1.00	0.04	4.30
F5	−1.41	0.00	0.00	2.50
F6	1.41	0.00	0.05	2.50
F7	0.00	−1.41	0.03	0.00
F8	0.00	1.41	0.03	5.00
F9(c)	0.00	0.00	0.03	2.50
F10(c)	0.00	0.00	0.03	2.50
F11(c)	0.00	0.00	0.03	2.50

(c) Central points.

Results and Discussion

3.1. Characterization of the Films

Table 2 shows the results of the thickness (t), water vapor permeability (WVP), mechanical properties (tensile strength and percent elongation at break), and thermogravimetric data of TGA (mass loss) and DTG (rate of mass loss), of the 11 film formulations, as responses (Y) according to a 2^2 central composite experimental design with 3 central points.

The thickness of the starch films with different quantities of two additives varied of 0.164 to 0.212 mm and was 24.39% higher and 9.92% smaller when compared to control film, (C2, Table 2). Control of the thickness in the films produced by casting is a step that requires great attention since variations here can affect their properties, including the mechanical and barrier properties, which certainly compromise the performance of the package [39], [40].

The incorporation of two additives into cassava bio-based films caused significant difference (p>0.05) between thickness of the 11 formulations and those in relation C2 control, according to Tukey's test (Table 2). For this parameters, ANOVA also indicated that the differences between the formulations were not statistically significant (p>0.05).

3.1.1. Mechanical Properties. The effect of the extract concentration on the mechanical properties was evaluated by the tensile strength (TS) and elongation at the break percentage (ε) of the films. For all films, the values of TS varied from 0.730±0.10 to 4.360±0.58 MPa, and the ε varied from 40.33±1.06 to 157.00±0.22% (Table 2).

A comparison between TS of the films containing additives with C2 control film showed greater tensile strength in all formulations, except F2. However, all 11 formulations showed lower percent elongation at break (ε) than C2 control (Table 2).

The formulation F7 (0.03% of carotenoic colorant and 0.00% of green tea extract) showed a higher TS and a lower percent elongation at break (Table 2), which characterized this material as more rigid than the control. These differences on the mechanical behavior of the formulated films could be explained by the carotenoic colorant interacting with films constituents and changing the properties of the continuous phases and the effect of crystallinity formed after the processing and storage of starch

film. Furthermore, the molecule regularity of amylose provided for the formation of crystalline regions in formulation F7 and, together with a greater number of points of contact, contributed to a behavior similar to conventional semi-crystalline polymers, for instance, a higher tensile strength and lower elongation at break percentage.

Between the formulations with 0.03% carotenoic colorant, the formulations F10, F11 and F12 (central points) were less affected in elongation at break percentage by comparison with control (Table 2).

The incorporation of additives in certain concentrations tested confer greater resistance to films (Table 2), an important characteristic for use in the packaging sector in general, where large deformation (ε) of the films is not required. Figure 1 shows the response surface and Pareto chart to tensile strength (Fig. 1A) and elongation at break percentage (Fig. 1A).

The ANOVA statistical analysis applied to the results indicated that the addition of green tea extract (X_2) negatively affected the tensile strength, while both additives (X_1 and X_2) the elongation at break percentage (Fig. 1).

The properties of the starch films depends a series factors, such as, the nature of starch and its cohesive structure, type of processing, environmental conditions, type of plasticizer, their thickness among others [39].

The film base already had added plasticizers (glycerol), and the concentration in the final material may have been too high, resulting in excessive interactions between the film network and the plasticizers and lower film flexibility. This variation could be caused by some of the natural compounds present in the extracts, which could greatly affect a starch film network and the mechanical performance [40], [41].

The humidifying ability of such components can alter the mechanical resistance of the biodegradable materials. Natural components that absorb water can increase the hydrophilicity of cassava starch biobased films, which are already highly hydrophilic materials, as function the type of processing and the environmental conditions of storage [42].

Correlation inversely proportional of the thickness (0.164–0.212 mm) with the tensile strength (0.73–4.36 MPa) and break elongation (40.33–157.00%) (Table 2), of $r^2 = 0.594$ and

Table 2. Thickness (t), mechanical properties (TS and ε), water vapor permeability (WVP), and thermal analysis (thermic events) of cassava starch films with addition of carotenoid colorant and green tea extract according to a (2^2) second order experimental design with 3 central points.

| Formulation Films | Real Values | | t (mm) | TS (MPa) | ε (%) | WVP × 10⁻⁸ (g_{H2O}.mm/m².h.kPa) | 1st event | | 2nd event | |
	Carotenoic Colorante (X_1 %)	Green Tea Extract (X_2 %)					T_{onset} (°C)	Mass loss (%)	T_{onset} (°C)	Mass loss (%)
C2	–	–	0.204±0.015[a]	0.83±0.22[c]	253.30±0.05[a]	15.86±0.18[a]	38.72[c]	8.87[d]	315.23[a]	81.43[d]
F1	0.01	0.70	0.177±0.021[b]	2.45±0.18[d]	66.31±0.07[b]	2.94±0.19[b]	31.55[a]	13.12[a]	323.58[e]	66.87[a]
F2	0.01	4.30	0.212±0.012[c]	0.73±0.10[e]	90.61±0.08[a]	2.95±0.13[b]	29.35[d]	13.63[a]	301.10[b]	66.06[b]
F3	0.04	0.70	0.165±0.005[d]	1.71±0.30[f]	78.21±0.17[b]	5.65±0.30[c]	33.16[e]	11.79[b]	327.41[f]	70.65[e]
F4	0.04	4.30	0.196±0.010[e]	0.96±0.22[g]	131.10±0.11[a]	0.29±0.93[d]	39.26[f]	12.91[a]	301.97[b,c]	65.13[c]
F5	0.00	2.50	0.199±0.015[f]	1.66±0.07[h]	62.59±0.06[b]	2.05±0.65[e]	26.50[g]	13.06[a]	308.92[g]	72.93[f]
F6	0.05	2.50	0.193±0.012[g]	1.16±0.17[a]	105.80±0.13[a]	0.68±0.12[f]	32.04[b]	13.12[a]	313.93[a]	64.94[a]
F7	0.03	0.00	0.164±0.012[h]	4.36±0.58[i]	40.33±1.06[c]	7.60±0.19[g]	29.82[d]	11.27[b,c]	334.45[h]	76.62[g]
F8	0.03	5.00	0.193±0.01[i]	1.55±0.28[b]	84.54±3.17[a]	8.56±0.47[h]	46.68[h]	11.83[b]	298.14[i]	66.59[a,b]
F9 (c)	0.03	2.50	0.202±0.027[j]	1.58±0.24[b]	142.10±0.21[a]	6.43±0.20[i]	31.84[a,b]	11.12[b]	303.13[c]	55.54[c]
F10 (c)	0.03	2.50	0.200±0.07[l]	1.40±0.30[j]	157.00±0.22[a]	6.51±0.12[j]	32.26[b]	10.92[c]	306.78[d]	55.01[c]
F11 (c)	0.03	2.50	0.210±0.02[m]	1.12±0.4[a]	126.90±0.30[a]	7.63±0.85[g]	31.64[b]	11.03[c]	305.93[d]	54.94[c]

Control C2 = additives without film; (c) Central points. TS = Tensile strength (MPa); ε = percent elongation at break (%); T_{onset} = degradation onset temperature (°C). Means with the same letters in the same columns were not statistically different (p>0.05) according to Tukey's test.

A

B

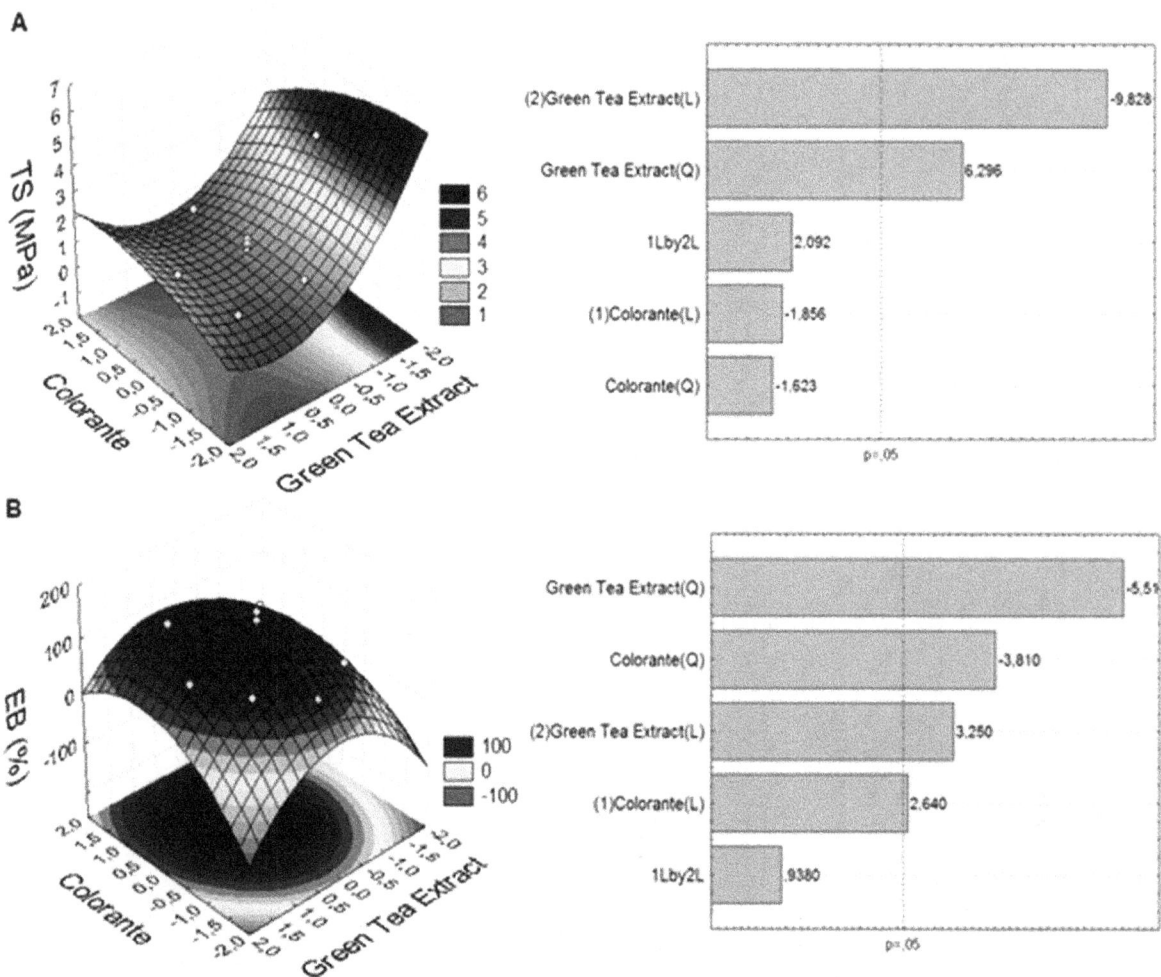

Figure 1. Response surface plot and Pareto chart for understanding the tensile strength (TS) and elongation at break percentage (EB) (Y response), as function on the content of carotenoic colorant and green tea extract (coded values of X_1 e X_2).

$r^2 = 0.391$ respectively, indicate that this parameter can exert influence on the mechanical properties of the films analyzed. Janson and Thuvander [43] also observed same effect, a decrease from 100 to 20% in elongation with increasing thickness 0.3 to 2.5 mm. According to authors, the effect in mechanical resistance could be too explained by the difference in the thickness among the samples, although it was also not statistically significant (p> 0.05).

3.1.2. Water Vapor Permeability. The water vapor permeability coefficient of a film is a constant value for permeation of the water vapor at a given temperature. The permeability of a film depends on the chemical structure and morphology, the nature of permeate and the temperature of the environment [44]. The water vapor permeability of the cassava starch bio-based films incorporated with different concentrations of carotenoic colorant and green tea was examined.

The data indicate (Table 2) that the incorporation of additives resulted a decrease of 50 times in water vapor permeability when compared to C2 control (additives without film). The incorporation of additives, (independent variables) in different percentages in cassava starch plasticized with glycerol matrix resulted in favorable statistically significant effect (p<0.05) in WVP (Table 2; Fig. 2). These results can be explained mainly by the nature of the

carotenoic colorant and the films constituents, which can give increased cohesion in the matrix.

The lower WVP of the films incorporated with additives might result from the interactions of mainly of carotenoic compounds with hydrophobic regions of leached amylose and with amylopectin side chains through Van der Waals forces, which limit the availability of hydrogen groups to form hydrophilic bonding with water [45]. This leads to a decrease in the affinity of the cassava starch film towards water.

Although the effect of incorporation of green tea extract was not significant, may there is a contribution of its components to the reduction of WVP. Siripatrawan and Harte [17] reported the influence the green tea extracts in chitosan films. The permeability coefficient decreased from 0.256 ± 0.023 to 0.087 ± 0.012 g mm m^{-2} d^{-1} kPa^{-1}, while the density increased from 1.21 ± 0.03 to 1.67 ± 0.03 g cm^{-3}, as the concentration of green tea increased from 0 to 20%. Incorporation of green tea into chitosan film caused the resulting films to become denser, with less water vapor permeability. Curcio and others [46] also observed the formation of covalent bonds between gallic acid antioxidant and chitosan, as verified by FTIR.

3.1.3. Thermogravimetric analyze. TG is a technique in which the change in mass of the sample is determined by the temperature and/or time [47]. Schlemmer investigated the

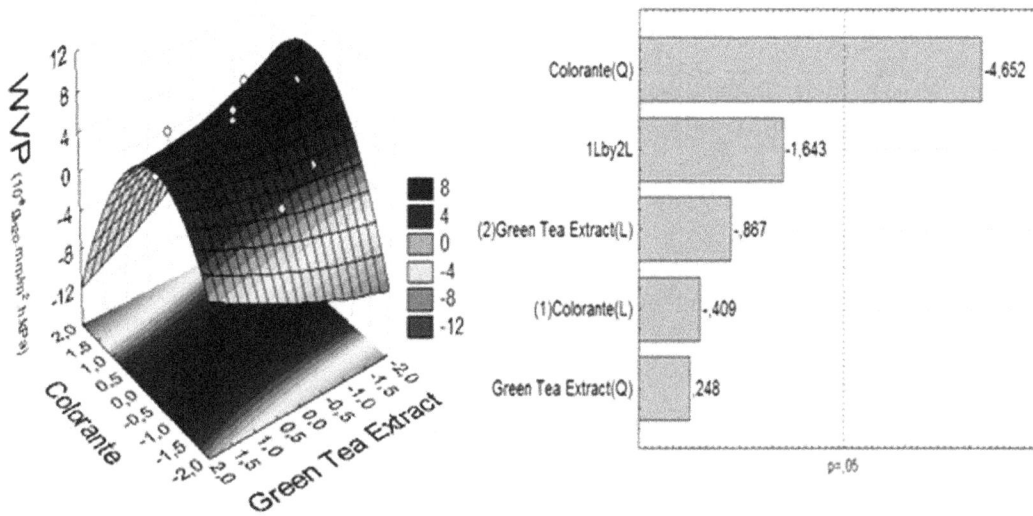

Figure 2. Response surface plot and Pareto chart for understanding water vapor permeability (WVP), as function on the content of colorant and GTE, with additives coded values (X_1 e X_2).

behavior of pure cassava starch by TGA [48]. The starch had only one stage of decomposition, with a Td thermal decomposition value of approximately 300°C. As illustrated in Table 2, we observed only two events: a very slight weight reduction approximately 100°C, attributed to lost water, reducing the mass of the material at least 12%; and a second weight reduction at approximately 300°C, which corresponds to the degradation of the starch and the highest percentage of mass loss at approximately 70% in the formulated films and in the control films at 600°C.

The analysis facilitates the knowledge about possible interactions between the matrix and the additives by providing information about the stability and applicability of the developed film. Figures 3a e b presents the TG and DTG curves of the starch films. Despite some differences in terms of the presence of bound water and residual mass, similar profiles in the curves of mass loss were observed of the analyzed material presents with the pure starch and control film, indicating good interactions in the polymer matrix.

The percentage of bound water varies from 9% to 13% and the main stage of degradation, which corresponds to ~70% of mass loss, starts at ~ 200°C and the residual mass, is 15% at 600°C. It is noteworthy that the humidity of the films ranged from 7 to 10%. Apparently the residual mass is related to the nature of the additives, impurities and inorganic components, and the conditions for analysis in an inert atmosphere (N_2), so there is no complete burning, even organic substances.

In the first event, when compared to the control, formulations with different concentrations of additives, showed a decrease in temperature beginning at 20%, except for the F8 formulation, and

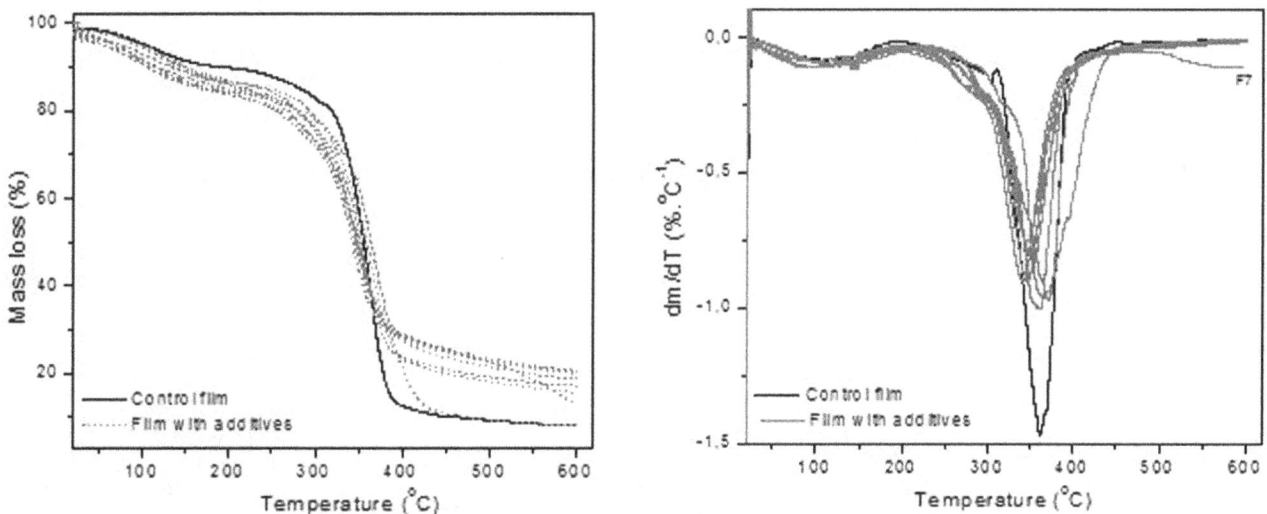

Figure 3. TG (A) and DTG (B) curves for the control film and the film with different additives concentrations.

formulations with high concentrations of additives (F3, F4 and F8) this temperature increased by 10% and the weight loss was over 60% compared to the mass loss of control. In the latter event, there was a reduction T_{onset}, Td_2 and 5%. However for the formulation F7 (0.03% colorant), the behavior was different from the others, where T_{onset} concerning the decomposition event was shifted to higher temperature suggesting greater interaction between the additive and the constituents of polymer matrix. The formulation F7 (0.03% CCN) when compared to control and other formulations in different parameters analyzed always showed characteristics that confirmed the cohesiveness of the matrix under these conditions.

Some discrepancies can also be observed in the method thermometric, as function probably due to small changes in mass of the samples used in the analysis and the high sensibility.

3.2. Stability of Additives Incorporated into Bio-based Films during the Storage of the Packaged Product

The green tea extract and colorant were incorporated in films of cassava starch as a source of active compounds, as evidenced in other studies, carotenoids and phenolic compounds are the two major groups of bioactive compounds with antioxidant activity [36], [49], [50]. According to the experimental design, different concentrations of these compounds were incorporated into the polymer matrix, and these compounds in films were monitoring as total carotenoids (TC), total polyphenols (PT) and flavonoids (TF) during storage of the packaged butter.

All the parameters evaluated in the films (TP, TF and TC) showed changes during the storage of the packaged product for 45 days, suffering significant reductions, the storage conditions can be attributed to migration to packaged product by oxidation and decomposition, can protect the packaged product (Table 3).

The formulations films showed reductions in levels of TP, TF and TC ranging from 11.10 to 53.00%, from 14.61 to 68.56% and from 11.77 to 91.52%, respectively (Table 3), demonstrating that even after 45 days of storage, part of the bioactive compounds of additions remained viable in the films. It is observed that the formulation F5 (only green tea extract) showed the greatest reduction in the content of TC (0.007 mg/g, 91.51%) of the formulation F7 (only carotenoic colorant) (9.84 mg/g 40.09%),

probably due to the stability of the pigment found in colorant compared to the extract, whereas for the content of polyphenols and flavonoids for the F7 formulation showed the greatest reductions in baseline (53.00% and 68.56%).

The formulations of the central points, on average, showed the lowest reduction in the content of TP (11.63%) followed by F6 (28.30%). The formulation F8 showed a smaller reduction in the content of TF (14.61%) and F1 showed high reduction (42.82%). The content of TC, F1 showed a lower reduction (11.76%) and F3 had the second highest (41.67%) of the initial content after 45 days of storage of the packed product (Table 3).

A similar behavior was observed in films containing mango and acerola pulps added as antioxidants, which were used to pack palm oil. In this case, a decrease after 45 days of storage ranged from 24.53 to 43.60% for TC, while decreases in polyphenols and vitamin C ranged from 17.80 to 36.12% and from 69.50 to 85.00%, respectively [8].

According to Wessling and others [13] tocopherol incorporated in polyethylene materials showed a resistance during 4 weeks of storage than those film controls containing butylated hydroxytoluene (BHT), which degraded in just in 1 week.

Experimental results for the different formulations of films used to package butter showed significant ($p<0.05$) differences in TP, TF and TC after 45 days of storage (Fig. 4). This resulted in a second-order polynomial equation, which represents the model equation used to evaluate the increase of TP (eq. 4) in films as a function of the concentrations of colorant (%, X_1) and green tea extract (%, X_2) and the interaction between them (X_1 and X_2). According to eq. 4, the increase in the TP concentration in films depends upon the interaction between both independent variables (X_1 and X_2), while the decrease in the TP concentration only depends upon these variables independently.

$$TP = 21.291 + 2.295X_1 + 10.650X_1^2 + 8.572X_2 \\ -4.676X_2^2 - 2.120X_1X_2 \tag{4}$$

$$R^2 = 0.88.$$

Table 3. Decrease of antioxidants in the film formulations with different concentrations of green tea extracts and carotenoic colorant after butter storage by 45 days.

Formulation Films	Real Values		Decrease TP (mg/100 g)	Decrease TF (mg/g)	Decrease TC (µg/g)
	Colorante (%)	Green Tea Extract (%)			
F1	0.01	0.70	17.10	16.43	0.55
F2	0.01	4.30	33.72	35.69	1.95
F3	0.04	0.70	17.89	5.55	11.44
F4	0.04	4.30	22.03	36.82	11.63
F5	0.00	2.50	26.24	20.00	0.83
F6	0.05	2.50	49.11	28.24	13.32
F7	0.03	0.00	0.44	0.49	6.59
F8	0.03	5.00	28.60	20.28	2.88
F9(c)	0.03	2.50	19.06	37.17	3.38
F10(c)	0.03	2.50	19.41	31.34	2.53
F11(c)	0.03	2.50	25.40	34.95	3.98

(c): Central Points. TC: Total Carotenoids. TP: Total Polyphenols. TF: Total Flavonoids.

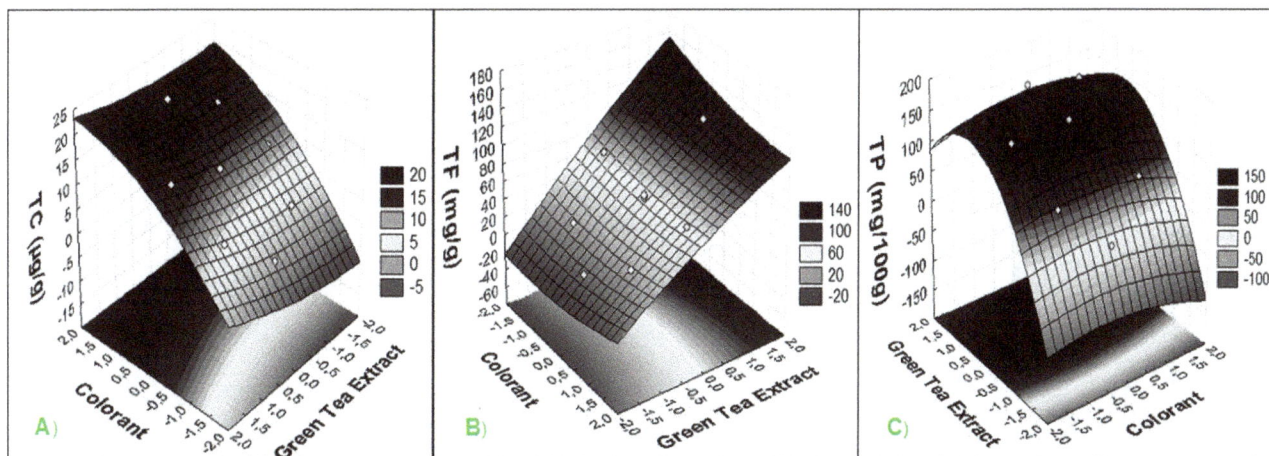

Figure 4. The response surface plot demonstrating the effect of incorporated additives to illustrate the effect of the increased factor on TP (mg/100 g), TF (mg/g) and TC (µg/g) film preparation after 45 days of butter storage.

$$TF = 34.487 + 0.237X_1 - 3.592X_1^2 + 9.813X_2$$
$$- 10.458X_2^2 + 3.001X_1X_2 \tag{5}$$

$$R^2 = 0.86.$$

$$TC = 3.295 + 4.780X_1 + 2.011X_1^2 - 0.456X_2$$
$$+ 0.842X_2^2 - 0.300X_1X_2 \tag{6}$$

$$R^2 = 0.96.$$

3.3. Monitoring Packaged butter during Storage

As expected, the butter packaged in different formulations of biodegradable films incorporated with different concentrations of carotenoic colorant and green tea extract (F1 through F11) showed a change in the initial content of total carotenoids and an differentiate increase in the peroxide index. Butter packaged in the active films had a smaller increase in PI ($p < 0.05$) when compared with the three controls, the peroxide index increased 692.00% in exposed butter (C1, unpackaged), 559.21% in packaged formulation without additives (C2) and 583.73% in LDPE (C3), indicating the effectiveness of the additives as antioxidants (Table 4). This effect appears to be dependent on the additives concentrations because the butter packed in films (F8, F2 and F4 - high concentrations of green tea extract) had a greater oxidation (PI 4.330, 4.307 and 4.221 meq/kg; 456.67, 451.45 e 451.14%) compared to F6-butter (average concentration of additives) (PI 2.576 meq/kg; 231.57%) ($p < 0.05$), under the same storage conditions after 45 days (Table 4).

The peroxide index (PI) of the butter packaged in formulation F8 (0.03% colorant and 5% green tea extract) increased in 456.67% after 45 days of storage, while the butter packed in the formulation F6 (0.05% colorant and 2.5% green tea extract) showed lower PI (231.57%) ($p < 0.05$), demonstrating that as increase the concentration of active compounds incorporated in

packaging, enhances the oxidation of the packed product (Table 4).

The butter packed in film formulation F7 (only colorant) had a higher PI (PI = 3.980 meq/kg; 412.27%) than butter packed in formulation F5 (only green tea extract) (PI = 3.294 meq/kg; 330.34%). This demonstrates that a greater protective effect of phenolic compounds exists, and that the green tea extract was more effective than the colorant ($p < 0.05$) when used individually (Table 4).

The packaged butter showed an increase in total carotenoids values after 45 days of storage, that was proportional to the concentration of colorant incorporated in the films ($r^2 = 0.87$). For example, butter packed in F6 (maximum concentration of colorant) showed greater increase in TC (154.9%) over the butter contained in the formulations F1 and F2 (0.01% only colorant) of 60 and 41%, respectively, under the same storage conditions.

In this context, the colorant (%, X_1) and green tea extract (%, X_2) incorporated into the films in different proportions, and increase in the PI content (meq/kg) of butter after 45 days of storage can be expressed using a second-order polynomial equation (eq. 7). The increase in this parameter depends upon the concentration of the colorant and green tea extract and the interaction of these two factors. Figure 5 shows the response surface graph, illustrating the increase of butter PI (Fig. 5A) and TC (Fig. 5B) packaged with different film formulations. The graph indicates that the point of maximum PI increase corresponds to the maximum concentration of both additives, and the point of minimum PI occurs at the average concentration of both additives ($p < 0.05$).

The Pareto chart of standardized effect estimates (absolute values) showed the magnitude of the positive effect of green tea extract concentration (quadratic and linear function) on increasing PI, demonstrating that depending on concentration, this additive has a favorable influence on that variable. However, for the day 45, the Pareto chart demonstrated that the linear function of the colorant variable showed a negative effect on the increase in PI, (Fig. 5A).

Table 4. Alterations of the butter packed in different film formulations after 7, 15, 30, and 45 days of storage.

Formulation Films	Real Values		Increase PI (%)				TC, Increase (+), decrease (−), (%)
	Colorante (%)	Green Tea Extract (%)	0-7	0-15	0-30	0-45	0-45
C1	–	–	271.1	424.9	474.4	692.1	−12.01
C2	–	–	208.1	327.8	379.3	559.2	−3.72
C3	–	–	244.1	374.6	442.8	583.7	−15.13
F1	0.01	0.70	90.4	204.2	286.8	365.4	+60.00
F2	0.01	4.30	96.2	261.6	315.7	451.4	+41.62
F3	0.04	0.70	79.6	184.2	261.0	328.2	+111.90
F4	0.04	4.30	96.4	227.9	305.2	451.1	+86.51
F5	0.00	2.50	79.4	147.1	253.4	330.3	−15.22
F6	0.05	2.50	38.5	102.3	205.6	231.6	+154.91
F7	0.03	0.00	88.9	209.1	295.1	412.3	+66.00
F8	0.03	5.00	113.4	258.9	342.3	456.7	+111.80
F9(c)	0.03	2.50	81.2	150.7	258.9	319.3	+90.61
F10(c)	0.03	2.50	76.6	134.3	236.5	284.8	+78.70
F11(c)	0.03	2.50	77.6	133.9	241.9	297.1	+109.90

(c) Central points. C1 = unpackaged; C2 = starch films without additives, C3 = package in LPDE.

A

B

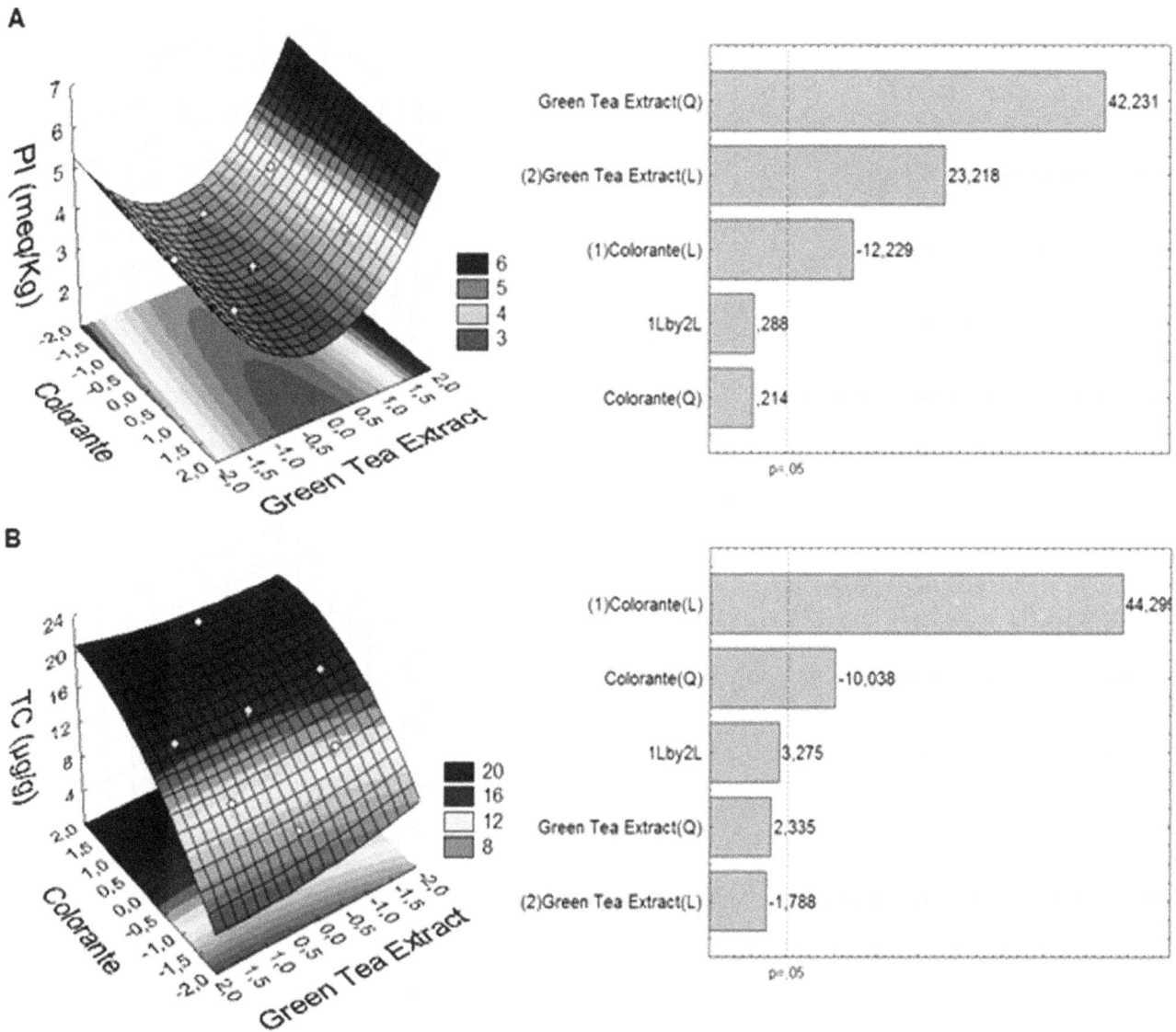

Figure 5. Response surface plot and Pareto chart for understanding the increase of peroxide index (PI) values and total carotenoids (TC) after 45 days of butter storage.

$$PI(day\ 45) = 2.28 - 0.109X_1 - 0.0003X_1^2 + 0.387X_2 \\ + 0.697X_2^2 + 0.0005X_1X_2 \tag{7}$$

$$R^2 = 0.98.$$

$$TC(day\ 45) = 7.53 + 2.916X_1 - 0.506X_1^2 + 0.137X_2 \\ - 0.181X_2^2 - 0.029X_1X_2 \tag{8}$$

$$R^2 = 0.86.$$

In equation 8, it is verified that the carotenoid content of the butter only depends on the linear term of the colorant concentration ($p < 0.05$) with a correlation coefficient of 86%, obtaining a first order model, which is suitable for describe the

results shown in the response surface. This demonstrates the migration of carotenoid packaging for butter. This result was desired, as for the mechanism of active packaging is efficient active compound must migrate to the packaged product thus exerting the desired action.

These results are in agreement with some studies that have shown that oxygen can permeate through the film, reacting preferentially with some compounds present in the film formulation. This allows the packaged product to be preserved for a longer period of time [16]. Justifying the incorporation of antioxidant additives in packaging, the colorant and green tea extract, as additives in film-forming dispersions, are effective in preserving a packaged product against oxidation.

3.4. Correlations between parameters of the Bio-based Films and Packing Butter during Storage

The results of this study suggest that the protection of packaged products against oxidation can be attributed to the concentration dependent radical scavenging activity of antioxidant compounds present in film forming dispersions.

Considering the results for the PI values from the packaged product, the protective effect against lipid oxidation is likely a consequence of a physical process because of the carotenoids content. This is especially true of the colorant because the butter packed in films containing this additive showed a lower rate of oxidation compared to the controls.

When comparing the results of the 11 formulations, it appears that the butter wrapped in the formulations that had higher levels of green tea had a significant increase in peroxide index (about 450%, $p<0.05$). This result is likely a result of pro-oxidant actions of the bioactive compounds (Table 4), as other studies have reported that high concentrations of these compounds can result in pro-oxidant activity under certain conditions [49], [50] when there is an imbalance between oxidant and antioxidant molecules, which results in the induction of oxidative damage by free radicals, or if there is no equivalent concentration of the other antioxidants to regenerate the radical. These oxidative stress mechanisms are not yet understood [49], [50], [51], [52].

3.5. Films Applications

The applications of starch-based plastics can include biodegradable packaging in the food industry and pharmaceutical, particularly as coatings to enhance the shelf life of fatty products [53], [54], [55]. The commercial application of films containing food starch depends of improvements on the mechanical properties and the water vapor permeability of these materials, which are limiting properties [53]. Researches are needed to proposing improvements, development new products by combination this biopolymer with new materials, biopolymeric blends and composites, as biodegradable nanoparticles, for to improve coating properties and increase potential applications of food packaging [1], [53], [55], [56]. Biodegradable films can be applied to food products, such as a barrier for fat, butter and margarine [54], as active biodegradable films that can protect the packaged

product. The functionality of the product is related also to the how well the food product resists deterioration.

Conclusions

This study indicated that an active film from cassava starch films could be achieved by incorporation with colorant and green tea extract, as a source of bioactive compounds, into a cassava starch packaging for lipid foods. Addition of colorant and green tea extract improved significantly functional properties mechanical, water vapor barrier and antioxidant properties of the resulting films. These changes, as verified by results, could be attributed to the interactions between functional groups of starch, colorant and green tea extract polyphenol compounds, cohesive molecular reorganization, which resulted in a reduction in permeability to water vapor and providing greater rigidity to the material compared to the controls. The results provide an oxidative protection in packaged butter, by decrease peroxide index, when using these film additives. However, high colorant and green tea contents can act as pro-oxidant agents, which suggest that these additives should be used at low concentrations. Nevertheless, further studies are required before using this film as an active packaging for food products, to verify this activity and the effects of incorporating antioxidant additives into other matrixes.

Acknowledgments

The authors would like to acknowledge Cargill Agrícola SA and Chr Hassen SA for the donated ingredients and the Brazilian Committee for Postgraduate Courses in Higher Education (FAPESB) for the scholarships granted, and National Council for Scientific and Technological Development for the financial incentive (CNPq 505831/2008-2).

Author Contributions

Conceived and designed the experiments: KKNCLP. Performed the experiments: ACVC JCPS DJA. Analyzed the data: KKNCLP COS. Contributed reagents/materials/analysis tools: KKNCLP JID. Wrote the paper: KKNCLP.

References

1. Khwaldia K, Arab-Tehrany E, Desobry S (2010) Biopolymer coatings on paper packaging materials. Comprehensive Reviews in Food Science and Food Safety 9: 82–91.
2. Moradi M, Tajik H, Rohani SMR, Oromiehie AR, Malekinejad H, et al. (2012) Characterization of antioxidant chitosan film incorporated with *Zataria multiflora* Boiss essential oil and grape seed extract. Food Science Technology 46: 477–84.
3. Vieira MGA, da Silva MA, Santos LO, Beppu MM (2011) Natural-based plasticizers and biopolymer films: A review. European Polymer Journal 47: 254–63.
4. Bonilla J, Atarés L, Vargas M, Chiralt A (2013) Properties of wheat starch film-forming dispersions and films as affected by chitosan addition. Journal Food Engineering 114: 301–12.
5. FAOSTAT - Food and Agriculture Organization of The United Nations, Production Crops FAO Codes (2012) http://faostat.fao.org/site/567/DesktopDefault.aspx?PageID=567.
6. Gimenez B, Lacey AL, Pérez-Santín E, López-Caballero ME, Montero P (2013) Release of active compounds from agar and agar-gelatin films with green tea extract. Food Hydrocolloids 30: 264–71.
7. Vermeiren L, Devlieghere F, Van Beest M, de Kruijf N, Debevere J (1999) Developments in the active packaging of foods. Trends Food Science Technology 10: 77–86.
8. Souza CO, Silva LT, Silva JR, Lopez JA, Veiga-Santos P, et al. (2011) Mango and acerola pulps as antioxidant additives in cassava starch bio-based film. Journal Agricultural Food Chemistry 59: 2248–54.
9. Falguera V, Quintera JP, Jimenez A, Munoz JA, Ibarz A (2011) Edible films and coatings: structures, active functions and trends in their use Trends. Food Science Technology 22: 292–303.
10. Miller KS, Krochta JM (1997) Oxygen and aroma barrier properties of edible films: A review. Trends Food Science Technology 8: 228–37.
11. Finley JW, Given Jr P (1986) Technological necessity of antioxidants in the food industry. Food Chemical Toxicology. 24: 999–1006.
12. Nerín C, Tovar L, Salafranca J (2008) Behaviour of a new antioxidant active film versus oxidizable model compounds. Journal Food Engineering 84: 313–20.
13. Wessling C, Nielsen T, Andres L (2000) The influence of atocopherol concentration on the stability of linoleic acid and the properties of low-density polyethylene. Packaging Technology and Science 13: 19–28.
14. Heumann BF (1990) Antioxidants: Firms seeking products they can label as 'natural'. INFORM. 1(12) http://aocs.files.cms-plus.com/inform/1990/12/1002.pdf.
15. Song HS, Bae JK, Park I (2013) Effect of heating on DPPH radical scavenging activity of meat substitute. Preventive Nutrition and Food Science 18(1): 80–84.
16. Grisi CVB, Veiga-Santos P, Silva LT, Cabral-Albuquerque EC, Druzian JI (2008) Evaluation of the viability of incorporating natural antioxidants in bio-based packagings. In Food Chemistry Research Developments; Nova Science Publishers 1: 1–11.
17. Siripatrawan U, Harte BR (2010) Physical properties and antioxidant activity of an active film from chitosan incorporated with green tea extract. Food Hydrocolloids 24: 770–5.
18. Giménez B, Moreno S, López-Caballero ME, Montero P, Gómez-Guillén MC (2013) Antioxidant properties of green tea extract incorporated to fish gelatin films after simulated gastrointestinal enzymatic digestion. LWT - Food Science and Technology 53(2): 445–451.
19. Martín-Diana AB, Rico D, Barry-Ryan C (2008) Green tea extract as a natural antioxidant to extend the shelf-life of fresh-cut lettuce. Innovative Food Science and Emerging Technologies 9: 593–603.
20. He Y, Shahidi F (1997) Antioxidant activity of green tea and its catechins in a fish meat model system. Journal Agricultural Food Chemistry 45: 4262–6.
21. Anesini C, Ferraro GE, Filip R (2008) Total polyphenol content and antioxidant capacity of commercially available tea (*Camellia sinensis*) in Argentina. Journal Agricultural Food Chemistry 56: 9225–9.
22. Gramza A, Khokharb S, Yokob S, Gliszczynska-Swigloc A, Hesa M, et al. (2006) Antioxidant activity of tea extracts in lipids and correlation with

polyphenol content. European Journal of Lipid Science and Technology 108: 351–62.

23. Wanasundara UN, Shahidi F (1998) Antioxidant and pro-oxidant activity of green tea extracts in marine oils. Food Chemistry 63: 335–42.

24. Rice-Evans CA, Miller NJ, Paganga G (1996) Structure-antioxidant activity relationships of flavonoids and phenolic acids. Free Radical Biology and Medicine 20: 933–956.

25. Re R, Pellegrini N, Proteggente A, Pannala A, Yang M, et al. (1999) Antioxidant activity applying an improved ABTS radical cation decolorization assay. Free Radical Biology and Medicine 26: 1231–7.

26. Montenegro MA, Rios AO, Mercadante AZ, Nazareno MA, Borsarelli CD (2004) Model studies on the photosensitized isomerization of bixin. J Agric Food Chem 52: 367–73.

27. Boon CM, Ng MH, Choo YM, Mok SL (2013) Super, red palm, and palm oleins improve the blood pressure, heart size, aortic media thickness and lipid profile in spontaneously hypertensive rats. Plos One 8: 2, e55908.

28. Montenegro MA, Nunes IL, Mercadante AZ, Borsarelli CD (2007) Photo-protection of vitamins in skimmed milk by an aqueous soluble lycopene-gum arabic microcapsule. Journal Agricultural Food Chemistry 55: 323–9.

29. Azeredo HMC, Mattoso LHC, Wood D, Williams TG, Avena-Bustillos RJ, et al. (2009) Nanocomposite edible films from mango Puree reinforced with cellulose Nanofibers. Journal Food Science 74: 31–5.

30. Nishiyama MF, Costa MF, Costa AM, Souza CG, Bôer CG, et al. (2010) Brazilian green tea (Camellia sinensis var assamica): effect of infusion time, mode of packaging and preparation on the extraction efficiency of bioactive compounds and on the stability of the beverage. Food Science Technology 30: 191–6.

31. ASTM Standards (1989) E96–80: Standard test methods for water vapor transmission of materials. West Conshohocken, Pennsylvania, USA.

32. Veiga-Santos P, Suzuki CK, Cereda MP, Scamparini ARP (2007) Microstructure and color of starch-gum films: Effect of additives and deacetylated xanthan gum. Food Hydrocolloids 19: 1064–73.

33. Gontard N, Gilbert S, Cuq JL (1993) Water and glycerol as plasticizer effect mechanical and water vapor barrier properties of an edible wheat gluten film. Journal Food Science 58: 206–11.

34. ASTM - American Society for Testing and Materials (1995) Designation E96–95: Standard Method for Water Vapor Transmission of Materials. Philadelphia: ASTM. (Annual Book of ASTM Standards).

35. Association of Official Analytical Chemists (2000) Official Methods of Analysis Cd 8b-90; AOAC: Gaithersburg, MD.

36. Passoto JA, Penteado MVC, Mancini-Filho J (1998) Activity of β-carotene and vitamin A: A comparative study with synthetic antioxidant. Food Science Technology 18: 3: 624–32.

37. Silva SR, Mercadante AZ (2002) Carotenoid composition of fresh yellow passion fruit (Passiflora edulis). Food Science and Technology 22: 254–8.

38. Lee ES, Lee HE, Shin JY, Yoon S, Moon JO (2003) The flavonoid quercetin inhibits dimethylnitrosamine-induced liver damage in rats. Journal Pharmacy Pharmacology 55(8): 1169–1174.

39. Xiong HG, Tang S, Tang H, Zou P (2008) The structure and properties of a starch-based biodegradable film. Carbohydrate Polymers 71 (2): 263–268.

40. Jansson A, Thuvander F (2004) Influence of thickness on the mechanical properties for starch films. Carbohydrate Polymers 56: 499–503.

41. Zhu F, et al. (2009) Effect of phytochemical extracts on the pasting, thermal, and gelling properties of wheat starch. Food Chemistry 112(4): 919–923.

42. Avérous L, Fringant C, Moro L (2001) Starch-based biodegradable material suitable for thermoforming packaging. Starch 53: 368–71.

43. Jansson A, Thuvander F (2004) Influence of thickness on the mechanical properties for starch films. Carbohydrate Polymers 56: 499–503.

44. Siracusa V (2012) Food packaging permeability behaviour: A report. International Journal of Polymer Science 2012: 1–11.

45. Immel S, Lichtenthaler FW (2000) The Hydrophobic Topographies of Amylose and its Blue Iodine Complex. Starch/Stärke 52(1): 1–8.

46. Curcio M, Puoci F, Iemma F, Parisi OI, Cirillo G, et al. (2009) Covalent insertion of antioxidant molecules on chitosan by a free radical grafting procedure. Journal Agricultural Food Chemistry 57: 5933–8.

47. Gumel AM, Annuar MSM, Heidelberg T (2012) Biosynthesis and character-ization of polyhydroxyalkanoates copolymers produced by Pseudomonas putida Bet001 isolated from palm oil mill effluent. Plos One, 7(9): e45214.

48. Schlemmer D, Oliveira ER, Sales MJA. Polystyrene/thermoplastic starch blends with different plasticizers. Journal of Thermal Analysis and Calorimetric 87(3): 635–638.

49. Cao G, Sofic E, Prior RL (1997) Antioxidant and prooxidant behavior of flavonoids: structure-activity relationships. Free Radical Biology and Medicine 22: 749–60.

50. Heim K, Tagliaferro AR, Bobilya DJ (2002) Flavonoid antioxidants: chemistry, metabolism and structure-activity relationships. Journal Nutrition Biochemistry 13: 572–84.

51. Xiao X, Shi D, Liu L, Wang J, Xie X, et al. (2011) Quercetin suppresses cyclooxygenase-2 expression and angiogenesis through inactivation of P300 signaling. Plos One 6(8): e22934.

52. Munoz-Munoz JL, García-Molina F, Molina-Alarcón M, Tudela J, García-Cánovas F, et al. (2008) Kinetic characterization of the enzymatic and chemical oxidation of the catechins in green tea. Journal Agricultural Food Chemistry 56: 9215–24.

53. Silva JBA, Pereira FV, Druzian JI (2012) Cassava starch-based films plasticized with sucrose and inverted sugar and reinforced with cellulose nanocrystals. Journal of Food Science 77: 14–19.

54. Haugaard VK (2001) Potential food applications of biobased materials. An EU-concerted action project. Starch/Stärke 5: 189–200.

55. Singh A, Sharma PK, Malviya R (2011) Eco friendly pharmaceutical packaging material. World Applied Sciences Journal 14 (11): 1703–1716.

56. Wei B, Xu X, Jin Z, Tian Y (2014) Surface Chemical Compositions and Dispersity of Starch Nanocrystals Formed by Sulfuric and Hydrochloric Acid Hydrolysis. Plos One 9(2): e86024.

11

Modulation of PPARγ Provides New Insights in a Stress Induced Premature Senescence Model

Stefania Briganti[◦], Enrica Flori[◦], Barbara Bellei, Mauro Picardo*

Laboratory of Cutaneous Physiopathology, San Gallicano Dermatologic Institute, Istituto di Ricovero e Cura a Carattere Scientifico, Rome, Italy

Abstract

Peroxisome proliferator-activated receptor gamma (PPARγ) may be involved in a key mechanism of the skin aging process, influencing several aspects related to the age-related degeneration of skin cells, including antioxidant unbalance. Therefore, we investigated whether the up-modulation of this nuclear receptor exerts a protective effect in a stress-induced premature senescence (SIPS) model based on a single exposure of human dermal fibroblasts to 8-methoxypsoralen plus + ultraviolet-A-irradiation (PUVA). Among possible PPARγ modulators, we selected 2,4,6-octatrienoic acid (Octa), a member of the parrodiene family, previously reported to promote melanogenesis and antioxidant defense in normal human melanocytes through a mechanism involving PPARγ activation. Exposure to PUVA induced an early and significant decrease in PPARγ expression and activity. PPARγ up-modulation counteracted the antioxidant imbalance induced by PUVA and reduced the expression of stress response genes with a synergistic increase of different components of the cell antioxidant network, such as catalase and reduced glutathione. PUVA-treated fibroblasts grown in the presence of Octa are partially but significantly rescued from the features of the cellular senescence-like phenotype, such as cytoplasmic enlargement, the expression of senescence-associated-β-galactosidase, matrix-metalloproteinase-1, and cell cycle proteins. Moreover, the alterations in the cell membrane lipids, such as the decrease in the polyunsaturated fatty acid content of phospholipids and the increase in cholesterol levels, which are typical features of cell aging, were prevented. Our data suggest that PPARγ is one of the targets of PUVA-SIPS and that its pharmacological up-modulation may represent a novel therapeutic approach for the photooxidative skin damage.

Editor: Andrzej T. Slominski, University of Tennessee, United States of America

Funding: This work has been partly supported by a research grant provided by Giuliani Pharma, Milan, Italy. The funders had no role in study design, data collection and analysis, decision to publish, or preparation of the manuscript. No additional funding received for this study.

Competing Interests: The authors received an unrestricted research grant from Giuliani Pharma Milan, thus they declare a financial competing interest.

* Email: picardo@ifo.it

◦ These authors contributed equally to this work.

Introduction

Ultraviolet (UV) radiation elicits premature aging of the skin and cutaneous malignancies [1]. UVA rays generate reactive oxygen species (ROS) via photodynamic actions [2], resulting in skin degeneration and aging [3,4] and, in particular, oxidative damage to lipids, proteins, and DNA [5–7]. Moreover, UVA-induced ROS regulate the gene expression of matrix metallo-proteinases (MMPs), which are the main enzymes responsible for dermal extracellular matrix degradation [8–10]. As a result, the incidence of skin photoaging and skin cancer dramatically increases with increased exposure to UVA rays [11]. To protect its structure against UV, skin has developed several defence systems which include pigmentation, antioxidant network and neuro-immune-endocrine functions, which are tightly networked to central regulatory system and are involved in the protection and in the maintenance of global homeostasis, through the production of cytokines, neurotransmitters, neuroendocrine hormones [12]. Thus, UV would stimulate production and secretion of α-melanocyte-stimulating hormone, proopiomelanocortin-derived β-endorphin, adrenocorticotropin, corticotrophin releasing factor, and glucocorticoids [13]. An unbalance between pro-inflammato-ry or anti-inflammatory responses activated by these mediators may be related to cellular degeneration in aged skin.

A way to investigate *in vitro* aging process is the study of cellular senescence, a loss of proliferative capacity attributed to telomere shortening during cell replication or after exposure to pro-oxidant stimuli and closely interconnected with aging, longevity and age-related disease [14,15]. Due to the key role of oxidative stress in the photoaging process, the change of proliferating skin cells to photo-aged cells resembles premature senescence under conditions of artificially increased ROS levels. Consistently, stress-induced premature senescence (SIPS) models can represent useful tools with which to investigate the biological and biochemical mecha-nisms involved in photo-induced skin damage and photocarcino-genesis and to evaluate the potential protective effects of new molecules. SIPS can be induced in human skin dermal fibroblasts (HDFs) by a single subcytotoxic exposure to UVA-activated 8-methoxypsoralen (PUVA) [16], widely used in the treatment of different skin disorders like psoriasis, T-cell lymphoma and other inflammatory skin disorders. We previously reported that oxida-tive stress and cell antioxidant capacity are involved in both the induction and maintenance of PUVA-SIPS and supplementation with low-weight antioxidants abrogated the increased ROS generation and rescued fibroblasts from the PUVA-dependent changes in the cellular senescence phenotype [17]. Moreover, PUVA treatment induced a prolonged expression of interstitial collagenase/MMP-1, leading to connective tissue damage, a hallmark of premature aging [17], confirming this experimental model as a useful tool to investigate in vitro the mechanisms of skin ageing. The function of nuclear receptors has been reported to be involved in the molecular mechanisms controlling the aging process. The peroxisome proliferator-activated receptor (PPAR) family regulates the function and expression of complex gene networks, especially involved in energy homeostasis and inflammation [18–20], and modulate the balance between MMP activity and collagen expression to maintain skin homeostasis [21]. In particular, PPARγ has been implicated in the oxidative stress response, an imbalance between antithetic pro-oxidation and antioxidation, and in this delicate and intricate game of equilibrium, PPARγ stands out as a central player specializing in the quenching and containment of damage and fostering cell survival. Moreover, PPARγ activation has been reported to restore the "youthful" structure and function of mitochondria that are structurally and functionally impaired by excessive oxidant stress [22]. However, PPARγ does not act alone, but is interconnected with various pathways, such as the nuclear factor erythroid 2-related factor 2 (NRF2), Wnt/β-catenin, and forkhead box protein O (FoxO) pathways [23]. PPARγ activation has been reported to be a link to melanocyte differentiation pathways, as suggested by the ability of PPARγ ligands to regulate Microphthalmia-associated transcription factor gene and Wnt/β-catenin levels, promoting differentiation and growth arrest of melanoma cells [24]. Given these features, PPARγ is emerging as an important regulator of skin photodamage.

Among anti-aging agents, topical all-trans-retinoic acid (AtRA) inhibits MMP expression [25] and has a significant diminishing effect on UV-induced photoaging, such as wrinkles, water loss, and reduced wound healing [26]. However, irritant reactions, such as burning, scaling, or dermatitis, limit the acceptance of AtRA by patients [27]. To minimize these side effects, various novel drug delivery systems have been developed; in addition, screening to discover new natural or synthetic retinoid-like molecules has been conducted. Psittacofulvins are a mixture of polyenals identified exclusively in the red plumage of the Ara macao [28], indicating that these compounds are produced at the feather bulb for defense against environmental insults. Parrodienes, congeners of psittaco-fulvins that are considered retinoid-like molecules, as they possess a polyene structure and an alcohol functional group, have been synthesized to investigate the biological effects of psittacofulvins. Studies have shown that parrodienes possess antioxidant [29] and anti-inflammatory activities and are able to inhibit the lipoperox-idation of cell membranes induced by CCl_4 [30]. Among the parrodiene family members, 2,4,6-octatrienoic acid (Octa) pro-motes melanogenesis and antioxidant defense in normal human melanocytes, and its mechanism of action involves the modulation of PPARγ [31]. We added Octa to PUVA-treated HDFs to evaluate Octa's ability to counteract PUVA-SIPS and to investigate whether PPARγ is involved in photo-induced cell senescence.

Materials and Methods

Standards and reagents

Dulbecco's modified Eagle's medium (DMEM), penicillin and streptomycin were purchased from Gibco, Life Technologies Italia, Milan, Italy. Octa was furnished by Giuliani Pharma, Milan, Italy. Crystalline 8-methoxypsoralen (8-MOP), dimethyl-sulfoxide (DMSO), 3-(4,5 dimethylthiazol)-2,5-diphenyl tetrazoli-

um bromide (MTT), butylated hydroxytoluene (BHT), 6-hydroxy-2,5,7,8-tetramethylchromane-2-carboxylic acid (Trolox), N-ethyl-maleimide (NEM), thiosalicylic acid (TSA), sodium methoxide, potassium hydroxide (KOH), retinol (ReOH) and all-trans retinoic acid AtRA were from Sigma-Aldrich, Milan, Italy. 2′,7′-dichlor-odihydrofluorescein diacetate ($DCFH_2$-DA) was from Molecular Probes (Eugene, OR, USA). All organic solvents used were of HPLC-grade.

Cell culture and treatments

Human Dermal Fibroblasts (HDFs) were derived from neonatal foreskin of healthy male caucasian individuals (n = 3), phenotype III, ranged from 4 to 7 years old and were isolated as previously described [16]. Cells were grown in DMEM supplemented with 10% FBS, penicillin (100 U/ml) and streptomycin (100μg/ml) and used between passage 2 and 8. Institutional Research Ethic committee (Istituti Fisioterapici Ospitalieri) approval was obtained to collect sample of human material for research. The Declaration of Helsinki Principle was followed and due to the fact that the study included children participants their parents gave written informed consent. Stock solutions (10 mM) of Octa was prepared in DMSO. The maximum concentration of Octa, without affecting cell viability or proliferation, was determined by MTT assay and Trypan blue exclusion test (data not shown). Moreover we did not observe any relevant modification of protein content in Octa treated cells (data not shown).

PUVA treatment

8-MOP (25 ng/ml) was added to the cell culture medium overnight. Cell were washed twice with phosphate-buffered saline (PBS) containing 8-MOP 25 ng/ml. HDFs were irradiated at a dose of 6 J/cm^2 using a Bio-Sun irradiation apparatus (Vilbert Lourmat, Marnè-la-Vallée, France) with maximum emission at 365nm in the UVA spectral region (340 to 450 nm). Following irradiation, PBS was replaced by fresh medium which was changed every three days. Octa was diluted in cell culture medium at a final concentration of 2 μM and added to HDFs immediately following PUVA and twice a week thereafter.

Cell morphology

To monitor fibroblast morphology after PUVA treatment, fibroblasts were fixed and stained with Comassie brilliant Blue as previously described [32].

Senescence associated beta-galactosidase (SA-β-gal) staining

SA-β-gal staining was performed as previously described [33]. The proportion of cells positive for SA-β-gal activity are given as percentage of the total number of fibroblasts counted in each dish. Triplicates were performed. The stained dishes were photo-graphed, positive fibroblasts counted and the results expressed as mean ± S.D. of SA-β-gal positive fibroblasts in% of total fibroblast number.

MMP-1 ELISA

MMP-1 total release (proMMP-1, active MMP-1 and MMP-1/TIMP-1 complex) was measured using an Human, Biotrack ELISA immunoassay (Amersham Pharmacia Biotech, Milan, Italy), according to the manufacturer's instructions, and was normalized against protein concentration, determined by Quick Start Bradford Dye Reagent (Bio-Rad, Hercules, CA, USA). The results are the mean ± S.D. of experiments performed in each donor (n = 3) in triplicate.

Determination of ROS generation

The generation of intracellular ROS was determined by employing the cell-permeable fluorogenic probe DCFH-DA. In brief, DCFH-DA is diffused into cells and deacetylated by cellular esterases to non fluorescent $2', 7'$-dichlorofluorescin (DCFH), which is rapidly oxidazed to highly fluorescent $2', 7'$-dichlorofluorescein (DCF). The fluorescence intensity of the supernatant was measured with a multiplate reader (DTX 880 Multimode Detector; Beckman Coulter Srl, Milan, Italy) at 485nm excitation and 535 nm emission. Cellular oxidant levels were expressed as relative DCF fluorescence per microgram of protein. The results are the mean $\pm$ S.D. of experiments performed in each donor (n = 3) in triplicate.

JC-1 assay for mitochondrial membrane potential

Mitochondrial trans-membrane potential $(\Delta \Psi_m)$ was assessed in live HDFs using the lipophilic cationic probe $5,5',6,6'$-tetrachloro-$1,1',3,3'$-tetraethylbenzimidazolcarbocyanine iodide (JC-1, Molecular Probes). For quantitative fluorescence measurements, cells were rinsed once after JC-1 staining and scanned with a Flow cytometer (FACS-Calibur, Becton Dickinson, San José, CA, USA) at 485 nm excitation, and 530 and 570 nm emission, to measure green and orange-red JC-1 fluorescence, respectively. Results of experiments performed in each donor (n = 3) in triplicate are expressed as percentage of variation ($\pm$ SD) respect to control values of the orange-red/green fluorescence intensity ratio.

Catalase (Cat) activity

Fibroblasts were lysed in PBS by repeated freezing and thawing, in the presence of protease inhibitors. Cat activity was determined by spectrophotometric monitoring the rate of disappearance of H_2O_2 at 240 nm [34]. A standard curve was obtained with bovine catalase (Sigma-Aldrich, Srl Milan, Italy). Units were normalized for protein content. Results of experiments performed in each donor (n = 3) in triplicate are given as % of relative units of Cat per mg protein $\pm$ S.D.

Biological Antioxidant Potential (BAP) Assay

BAP was measured with a commercially available assay kit (Diacron srl, Grosseto Italy). The principle of the test is to measure the color change upon reduction of Fe^{3+} to Fe^{2+} by the reducing components in the sample. The optical density was measured at 505 nm by a microplate reader. The data were obtained by interpolating the absorbance on a calibration curve obtained with Trolox (30–1000 µM). Results of experiments performed in each donor (n = 3) in triplicate are expressed as medium percentage of variation ($\pm$ S.D.) respect to control values of untreated cells.

Glutathione (GSH) measurement

GSH levels were determined in cell lysates by high-performance liquid chromatography-mass spectrometry (HPLC-MS) as previously described [35]. The mean value of experiments performed in each donor (n = 3) in triplicate is given as GSH in nmol/mg of total protein $\pm$ S.D.

Alpha-tocopherol (α-Toc) analysis

Cells were extracted in hexane:ethanol 3:1 in the presence of γ and δ tocopherol (Sigma-Aldrich, Milan, Italy), as internal standards, and the tocopherols were analysed by gas chromatography-mass spectrometry (GC-MS) as previously described [36]. The mean value of experiments performed in each donor (n = 3) in duplicate is expressed as nanogram per milligram of proteins $\pm$ S.D.

Assessment of cell membrane phospholipids polyunsaturated fatty acids

Cell pellets were extracted twice in chloroform/methanol (2:1, v:v) in the presence of tricosanoic acid methyl ester (Sigma Aldrich, Milan Italy), as internal standard. Fatty acids of cell total lipid extract were analysed by GC-MS on a capillary column (FFAP, 60 m×0.32 µm×0.25 mm, Hewlett Packard, Palo Alto, CA, USA), as previously reported [36]. Results of experiments performed in each donor (n = 3) in triplicate are given as mean percentage $\pm$ S.D.

Conjugated Dienes

Conjugated diene level was evaluated as described by Kurien and Scofield [37] with modification. Cells were extracted with 3 ml chloroform/methanol (2:1, v/v). After centrifugation at 3,000 rpm for 15 min, 2 ml of organic phase was transferred into another tube and dried at 45°C. The dried lipids were dissolved in 2 ml of methanol and absorbance at 234 nm was determined. It corresponds to the maximum absorbance of the extracted compounds. Results of experiments performed in each donor (n = 3) in triplicate are given as mean percentage $\pm$ S.D.

Lipid peroxidation (LP) evaluation

After treatment with PUVA, cells were trypsinized and collected. Suspensions with approximately $1,5 \times 10^6$ cells ml^{-1} were centrifuged (8000 rpm for 5 min) and the pellet was suspended in 0,5 ml of PBS and extracted twice in chloroform/methanol (2:1, v:v). Measurement of LP was assessed according to the thiobarbituric acid (TBA) method [38] with slight modifications. The spectrum was recorded in the 400–600 nm range showing a maximum at 532 typical for the MDA-TBA complex. Optical density at 532 nm was corrected for background absorption by interpolation. The standard curve was constructed using 1,1,1,3-tetraethoxypropane, after hydrolysis with 1% H_2SO_4, as external standard. The levels of lipid peroxides were expressed as nmol of TBA reactive species (TBARS)/mg protein. The results are the means of three different assays performed in each donor (n = 3).

Analyses of cell membrane cholesterol and oxysterols

HDFs were suspended extracted with methanol containing BHT 100 µM and 5-α-cholestane 100 ng (Sigma-Aldrich, Milan, Italy) as internal standard. Cholesterol (CH) was measured by GC-MS as previously described [39]. Selected ion monitoring (SIM) was carried out by monitoring m/z 329 and 458 for CH, 454 for 7β-OH-cholesterol (7β-OH-CH), 456 for 7-keto-cholesterol (7-keto-CH), 217 and 357 for 5α-cholestane (IS). The mean value of experiments performed in each donor (n = 3) in duplicate is expressed as microgram (for CH) or as nanogram (for 7β-OH-CH and 7-keto-CH) for per milligram of proteins $\pm$ S.D.

Western Blot analysis of cell cycle proteins

Samples were lysed in RIPA buffer with protease inhibitors. Aliquotes of cell proteins (30 µg) were resolved on SDS-polyacrilamide gel and transferred to nitrocellulose membrane and then treated with anti-p53 (clone DO-1, Dako, Milan, Italy; diluted 1:3000 in TBS-T), anti p21 (Santa Cruz Biotechnology Inc., Santa Cruz, CA, USA; diluted 1:3000 in TBS-T), anti phospho-p38 (Cell Signaling Technology Inc., Danvers, MA, USA; diluted 1:3000 in TBS-T), or anti IκB-alpha (Santa Cruz Biotechnology Inc., Santa Cruz, CA, USA; diluted 1:1000 in TBS-T) overnight at 4°C. Horseradish-peroxidase-conjugated goat anti-mouse or anti-rabbit immuglobulins (Santa Cruz, Biotechnology

Inc., Santa Cruz, CA, USA) were used as secondary antibodies. Antibodies complexes were visualized using the ECL Chemiluminescence Luminol Reagent (Santa Cruz Biotechnology Inc., Santa Cruz, CA, USA). As a loading control, the blots were reprobed with an anti-β-tubulin or anti- glyceraldehyde-3-phosphate dehydrogenase (GAPDH) antibody (Sigma-Aldrich, Milan, Italy).

RARE Transfection and luciferase assays

Cells were plated in a 24-well plate at a density of 2×10^4 cells/well and left to grow overnight. Afterwards cells were transfected with retinoid responsive element (RARE) reporter, negative control and positive control (CignalTM RARE Reporter Assay Kit; Superarray Bioscience Corp., Frederick, USA). After 24 h, cells were treated with 5µM ReOH for 6 h, 5µM AtRA for 6–48 h, and 4µM Octa for 6–48 h. Measurement of luciferase activity was carried out at the end of the treatments. Cells were harvested in 100 µl of lysis buffer and soluble extracts assayed for luciferase and Renilla activities by using Dual-Luciferase Reporter Assay System (Promega Corp., Madison, USA) according to the manufacturer's procedure.

RNA extraction and real time RT-PCR

Total RNA was isolated using an RNeasy Mini kit (Qiagen, Hilden, Germany). Following DNAse I treatment, cDNA was synthesized from 1 µg of total RNA using ImProm-II Reverse Transcriptase (Promega Corporation, Madison, WI) according to the manufacturer's instructions. Real time RT-PCR was performed with SYBR Green PCR Master Mix (Bio-Rad, Hercules, CA) and 200 nM concentration of each primer. Sequences of all primers used are indicated in Table S1. Reactions were carried out in triplicates using the Real-Time Detection System (iQ5 Bio-Rad, Milan, Italy) supplied with iCycler IQ5 optical system software version 2.0 (BioRad). The thermal cycling conditions comprised an initial denaturation step at 95°C for 3 minutes, followed by 40 cycles at 95°C for 10 seconds and 60°C for 30 seconds. Levels of gene expression in each sample were quantified applying the $2^{-\Delta\Delta C_T}$ method, using GADPH as an endogenous control.

PPARγ transactivation assay

HDFs were transfected with pGL3-(Jwt)3TKLuc reporter construct [40] using Amaxa human fibroblasts Nucleofector kit (Lonza, Basel, Switzerland) according to the manufacturer's instructions. Twenty-four and forty-eight hours after treatment with PUVA ± Octa, cells were harvested and assayed for luciferase activity using Promega's Dual Luciferase (Promega) according to the manufacturer's protocol. The renilla luciferase plasmid was also transfected as an internal control for monitoring transfection efficiency and for normalizing the firefly luciferase activity. The mean value of luciferase activity performed in each donor (n = 3) in duplicate is expressed as fold of the activity ± S.D. obtained in cells treated divided by luciferase activity from non-stimulated cells.

RNA interference experiments

For the RNA interference experiments, HDFs were transfected with 100 pmol (h) siRNA specific for PPARγ (sc-29455; Santa Cruz Biotechnology). An equivalent amount of non-specific siRNA (sc-44234; Santa Cruz Biotechnology) was used as a negative control. Cells were transfected using the Amaxa human fibroblasts Nucleofector kit (Lonza) according to manufacturer's instructions. To ensure identical siRNA efficiency among the plates, cells were transfected together in a single cuvette and plated immediately after nucleofection. Twenty-four hours following

transfection, HDFs were treated with PUVA and post-incubated with 2µM Octa in agreement with the experimental design.

Statistical analysis

Statistically significant differences were calculated using Student's t-test. The minimal level of significance was p≤0.05.

Results

Identification of a specific PPARγ modulator as a useful tool to study possible interference with PUVA-induced damage

To investigate the role of PPARγ modulation in PUVA-SIPS, we used Octa, a compound we previously reported to activate PPARγ in human melanocytes [31]. Because the chemical structure of Octa resembles the polyene chain of carotenoids, we evaluated the ability of this molecule to modulate the retinoid-mediated signaling in HDFs to study the activation of retinoic acid receptor (RAR) and the subsequent transcriptional activation of RARE. We compared the effects with those caused by the specific retinoid receptor ligands AtRA and ReOH. Both ReOH and AtRA induced an early (6 h) and relevant enhancement of the expression of the RARE-driven reporter, whereas Octa showed a mild capacity to transactivate RARE only after 48 h (Fig. 1A). Moreover, Octa treatment did not exhibit any ability to induce the mRNA expression of cellular retinoic acid binding protein 2 (CRABPII) or cytochrome P450 hydroxylase (CYP26), two genes that contain RARE reporter promoters, which are directly involved in the proliferative response elicited by retinoid-like molecules, whereas atRA induced a relevant up-regulation of both genes (Fig. 1B). In contrast, Octa was more effective than atRA in inducing the expression of PPARγ and fatty acid binding protein-5 (FABP5), a carrier protein for PPAR ligands, at the evaluated time points (6 and 24 h) (Fig. 1C). Consistently, a luciferase assay using the pGL3-(Jwt)TKLuc reporter construct [40] showed that Octa enhanced luciferase expression at 24 and 48 h (Fig. 1D).

PUVA induced a significant reduction of PPARγ expression and activity

A reduction of PPARγ expression in H_2O_2-SIPS HDFs has been reported to reflect age-related inflammation and aging progression [41]. We previously demonstrated that azelaic acid, a natural compound that is able to act as a ligand of PPARγ, was able to revert, at least in part, PUVA-induced decrease in PPARγ activation [42]. To confirm that PPARγ represents a main biological target of PUVA-SIPS, we performed photo-irradiated HDFs RT-PCR analysis and luciferase assay to evaluate the changes in PPARγ expression and/or activity induced by PUVA treatment. Our results showed that PUVA exposure induced an early reduction of PPARγ expression (at 6 and 24 h) as well as a significant decrease in transcriptional activity (at 24 and 48 h) (Fig. 2A and 2B). Octa treatment significantly counteracted the decreased expression and inhibition of PPARγ (Fig. 2C and 2D).

PUVA-induced ROS production and mitochondria damage are counteracted by PPARγ modulation

Exposure of HDFs to PUVA induces mitochondrial membrane damage with a persistent intracellular ROS accumulation [17]. To determine whether PPARγ modulation has a protective effect, ROS generation was determined at 24 h, 48 h, and 1 week after PUVA exposure using the DCFH2-DA assay. PUVA led to a significant time-dependent ROS increase in HDFs, and post-incubation with Octa significantly decreased (p<0.01) ROS

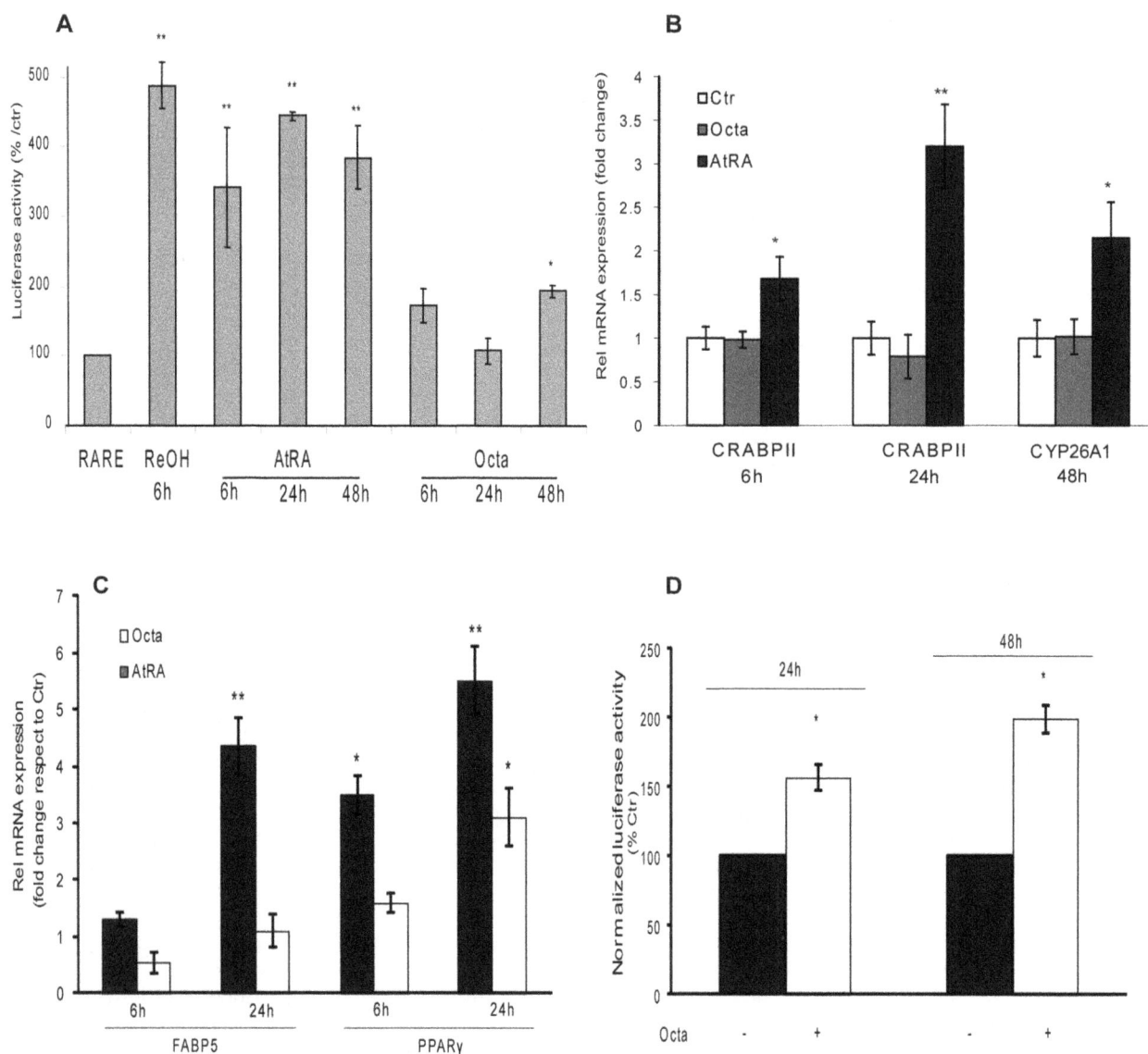

Figure 1. Evidence for Octa-mediated activation of PPARγ-linked signal transduction. (A) Activation of RARE. Cells (2×10^4cells/well) were plated in a 24-well plate and after 24 h they were transfected with RARE. After 24 h, cells were treated with 5μM ReOH for 6 h, 5μM AtRA for 6–48 h, and 2μM Octa for 6–48 h. Measurement of luciferase activity was assessed as reported in **Materials and Methods**. (B) Quantitative real-time RT-PCR was performed to measure the expression of CRABPII and CYP26A1 mRNA at various time points after treatment with 2μM Octa or 5 μM AtRA. The values were normalized to GAPDH mRNA levels. (C) Quantitative real-time RT-PCR was performed to measure the expression of FABP5 and PPARγ mRNA at various time points after treatment with 2μM Octa or 5μM AtRA. The values were normalized to GAPDH mRNA levels. (D) Luciferase activity analysis of cells transfected with pGL3-(Jwt)3TKLuc reporter construct. After 24 h of transfection, cells were treated with 2μM Octa. The measurement of luciferase activity was carried out 24 h and 48 h after treatment. *p<0.05; **p<0.001 respect to untreated control cells.

production at all evaluated time points (Fig. 3A). ROS generation was correlated with a decrease in mitochondrial $\Delta\Psi_m$ based on JC-1 staining. Consistent with the literature [43], PUVA determined a progressive decline in the ratio of orange-red/green fluorescent JC-1 density compared with sham-irradiated fibroblasts after 24 h, 48 h, and 1 week. PPARγ modulation induced a significant improvement of mitochondrial $\Delta\Psi_m$ (Fig. 3B).

PPARγ modulation counteracted the imbalance of the redox system in PUVA-treated HDF

The PUVA-induced imbalance in the intracellular redox environment was investigated by analyzing the following: a) BAP, an index of overall antioxidant status; b) Cat activity, which

is directly involved in the persistent accumulation of hydrogen peroxide in senescent cells; c) GSH, a major endogenous antioxidant; and d) α-Toc, which protects the cell membrane lipid layer by acting as a chain anti-breaking antioxidant. Because our aim was to investigate the ability of PPARγ modulation to interfere with already activated cell senescence, we did not incubate fibroblasts with Octa before PUVA exposure and we considered untreated fibroblasts as the controls. Up to 1 week, PUVA caused a decline in BAP (p<0.01), GSH levels, and α-Toc content, which were recovered by post-treatment with Octa (Fig. 4A, 4C, 4D). Moreover, PUVA led to a relevant and long-lasting decrease in Cat activity to 50% of the baseline value after 24 h, which was still reduced to 47% after 1 week. Octa protected

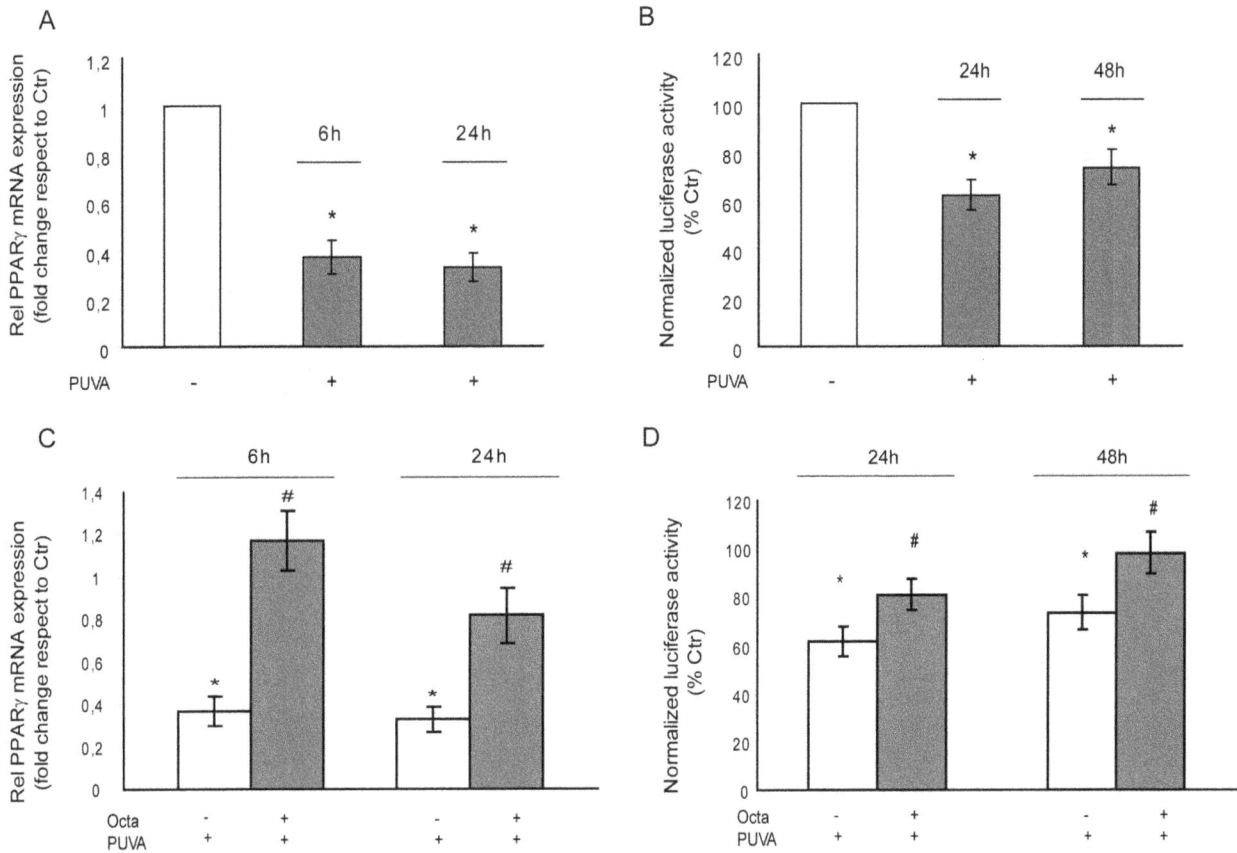

Figure 2. Evaluation of PUVA induced effects on PPARγ expression and activity. (A) Real-time RT-PCR was performed to measure the expression of PPAR-γ mRNA 6 h and 24 h after PUVA exposure. The level of PPAR-γ mRNA was normalized to the expression of GAPDH and is expressed relative to untreated control cells (*p<0.05 respect to Ctr). (B) Luciferase activity analysis of cells transfected with pGL3-(Jwt)3TKLuc reporter construct. After 24 h of transfection, cells were treated with PUVA. The measurement of luciferase activity was carried out 24 h and 48 h after treatment (*p<0.05 respect to Ctr). (C) Real-time RT-PCR was performed to measure the effect of Octa post-treatment on the expression of PPAR-γ mRNA 6 h and 24 h after PUVA exposure. The level of PPAR-γ mRNA was normalized to the expression of GAPDH and is expressed relative to untreated control cells (*p<0.05 respect to Ctr; #p<0.05 respect to PUVA). (D) Luciferase activity analysis of cells transfected with pGL3-(Jwt) 3TKLuc reporter construct. After 24 h of transfection, cells were treated with PUVA and post-incubated with Octa. The measurement of luciferase activity was carried out 24 h and 48 h after treatment (*p<0.05 respect to Ctr; #p<0.05 respect to PUVA).

against enzyme damage, leading to a recovery of Cat activity within 1 week (Fig. 4B).

PPARγ activation is needed to promote cell antioxidant defense

In parallel, we treated non-irradiated fibroblasts with Octa to evaluate its capacity to enhance basal antioxidant defense. Endogenous antioxidants and, in particular, total antioxidant capacity and Cat activity in sham-irradiated cells were significantly increased by supplementation with Octa (Fig. 5A). Considering that PPARγ regulates the expression of catalase via functional PPREs identified in its promoter [44], we investigated the implication of PPARγ in the activation of this endogenous antioxidant by Octa in both sham-irradiated and PUVA exposed HDFs, that were transiently transfected with PPARγ siRNA (siPPARγ) (Fig. 5B). As expected, Octa significantly increased Cat activity in siCtr cells but failed to up-regulate Cat in PPARγ-silenced HDFs (Fig. 5C). Furthermore, in PPARγ-deficient HDFs, Octa failed to counteract the decrease in Cat activity caused by PUVA (Fig. 5C), indicating that the increase in the antioxidant enzyme was PPARγ dependent.

Possible interference of PPARγ against the PUVA-induced modulation of the cellular stress response system

The activation of nuclear factor erythroid-related factor 2 (NRF2) and subsequent induction of NRF2-dependent genes are part of an efficient adaptive response mechanism to electrophilic and oxidant stress, as occurs upon UVA irradiation [45]. Quantitative PCR results indicated that the copy of the cellular NRF2 mRNA increased 2.3 and 5.1-fold, 6 h and 24 h, respectively, after PUVA treatment (Fig. 6A). Cells supplemented with Octa after PUVA exposure showed a significant reduction (p<0.01) of NRF2 mRNA, and no significant modifications of basal level of NRF2 mRNA were observed in HDFs treated with Octa (Fig. 6A). Moreover, NRF2 plays a key role in the UVA-induced up-modulation of heme oxygenase 1 (HO-1), which is considered an immediate cellular response to oxidative insults [46,47]. However, whereas modest HO-1 expression is cytoprotective, the exacerbation of oxidative injury correlates with high HO-1 expression [48]. In HDFs, the basal level of HO-1 expression was low and PPARγ modulation did not induce relevant changes (Fig. 6B). In response to PUVA, HO-1 expression increased significantly in a time-dependent manner up to 40-fold after 24 h (Fig. 6B), indicating a promotion by the persistent oxidative stress, and Octa treatment significantly reduced (p<0.001) this effect.

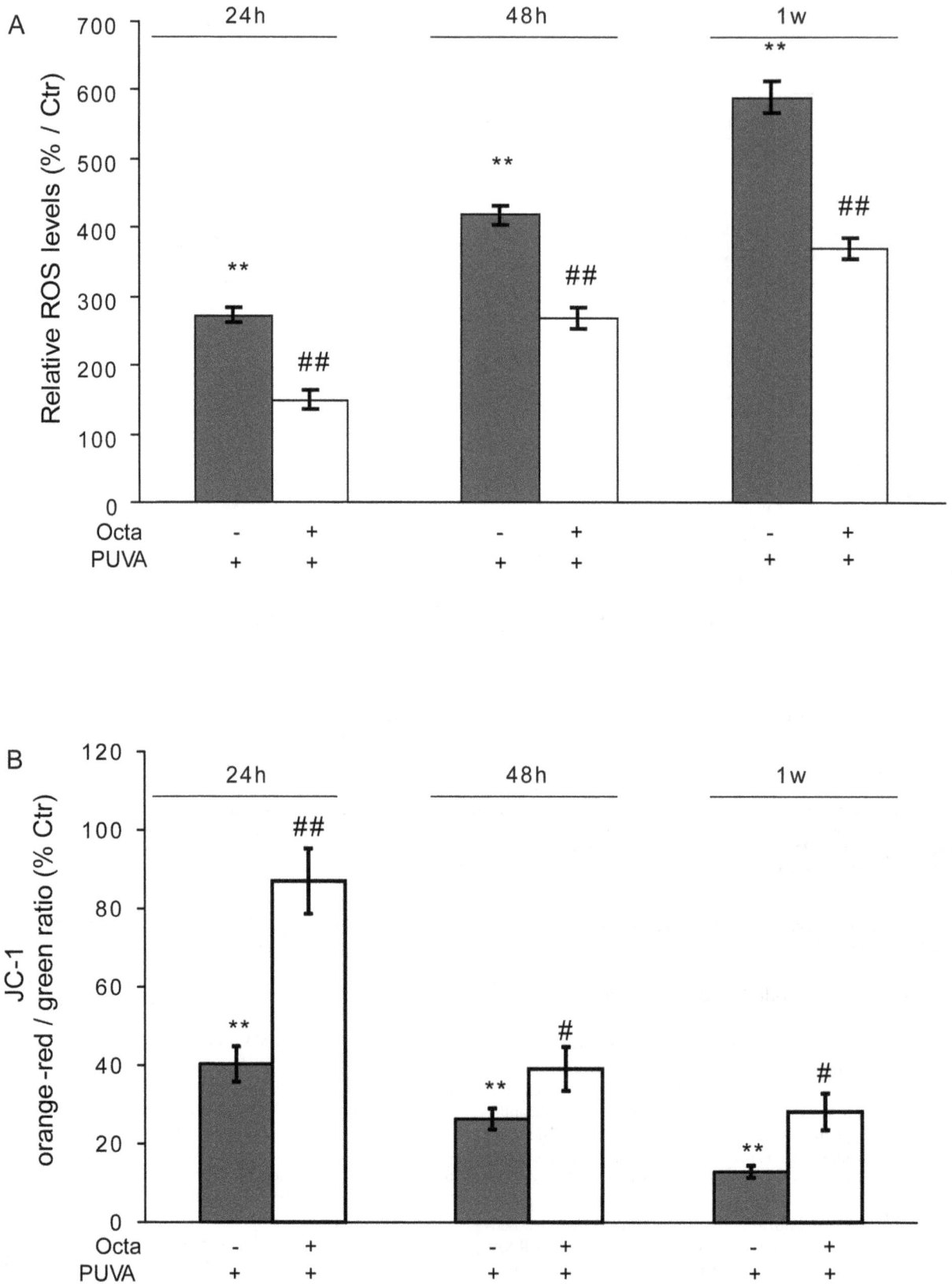

Figure 3. Effects of PPARγ modulation against PUVA-induced intracellular ROS accumulation and mitochondria damage. HDFs were treated with PUVA or left untreated (Ctr). Immediately after irradiation PBS was replaced by fresh medium with or without Octa 2μM for 24 h, 48 h or 1 week. (A) Intracellular oxidative stress was assessed by Flow cytometry using the fluorescent probe $DCFH_2$-DA. The median value of fluorescence was used to evaluate the intracellular content of DCF as a measure of the ROS formation. (B) $\Delta\Psi_m$ was assessed in live HDFs using the lipophilic cationic probe JC-1. For quantitative fluorescence measurements, cells were rinsed once after JC-1 staining and scanned with a Flow cytometer **p< 0.001 statistically different from unirradiated cells; ##p<0.001 compared with PUVA-treated fibroblasts.

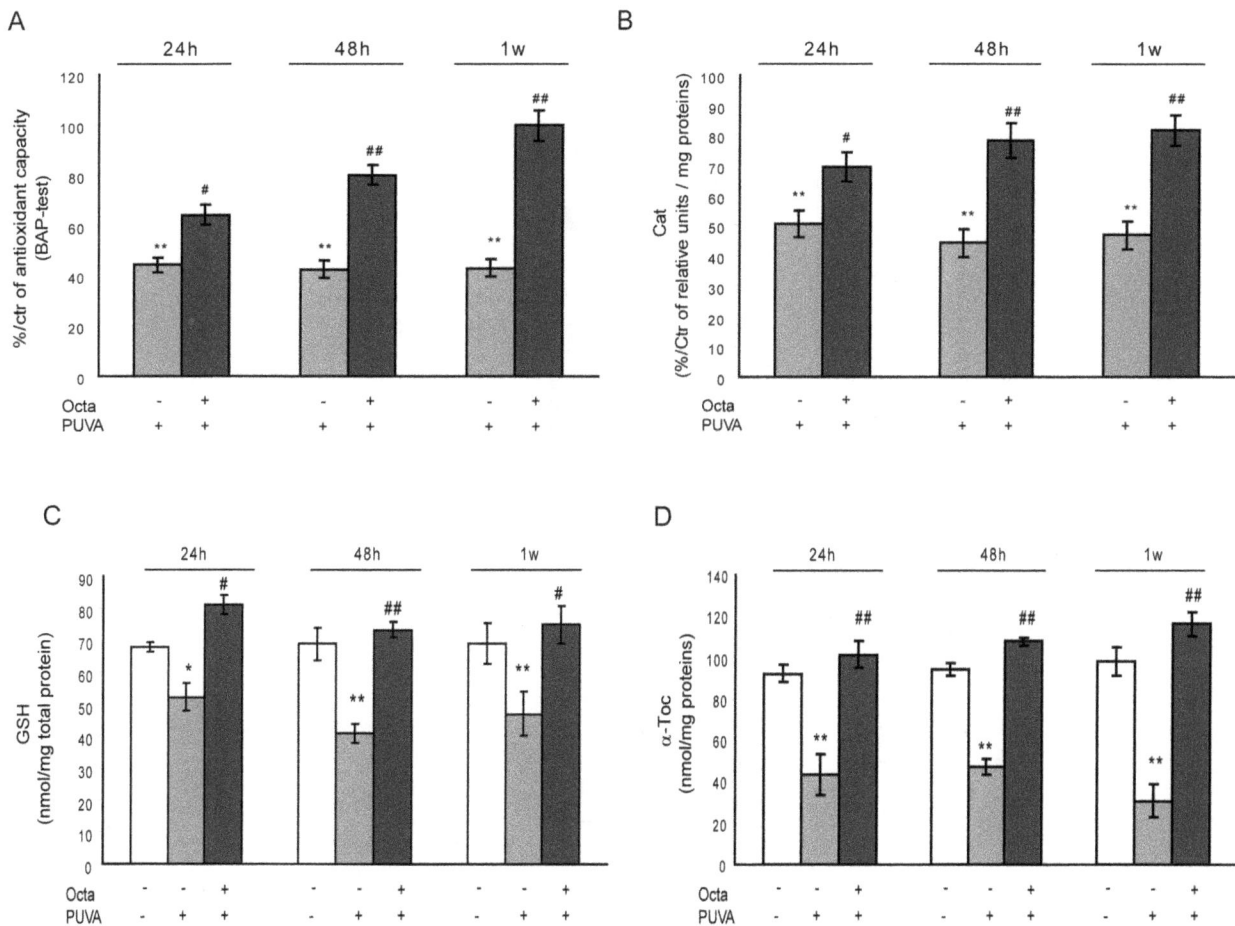

Figure 4. Protective action of PPARγ modulation on PUVA-induced imbalance of cell antioxidant system. HDFs (1×10^6) were lysed in PBS and protease inhibitor cocktail. Cell lysates were used for analytical determinations. (A) Total antioxidant capacity (TAC) was assessed by BAP-test as described under **Materials and Methods** section. (B) Cat enzyme activity was determined by spectrophotometry as described under **Materials and Methods**. (C) GSH concentrations were determined by HPLC-MS as described in **Materials and Methods**. (D) α-Toc is measured by GC-MS as described in **Materials and Methods**. *$p<0.05$; **$p<0.001$ respect to control fibroblasts; #$p<0.05$; ##$p<0.001$ compared with PUVA-treated fibroblasts.

Consistent with the incapacity of Octa to increase the basal level of NRF2 mRNA, we observed only a slight increase in basal intracellular GSH, which is synthesized by glutamate cysteine ligase, an NRF2-dependent gene (Fig. 6C). As discussed above, PUVA exposure caused a strong and long-lasting GSH depletion and Octa significantly counteracted this effect (Fig. 4C).

The FoxO1 is a transcription factor that is directly involved in cell responses to ROS [49], and it plays a substantial role in skin photoaging [50]. Moreover, a regulatory feedback loop involving PPARγ and FoxO and characterized by a transrepression mechanism has been described [51]. In this set of experiments, PUVA-treated HDFs showed a significant increase in FoxO1a mRNA expression, and PPARγ stimulation was able to reverse this effect (Fig. 6D), suggesting that this molecule promotes an antioxidant defense response by also interfering with the FoxO-induced repression of PPARγ.

PPARγ modulation reduced the senescence-like phenotype in PUVA-treated HDFs

We showed that an altered expression and activity of PPARγ is an early effect determined by PUVA and may be implicated in the appearance of the PUVA-induced cell-senescent phenotype. The

interference of PPARγ in PUVA-induced cell senescence was also investigated by examining its effect on typical senescence features, such as cell morphology, SA-β-gal expression, MMP-1 release, and regulatory cell cycle protein expression. PPARγ modulation was able to rescue, at least in part, the enlarged and flattened senescent fibroblast morphology observed 4 weeks after PUVA exposure (Fig. 7A). SA-β-gal is a β-galactosidase whose activity is detectable at pH 6.0 in cultured cells undergoing replicative or induced senescence but whose activity is absent from proliferating cells [33]. In HDFs exposed to PUVA, SA-β-gal activity was detected after 1 week followed by a steady increase up to 4 weeks, when virtually all of the fibroblasts exhibited *de novo* activity of SA-β-gal (insert in Fig. 7B). The total number of cells was not significantly different, but the percentage of SA-β-gal-positive fibroblasts was significantly suppressed (approximately 40%) by post-treatment with Octa (Fig. 7B). In HDFs, PUVA induced a strong and persistent release of MMP-1, the main metalloproteinase induced by UV exposure [52,53], with a maximum at 48 h after photo-irradiation and an approximately 10-fold (SE ± 0.42) higher amount compared to that of non-irradiated control cells (Fig. 7C). Octa caused a mild but significant decrease of MMP-1 release with a maximum reduction of 21% at 48 h (Fig. 7C);

A

B

C

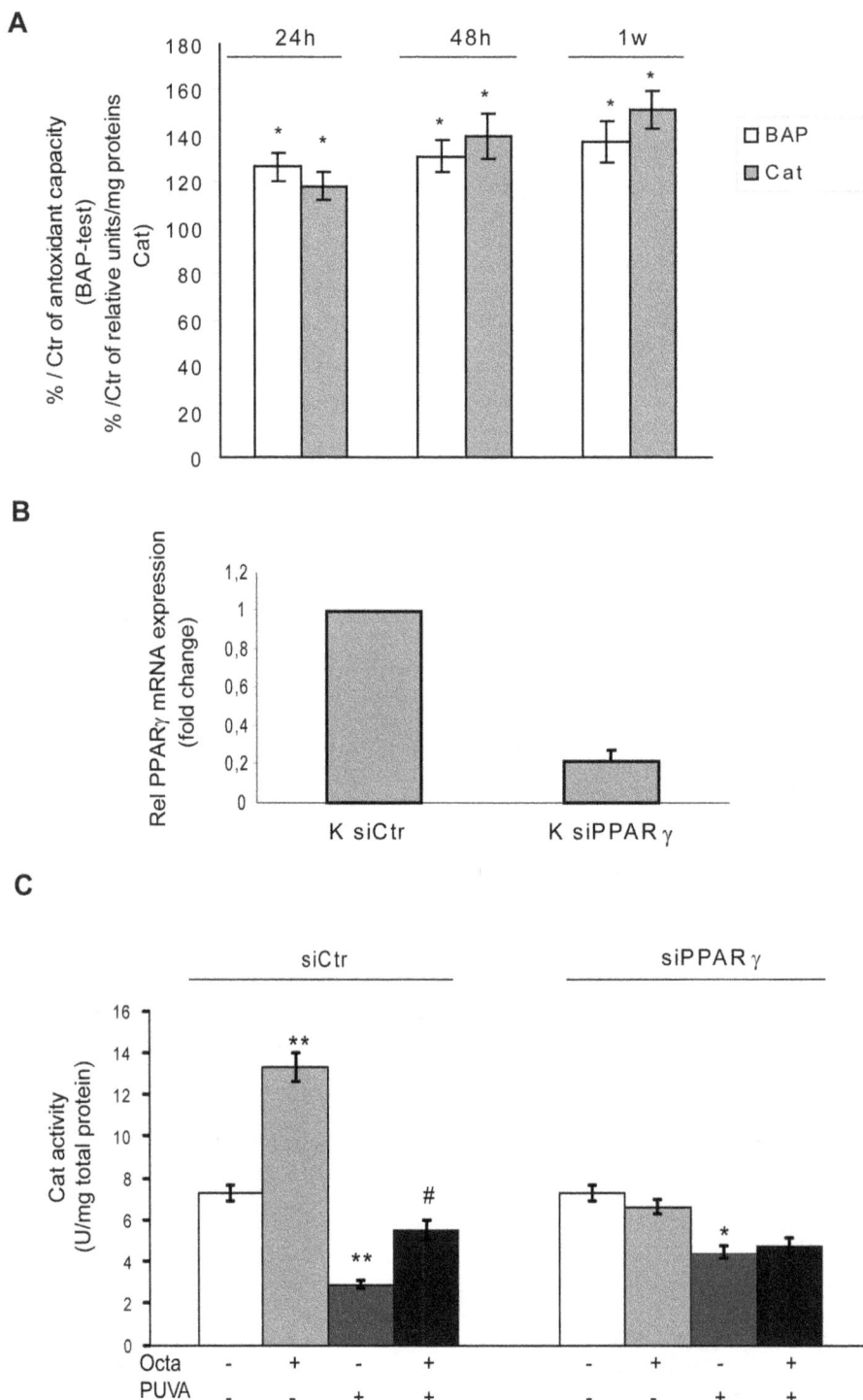

Figure 5. Evidence for PPARγ-induced promotion of cell antioxidant defence. (A) Octa treatment for 24 h, 48 h and 1 week determined a significant increase of antioxidant cell response. TAC was assessed by BAP-test and Cat enzyme activity was determined by spectrophotometry as described under **Materials and Methods** section. (B) HDFs were transfected with siRNA specific for PPARγ (siPPARγ) or non-specific siRNA (siCtr). PPARγ level was evaluated by real-time RT-PCR (C) The activity of Cat was assessed in HDFs transfected with siPPARγ or siCtr and exposed to 2μM Octa for 6 h. In parallel Cat activity was measured in HDFs transfected with siPPARγ or siCtr and exposed to PUVA w/o post-incubation with 2μM Octa. *$p < 0.05$; **$p < 0.001$ respect to control fibroblasts; #$p < 0.05$ compared with PUVA-treated fibroblasts.

however, it had no effect on the basal secretion of MMP-1 (data not shown). Growth arrest is an important feature of cellular senescence and stress-induced premature senescence. We observed a strong expression of p53 and p21 proteins starting from 24 h

after PUVA that was still elevated after 1 week (Fig. 7D). p53 and p21 were not detectable in untreated and Octa-treated control cells. Octa significantly reversed the up-regulation of p53 protein expression after 24 h and p21 after 1 week (0.65-fold and 0.7-fold

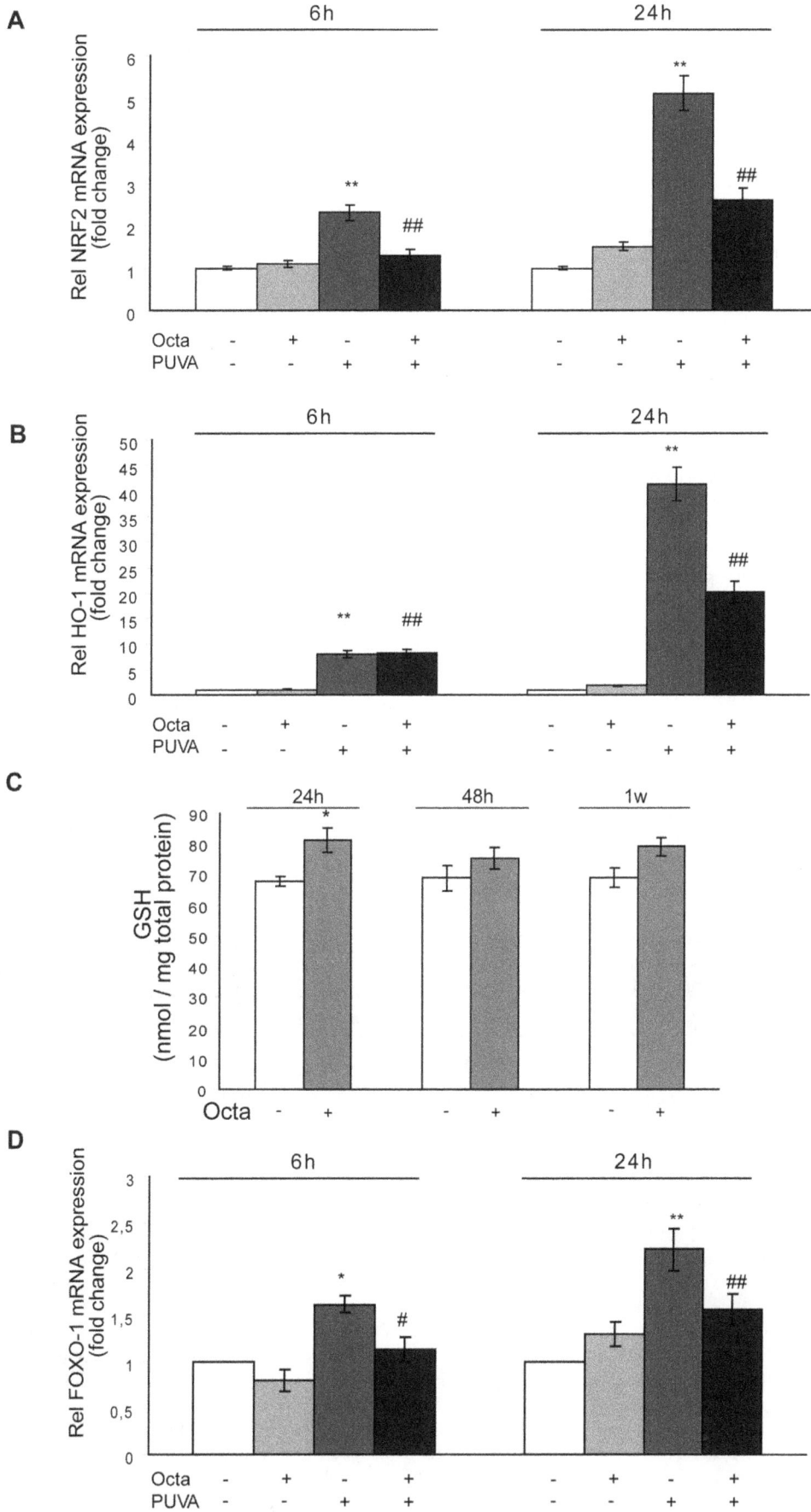

Figure 6. Possible interference of PPARγ against PUVA induced modulation of the cellular stress response system. (A) RT-PCR was performed to measure the expression of NRF2 mRNA 6 and 24 h after PUVA exposure, w/o Octa post-incubation. The level of NRF2 mRNA was normalized to the expression of GAPDH and is expressed relative to untreated control cells (**p<0.001 respect to Ctr; ##p<0.001 compared with PUVA-treated fibroblasts). (B) RT-PCR was performed to measure the expression of HO-1 mRNA 6 and 24 h after PUVA exposure, w/o Octa post-incubation. The level of HO-1 mRNA was normalized to the expression of GAPDH and is expressed relative to untreated control cells (**p<0.001 respect to Ctr; ##p<0.001 compared with PUVA-treated fibroblasts). (C) GSH concentrations were determined by HPLC-MS) as described in **Materials and Methods** (*p<0.05 respect to control fibroblasts). (D) RT-PCR was performed to measure the expression of FoxO1 mRNA 6 and 24 h after PUVA exposure, w/o Octa post-incubation. The level of FoxO1 mRNA was normalized to the expression of GAPDH and is expressed relative to untreated control cells. *p<0.05 respect to control fibroblasts; #p<0.05 compared with PUVA-treated fibroblasts.

compared to PUVA-treated samples, respectively) (Fig. 7D). In addition, we detected a moderate expression of p16 in PUVA-treated cells, but Octa post-treatment did not induce a significant reduction (data not shown).

PPARγ modulation interferes with changes in cellular membrane lipids in PUVA-treated HDFs

Unsaturated lipids in cell membranes, including phospholipids and cholesterol, are well-known targets of oxidative modification, which can be induced by a variety of stresses, including UVA-induced photodynamic stress. To evaluate the modifications of the plasma membrane induced by PUVA oxidative damage, we assessed the content of polyunsaturated fatty acids of membrane phospholipids (Pl-PUFA) and the level of CH as the main lipid component of raft domains of cell membranes. PUVA induced a significant modification of the fatty acid composition of cell membrane lipids, with a strong reduction in the Pl-PUFA percentage, which was detectable immediately after irradiation (data not shown) and was still reduced at 1 week (Fig. 8A); this was accompanied a bi-modal alteration in the CH level, with an early (up to 48 h) reduction followed by a relevant accumulation 1 week after photo-irradiation (Fig. 8B). PUVA-induced lipid alterations were almost completely reversed by Octa (Fig. 8A and 8B). Moreover, we evaluated the formation of oxidative products, such as conjugated dienes, fatty acid hydroperoxides, TBARS, and oxysterols (7β-hydroxycholesterol (7-β-OH-CH) and 7-ketocholesterol (7-Keto-CH)), in photo-irradiated cells. PUVA-treated HDFs showed a time-dependent accumulation of lipid peroxidation products. In particular, conjugated dienes were the early products of PUVA-induced lipoperoxidation, and their levels peaked after 3 h (Fig. 8C), whereas both TBARS and oxysterols constantly increased up to 1 week after photo-irradiation (Fig. 8D, 8E, and 8F). Octa significantly reduced the PUVA-induced generation and accumulation of these cell membrane oxidation products (Fig. 8C, 8D, 8E, and 8F). Interestingly, oxysterols were reported to induce the expression of p21 and modulation of the phosphorylation signaling involved in the activation of nuclear factor κB (NF-κB), a transcription factor involved in the induction of pro-inflammatory cytokines. [54]. Considering that these processes are implicated in the aging process and age-related inflammatory responses, we hypothesize that the Octa-mediated reduction of oxysterols plays a key role in the disruption of PUVA-SIPS by Octa.

PPARγ interfered with the PUVA-induced phosphorylation pathway and NF-kB activation

The UV-induced inflammatory process in the skin is characterized by ROS-mediated phosphorylation of mitogen-activated proteins kinases (MAPKs), including p38 kinase, and the subsequent activation of NF-κB, To determine the alterations of the phosphorylation pathway and NF-κB activation in the PUVA-SIPS model and the possible protective effect of PPARγ modulation, phosphorylation of p38 and expression of IκBα were

evaluated by Western Blot. PUVA-treated HDFs showed an increased phosphorylation of p38 (Fig. 9A) and a decreased expression of IκBα (Fig. 9B), indicating that the activation of the pro-inflammatory response is involved in the senescence-like phenotype. Octa post-treatment inhibited p38 phosphorylation and decreased IκBα expression at 24 h and 48 h after PUVA treatment, respectively (Fig. 9A and 9B). The ability of Octa to interfere with the PUVA-induced activation of the phosphorylation pathway and the activation of NF-κB at later time points compared to its effects on PPARγ activation and generation of cell membrane lipid peroxidation products suggests that p38 and NF-κB are not direct targets of Octa, but they can be modified by the Octa-induced activation of PPARγ and a reduction of the ROS-induced lipoperoxidation process.

Discussion

PUVA-SIPS is characterized by the persistent induction of ROS and stable alteration of the cell redox system inducing robust aging markers, including morphological changes, increased staining of SA-β galactosidase, and MMP-1 release, thereby representing a suitable tool for the analysis of photoaging-related mechanisms *in vitro* [17]. The imbalance of the antioxidant network is crucial for propagating PUVA-induced oxidative stress, as demonstrated by the ability of antioxidant molecules to counteract the phenomenon [17], most likely not exclusively due to scavenging ROS but also to the modulation of cell signaling pathways.

To further investigate the mechanism mediating the imbalance of the cell-redox system, we focused on possible cell targets and transcription factors involved in the induction of PUVA-SIPS. We previously reported that azelaic acid, a modulator of PPARγ, interfered with PUVA-induced cell responses, and here we sought to determine whether this nuclear receptor represents a "conductor" of PUVA-SIPS. PPARs regulate the expression of genes involved in multiple biological pathways, including cellular lipid metabolism, inflammation, differentiation, and proliferation [18–20]. Therefore, these nuclear receptors are possible regulators of mitochondrial functions, inflammatory responses, and antioxidant imbalances observed in premature cell senescence. Reduced activity of the proteins PPARγ coactivator-1α (PGC-1α) and PPARγ coactivator-1β (PGC-1β), which are master regulators of PPARγ, is associated with mitochondrial dysfunction and reduced expression of numerous ROS-detoxifying enzymes [22]. We investigated the possible interplay among PPARγ modulation and the PUVA-induced senescence-like phenotype by employing Octa, a polyunsaturated acid with retinoid-like molecular features. Despite its reported features in common with retinoids [29], the molecule caused only a weak activation of RARE and was not associated with the modulation of RA target genes, such as CYP26, which is a cytochrome P450 isoenzyme that specifically metabolizes RA [55], or CRABPII, which transports retinoids to the nucleus [56]. In contrast, Octa significantly activated PPARγ and FABP5, which is an intracellular protein that binds lipid molecules and transports them to PPARs [57]. Consistent with the

A

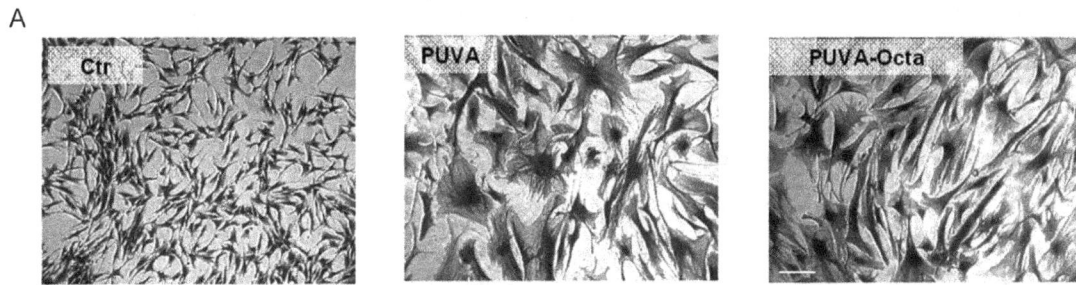

B

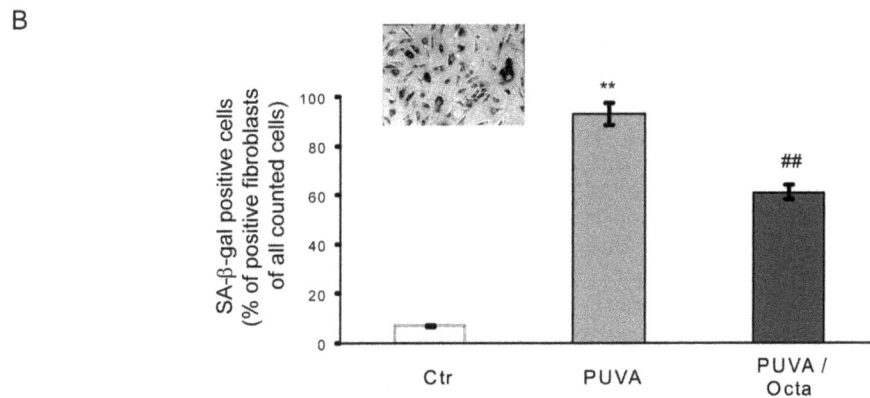

C

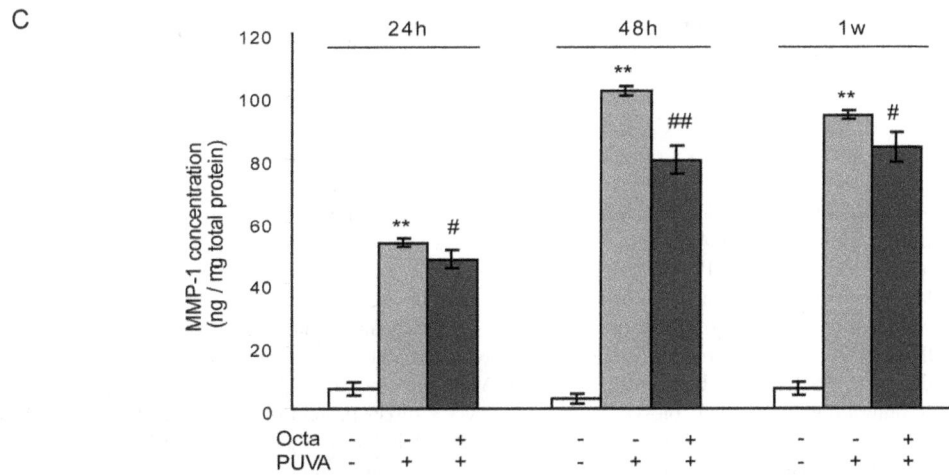

D

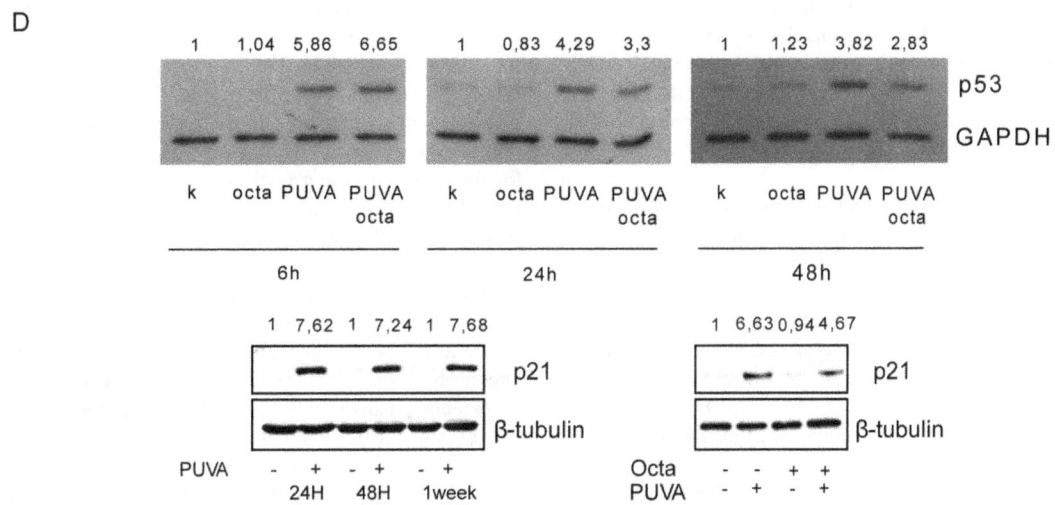

Figure 7. Effect of PPARγ modulation on PUVA-induced expression of senescence-like phenotype in HDFs. After PUVA treatment, HDFs were cultured in the absence or in the presence of 2μM Octa. The medium was changed every 3 days to ensure efficient antioxidant capacity. (A) To evaluate fibroblast morphology, 2 weeks after PUVA in the absence or presence of Octa treatment, cells were fixed and stained with Comassie Brilliant Blue. Scale bar 50 μm. (B) SA-β-gal expression was detected as described in **Materials and Methods**. The *inset* represents fibroblasts after PUVA-treatment revealing a senescent phenotype with enlarged cytoplasmic morphology and SA-β-gal expression. The number of SA-β-gal positive fibroblasts is shown as mean ± SD of three independent experiments. **p<0.001 as compared with mock treated controls; ##p<0.001 as compared with PUVA-treated fibroblasts. (C) Supernatants were collected from mock-treated fibroblasts, at 24 h, 48 h and 1 week post PUVA-treatment. MMP-1 release was assessed by ELISA-kit. Three independent experiments in each donor (n = 3) were performed to determine specific MMP-1 protein concentrations in the supernatants. **p<0.001 as compared with mock-treated fibroblasts; #p<0.05; ##p<0.001 as compared with PUVA-treated fibroblasts. (D) Total cellular proteins (30μg/lane) were subject to 10% SDS-PAGE. Variation of protein loading was determined by reblotting membrane with an anti-β-tubulin antibody. Western Blot assays are representative of at least three experiments. Increase of p53 and p21 proteins expression is remarkable 24 h after irradiation as well as until 7 days. Octa treatment decreased PUVA-induced expression of p53 protein (at 24 and 48 h) and of its target gene p21 (at 1 week).

results reported for H₂O₂-SIPS [41], PUVA-treated HDFs showed an immediate decrease in the expression and activity of PPARγ, indicating a relevant role of this receptor in the biological modifications induced by senescence-like phenotype. Octa mitigated the PUVA effects, indicating that PPARγ modulation may be responsible for the protective mechanism. Because PPARγ promotes mitochondrial function and endogenous antioxidants, we evaluated the effects of the nuclear receptor modulation against PUVA-induced damage to these cellular targets. Mitochondrial oxidative stress, characterized by the reduction of the oxidative phosphorylation efficiency and ΔΨm, promotes the senescence of skin cells both in vitro [58] and in vivo [59]. In PUVA-treated HDFs, we observed a progressive accumulation of intracellular ROS and a decline in ΔΨm, indicating that mitochondria are involved in the senescence-like phenotype. However, the excessive ROS generation induced by PUVA overwhelmed the cell redox system. Because antioxidant enzymes are themselves targets of oxidative modifications [60], PUVA-SIPS mimics the alterations observed in photoaged cells [61]. In particular, PUVA-treated HDFs showed a dramatic decline in Cat activity and a significant reduction in intracellular GSH, which are both critical for preserving cellular redox balances, with a very low recovery to basal values. Despite the reported antioxidant action of Octa [29], the compound reduced but did not abrogate PUVA-induced intracellular ROS accumulation and the alteration of mitochondrial integrity, suggesting that scavenging ability is only partly involved in the protective effect of Octa. Octa treatment promoted the increase of both Cat activity and GSH levels in both untreated and PUVA-exposed HDFs, interfering with their biosynthetic pathways.

PPARγ is directly involved in the regulation of the expression of Cat via functional PPREs identified in its promoter [44], and the activation of PPARγ by Octa was functionally relevant for the induction of catalase activity, as the use of a specific PPARγ siRNA abolishes this effect. Moreover, silencing the PPARγ receptor significantly reduced the PUVA-induced decrease in Cat activity and completely abrogated the protection of Octa against this damage.

PPARγ regulates antioxidant defense and counteracts mitochondrial damage in close connection with other transcription factors involved in the oxidative stress response [23]. In the activation of cellular defense against the oxidative stress antioxidant response, PPARγ cooperates with NRF2, a transcription factor that regulates the expression of antioxidant genes, including HO-1 and the glutamate cysteine ligase, which is the rate-limiting enzyme for the cellular biosynthesis of GSH [23]. PUVA induced an increased expression of NRF2, indicating the attempt of the cells to activate an adaptive response against oxidative stress. Among the target genes of NRF2, HO-1 acts as a general marker of oxidative stress [47]. The activation by UVA is an emergency

stress response that results in the clearance of excess heme levels. However, HO-1 overexpression has deleterious consequences if the excess free heme is not quickly catabolized [48]. The balance of expression is particularly delicate for UVA, which itself damages heme-containing proteins and releases labile iron. Moreover, the induction of HO-1 by the ROS-generating system occurs in association with the depletion of intracellular GSH and may be enhanced by the chemical depletion of GSH [62,63]. Octa significantly reduced NRF2 and HO-1 mRNA expression in PUVA-treated HDFs, suggesting an attempt to interrupt the persistent activation of detoxifying genes, which may indicate a compromised redox homeostasis in photo-irradiated cells. Although the mRNA expression of NRF2 was increased in photo-irradiated cells, a stable decline in intracellular GSH was observed, whereas Octa effectively counteracted this damage, indicating its ability to promote the maintenance of the NRF2 signaling pathway, leading to the up-modulation of the GSH level. These findings strongly suggest a relationship between NRF2 and PPARγ in the PUVA-induced senescence-like phenotype. However, the mechanisms that regulate the reciprocal feedback circuit between these transcription factors require further investigation. Moreover, PPARγ acts at an intersection of the intracellular signaling pathways activated by FoxO1, a transcription factor that plays a pivotal role in cell fate decisions because it regulates and is regulated by oxidative stress [49]. FoxO1 may modulate PPARγ at the mRNA and protein levels [51], acting as a transcriptional repressor binding to the PPARγ promoter [64] and reducing PPARγ activity through a transrepression mechanism that involves a direct protein-protein interaction [65]. Octa decreases the PUVA-induced nuclear concentration of FoxO1, ROS accumulation, and mitochondrial damage, suggesting an interference with the regulatory feedback loop between PPARγ and FoxO proteins. Moreover, due to their ability to cross talk with the p53 tumor suppressor gene, FoxOs can participate in ROS-induced cell cycle arrest, a typical feature of cell senescence [50]. PUVA activates p53 stabilization, phosphorylation, and nuclear localization as well as the induction of p21 (Waf/Cip1), which is needed for the entry into the growth arrest state [66,67]. Octa interfered with the increase of p53 and p21, interrupting the positive axis between FoxO1 and cell cycle proteins. The evidence that the molecule did not interfere with immediate (up to 6 h) PUVA-induced ROS generation (data not shown) and p53 expression indicates that scavenger ability is not relevant for Octa interference with the senescence-like phenotype. In contrast, the compound effectively counteracted typical features of PUVA-induced cell senescence, such as enlarged cell shape, the up-modulation of MMPs and the subsequent malfunction of the connective tissue remodeling process, and a steady increase in SA-β-gal expression, suggesting that the up-modulation of PPARγ can effectively contribute to its "anti-senescence" action.

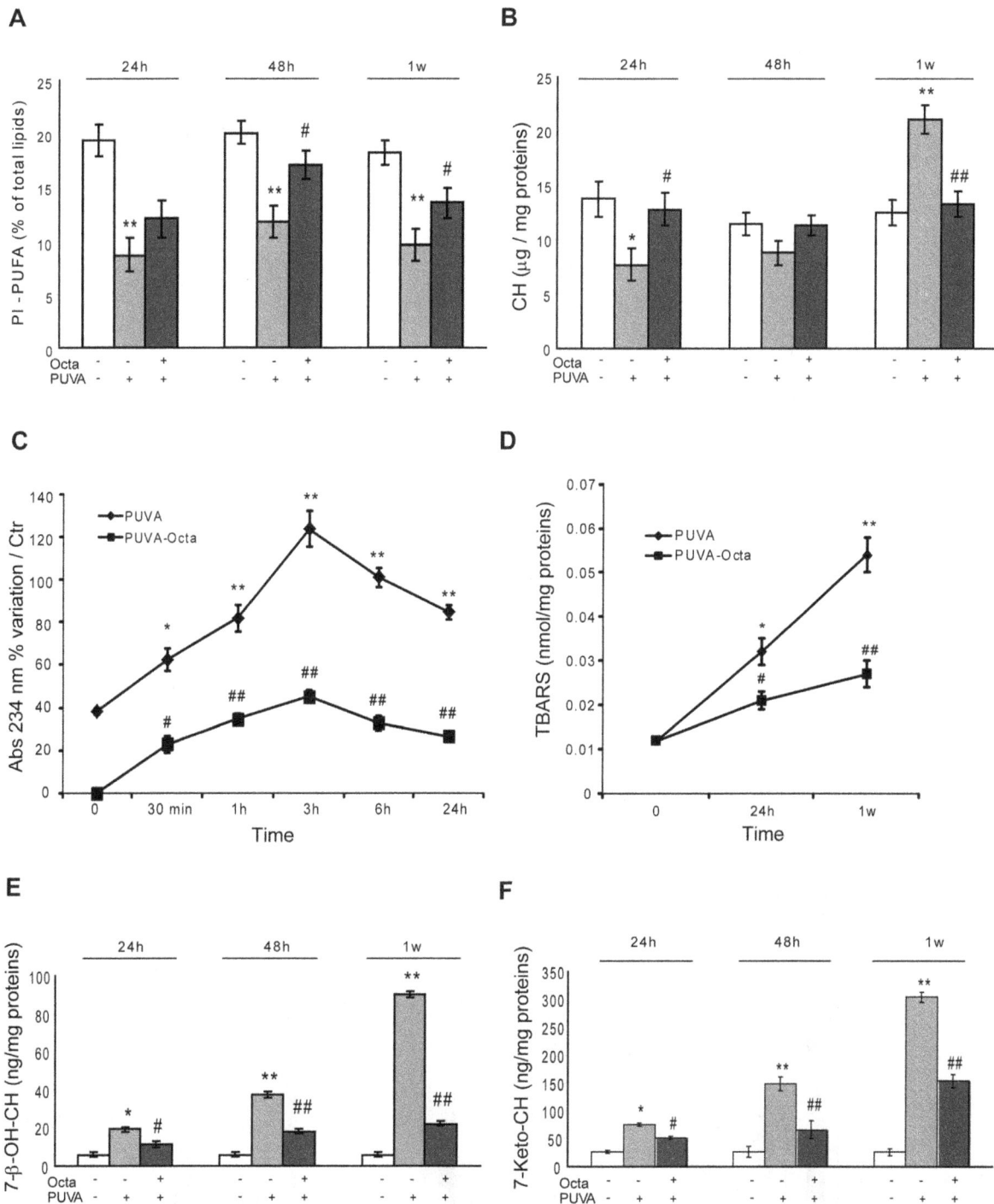

Figure 8. Octa counteracts alteration of lipid cell membrane homeostasis in PUVA treated HDFs. (A) Polyunsaturated fatty acids of membrane phospholipids (Pl-PUFA) in PUVA-treated HDFs were assessed GC-MS as described in **Materials and Methods**. (B) Chol content was analyzed by GC-MS as described in **Materials and Methods**. (C) Early lipid peroxidation products were assessed by the spectrophotometric evaluation of conjugated diene levels as described in **Materials and Methods**. (D) End products of lipid peroxidation were measured according to TBA assay as described in **Materials and Methods**. (E) and (F) Chol oxidation was evaluated by assessing 7β-OH-CH and 7-keto-CH as described in **Materials and Methods**. *p<0.05; **p<0.001 respect to control fibroblasts; #p<0.05; ##p<0.001 compared with PUVA-treated fibroblasts.

Since PPARγ is a key player in lipid metabolism and because damage to cellular lipids is involved in the imbalance of the antioxidant network, we investigated the consequences of PUVA treatment for lipid composition and the possible interference of Octa against this damage. Among the cell compartments,

membrane phospholipids play a causal role in the aging process by modulating oxidative stress and molecular integrity [68,69]. 8-MOP can permeate cell membranes and establish photochemical cross-links between its furan or pyrone ring and unsaturated lipid molecules [70], and the subsequent UVA exposure disturbs the

A

B

C

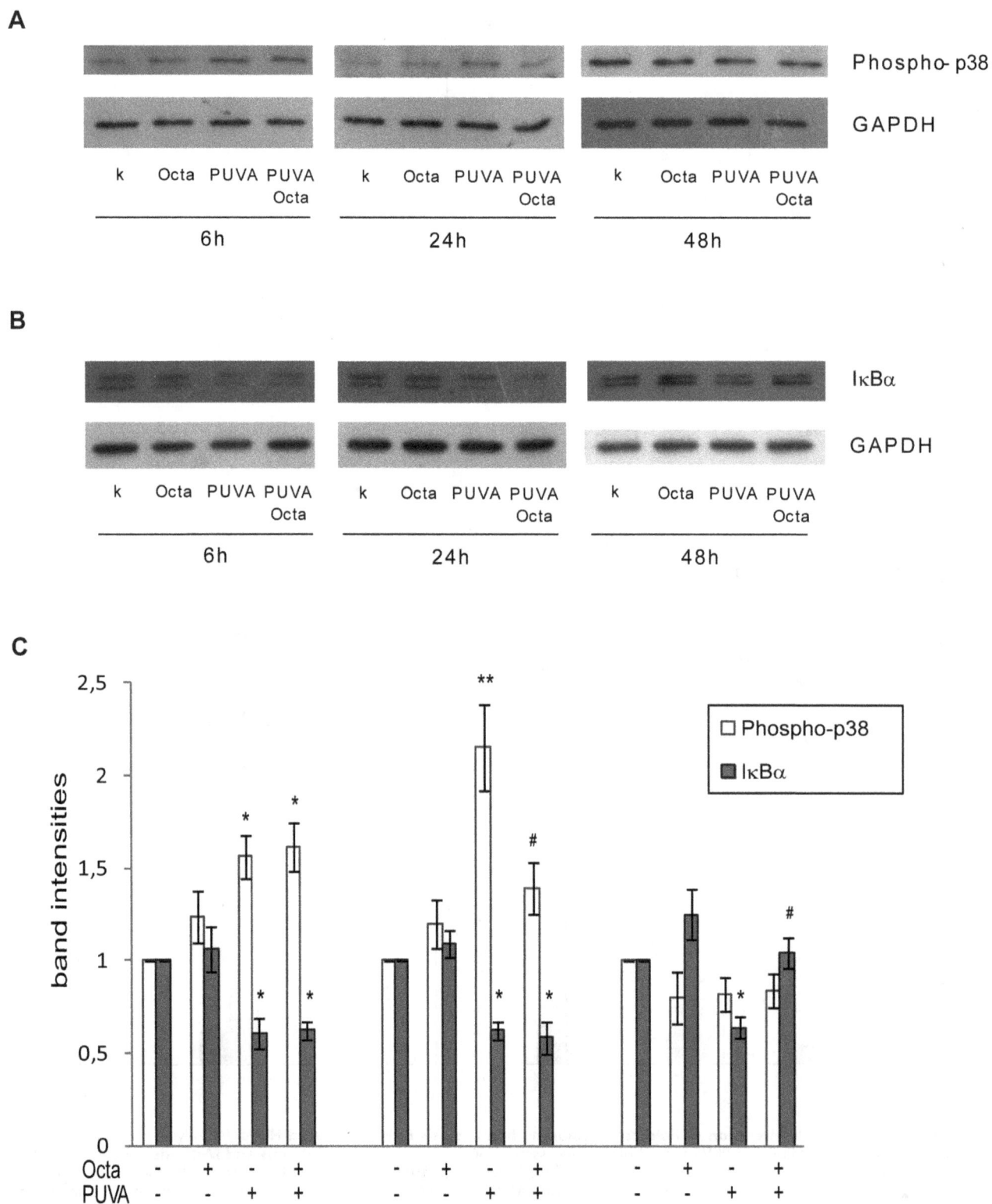

Figure 9. PPARγ interference with PUVA-induced phosphorylation pathway and NF-κB activation. Total cellular proteins (30μg/lane) were subject to 10% SDS-PAGE. Variation of protein loading was determined by reblotting membrane with an anti-GADPH antibody. PUVA-treated HDFs showed an increased phosphorylation of p38 (A) and a decreased expression of IκBα (B). Octa post-treament inhibited p38 phosphorylation (A) as well as decrease of IκBα expression (B) 24 h and 48 h after PUVA treatment, respectively. (C) Densitometric scanning of band intensities obtained from two separate experiments performed in each donor were used to quantify change of protein expression (control value taken as 1-fold in each case). *p<0.05; **p<0.001 respect to control fibroblasts; #p<0.05 compared with PUVA-treated fibroblasts.

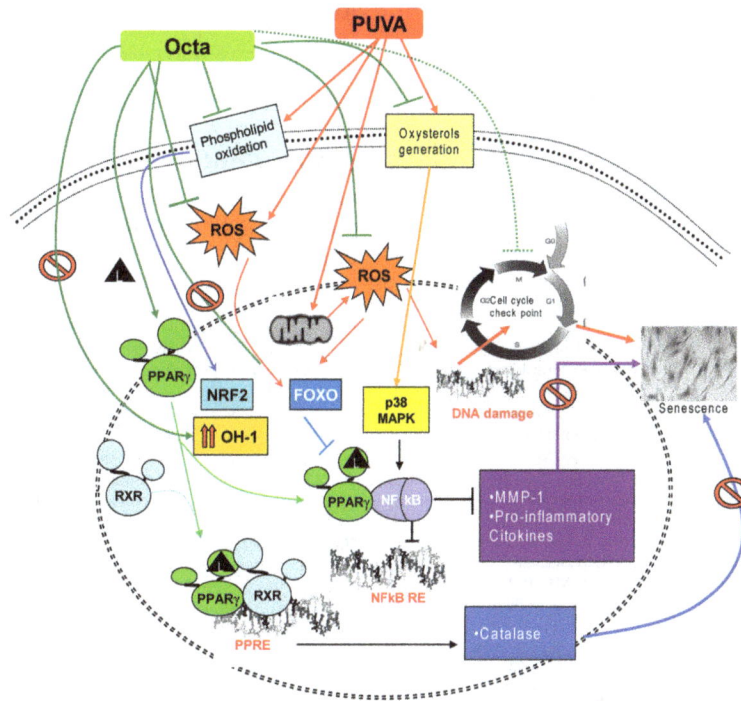

Figure 10. Summary scheme of possible role of PPARγ modulation in counteracting PUVA-SIPS of HDFs. PUVA exposure induced intracellular generation of ROS, alteration of mitochondria function, activation of antioxidant stress response and MAPK phosphorylation pathway, dysregulation of membrane lipid metabolism, DNA-oxidative damage and altered expression of cell cycle regulators. PPARγ modulation by Octa may counteract PUVA-induced senescence-like phenotype. Moreover, Octa ability to reduce phospholipid oxidation and oxysterol generation contributes to the reduction of PUVA-induced inflammatory response and redox imbalance.

integrity of HDF membrane lipids, as demonstrated by the early and permanent decrease in the Pl-PUFA content and the relevant generation of both early and end-products of lipid peroxidation. The oxidative products of cellular lipids diffuse in the cytosol, interacting with intracellular organelles and determining a propagation of the oxidative stress reaction. Phospholipid oxidation products have been reported to activate NRF2 and HO-1 as a compensatory reaction of cells against oxidative stress [71]. However, the accumulation of lipoperoxidation products induced by PUVA can lead to an excessive over-expression of HO-1, shifting the emergency stress response to a deleterious effect against the cell structure. Therefore, the Octa-induced reduction of PUVA-induced phospholipids oxidation products may contribute to the regulation of NRF2 and HO-1 and the subsequent preservation of cell integrity. In addition to phospholipids, CH plays an indispensable role in regulating the properties of cell membranes and the fluidity and the integrity of lipid rafts [72,73]. CH accumulation has been observed in fibroblasts obtained from aged skin [74] as well as *in vitro* senescent cells [75]. The pro-oxidant effect of PUVA caused an early decrease in CH and the immediate generation of oxysterols, peroxidation products of CH metabolism representing reliable markers of oxidative stress *in vivo* [76]. Moreover, the stable appearance of the senescence-like phenotype was associated with a time-dependent accumulation of CH and oxysterols. The observed effect of PUVA on CH metabolism prompted us to investigate the role of PPARγ in controlling the activation of the inflammatory response by chronic oxidative stress which is associated with the induction of cell senescence. The age-related inflammatory chronic state has been associated with a reduction of PPARγ function and an increased generation of oxysterols, which act as secondary messengers in

MAPK signaling pathways [77], an important component of the pathway that regulates cellular senescence as well as the inflammatory response [78]. PUVA-SIPS was characterized by a progressive generation of oxysterols and the up-modulation of phosphorylation signaling involved in NF-κB activation and, in particular, the increase in phosphorylated p38 and the decrease in IκBα, leading to NF-κB activation. In PUVA-exposed cells, the ability of Octa to counteract the accumulation of oxysterols and the changes in the level of CH may contribute to the observed interference with the phosphorylation pathway. It has been suggested that oxysterols act as signaling molecules [79] by influencing lipid membrane integrity as well as the structure and function of PPAR and RXR receptors and their subsequent modulation of the antioxidant response and inflammation [80]. Therefore, PUVA-SIPS contributes to the identification of how biochemical modulators are integrated in the induction of the chronic inflammation state that is typical of aged skin and provides new insights in the activation of nuclear receptors as novel therapeutic approaches for photo-aging (Fig. 10).

Conclusions

Taken together, our data suggest that PUVA-SIPS involves a complex interplay of various cellular transcription factors activated by sustained and long-lasting oxidative stress. Mitochondria are the most probable cell targets, and the modulation of PPARγ provides relevant insights into the mechanism of PUVA-SIPS. The reciprocal influences of PUVA-induced signaling pathways have been investigated by employing Octa due to its ability to increase the trans-activation of PPARγ by acting as a partial agonist and interfering with ROS-dependent cellular signaling mechanisms. Interestingly, Octa counteracts certain molecular markers of

PUVA-SIPS by improving physiological defense mechanisms without significant changes to the cell redox environment.

Supporting Information

Table S1 List of primers used for quantitative real time PCR. Sequences of primers indicated with an F correspond to sense strands and with an R correspond to anti-sense.

References

1. Halliday GM (2005) Inflammation, gene mutation and photoimmunosuppression in response to UVR-induced oxidative damage contributes to photocarcinogenesis. Mutat Res 571: 107–120.
2. Bruls WA, Van Weedlden H, Van der Leun JC (1984) Transmission of UV-irradiation through human epidermal layers as a factor influencing the minimal erythema dose. Photochem Photobiol 39: 63–67.
3. El-Domyati M, Attia S, Saleh F, Brown D, Birk DE, et al. (2002) Intrinsic aging vs. photoaging: a comparative histopathological, immunohistochemical, and ultrastructural study of skin. Exp Dermatol 11: 398–405.
4. Yasui H, Sakurai H (2002) Age-dependent generation of reactive oxygen species in the skin of live hairless rats exposed to UVA light. Exp Dermatol 12: 655–661.
5. Cunningham ML, Krinsky NI, Giovanazzi SM, Peak MJ (1985) Superoxide anion is generated from cellular metabolites by solar radiation and its components. Free Radic Biol Med 1: 381–385.
6. Hanson KM, Clegg RM (2002) Observation and quantification of ultraviolet-induced reactive oxygen species in ex vivo human skin. Photochem Photobiol 76: 57–63.
7. Vile GF, Tyrrell RM (1995) UVA radiation-induced oxidative damage to lipids and proteins in vitro and in human skin fibroblasts is dependent on iron and singlet oxygen. Free Radic Biol Med 18: 721–730.
8. Berneburg M, Grether-Beck S, Kurten V, Ruzicka T, Briviba K, et al. (1999) Singlet oxygen mediates the UVA-induced generation of the photoaging-associated mitochondrial common deletion. J Biol Chem 274: 15345–15349.
9. Scharffetter-Kochanek K, Brenneisen P, Wenk J (2000) Photoaging of the skin from phenotype to mechanisms. Exp Gerontol 35: 307–316.
10. Wenk J, Brenneisen P, Meewes C (2001) UV-induced oxidative stress and photoaging. Curr Probl Dermatol 29: 74–82.
11. Pinnel SR (2003) Cutaneous photo-damage, oxidative stress and topical antioxidant protection. J Am Acad Dermatol. 48: 1–22.
12. Slominski AT, Zmijewski MA, Skobowiat C, Zbytek B, Slominski RM, et al. (2012) Sensing the Environment: Regulation of Local and Global Homeostasis by the Skin's Neuroendocrine System. Advances in Anatomy, Embriology and Cell Biology. New York: Springer-Verlag Berlin Heidelberg. 115p.
13. Nejati R, Kovacic D, Slominski A (2013) Neuro-immune-endocrine functions of the skin: an overview. Expert Rev Dermatol 8: 581–583.
14. Chen JH, Hales NC, Ozanne SE (2007) DNA damage, cellular senescence and organismal ageing: causal or correlative? Nucleic Acids Res 35: 7417–7428.
15. Herbig U, Ferreira M, Condel L, Carey D, Sedivy JM (2006) Cellular senescence in aging primates. Science 311: 1257.
16. Hermann G, Brenneisen P, Wlaschek M, Wenk J, Faisst K, et al. (1998) Psoralen photoactivation promotes morphological and functional changes in fibroblasts in vitro reminiscent of cellular senescence. J Cell Sci 111: 759–767.
17. Briganti S, Wlaschek M, Hinrichs C, Bellei B, Flori E, et al. (2008) Small molecular antioxidants effectively protect from PUVA-induced oxidative stress responses underlying fibroblast senescence and photoaging. Free Radic Biol Med 45: 636–644.
18. Desvergne B, Wahli W (1999) Peroxisome proliferator-activated receptors: nuclear control of metabolism. Endocrine Reviews 20: 649–688.
19. Varga T, Czimmerer Z, Nagy L (2011) PPARs are a unique set of fatty acid regulated transcription factors controlling both lipid metabolism and inflammation. Biochimica Biophysica Acta 1812: 1007–1022.
20. Qq Kwak BR, Mulhaupt F, Mach F (2002) The role of PPARγ ligands as regulators of the immune response. Drug News Perspectives 15: 325–332.
21. Ham SA, Kang ES, Lee H, Hwang JS, Yoo T, et al. (2013) PPARδ inhibits UVB-induced secretion of MMP-1 through MKP-7-mediated suppression of JNK signaling. J Invest Dermatol 133: 2593–2600.
22. McCarty MF, Barroso-Aranda J, Contreras F (2009) The "rejuvenatory" impact of lipoic acid on mitochondrial function in aging rats may reflect induction and activation of PPAR-γ coactivator-1α. Medical Hypotheses 72: 29–33.
23. Polvani S, Tarocchi M, Galli A (2012) PPARγ and Oxidative Stress: Con(β) Catenating NRF2 and FOXO. PPAR Res 2012: 641087.
24. Grabacka M, Placha W, Urbanska K, Laidler P, Płonka PM, et al. (2008) PPAR gamma regulates MITF and beta-catenin expression and promotes a differentiated phenotype in mouse melanoma S91. Pigment Cell Melanoma Res 21: 388–396.
25. Jurzak M, Latocha M, Gojniczek K, Kapral M, Garncarczyk A, et al. (2008) Influence of retinoids on skin fibroblasts metabolism in vitro. Acta Pol Pharm 65: 85–91.
26. Weiss JS, Ellis CN, Headington JT, Voorhees JJ (1988) Topical tretinoin in the treatment of aging skin. J Am Acad Dermatol 19: 169–175.
27. Kim BH, Lee YS, Kang KS (2003) The mechanism of retinol-induced irritation and its application to anti-irritant development. Toxicol Lett 146: 65–73.
28. Stradi R, Pini E, Celentano G (2001) The chemical structure of pigments in Ara macao plumage. Comp Biochem Physiol Part B 130: 57–63.
29. Morelli R, Loscalzo R, Stradi R, Bertelli A, Falchi M (2003) Evaluation of the antioxidant activity of new carotenoid-like compounds by electron paramagnetic resonance. Drugs Exp Clin Res 29: 95–100.
30. Pini E, Bertelli A, Stradi R, Falchi M (2004) Biological activity of parrodienes, a new class of polyunsaturated linear aldehydes similar to carotenoids. Drugs Exp Clin Res 30: 203–206.
31. Flori E, Mastrofrancesco A, Kovacs D, Ramot Y, Briganti S, et al. (2011) 2,4,6-Octatrienoic acid is a novel promoter of melanogenesis and antioxidant defence in normal human melanocytes via PPAR-γ activation. Pigment Cell. Melanoma Res. 24: 618–630.
32. Bayreuther K, Francz PI, Rodemann HP (1992) Fibroblasts in normal and pathological terminal differentiation, aging, apoptosis and transformation. Arch. Geront. Geriatr. Suppl 3: 47–74.
33. Dimri GP, Lee X, Basile G, Acosta M, Scott G, et al. (1995) A biomarker that identifies senescent human cells in culture and in aging skin in vivo. Proc Nat Acad Sci USA 92: 9363–9367.
34. Claiborne A (1985) Catalase activity. In: Greewald RA, editors. Handbook of Methods for Oxygen Radical Research. Boca Raton, FL: CRC. pp. 283–284.
35. Camera E, Rinaldi MR, Briganti S, Picardo M, Fanali S (2001) Simultaneous determination of reduced and oxidized glutathione in peripheral blood mononuclear cells by liquid chromatography-electrospray mass spectrometry. J Chromatogr B Biomed App 757: 69–78.
36. Picardo M, Grammatico P, Roccella F, Roccella M, Grandinetti M, et al. (1996) Imbalance in the antioxidant pool in melanoma cells and normal melanocytes from patients with melanoma. J Invest Dermatol 107: 322–326.
37. Kurien BT, Scofield RH (2003) Free radical mediated peroxidative damage in systemic lupus erythematosus. Life Sciences 73: 1655–1666.
38. Stocks J, Dormandy TL (1971) The autooxidation of human red cell lipids induced by hydrogen peroxide. British J Haematol 20: 95–111.
39. Saito Y, Yoshida Y, Niki E (2007) Cholesterol is more susceptible to oxidation than linoleates in cultured cells under oxidative stress induced by selenium deficiency and free radicals. FEBS Lett. 581: 4349–4354.
40. Rocchi S, Picard F, Vamecq J, Gelman L, Potier N, et al. (2001) A unique PPAR-gamma ligand with potent insulin-sensitizing yet weak adipogenic activity. Mol Cell 8: 737–747.
41. Lee YH, Lee NH, Bhattarai G, Yun JS, Kim TI, et al. (2010) PPARgamma inhibits inflammatory reaction in oxidative stress induced human diploid fibroblast. Cell Biochem Funct 28: 490–496.
42. Briganti S, Flori E, Mastrofrancesco A, Kovacs D, Camera E, et al. (2013) Azelaic acid reduced senescence-like phenotype in photo-irradiated human dermal fibroblasts: possible implication of PPARγ. Exp Dermatol 22: 41–47.
43. Canton M, Caffieri S, Dall'Acqua F, Di Lisa F (2002) PUVA-induced apoptosis involves mitochondrial dysfunction caused by the opening of the permeability transition pore. FEBS Lett 522: 168–172.
44. Okuno Y, Matsuda M, Miyata Y, Fukuhara A, Komuro R, et al. (2010) Human catalase gene is regulated by peroxisome proliferator activated receptor gamm through a response element distinct from that of mouse. Endocr J 57: 303–309.
45. Tian FF, Zhang FF, Lai XD, Wang LJ, Yang L, et al. (2011) Nrf2-mediated protection against UVA radiation in human skin keratinocytes. Biosci Trends 5: 23–29.
46. Raval CM, Zhong JL, Mitchell SA, Tyrrell RM (2012) The role of Bach1 in ultraviolet A-mediated human heme oxygenase 1 regulation in human skin fibroblasts. Free Radic Biol Med 52: 227–236.
47. Zhong JL, Edwards GP, Raval C, Li H, Tyrrell RM (2010) The role of Nrf2 in ultraviolet A mediated heme oxygenase 1 induction in human skin fibroblasts. Photochem Photobiol Sci 9: 18–24.
48. Suttner DM, Dennery PA (1999) Reversal of HO-1 related cytoprotection with increased expression is due to reactive iron. Faseb J 13: 1800–1809.
49. Essers MA, Weijzen S, de Vries-Smits AM, Saarloos I, de Ruiter ND, et al. (2004) FOXO transcription factor activation by oxidative stress mediated by the small GTPase Ral and JNK. The EMBO Journal 23: 4802–4812.

Acknowledgments

The pGL3-(Jwt)3TKLuc reporter construct was kindly provided by Dr R. Ballotti and Dr S. Rocchi (INSERM U895, Centre Méditerranéen de Médecine Moléculaire, Nice, France).

Author Contributions

Conceived and designed the experiments: SB MP. Performed the experiments: SB EF BB. Analyzed the data: SB EF MP. Wrote the paper: SB.

50. Tanaka H, Murakami Y, Ishi I, Nakata S (2009) Involvement of a forkhead transcription factor, FOXO1a, in UV-induced changes of collagen metabolism. J Invest Dermatol Symposium Proceedings 14: 60–62.

51. Dowell P, Otto CT, Adi S, Lane MD (2003) Convergence of peroxisome proliferator-activated receptor γ and Foxo1 signaling pathways. J Biol Chem 278: 45485–45491.

52. Naru E, Suzuki T, Moriyama M, Inomata K, Hayashi A, et al. (2005) Functional changes induced by chronic UVA irradiation to cultured dermal fibroblasts. Br J Dermatol 153: 6–12.

53. Brenneisen P, Sies H, Scharffetter-Kochanek K (2002) Ultraviolet-B irradiation and matrix metalloproteinases: from induction via signalling to initial events. Ann N Y Acad Sci 973: 31–43.

54. McCubrey JA, Lahair MM, Franklin RA (2006) Reactive oxygen species-induced activation of the MAP-kinase signalling pathways. Antiox Redox Signal 8: 1775–1789.

55. Thatcher JE, Isoherranen N (2009) The role of CYP26 enzymes in retinoic acid clearance. Expert Opin Drug Metab Toxicol 5: 875–886.

56. Mongan NP, Gudas LJ (2007) Diverse actions of retinoid receptors in cancer prevention and treatment. Differentiation 75: 853–870.

57. Furuhashi M, Hotamisligil GS (2008) Fatty acid-binding proteins: role in metabolic diseases and potential as drug targets. Nat Rev Drug Discov 7: 489–503.

58. Chiba Y, Yamashita Y, Ueno M, Fujisawa H, Hirayoshi K, et al. (2005) Cultured murine dermal fibroblast-like cells from senescence-accelerated mice as in vitro model for higher oxidative stress due to mitochondrial alterations. J Gerentol A Biol Sci Med Sci 60: 1087–1098.

59. Koziel R, Greussing R, Maier AB, Declercq L, Jansen-Dürr P (2011) Functional interplay between mitochondrial and proteasome activity in skin aging. J Invest Dermatol 131: 594–603.

60. Afaq F, Mukhtar H (2001) Effects of solar radiation on cutaneous detoxification pathways. J Photochem Photobiol B 63: 61–69.

61. Shin MH, Rhie GE, Kim YK, Park CH, Cho KH, et al. (2005) H₂O₂ accumulation by catalase reduction changes MAP kinase signaling in aged human skin in vivo. J Invest Dermatol 125: 221–229.

62. André M, Felley-Bosco E (2003) Heme oxygenase-1 induction by endogenous nitric oxide: influence of intracellular glutathione. FEBS Lett 546: 223–227.

63. Lehmann JC, Listopad JJ, Rentzsch CU, Igney FH, von Bonin A, et al. (2007) Dimethylfumarate induces immunosuppression via glutathione depletion and subsequent induction of heme oxygenase 1. J Invest Dermatol 127: 835–845.

64. Armoni M, Harel C, Karni S, Chen H, Bar-Yoseph F, et al. (2006) FOXO1 represses peroxisome proliferator-activated receptor-gamma1 and -gamma2 gene promoters in primary adipocytes. A novel paradigm to increase insulin sensitivity. J Biol Chem 281: 19881–19891.

65. Fan W, Yanase T, Morinaga H, Okabe T, Nomura M, et al. (2007) Insulin-like growth factor 1/insulin signaling activates androgen signaling through direct interactions of Foxo1 with androgen receptor. J Biol Chem 282: 7329–7338.

66. Santamaria AB, Davis DW, Nghiem DX, McConkey DJ, Ullrich SE, et al. (2002) p53 and Fas ligand are required for psoralen and UVA-induced apoptosis in mouse epidermal cells. Cell Death Differ 9: 549–560

67. Waldman T, Kinzler KW, Vogelstein B (1995) p21 is necessary for the p53-mediated G1 arrest in human cancer cells. Cancer Res 55: 5187–5190.

68. Pamplona R (2008) Membrane phospholipids, lipoxidative damage and molecular integrità: A causal role in aging and longevity. Biochim Biophys Acta 1777: 1249–1262.

69. Park HY, Youm JK, Kwon MJ, Park BD, Lee SH, et al. (2008) K6PC-5, a novel sphingosine kinase activator, improves long-term ultraviolet light-exposed aged murine skin. Exp Dermatol 17: 829–836.

70. dos Santos DJ, Eriksson LA (2006) Permeability of psoralen derivatives in lipid membranes. Biophys J 91: 2464–2474.

71. Gruber F, Mayer H, Lengauer B, Mlitz V, Sanders JM, et al. (2010) NF-E2-related factor 2 regulates the stress response to UVA-1-oxidized phospholipids in skin cells. FASEB J 24: 39–48.

72. Brown DA, London E (2000) Structure and function of of sphingolipid- and cholesterol-rich membrane rafts. J Biol Chem 275: 17221–17224.

73. Simons K, Toomre D (2000) Lipid raftes and signal transduction. Mol Cell Biol 1: 31–39.

74. Park WY, Park JS, Cho KA Kim DI, Ko YG, et al. (2000) Up-regulation of caveolin attenuates epidermal growth factor signaling in senescent cells. J Biol Chem 275: 20847–20852.

75. Maeda M, Scaglia N, Igal RA (2009) Regulation of fatty acid synthesis and Delta9-desaturation in senescence of human fibroblasts. Life Sci 84: 119–124.

76. Schroepfer GJ (2000) Oxysterols: Modulators of cholesterol metabolism and other processes. Physiol Rev 80: 361–554.

77. Anticoli S, Arciello M, Mancinetti A, De Martinis M, Ginaldi L, et al. (2010) 7-ketocholesterol and 5,6-secosterol modulate differently the stress-activated mitogen-activated protein kinases (MAPKs) in liver cells. J CellPhysiol 222: 586–595.

78. Wada T, Stepniak E, Hui L, Leibbrandt A, Katada T, et al. (2008) Antagonistic control of cell fates by JNK and p38-MAPK signaling. Cell Death Differ 15: 89–93.

79. Feingold KR, Jiang YJ (2011) The mechanisms by which lipids coordinately regulate the formation of the protein and lipid domains of the stratum corneum: Role of fatty acids, oxysterols, cholesterol sulfate and ceramides as signaling molecules. Dermatoendocrinol 3: 113–118.

80. Palozza P, Simone R, Catalano A, Monego G, Barini A, et al. (2011) Lycopene prevention of oxysterol-induced proinflammatory cytokine cascade in human macrophages: inhibition of NF-kB nuclear binding and increase in PPARγ expression. J Nutr Biochem 22: 259–268.

12

Comparative Physiological and Proteomic Analyses of Poplar (*Populus yunnanensis*) Plantlets Exposed to High Temperature and Drought

Xiong Li[1,2,3◊], Yunqiang Yang[1,2,3◊], Xudong Sun[1,2], Huaming Lin[1,2,3], Jinhui Chen[4], Jian Ren[5], Xiangyang Hu[1,2], Yongping Yang[1,2]*

1 Key Laboratory for Plant Biodiversity and Biogeography of East Asia, Kunming Institute of Botany, Chinese Academy of Sciences, Kunming, China, 2 Plant Germplasm and Genomics Center, The Germplasm Bank of Wild Species, Kunming Institute of Botany, Chinese Academy of Sciences, Kunming, China, 3 University of Chinese Academy of Sciences, Beijing, China, 4 Key Laboratory of Forest Genetics & Biotechnology, Nanjing Forestry University, Nanjing, China, 5 Department of Grassland Science, Yunnan Agricultural University, Kunming, China

Abstract

Plantlets of *Populus yunnanensis* Dode were examined in a greenhouse for 48 h to analyze their physiological and proteomic responses to sustained heat, drought, and combined heat and drought. Compared with the application of a single stress, simultaneous treatment with both stresses damaged the plantlets more heavily. The plantlets experienced two apparent response stages under sustained heat and drought. During the first stage, malondialdehyde and reactive oxygen species (ROS) contents were induced by heat, but many protective substances, including antioxidant enzymes, proline, abscisic acid (ABA), dehydrin, and small heat shock proteins (sHSPs), were also stimulated. The plants thus actively defended themselves against stress and exhibited few pathological morphological features, most likely because a new cellular homeostasis was established through the collaborative operation of physiological and proteomic responses. During the second stage, ROS homeostasis was overwhelmed by substantial ROS production and a sharp decline in antioxidant enzyme activities, while the synthesis of some protective elements, such as proline and ABA, was suppressed. As a result, photosynthetic levels in *P. yunnanensis* decreased sharply and buds began to die, despite continued accumulation of sHSPs and dehydrin. This study supplies important information about the effects of extreme abiotic environments on woody plants.

Editor: Jin-Song Zhang, Institute of Genetics and Developmental Biology, Chinese Academy of Sciences, China

Funding: This article was supported by Major State Basic Research Development Program (2010CB951700), the Young Academic and Technical Leader Raising Foundation of Yunnan Province (2012HB041) and the National Science Foundation of China (31170256; 31260167). The funders had no role in study design, data collection and analysis, decision to publish, or preparation of the manuscript.

Competing Interests: The authors have declared that no competing interests exist.

* Email: yangyp@mail.kib.ac.cn

◊ These authors contributed equally to this work.

Introduction

Global warming, the most typical manifestation of worldwide climate change, is a focus of increasing attention. Although warming experiments have often been used to simulate future climate conditions, this approach is limited by the unproven assumption that plant responses to experimental warming match their long-term responses to global warming [1]. Within natural habitats, however, plants are often subjected to a combination of different abiotic stresses, each with the potential to exacerbate the damage caused by the others. Recent studies have provided evidence that the molecular, biochemical, and physiological responses of plants to a combination of abiotic stresses are unique and cannot be directly extrapolated from their responses to each stress applied separately [2]. Because high temperatures can increase evapotranspiration rates [3], warming is usually accompanied by drought; plant growth is thus limited directly by heat stress or indirectly via water shortage. In fact, drought and heat shock are common stress factors that often reduce crop yield by more than 50% [4]. They are also two of the most important abiotic stress factors impacting the natural distributions of woody plants and limiting global ecosystem production [5].

Research on plant responses to heat, drought, or their combination has mainly focused on model plants and crops that are herbaceous, such as wheat [4,6–9], sorghum [7], potato [10], pea [11], bean [12,13], and tobacco [14]. The effects of high temperature and drought on the growth and development of woody plants have rarely been studied, and little is known regarding how the combination of these two factors impacts woody plants.

Yunnan poplar (*Populus yunnanensis* Dode), native to high altitude areas of southwestern China, is one of the woody plants most commonly used in stress resistance studies [15]. This plant plays an important role in forestry production, afforestation, and environmental conservation because of its fast growth rate, high biomass, and large populations [16]. Because *P. yunnanensis* populations have recently experienced climate warming and continuous drought stress in southwestern China, an understanding of the combined effects of heat and drought on this species is a research priority. Previous studies of *P. yunnanensis* have involved greenhouse experiments to determine the effects of abiotic stresses on its growth and physiology. The applied stresses have included heavy metals, salinity, acid rain, elevated CO_2, warming, drought, UV-B, and their various combinations [16–23], but the combined effects of heat and drought on *P. yunnanensis* are still largely unknown. In the present study, we performed experiments to explore the response of *P. yunnanensis* to sustained high temperature and drought using comparative proteomic and physiological analyses. To precisely determine the effect of combined heat and drought stress, we also performed comparison experiments involving the application of separate high temperature and drought treatments. We aimed to understand how global climate change may affect woody plants, with *P. yunnanensis* used as a model.

Results

Changes in phenotype and physiological status

Populus yunnanensis plantlets exhibited various phenotypes under different treatments. When plantlets were exposed to either high temperature or drought, a weak morphological change was detected throughout the 48-h stress (Figure 1A). During early stages (0–12 h) of combined heat and drought stress, no significant changes were observed in morphology. However, the buds of *P. yunnanensis* exhibited apparent withering by 24 h, which was even more pronounced at 48 h (Figure 1A). The number of withered leaves and leaf water content displayed little change over 48 h of exposure to high temperature stress (Figure 1B), and remained relatively stable under single drought stress (Figure 1B). When exposed to a combination of high temperature and drought, treated plants had a greater number of withered leaves and slightly decreased leaf water content from 0 h to 24 h compared with the controls (0 h), with both of these parameters changing drastically after 24 h (Figure 1B).

Maximum quantum yield (the ratio of variable to maximum fluorescence; F_v/F_m) and electron transport rates (ETRs) of photosystem II (PSII), which can indicate plant photosynthetic capacity [24], also showed various changes under different stresses. The ratio of F_v to F_m of plantlets exposed to high temperature changed significantly after 24 h of stress (Figure 2A and B), but was only slightly changed in plantlets exposed only to drought (Figure 2A and B). In contrast, F_v/F_m and ETR both decreased significantly in plantlets treated to 40°C without watering (Figure 2). More specifically, F_v/F_m values of samples treated for 6, 12, 24, and 48 h were 16.5, 29.3, 40.4, and 53.0% lower, respectively, compared with the controls (0 h) (Figure 2A and B), and the respective ETRs of these treated samples were 20.1, 38.9, 53.0, and 58.2% lower than the controls (Figure 2C).

Changes in proline, malondialdehyde (MDA), and reactive oxygen species (ROS) contents

Proline content is an important indicator of plant response to abiotic stress, especially drought. The proline content of *P. yunnanensis* rose gradually over time during individual heat or drought stress treatments (Figure 3A). Under combined heat and drought stress, proline content increased from 0 h to 12 h; after 24 h of stress, however, it decreased significantly (Figure 3A).

MDA and ROS such as H_2O_2 and $O_2{}^-$, which reflect grades of cellular oxidation [5], both gradually accumulated when plants were exposed to single or combined heat and drought stress treatments (Figure 3B–D). However, the degree to which MDA and ROS accumulated differed drastically under various treatments, with only slight accumulation under single drought stress and much greater accumulation under single heat stress (Figure 3B–D). When plants were stressed by heat and drought simultaneously, MDA and ROS were produced more rapidly and to greater degrees from 24 h to 48 h than from 0 h to 24 h (Figure 3B–D).

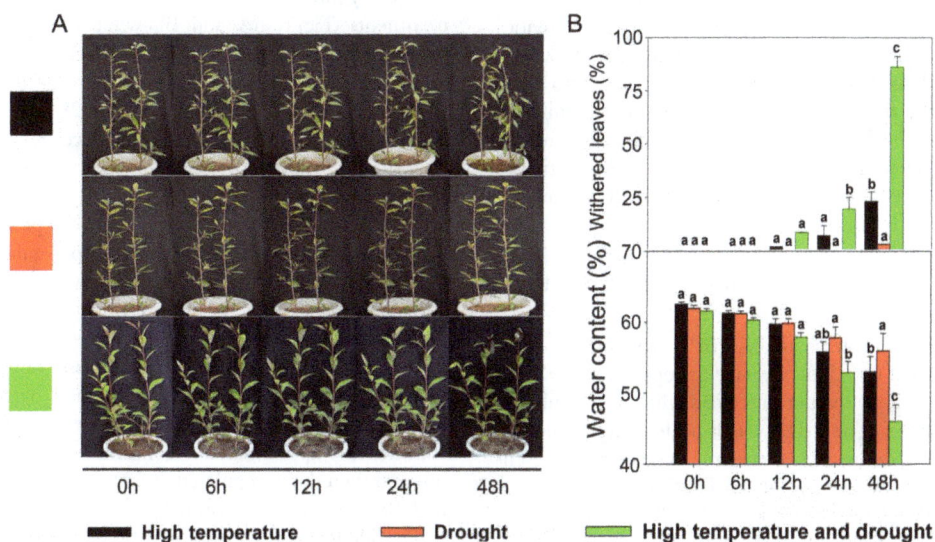

Figure 1. Effects of sustained heat, drought, and combined heat and drought on the morphology and relative water content of leaves of *Populus yunnanensis* plantlets. (**A**) Changes in plantlet morphology. (**B**) Number of withered leaves on plantlets. Data represent the means of five replicate experiments (± SE). Means labeled with different letters were significantly different according to Tukey's test (*P*<0.05). (**C**) Changes in plantlet leaf water content. Data represent the means of five replicate experiments (± SE). Means labeled with different letters were significantly different according to Tukey's test (*P*<0.05).

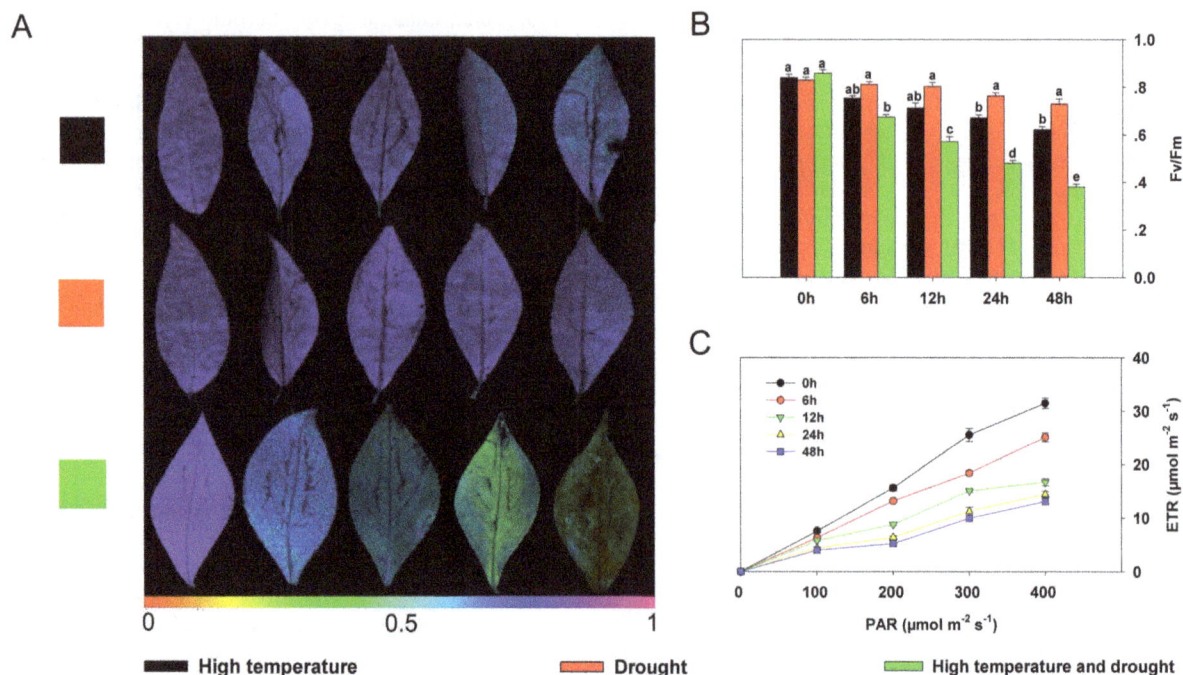

Figure 2. Effects of different treatment durations on leaf photosynthesis in *Populus yunnanensis* **plantlets under different stresses.**
(**A**) F_v/F_m images (bottom). The pseudocolor code depicted at the bottom of the image ranges from 0 (red) to 1.0 (purple). The experiment was replicated three times with similar results. One representative result is shown. (**B**) Average F_v/F_m values. F_v/F_m was determined for whole leaves exposed to different treatments. Data represent the means of five replicate experiments ($\pm$ SE). Means labeled with different letters were significantly different according to Tukey's test ($P<0.05$). (**C**) Electron transport rates determined after different durations of exposure to heat and drought stress. The data represent the means of five replicate experiments ($\pm$ SE).

Dynamics of antioxidant enzyme activities

Activities of antioxidant enzymes, which play essential roles in maintaining ROS homeostasis in plants, were affected differently by heat, drought, and a combination of the two stresses (Figure 4). When exposed to either heat or drought stress, all four antioxidant enzyme activities rose by different degrees with increasing stress duration. Heat triggered greater increases in enzyme activities than did drought. Under the double-stress treatment, however, catalase (CAT) and ascorbate peroxidase (APX) activities were stimulated during the first 12 h and then displayed a significant decline from 24 to 48 h (Figure 4). In a similar fashion, superoxide dismutase (SOD) and glutathione reductase (GR) activities increased from 0 to 24 h, and then decreased markedly after 48-h stress (Figure 4).

Protein profiling of the response of *P. yunnanensis* to different stresses

To obtain a profile of proteins involved in *P. yunnanensis* stress response, we performed two-dimensional electrophoresis (2-DE) of samples subjected to high temperature, drought, and combined heat and drought stress. The 2-DE was repeated three times on each sample with similar results; therefore, one set of representative gels per treatment was visualized by Coomassie Brilliant Blue (CBB) staining (Figures S1, S2, and S3). After staining, more than 600 protein spots were detected within each treatment. Of these, we observed 47, 24, and 90 proteins whose expressions varied by at least 1.5-fold ($P<0.05$) among samples subjected to heat, drought, and combined heat and drought treatment, respectively. Fifty-seven of these differentially expressed proteins (Table S1) were unambiguously identified by matrix-assisted laser desorption/ionization-tandem time-of-flight mass spectrometry

(MALDI-TOF-MS/MS) and screened against the NCBI nonredundant protein database (Table 1). Among these identified proteins, the expressions of 39, 11, and 57 were altered by high temperature, drought, and double-stress treatments, respectively (Figure 5A). Interestingly, the 11 proteins whose expressions were altered by drought stress were also affected by the other two treatments (Figure 5A and B), while the expressions of the 39 differentially expressed proteins observed under high temperature stress were also changed by combined heat and drought stress (Figure 5A). In addition, expressions of 28 proteins were altered by high temperature stress alone and in combination with drought (Figure 5A), while expressions of 18 proteins were changed only when subjected to combined heat and drought stress (Figure 5A and B). Under high temperature stress, 39 protein spots displayed five types of changes (17 continuously up-regulated, 5 continuously down-regulated, 8 first up- and then down-regulated, 3 first down- and then up-regulated, and 6 otherwise) (Figure 5C). Under drought stress, 11 protein spots exhibited three types of changes (8 continuously up-regulated, 2 continuously down-regulated, and 1 otherwise) (Figure 5C). Under combined high temperature and drought stress, 57 protein spots reflected five types of changes (9 continuously up-regulated, 7 continuously down-regulated, 22 first up- and then down-regulated, 3 first down- and then up-regulated, and 16 otherwise) (Figure 5C).

The identified proteins could be classified into nine functional groups, namely, enzyme system (15, 4, and 19 under heat, drought, and combined stresses, respectively), defense-related (4, 2, and 7), cell structure and division (4, 1, and 5), nucleic acid metabolism (5, 0, and 5), redox metabolism (3, 1, and 7), photosynthesis (1, 0, and 3), signal transduction (2, 3, and 3), energy metabolism (3, 1, and 5), and other proteins (2, 0 and 3)

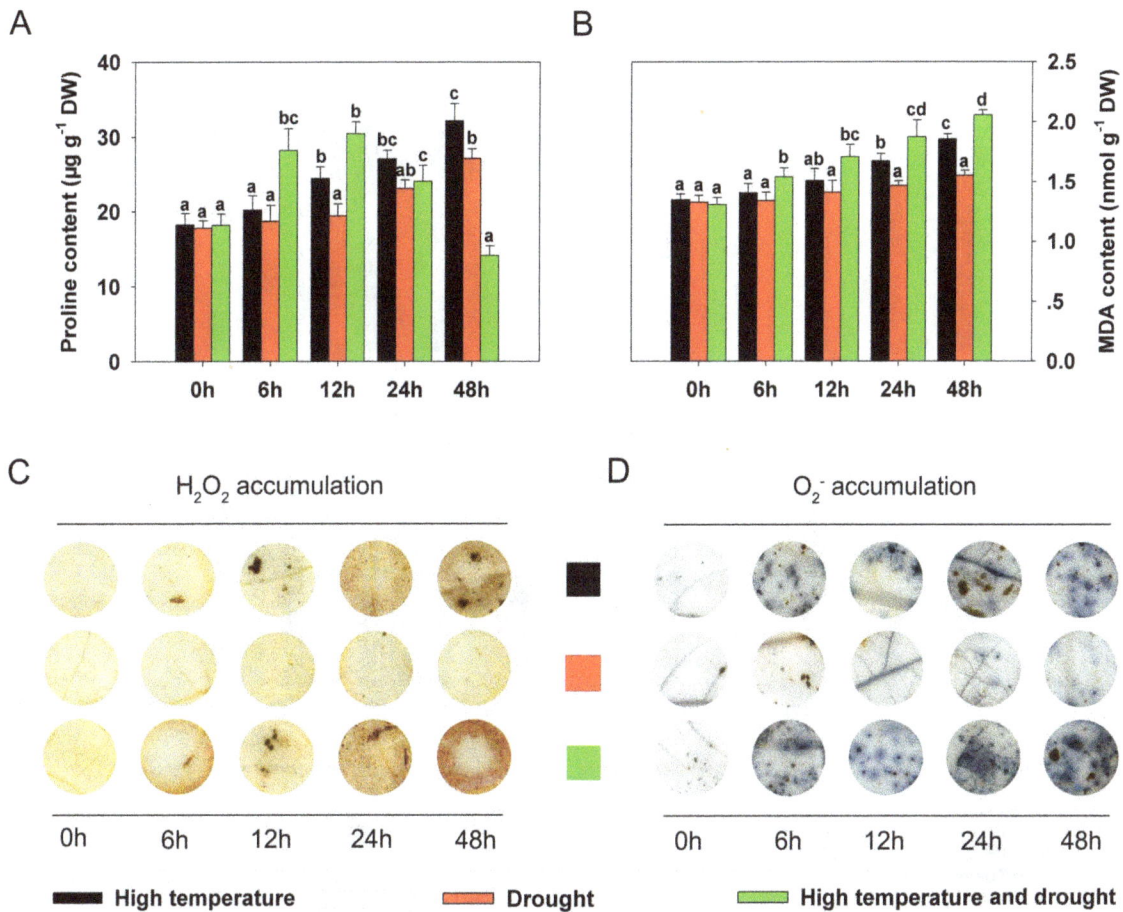

Figure 3. Accumulation of proline, malondialdehyde (MDA), and reactive oxygen species (ROS) (H_2O_2 and O_2^-) in *Populus yunnanensis* plantlets after different durations of exposure to different stresses. (A) Proline content at different times under heat, drought, or combined heat and drought. **(B)** MDA content at different times under heat, drought, or combined heat and drought. Data (B and C) represent the means of five replicate experiments ($\pm$ SE). Means labeled by different letters were significantly different according to Tukey's test ($P<0.05$). **(C)** *In situ* detection of changes in leaf H_2O_2 levels at different times under heat, drought, or their combination. **(D)** *In situ* detection of changes in leaf O_2^- levels at different times under heat, drought, or combined heat and drought.

(Figure 6), implying that these biological processes were affected by heat and drought stress. In particular, the enzyme system consisting of antioxidant enzyme (spots 1, 9, 14, 25, and 28), synthases (spots 4, 36, 38, and 91), kinases (spots 35, 74, 81, and 92), phosphatases (spots 11 and 26), transferases (spots 17, 77, and 99), and mutases (spot 72) accounted for nearly one-third (33%) of all differentially expressed proteins (Figure 6A). Moreover, proteins related to defense (spots 3, 6, 12, 16, 18, 22, and 82) and redox metabolism (10, 31, 33, 37, 39, 80, and 83) both constituted a large proportion of differentially expressed proteins, i.e., 12%, (Figure 6A), indicating their special roles during stress response. In addition, 18 proteins that were only differentially expressed under combined stress belonged to various functional categories with different expression patterns (Figures 5B and 6; Table 1), suggesting they were specifically induced or affected by combined stress but not by the individual stresses.

Differing expressions of abiotic stress-related proteins

To investigate the accumulation of some abiotic stress-related proteins during the course of both single and combined heat and drought treatments in *P. yunnanensis*, we performed protein immunoblot analysis with specific antibodies against plant mitogen-activated protein kinase 6 (MAPK6), heat shock protein

18.2 (HSP18.2), abscisic acid (ABA) synthase 9-cis-epoxycarotenoid dioxygenase (NCED), and dehydrin (Figure 7). Accumulations of these four proteins were induced to varying degrees by single high-temperature and drought stress treatments. Similar to the results of the 2-DE analysis, accumulations of proteins under high temperature stress were much larger than under drought stress (Figure 7). Under combined stress, however, these proteins showed different expressions. MAPK6 was induced from 0 to 24 h but inhibited after 24 h (Figure 7). Notably, the expression peak of NCED, which mediates the synthase of ABA, occurred 6 h after the start of the treatment (Figure 7). Unlike MPK6 and NCED, parallel changes occurred in the accumulation of HSP18.2 and dehydrin; they experienced sustained increases throughout the stress treatment (Figure 7).

Discussion

The effects of different treatments on *P. yunnanensis* plantlets

In plants, a series of integrated events at morphological, physiological, and proteomic levels are triggered by exposure to abiotic stresses [5]. Superoptimal temperatures can lead to changes in plant photosynthesis, protein synthesis, and cell

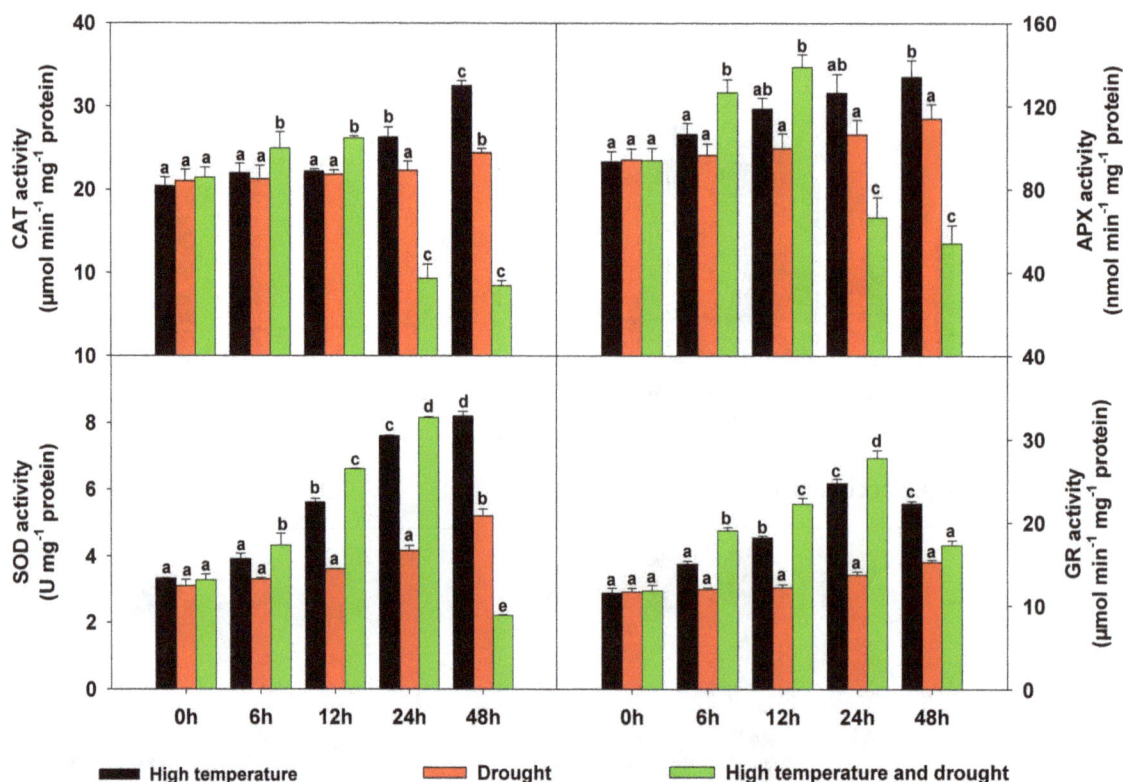

Figure 4. Changes in antioxidant enzyme activities in *Populus yunnanensis* **plantlets after different durations of exposure to different stresses.** The data represent the means of five replicate experiments (± SE). Means labeled with different letters were significantly different according to Tukey's test ($P<0.05$).

contents [25]. As high temperatures cause strong evapotranspiration that induces drought, plants typically suffer from combined heat and drought rather than heat stress alone. Because environmental factors interact synergistically [26,27] or antagonistically [26], plants can be influenced quite differently by a combination of stresses than by single factors. In the present study, we investigated the morphological and physiological changes of *P. yunnanensis* plantlets exposed for 48 h to high temperature, drought, and both stresses simultaneously. Our results consistently indicated that the plantlets were obviously influenced by sustained heat stress but barely affected by 48-h drought (Figure 1). Similar to the results of a previous study using the same species [18], we found that a combination of the two stresses directly damaged the plantlets and had an enhanced effect relative to the influence of either individual stress (Figure 1). To investigate potential plantlet response mechanisms, we further applied proteomics approaches to analyze the internal processes of plantlets treated by heat, drought, and combined heat and drought. The findings revealed by those analyses are discussed below.

Chlorophyll fluorescence and photosynthesis-related proteins

The photosynthetic apparatus associated with PSII is highly sensitive to heat, drought, and various other stresses [13,28] that usually reduce the photosynthetic rate and increase the rate of photorespiration [9,29]. In the present study, a continuous reduction in photosynthetic rate, which reflected the level of stress, was deduced from the change in chlorophyll fluorescence (F_v/F_m) under three different stress regimes (Figure 2A and B). This result, which clearly reveals the serious impact of high

temperature accompanied by drought on the photosynthetic system, was corroborated by a decline in ETRs (Figure 2). Three identified proteins related to photosynthesis – plasma membrane H+-ATPase (spot 13), oxygen-evolving enhancer protein 1 (spot 41), and phosphate import ATP-binding protein PstB (spot 93) – displayed distinctly reduced expression levels under combined heat and drought stress (Table S1). Under single heat or drought stress conditions, however, only the expression of phosphate import ATP-binding protein PstB was reduced significantly (Table S1). This result implies that a single stress, unlike combined stress, had a relatively minimal effect on plants and did not strongly interrupt the regulatory network of the photosynthetic system.

Antioxidant enzymes and related proteins

ROS comprising H_2O_2, O_2^-, OH, and $^1O_2$ are important signal molecules in plants [30]. Under normal conditions, ROS are maintained in homeostasis, with their excessive accumulation prevented by antioxidant enzymes and other substances located in different cell compartments. When plants are exposed to various stresses, ROS are typically induced in sufficient numbers to cause oxidative damage; as confirmed by several previous studies [29,31,32], corresponding antioxidant molecules are induced in response. In the present study, we tested the accumulation of H_2O_2 and O_2^- in conjunction with the activities of antioxidant enzymes (CAT, APX, SOD, and GR) and the expression of related proteins. Small amounts of H_2O_2 and O_2^- were detected under individual heat and drought stress conditions (Figure 3C and D), and the activities of the four antioxidant enzymes increased significantly over the course of the stress treatments (Figure 4). When plants were subjected to combined stress,

Table 1. Identification of differentially expressed proteins in leaves of *Populus yunnanensis* plantlets after different durations of heat and drought stress as analyzed by MALDI-TOF-MS/MS.

Spot	Protein name	Acc. No.[a]	Theo. M_w/pI[b]	Exp. M_w/pI[f]	SC[d] (%)	Score[e]	Organism
Defense related							
3	desiccation-related protein LbLEA3_3–06	gi|169159964	21.12/6.93	21.98/4.88	29.5	40	*L. brevidens*
6	resistance protein	gi|37221893	18.35/5.21	26.84/4.59	57.1	39	*A. stenosperma*
12	Heat shock 22 kDa protein	gi|3122228	23.97/6.34	21.89/5.02	29.9	23	*G. max*
16	26.7 kDa heat shock protein	gi|122247294	26.71/6.78	22.62/5.27	26.7	23	*O. sativa*
18	Late embryogenesis abundant protein D-113	gi|126075	17.48/5.81	17.08/5.29	20.7	22	*G. hirsutum*
22	23.1kDa heat-shock protein	gi|147225064	23.24/5.04	22.06/5.79	28.0	40	*T. monococcum*
82	3-isopropylmalate dehydratase small subunit	gi|166989796	24.23/5.66	27.88/5.23	58.1	43	*A. sp.*
Cell structure and division							
7	small GTP-binding protein	gi|1053067	22.78/5.27	26.72/4.71	54.2	38	*S. lycopersicum*
30	Xyloglucan endotransglucosylase/hydrolase protein A	gi|38605156	34.15/6.99	35.87/6.69	30.8	34	*P. angularis*
48	Ectoderm-neural cortex protein 2	gi|81901549	66.87/6.24	63.79/6.33	23.4	40	*M. musculus*
51	Tetratricopeptide repeat protein 30A2	gi|81918137	76.87/5.08	78.95/5.72	20.7	33	*R. norvegicus*
63	Leucine-rich PPR motif-containing protein	gi|123910179	15.75/6.00	16.81/5.66	19.6	38	*X. tropicalis*
Nucleic acid metabolism							
8	retrotransposon protein, putative	gi|78708153	21.70/5.88	23.50/4.52	45.3	39	*O. sativa*
67	Arginine – tRNA ligase	gi|238688807	62.46/5.72	47.77/5.04	21.1	34	*T. sp.*
76	UPF0042 nucleotide-binding protein Sala_2050	gi|118574110	34.77/6.35	37.36/5.94	18.1	33	*S. alaskensis*
84	Single-stranded DNA-binding protein	gi|6647824	19.09/5.04	17.67/5.38	35.6	32	*T. pallidum*
87	Vigilin	gi|218511884	14.20/6.43	15.00/6.42	10.8	33	*H. sapiens*
Redox metabolism							
10	precursor of dehydrogenase dihydrolipoamide dehydrogenase 1	gi|224099079	54.48/7.24	46.92/5.10	25.4	39	*P. trichocarpa*
31	glyceraldehyde 3-phosphate dehydrogenase	gi|255537011	32.11/7.72	33.79/6.17	32.5	49	*R. communis*
33	Glyceraldehyde-3-phosphate dehydrogenase	gi|122222108	56.56/6.61	35.72/5.81	19.9	29	*O. sativa*
37	flavin-containing monooxygenase YUCCA	gi|171362744	46.08/9.08	44.06/6.33	19.5	39	*O. sativa*
39	Alternative oxidase 1c	gi|3913142	37.91/6.90	42.43/5.62	37.7	39	*A. thaliana*
80	Enoate reductase 1	gi|52788252	44.81/5.60	41.93/5.13	30.8	34	*K. lactis*
83	Dihydrodipicolinate reductase	gi|166224179	27.82/5.15	29.26/5.29	34.1	37	*S. sanguinis*
Photosynthesis							
13	plasma membrane H+-ATPase	gi|2605909	26.44/5.91	19.20/5.83	26.6	34	*K. virginica*
41	Oxygen-evolving enhancer protein 1	gi|131384	35.10/6.25	47.14/5.55	26.6	31	*P. sativum*
93	Phosphate import ATP-binding protein PstB	gi|123748310	31.21/6.25	34.38/6.33	57.2	42	*P. fluorescens*
Signal transduction							
15	Ras-related protein RABH1d	gi|75337262	23.21/6.38	24.13/6.84	41.0	26	*A. thaliana*

Table 1. Cont.

Spot	Protein name	Acc. No.[a]	Theo. M_w/pI[b]	Exp. M_w/pI[f]	SC[d] (%)	Score[e]	Organism
45	Peptide chain release factor 3	gi|122269173	59.32/5.17	59.74/5.51	29.7	39	*L. brevis*
79	Ribosome-releasing factor 2	gi|261277887	81.88/5.91	78.83/4.92	17.6	33	*D. persimilis*
Energy metabolism							
19	ribulose-1,5-bisphosphate/carboxylase large subunit	gi|313758185	18.35/5.24	15.33/5.40	30.5	67	*S. dodecandra*
24	ribulose-1,5-bisphosphate carboxylase/oxygenase large subunit	gi|17224644	27.83/6.21	25.5/6.1	36.5	80	*D. pyrenaica*
29	ribulose-1,5-bisphosphate carboxylase/oxygenase large subunit	gi|67079082	25.60/6.23	35.56/6.75	39.2	47	*D. villosa*
44	ATP synthase subunit beta, mitochondrial	gi|114421	59.93/5.95	60.79/5.40	32.3	46	*N. plumbaginifolia*
65	ATP synthase subunit beta	gi|190358701	53.73/5.09	59.33/5.35	45.3	59	*P. trichocarpa*
Enzyme system							
Antioxidant enzyme							
1	2-cys peroxiredoxin	gi|224140038	29.71/6.44	18.70/4.87	29.3	40	*P. trichocarpa*
9	putative ascorbate peroxidase APX5	gi|31980502	28.90/8.84	22.89/4.74	40.2	48	*A. thaliana*
14	catalase 2	gi|215959344	25.20/6.10	18.99/6.56	17.7	37	*V. unguiculata*
25	Glutathione S-transferase 16	gi|330250548	24.11/6.25	23.84/6.42	23.1	26	*A. thaliana*
28	Peroxidase 43	gi|26397928	35.81/5.68	29.2/6.74	12.4	25	*A. thaliana*
Synthase							
4	5-enol-pyruvylshikimate-phosphate synthase	gi|63334403	47.81/5.76	36.33/4.58	46.2	52	*C. sumatrensis*
36	S-adenosylmethionine synthase	gi|1346524	43.71/5.59	41.73/6.43	21.3	37	*P. deltoides*
38	Indole-3-glycerol phosphate synthase	gi|27735264	44.84/6.99	46.72/6.20	24.3	34	*A. thaliana*
91	Biotin synthase	gi|123725422	38.79/6.19	39.73/6.43	24.7	44	*S. glossinidius*
Kinase							
35	PTI1-like tyrosine-protein kinase At3g15890	gi|75335398	41.34/5.36	38.67/6.37	36.1	28	*A. thaliana*
74	Protein kinase C-like 1B	gi|42560537	81.24/6.67	78.28/5.85	15.8	40	*C. elegans*
81	Acetate kinase	gi|259709978	43.72/5.30	40.82/5.25	38.4	40	*C. botulinum*
92	Acetylglutamate kinase	gi|122279744	31.36/6.27	34.52/6.38	29.2	29	*L. borgpetersenii*
Phosphatase							
11	Probable protein phosphatase 2C 15	gi|75131368	48.62/5.72	46.92/5.17	30.1	23	*O. sativa*
26	Phytochrome-associated serine/threonine protein phosphatase 1	gi|75314041	35.38/4.93	27.26/6.53	23.8	30	*A. thaliana*
Transferase							
17	Caffeoyl-CoA O-methyltransferase	gi|3023419	28.01/5.02	22.06/5.32	32.8	25	*E. gunnii*
77	Acetyl-coenzyme A carboxylase carboxyl transferase subunit alpha	gi|254800799	36.62/5.79	38.23/6.05	30.7	40	*B. anthracis*
99	Octanoyltransferase	gi|171769182	25.08/6.62	21.45/6.71	23.4	33	*A. citrulli*

Table 1. Cont.

Spot	Protein name	Acc. No.[a]	Theo. M_w/pI[b]	Exp. M_w/pI[c]	SC[d] (%)	Score[e]	Organism
Mutase							
72	Phosphoglucosamine mutase	gi|166990410	49.20/5.41	50.96/5.48	26.0	34	*C. botulinum*
Other proteins							
34	UPF0496 protein At4g34320	gi|75213510	42.49/8.47	37.91/6.33	13.9	22	*A. thaliana*
43	predicted protein	gi|224109888	42.77/4.92	46.95/5.25	30.6	62	*P. trichocarpa*
94	UPF0135 protein CPE2004	gi|20978811	29.17/5.02	27.16/4.96	45.6	41	*C. perfringens*

[a]Database accession numbers according to NCBInr; [b]Theoretical M_w/pI; [c]Experimental M_w/pI; [d]Sequence coverage; [e]Mascot search score against the NCBInr database.

however, obviously different results were obtained. During the first 24 h, only small quantities of H_2O_2 and O_2^- were induced (Figure 3C and D), with the antioxidant enzymes also stimulated (Figure 4). After 24 h of stress, ROS levels increased substantially (Figure 3C and D) while antioxidant enzyme activities gradually decreased (Figure 4), indicating that the plants' antioxidant systems may have been disrupted. Interestingly, proteomics analyses also revealed that several antioxidant proteins, namely 2-Cys peroxiredoxin (spot 1), putative ascorbate peroxidase APX5 (spot 9), catalase 2 (spot 14), glutathione S-transferase 16 (spot 25), and peroxidase 43 (spot 28), varied dramatically in expression level (Table S1) – a result generally consistent with observed changes in antioxidant enzyme activities. MDA, commonly used as an index of cellular oxidation levels [32], reflected the status of ROS equilibrium. In our study, the MDA content of *P. yunnanensis* plantlets gradually rose over the course of the different stress treatments, with the greatest increase recorded under combined stress (Figure 3B). Taken together, these results suggest that new equilibria were established under single heat or drought conditions to prevent oxidative damage. Under combined stress, a new equilibrium was also established during early stages (0–24 h); during late stages (24–48 h), however, severe oxidative damage occurred along with obvious phenotypic changes.

Proline and proteins involved in abiotic stress

Proline, an osmotic regulator, can protect cells against heat and other stresses during various stages of acclimation [14]. Proline helps plants avoid oxidative damage and is considered to be an indicator of stress response at the cellular level in many plants [28]. Proline also has been suggested to mediate osmotic adjustment, stabilize macromolecules, serve as a compatible solute to protective enzymes, and store carbon and nitrogen for use during stress regimes such as heat and drought [33]. In our experiments, proline content rose gradually by various degrees under single heat and drought stress conditions, indicating its important role in stress response (Figure 3A). Under combined stress, in contrast, proline content initially increased but then declined (Figure 3A), implying the occurrence of two successive response phases. Similar evidence for these two phases came from the expression of NCED (Figure 7), a synthase of ABA, which is an important plant hormone modulating responses to abiotic stresses including heat, cold, and drought [34]. The various expressions of NCED indirectly suggest the significant roles and different regulatory functions of ABA during different stresses. Dehydrins are present in plants and can be induced by ABA, cold, salt, drought, and heat stress [35]. Western blotting revealed that dehydrin accumulated at different levels throughout the three different stress treatments (Figure 7), implying its significant role in resistance to these stresses.

Another peculiarity of plant response to abiotic stress is the abundant synthesis of sHSPs (17–30 kDa) [28], which constitute an important class of the HSP family. Members of the HSP family protect cells from the deleterious effects of extreme temperatures [14]. We identified three sHSPs –22-kDa (spot 12), 26.7-kDa (spot 16), and 23.1-kDa (spot 22) HSPs – using 2-DE as well as HSP18.2 detected by western blotting. All sHSPs detected under each stress, especially heat-related stress, gradually accumulated over the course of the treatment (Figure 7 and Table S1), indicating their important roles in stress resistance. Nevertheless, the protection conferred by sHSPs was not effective during later periods of combined stress.

Several other defense-related proteins, including desiccation-related protein LbLEA3_3–06 (spot 3), resistance protein (spot 6), late embryogenesis abundant protein D-113 (spot 18), and 3-

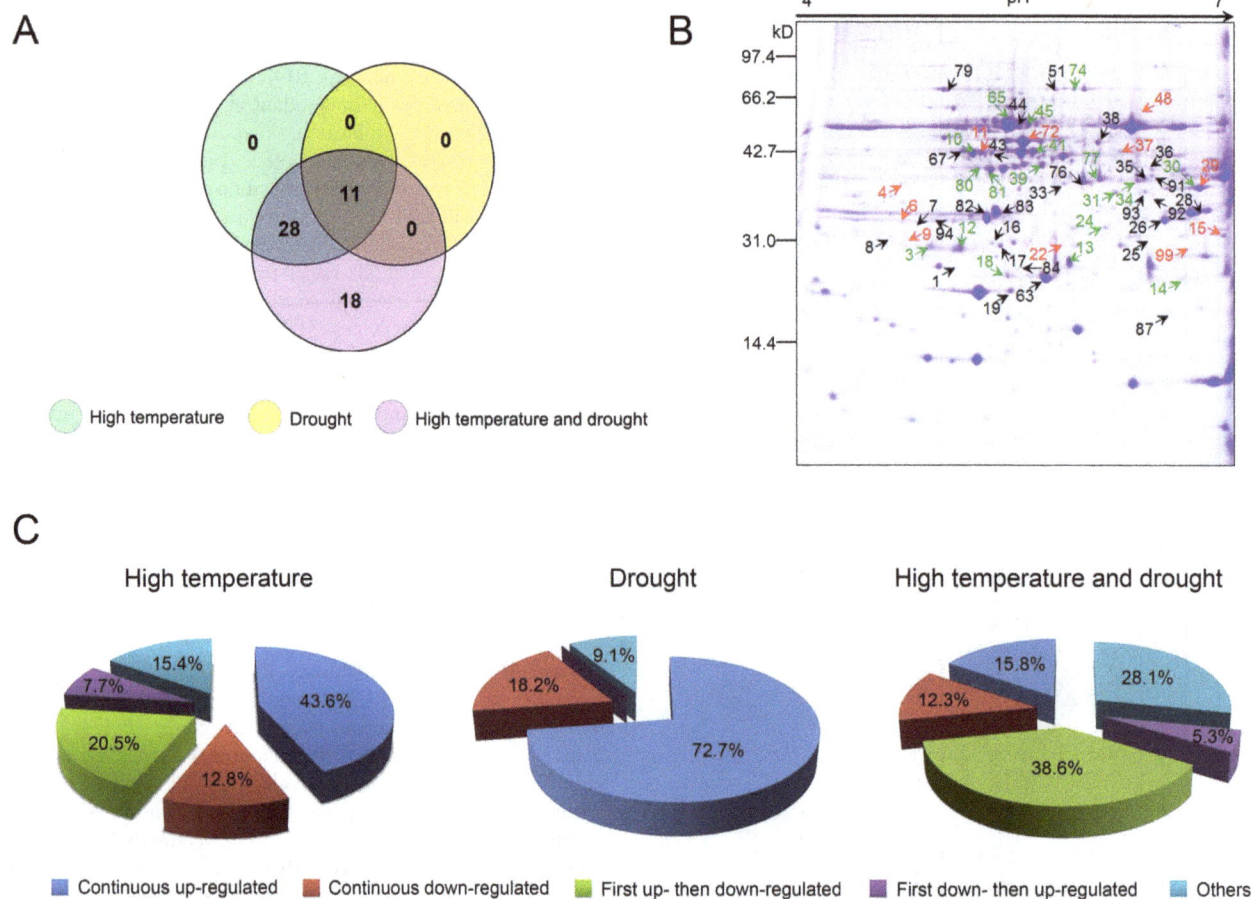

Figure 5. Results of comparative proteomics analyses of different treatments. (A) Representative 2-D gel showing spot numbers of identified proteins. Red spots represent common proteins differentially expressed under all three stresses. Green spots correspond to proteins specifically differentially expressed under combined stress. (B) Venn diagram of differentially expressed proteins under different treatments. (C) Expression patterns of differentially expressed proteins under different treatments.

isopropylmalate dehydratase small subunit (spot 82) were observed to be differentially expressed under different stress conditions (Table S1). Under single heat or drought stress conditions, these proteins were conformably up-regulated, indicating their roles in defense against these stresses (Table S1). Under combined stress, however, these proteins increased in early stages (0–12 or 24 h) and then declined (12 or 24–48 h) (Table S1). Changes in the expression of these abiotic stress-related proteins mirrored the physiological changes of the stress-treated plants.

Proteins involved in cell and nucleic acid activities

The stability of cell and DNA activities is a reflection of the status of plant stress response, as well as the basis of defense against stress. Plant cells can change their structures and division activity to respond to harsh environmental conditions [29]. We observed several cell structural and division-related proteins that were differentially expressed during either individual or combined heat and drought stress conditions. For example, levels of small GTP-binding protein (spot 7) and ectoderm-neural cortex protein 2 (spot 48) decreased in response to the stress treatments (Table S1). Tetratricopeptide repeat protein 30A2 (spot 51) and leucine-rich PPR motif-containing protein (spot 63) exhibited different degrees of increase in *P. yunnanensis* plantlets exposed to different stresses (Table S1), whereas xyloglucan endotransglucosylase/hydrolase

protein A (spot 30) was first induced but then decreased (Table S1). These results suggest that *P. yunnanensis* cell activities are actively regulated or passively influenced by high temperature and drought, similar to the reported response of *Portulaca oleracea* to high temperature conditions [29].

Although environmental stresses can cause nucleic acid damage, many preventative and damage-repair mechanisms exist to enable plant survival [36–38]. During heat or combined heat and drought stress, a putative retrotransposon protein (spot 8) and arginine-tRNA ligase (spot 67) were mainly decreased while nucleotide-binding protein Sala_2050 (spot 76) was mainly up-regulated; a positive correlation was observed between expression level and degree of stress (Table S1). However, the protein designated as single-stranded DNA-binding protein (spot 84) was induced by heat stress but decreased under combined stress (Table S1), supporting the tentative conclusion that the effects of heat and drought stress are exacerbated when the two stresses are combined. These results indicate that various related proteins, despite some degree of down-regulation, work in concert to maintain normal nucleic acid metabolism under stress.

Proteins involved in energy metabolism

On the basis of several proteomics analyses, proteins related to energy metabolism have been proven to play an important role in

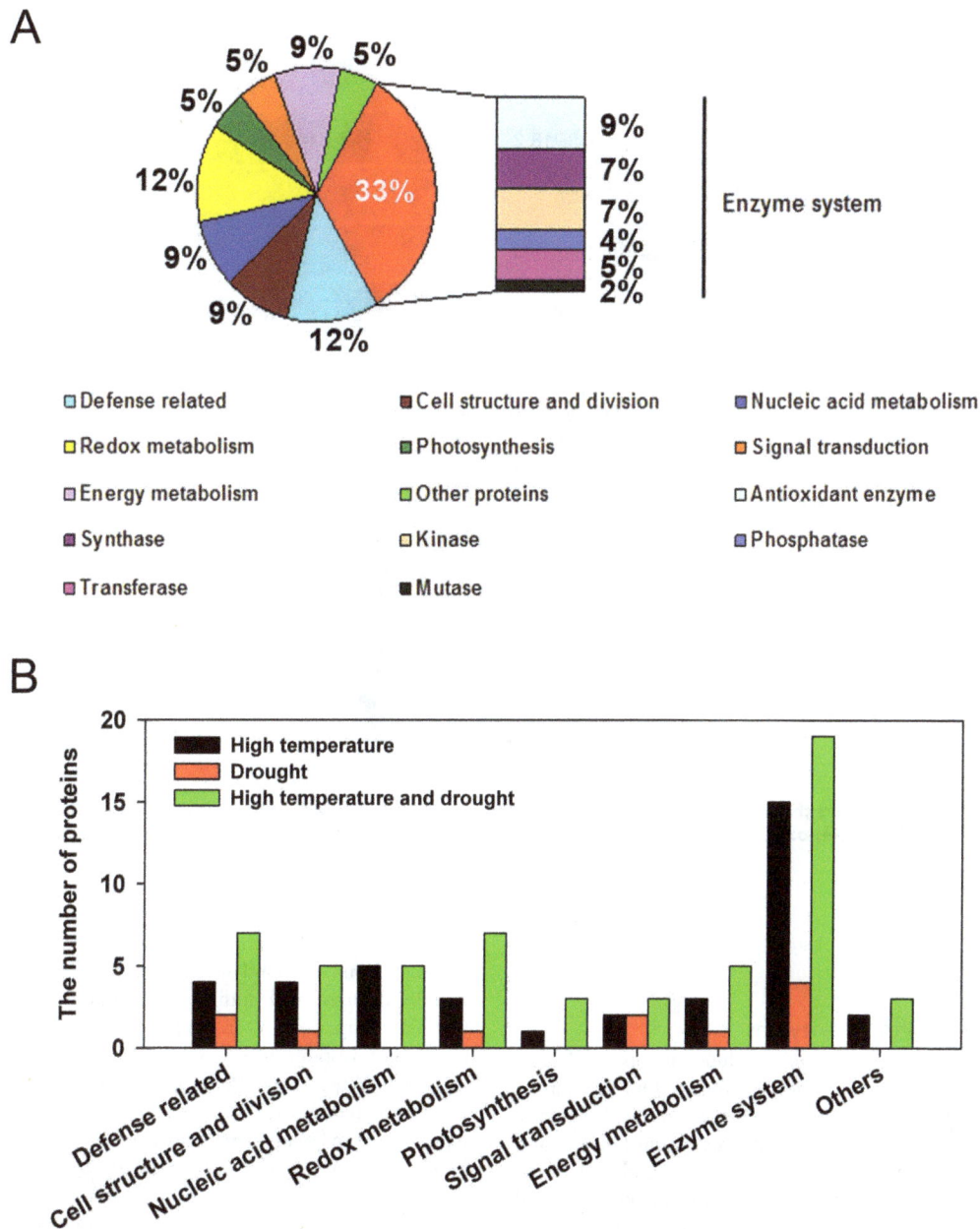

Figure 6. Functional classification of identified proteins and the number of proteins with various functions under different stresses.
(**A**) Functional classification of the identified proteins based on NCBI annotation. (B) The number of proteins with various functions under different stresses.

plant response to abiotic stress. Yang et al. [29] reported that several material- and energy-associated proteins in the thermotolerant plant *Portulaca oleracea* increase in response to high temperature stress, thereby contributing to its heat tolerance. Li et al. [32] found that the expression of two ATP synthases in *Kobresia pygmaea* were up-regulated along an elevational gradient corresponding to increasingly harsh environmental conditions. Because photosynthesis is greatly suppressed under stress, respiration, which is less susceptible and more adaptive than photosynthesis, can become a determinative factor for plant survival [39] by meeting the increased demand for ATP. In the present study, we obtained results in agreement with previous studies. One of two mitochondrial proteins of the ATP synthase

beta subunit (spots 44 and 65) was increased under high temperature stress despite showing no significant difference under drought conditions (Table S1). Three homologs (spots 19, 24, and 29) of the large subunit of ribulose-1,5-bisphosphate carboxylase/oxygenase, which is involved in carbon assimilation and photorespiration [40], were differentially up-regulated under individual drought or heat stress (Table S1), indicating their stress-response contributions. Stronger induction of a greater number of proteins related to energy metabolism was observed during early stages (0–24 h) of the combined stress treatment. Except for one protein whose expression remained at a high level, these proteins were then differentially decreased during later stages (24–24 h) (Table S1). These results support our conclusion that excessive

High temperature

Drought

High temperature and Drought

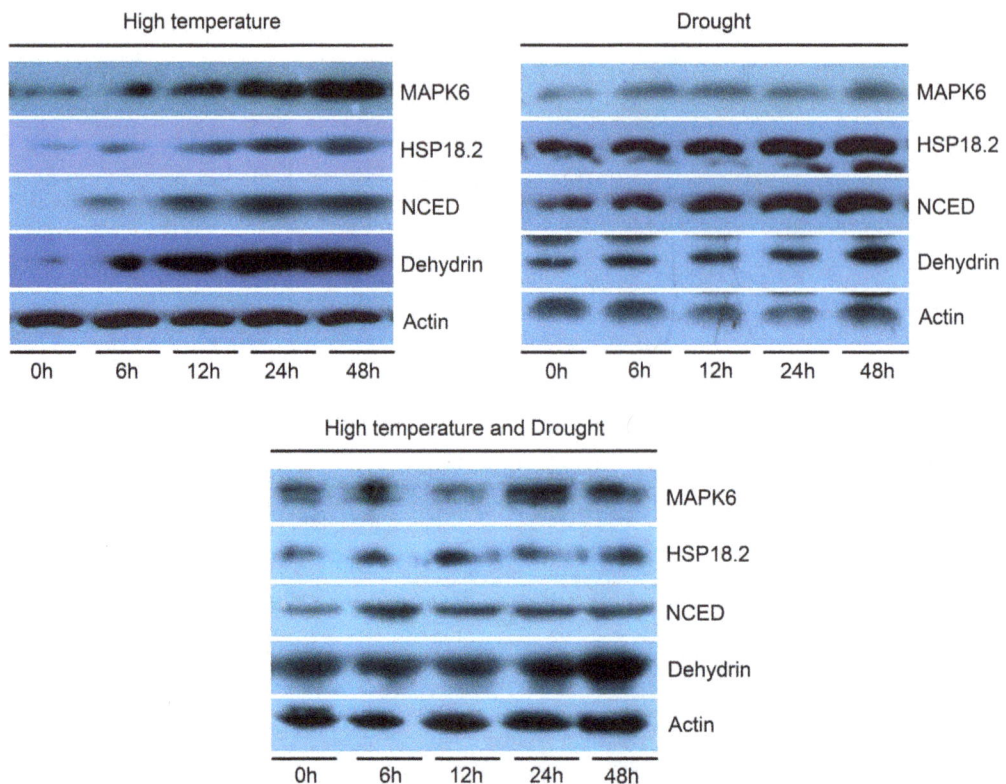

Figure 7. Western blot showing the effects of different stresses on plant mitogen-activated protein kinase 6 (MAPK6), heat shock protein 18.2 (HSP18.2), 9-cis-epoxycarotenoid dioxygenase (NCED), and dehydrin protein accumulation. Actin was included as a protein loading control.

exposure to heat and drought obstructs the defense system of *P. yunnanensis* plantlets.

Different types of protein enzymes

Protein enzymes, which primarily function as biological catalysts, participate in various plant life activities. Our proteomics analyses revealed that many enzymes besides antioxidant enzymes, including synthases, kinases, phosphatases, transferases and mutases, vary dramatically in expression levels under different stresses, with several specific changes observed under combined stress (Table S1). The observed expression changes suggest the importance or sensitivity of these enzymes in stress response. In particular, the basic post-translation protein modifications of phosphorylation and dephosphorylation modulate plant response to environmental stress [41]. We identified four protein kinases – PTI1-like tyrosine-protein kinase At3g15890 (spot 35), protein kinase C-like 1B (spot 74), acetate kinase (spot 81), and acetylglutamate kinase (spot 92) – and two phosphatases – probable protein phosphatase 2C 15 (spot 11) and phytochrome-associated serine/threonine protein phosphatase 1 (spot 26). Most of these enzymes were differentially up-regulated during different stresses until the combined stresses exceeded plant tolerance limits (after 24 h) (Table S1). These results suggest that the enzymes enhance plant response to single heat or drought stress and early stages of combined stress.

Conclusions

In this study, we performed comparative physiological and proteomic profiling to investigate how plantlets of poplar (*P.*

yunnanensis), a common broadleaved deciduous tree of southwestern China, respond to extreme high temperature accompanied by drought, with individual treatments of heat and drought used for comparison. Our results provide insight into how woody plants may respond to excessive heat, as is expected with global warming. When exposed to individual heat or drought stress, plantlets exhibited different levels of resistance, similar to results reported from many previous studies [7–9]. Nevertheless, as indicated in our proposed model (Figure 8), we detected two stages of response to combined stress. During the first stage, between 0 and 24 h, plants actively defended themselves to establish a new cellular homeostasis through both physiological and proteomic responses. This activity explains why plant morphology during this period barely changed. During the second stage, plants were overwhelmed by stress. ROS homeostasis was defeated by ROS overproduction, antioxidant enzyme activities declined, and the synthesis of some protective substances, such as proline and ABA, was suppressed. As a result, photosynthesis decreased sharply, and buds began to die despite continued accumulation of sHSPs and dehydrin. Our results indicate that extreme heat may threaten some non-resistant plants. Plant stress tolerance may be related to plant age [33], with differences existing between young plants and adults. Although excessive heat may not impact adult individuals, it can reduce population density and community structure by killing young plants. As a consequence, seedling fates are worthy of attention. At the same time, many previous studies have revealed significant sexual differences in abiotic stress responses in *P. yunnanensis*, with females usually experiencing greater negative effects than males [16–23]. This observation suggests that female plantlets may be more seriously damaged by exposure to

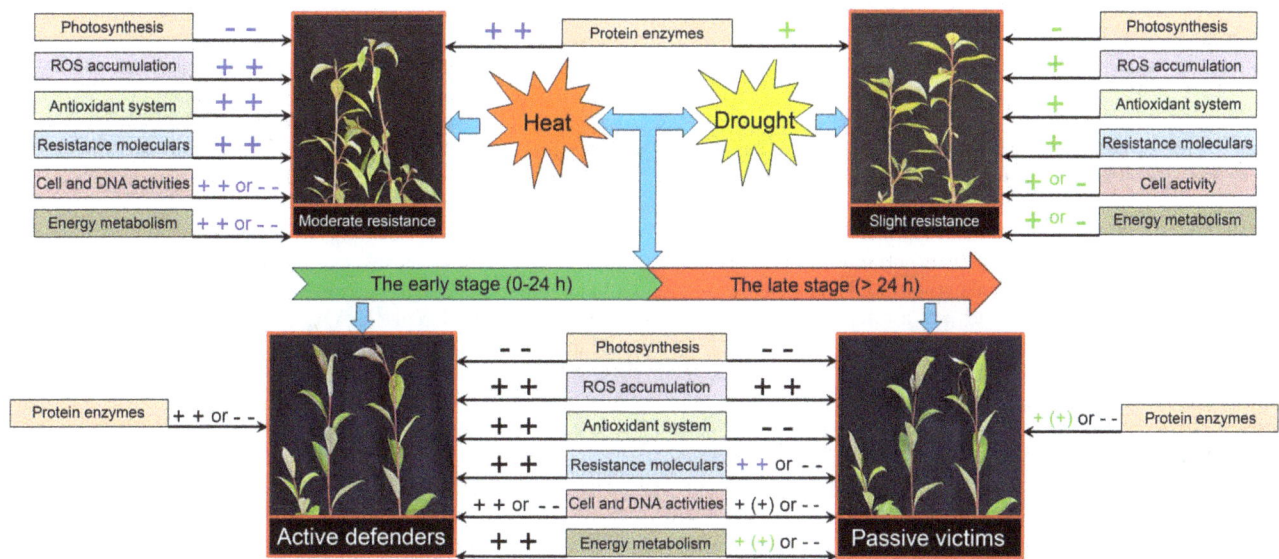

Figure 8. Schematic illustration of a proposed model for the process of *Populus yunnanensis* **plantlet response to high temperature, drought, and a combination of the two stresses.** The symbols "+" and "−" represent slight increases and decreases, respectively, while "+ +" and "− −" represent substantial changes. Information in parentheses is optional. Green, blue, and black symbols are used to show gradual increases in the amounts of proteins and substances involved in the process.

combined heat and drought stress, thus requiring more attention during extreme conditions.

Materials and Methods

Ethics statement

Plant materials used in this study were collected from the Kunming suburbs (E 102°44′24″, N 25°8′20″), Yunnan Province, China. No specific collecting permits were required for this location, as it was located adjacent to our institute, the Kunming Institute of Botany of the Chinese Academy of Sciences. Plant administration was under the auspices of the Kunming City Forestry Bureau, with any studies beneficial or non-damaging to plants permitted and supported by relevant departments. We confirm that the plant we used is a common native species that is neither endangered nor protected.

Plant materials and treatment

Yunnan poplar cuttings were obtained from male plants in March 2012. After survival in the field for 30 d, 150 healthy plantlets with an average of five nodes and a height of 15–20 cm were transplanted into plastic pots (15 cm ×20 cm) containing equal biomass in a greenhouse. The plants were grown in the greenhouse for 30 d under sunlight conditions (23–25°C day and 18–20°C night) and watered daily on a regular schedule with 100 ml of water per pot. The plantlets were then divided into three groups and subjected to different stresses in an incubator with a 12-h photoperiod (800 μmol photons m^{-2} s^{-1} light intensity). One of three stress treatments was applied to each group: (1) a constant temperature of 40°C with regular watering, (2) the normal pre-treatment temperature regime with no watering, or (3) a constant temperature of 40°C with no watering. Treatments were begun simultaneously during the day time, prior to the scheduled daily watering, with treatment continuing for 0, 6, 12, 24, or 48 h. After plant morphological changes were recorded, the fourth to sixth

leaves from the top were harvested to determine their physiological and biochemical properties for each treatment. Five replicates were performed per experiment, and samples from the 0-h treatment were used as controls for the data analysis.

Leaf-change observations and detection

Leaves showing obvious necrotic lesions and crinkling were considered to be withered. Before leaf harvesting, the number of withered leaves was recorded at each time point for all treatments. To measure water content of leaves subjected to stresses, the fourth to sixth leaves from the bottom were collected and any surface impurities removed. Fresh weights (FWs) were measured, with dry weights (DWs) recorded after drying at 80°C for 48 h [42].

Analysis of chlorophyll fluorescence

Chlorophyll fluorescence was analyzed as previously described [29,31] with a pulse-amplitude modulated chlorophyll fluorometer (Heinz Walz GmbH, Effeltrich, Germany). Briefly, *P. yunnanensis* plantlets after treatment were dark-adapted for 30 min to measure the maximal quantum yield of PSII (F_v/F_m), which was determined for each sample by analyzing a whole leaf. The maximal fluorescence (F_m) was recorded using a 0.8-s pulsed light at 4,000 μmol s^{-1} m^{-2}, and minimal fluorescence (F_o) was recorded during the weak measuring pulses. ETRs at a given actinic irradiance were calculated according to the instrument manual as follows: $(F_m' - F_s)/F_m' \times PAR \times 0.5 \times \alpha$, where $(F_m' - F_s)/F_m'$ is the quantum yield of PSII (φPSII) in light, PAR is the photosynthetically active irradiance, 0.5 is the assumed proportion of absorbed quanta used by PSII reaction centers, and α is the absorbance for poplar leaves.

Proline and MDA content measurements

Proline content was measured as previously reported [43]. Approximately 0.5 g of fresh leaves of each sample was homogenized in 8 ml of 3% aqueous sulfosalicylic acid, and the homogenate was centrifuged at 2,000×g for 10 min. Two milliliters each of the extract, acidic ninhydrin, and glacial acetic acid were heated for 1 h in a boiling water bath, with the reaction then terminated in an ice bath. The reaction mixture was extracted with 4 ml toluene, with vigorous mixing using a test-tube stirrer for 15–20 s. The chromophore-containing toluene was aspirated from the aqueous phase and warmed to room temperature, and its absorbance was read at 520 nm using toluene for a blank. The proline concentration was determined from a standard curve and calculated on a DW basis as follows:

(μg ml^{-1} proline × ml toluene) ×5 (g sample)$^{-1}$ = μg proline g^{-1} DW material.

MDA content was determined as described previously [44]. Approximately 0.5determined from a standard curve and g of fresh leaves per sample was homogenized in 10determined from a standard curve and ml of 10% trichloroacetic acid (TCA) and centrifuged at 12,000×g for 10determined from a standard curve and min. Two milliliters of 0.6% thiobarbituric acid in 10% TCA was then added to an aliquot of 2determined from a standard curve and ml of the supernatant. The mixture was heated in boiling water for 30determined from a standard curve and min and then quickly cooled in an ice bath. After centrifugation at 10 000×g for 10determined from a standard curve and min, the absorbance of the supernatant at 450, 532, and 600determined from a standard curve and nm was determined. The MDA concentration, which was expressed as nmol g^{-1} DW, was estimated from the formula: C (nmol ml^{-1}) = 6.45 (A$_{532}$−A$_{600}$) − 0.56A$_{450}$.

In situ H_2O_2 and O_2^- detection

In situ detection of H_2O_2 and O_2^- were performed using a previously reported method with some modifications [45]. To detect H_2O_2, three leaf discs drilled at specific time points during different treatments were vacuum-infiltrated in 10 ml of 1 mg ml^{-1} diaminobenzidine solution for 2 h, and were then cleared in boiling ethanol (95%) for 10 min. The samples were subsequently stored and examined in 95% ethanol. The amount of O_2^- in leaves was monitored by 10^{-2} M nitro-blue tetrazolium (NBT) reduction at specific time points. Three leaf pieces were vacuum-infiltrated with 10 ml NBT for 2 h, cleared in boiling ethanol (95%) for 10 min, and stored and examined in 95% ethanol.

Antioxidant enzyme activity assays

Approximately 0.5 g of leaves from each sample was homogenized in 10 ml extraction buffer (50 mM sodium phosphate [pH 7.0], 1 mM EDTA, 1 mM dithiothreitol [DTT], 1 mM glutathione, 5 mM MgCl$_2$·6H$_2$O, 1% [w/v] PVP-40, and 20% [v/v] glycerin). The homogenates were centrifuged at 12,000×g for 15 min at 4°C, and the total soluble protein content of the supernatants was measured by the Bradford method [46]. CAT (EC1.11.1.6), APX (EC1.11.1.11), SOD (EC1.15.1.1) and GR (EC1.8.1.7) activities were determined as previously described [47,48].

Protein extraction and 2-DE

Protein extraction and 2-DE were performed as reported previously [49], with some modifications. Approximately 10–20 g of leaves from samples exposed to different treatments for 0, 6, 12, 24, or 48 h were ground in liquid nitrogen, and the total soluble proteins were extracted on ice in acetone containing 10% (w/v) TCA and 0.07% (w/v) DTT. The homogenates were held at −20°C for 4 h and then centrifuged at 8,000×g for 30 min at 4°C. The pellets were washed with acetone containing 0.07% (w/v)

DTT at −20°C for 30 min and then centrifuged at 8,000×g for 20 min at 4°C; this step was performed a total of three times. Finally, the pellets were vacuum-dried and then dissolved in lysate (7 M urea, 2 M thiourea, 4% [w/v] CHAPS, and 60 mM DTT) for 2 h at room temperature with intermittent shocking, followed by centrifugation at 12,000×g for 20 min at 20°C. The supernatants were collected for the 2-DE experiments, which were performed in triplicate.

Extracted proteins (1,200 μg) were first separated by isoelectric focusing (IEF) using gel strips to build an immobilized non-linear pH gradient from 4 to 7 (Immobiline Dry Strip, pH 4–7 NL, 17 cm; Bio-Rad, Hercules, CA, USA) and then by sodium dodecyl sulfate-polyacrylamide gel electrophoresis (SDS-PAGE) using 12.5% polyacrylamide. The strips were rehydrated for 14 h in 320 μl of dehydration buffer and then focused at 20°C for a total of 64 kV-h with a PROTEAN IEF Cell system (Bio-Rad). After IEF, the strips were equilibrated for 20 min, first in equilibration buffer I (6 M urea, 0.375 M Tris [pH 8.8], 2% [w/v] SDS, 20% [v/v] glycerol, and 2% [w/v] DTT) and then in equilibration buffer II (6 M urea, 0.375 M Tris [pH 8.8], 2% [w/v] SDS, 20% [v/v] glycerol, and 2% [w/v] iodoacetamide). The equilibrated strips were placed over 12.5% (w/v) SDS-PAGE gels for 2-DE at 25 mA for 5 h. Gels were stained with colloidal CBB. After staining, gels were scanned using PDQuest 2D analysis software (Bio-Rad) on the basis of their relative volumes as described by Bai et al. [31]. To compensate for subtle differences in sample loading or gel staining/destaining during individual experiments, the volume of each spot was normalized [50].

Spot digestion and protein identification for MS analyses

Protein spots displaying significant changes in abundance following plant exposure to heat, drought, or their combination were excised manually from colloidal CBB-stained 2-DE gels using sterile pipette tips. Spots were transferred to 1.5-ml sterile tubes, destained with 50 mM NH$_4$HCO$_3$ for 1 h at 40°C, reduced with 10 mM DTT in 100 mM NH$_4$HCO$_3$ for 1 h at 60°C, and incubated with 40 mM iodoacetamide in 100 mM NH$_4$HCO$_3$ for 30 min. Gels were then minced, air-dried, and rehydrated in 12.5 ng μl^{-1} sequencing-grade modified trypsin (Promega, Fitchburg, WI, USA) in 25 mM NH$_4$HCO$_3$ overnight at 37°C. Tryptic peptides were extracted three times from the gel grains using 0.1% trifluoroacetic acid (TFA) in 50% acetonitrile. Supernatants were concentrated to approximately 10 μl using a SpeedVac (Thermo Fisher, Waltham, MA, USA) and then desalted using reversed-phase ZipTip pipette tips (C18, P10; Millipore, Billerica, MA, USA). Peptides were eluted with 50% acetonitrile and 0.1% TFA. Protein spots that differed in concentration by more than 1.5-fold and differed significantly (Student's t-test, $P<0.05$) compared with the control were analyzed by MS.

Lyophilized peptide samples were dissolved in 0.1% TFA, and MS analysis was conducted using a 4800 Plus MALDI-TOF/TOF Proteomics Analyzer (Applied Biosystems, Foster City, CA, USA). MS acquisition and processing parameters were set to reflector-positive mode and an 800–3,500-Da acquisition mass range, respectively. The laser frequency was 50 Hz, and each sample spectrum was acquired over 700 laser pulses. For secondary MS analysis, four to six ion peaks with signal-to-noise ratios exceeding 100 were selected from each sample as precursors. TOF/TOF signal data for each precursor were then accumulated from 2,000 laser pulses. Primary and secondary mass spectra were transferred to Excel files and compared against a non-redundant NCBI protein database (NCBI-nr 20101014) restricted to Viridiplantae (i.e., green plants) using the MASCOT search engine (www.matrixscience.com). The following search parameters were used:

no molecular weight restriction, one missed trypsin cleavage allowed, iodoacetamide-treated cysteine, oxidation of methionine, a peptide tolerance of 100 ppm, and an MS/MS tolerance of 0.25 Da. Protein identifications were validated manually based on at least three matching peptides. Keratin contamination was removed, and the MOWSE threshold was set above 20 ($P<0.05$). Only significant hits in the MASCOT probability analysis were accepted as protein identifications.

Western blotting

SDS-PAGE was performed as described previously [51] using 12% (w/v) polyacrylamide slab gels. For western blot analysis, the protein samples were electroblotted onto polyvinylidene difluoride membranes using a Trans-Blot cell (Bio-Rad). After transfer, the membranes were probed with the appropriate primary antibodies and HRP-conjugated goat anti-rabbit secondary antibody (Promega), and the signals were detected using an ECL kit (GE, Evansville, IN, USA). The primary antibodies were diluted as follows: polyclonal antibody against MAPK6 (1:1,000), HSP18.2 (1:2,000), NCED (1:3,000), dehydrin (1:3,000), and actin (1:2,000).

Statistical analysis

Statistical analyses were performed using SPSS version 12.0. ANOVA for all variables from measurements were used for testing the treatment differences. Differences were considered significant at the $P<0.05$ level.

Supporting Information

Figure S1 Representative set of 2-D gels of samples subjected to high temperature stress. Marked numbers represent differentially expressed proteins in the treatment.

Figure S2 Representative set of 2-D gels of samples subjected to drought stress. Marked numbers represent differentially expressed proteins in the treatment.

Figure S3 Representative set of 2-D gels of samples subjected to a combination of high temperature and drought. Marked numbers represent differentially expressed proteins in the treatment.

Table S1 Protein spot intensity ratios from different treatments at different treatment times (6, 12, 24, and 48 h) relative to the control (0 h).

Author Contributions

Conceived and designed the experiments: Yongping Yang XYH. Performed the experiments: XL Yunqiang Yang XDS HML JHC. Analyzed the data: XL Yunqiang Yang. Contributed reagents/materials/analysis tools: Yongping Yang XYH JR JHC. Contributed to the writing of the manuscript: XL Yunqiang Yang.

References

1. Wolkovich EM, Cook BI, Allen JM, Crimmins TM, Betancourt JL, et al. (2012) Warming experiments underpredict plant phenological responses to climate change. Nature 485: 494–497.
2. Barua D, Heckathorn SA (2006) The interactive effects of light and temperature on heat-shock protein accumulation in Solidago altissima (Asteraceae) in the field and laboratory. American Journal of Botany 93: 102–109.
3. Verlinden M, Van Kerkhove A, Nijs I (2013) Effects of experimental climate warming and associated soil drought on the competition between three highly invasive West European alien plant species and native counterparts. Plant Ecology 214: 243–254.
4. Grigorova B, Vassileva V, Klimchuk D, Vaseva I, Demirevska K, et al. (2012) Drought, high temperature, and their combination affect ultrastructure of chloroplasts and mitochondria in wheat (Triticum aestivum L.) leaves. Journal of Plant Interactions 7: 204–213.
5. Yang F, Wang Y, Miao LF (2010) Comparative physiological and proteomic responses to drought stress in two poplar species originating from different altitudes. Physiologia Plantarum 139: 388–400.
6. Hurkman WJ, DuPont FM, Altenbach SB, Combs A, Chan R, et al. (1998) BiP, HSP70, NDK and PDI in wheat endosperm. II. Effects of high temperature on protein and mRNA accumulation. Physiologia Plantarum 103: 80–90.
7. Machado S, Paulsen GM (2001) Combined effects of drought and high temperature on water relations of wheat and sorghum. Plant and Soil 233: 179–187.
8. Pradhan GP, Prasad PVV, Fritz AK, Kirkham MB, Gill BS (2012) Effects of drought and high temperature stress on synthetic hexaploid wheat. Functional Plant Biology 39: 190–198.
9. Shah NH, Paulsen GM (2003) Interaction of drought and high temperature on photosynthesis and grain-filling of wheat. Plant and Soil 257: 219–226.
10. Kim MD, Kim YH, Kwon SY, Yun DJ, Kwak SS, et al. (2010) Enhanced tolerance to methyl viologen-induced oxidative stress and high temperature in transgenic potato plants overexpressing the CuZnSOD, APX and NDPK2 genes. Physiologia Plantarum 140: 153–162.
11. Guilioni L, Wery J, Lecoeur J (2003) High temperature and water deficit may reduce seed number in field pea purely by decreasing plant growth rate. Functional Plant Biology 30: 1151–1164.
12. Stoyanova D, Yordanov I (1999) Influence of drought, high temperature, and carbamide cytokinin 4-PU-30 on photosynthetic activity of plants. 2. Chloroplast ultrastructure of primary bean leaves. Photosynthetica 37: 621–625.
13. Yordanov I, Velikova V, Tsonev T (1999) Influence of drought, high temperature, and carbamide cytokinin 4-PU-30 on photosynthetic activity of bean plants. 1. Changes in chlorophyll fluorescence quenching. Photosynthetica 37: 447–457.
14. Kuznetsov VV, Shevyakova NI (1997) Stress responses of tobacco cells to high temperature and salinity. Proline accumulation and phosphorylation of polypeptides. Physiologia Plantarum 100: 320–326.
15. Centritto M, Brilli F, Fodale R, Loreto F (2011) Different sensitivity of isoprene emission, respiration and photosynthesis to high growth temperature coupled with drought stress in black poplar (Populus nigra) saplings. Tree Physiology 31: 275–286.
16. Li L, Zhang YB, Luo JX, Korpelainen H, Li CY (2013) Sex-specific responses of Populus yunnanensis exposed to elevated CO2 and salinity. Physiologia Plantarum 147: 477–488.
17. Chen LH, Han Y, Jiang H, Korpelainen H, Li CY (2011) Nitrogen nutrient status induces sexual differences in responses to cadmium in Populus yunnanensis. Journal of Experimental Botany 62: 5037–5050.
18. Chen LH, Zhang S, Zhao HX, Korpelainen H, Li CY (2010) Sex-related adaptive responses to interaction of drought and salinity in Populus yunnanensis. Plant Cell and Environment 33: 1767–1778.
19. Duan BL, Xuan ZY, Zhang XL, Korpelainen H, Li CY (2008) Interactions between drought, ABA application and supplemental UV-B in Populus yunnanensis. Physiologia Plantarum 134: 257–269.
20. Duan BL, Zhang XL, Li YP, Li L, Korpelainen H, et al. (2013) Plastic responses of Populus yunnanensis and Abies faxoniana to elevated atmospheric CO2 and warming. Forest Ecology and Management 296: 33–40.
21. Jiang H, Korpelainen H, Li CY (2013) Populus yunnanensis males adopt more efficient protective strategies than females to cope with excess zinc and acid rain. Chemosphere 91: 1213–1220.
22. Jiang H, Peng SM, Zhang S, Li XG, Korpelainen H, et al. (2012) Transcriptional profiling analysis in Populus yunnanensis provides insights into molecular mechanisms of sexual differences in salinity tolerance. Journal of Experimental Botany 63: 3709–3726.
23. Peng SM, Jiang H, Zhang S, Chen LH, Li XG, et al. (2012) Transcriptional profiling reveals sexual differences of the leaf transcriptomes in response to drought stress in Populus yunnanensis. Tree Physiology 32: 1541–1555.
24. Yang LM, Tian DG, Todd CD, Luo YM, Hu XY (2013) Comparative Proteome Analyses Reveal that Nitric Oxide Is an Important Signal Molecule in the Response of Rice to Aluminum Toxicity. Journal of Proteome Research 12: 1316–1330.
25. Wu MT, Wallner SJ (1983) Heat-Stress Responses in Cultured Plant-Cells – Development and Comparison of Viability Tests. Plant Physiology 72: 817–820.
26. Bansal S, Hallsby G, Lofvenius MO, Nilsson MC (2013) Synergistic, additive and antagonistic impacts of drought and herbivory on Pinus sylvestris: leaf, tissue and whole-plant responses and recovery. Tree Physiology 33: 451–463.
27. Xue TL, Hartikainen H (2000) Association of antioxidative enzymes with the synergistic effect of selenium and UV irradiation in enhancing plant growth. Agricultural and Food Science in Finland 9: 177–186.
28. Mihailova G, Petkova S, Buchel C, Georgieva K (2011) Desiccation of the resurrection plant Haberlea rhodopensis at high temperature. Photosynthesis Research 108: 5–13.

29. Yang Y, Chen J, Liu Q, Ben C, Todd CD, et al. (2012) Comparative proteomic analysis of the thermotolerant plant Portulaca oleracea acclimation to combined high temperature and humidity stress. J Proteome Res 11: 3605–3623.

30. Chaves MM, Maroco JP, Pereira JS (2003) Understanding plant responses to drought – from genes to the whole plant. Functional Plant Biology 30: 239–264.

31. Bai X, Yang L, Yang Y, Ahmad P, Yang Y, et al. (2011) Deciphering the protective role of nitric oxide against salt stress at the physiological and proteomic levels in maize. J Proteome Res 10: 4349–4364.

32. Li X, Yang YQ, Ma L, Sun XD, Yang SH, et al. (2014) Comparative Proteomics Analyses of Kobresia pygmaea Adaptation to Environment along an Elevational Gradient on the Central Tibetan Plateau. Plos One 9.

33. Ashraf M, Foolad MR (2007) Roles of glycine betaine and proline in improving plant abiotic stress resistance. Environmental and Experimental Botany 59: 206–216.

34. Fujita M, Fujita Y, Noutoshi Y, Takahashi F, Narusaka Y, et al. (2006) Crosstalk between abiotic and biotic stress responses: a current view from the points of convergence in the stress signaling networks. Current Opinion in Plant Biology 9: 436–442.

35. Kosova K, Vitamvas P, Prasil IT (2011) Expression of dehydrins in wheat and barley under different temperatures. Plant Science 180: 46–52.

36. Boyko A, Kovalchuk I (2008) Epigenetic control of plant stress response. Environmental and Molecular Mutagenesis 49: 61–72.

37. Chinnusamy V, Zhu JK (2009) RNA-directed DNA methylation and demethylation in plants. Science in China Series C-Life Sciences 52: 331–343.

38. Chinnusamy V, Zhu JK (2009) Epigenetic regulation of stress responses in plants. Current Opinion in Plant Biology 12: 133–139.

39. Vassileva V, Signarbieux C, Anders I, Feller U (2011) Genotypic variation in drought stress response and subsequent recovery of wheat (Triticum aestivum L.). Journal of Plant Research 124: 147–154.

40. Ji J, Scott MP, Bhattacharyya MK (2006) Light is essential for degradation of ribulose-1.5-bisphosphate carboxylase-oxygenase large subunit during sudden death syndrome development in soybean. Plant Biology 8: 597–605.

41. Ichimura K, Shinozaki K, Tena G, Sheen J, Henry Y, et al. (2002) Mitogen-activated protein kinase cascades in plants: a new nomenclature. Trends in Plant Science 7: 301–308.

42. Rivero RM, Kojima M, Gepstein A, Sakakibara H, Mittler R, et al. (2007) Delayed leaf senescence induces extreme drought tolerance in a flowering plant. Proceedings of the National Academy of Sciences of the United States of America 104: 19631–19636.

43. Bates LS, Waldren RP, Teare ID (1973) Rapid Determination of Free Proline for Water-Stress Studies. Plant and Soil 39: 205–207.

44. Duan BL, Lu YW, Yin CY, Junttila O, Li CY (2005) Physiological responses to drought and shade in two contrasting Picea asperata populations. Physiologia Plantarum 124: 476–484.

45. Able AJ (2003) Role of reactive oxygen species in the response of barley to necrotrophic pathogens. Protoplasma 221: 137–143.

46. Barbosa H, Slater NKH, Marcos JC (2009) Protein quantification in the presence of poly(ethylene glycol) and dextran using the Bradford method. Analytical Biochemistry 395: 108–110.

47. Nakano Y, Asada K (1981) Hydrogen-Peroxide Is Scavenged by Ascorbate-Specific Peroxidase in Spinach-Chloroplasts. Plant and Cell Physiology 22: 867–880.

48. Jiang MY, Zhang JH (2001) Effect of abscisic acid on active oxygen species, antioxidative defence system and oxidative damage in leaves of maize seedlings. Plant and Cell Physiology 42: 1265–1273.

49. Damerval C, Devienne D, Zivy M, Thiellement H (1986) Technical Improvements in Two-Dimensional Electrophoresis Increase the Level of Genetic-Variation Detected in Wheat-Seedling Proteins. Electrophoresis 7: 52–54.

50. Wan XY, Liu JY (2008) Comparative proteomics analysis reveals an intimate protein network provoked by hydrogen peroxide stress in rice seedling leaves. Molecular & Cellular Proteomics 7: 1469–1488.

51. Laemmli UK, Beguin F, Gujerkel.G (1970) A Factor Preventing Major Head Protein of Bacteriophage T4 from Random Aggregation. Journal of Molecular Biology 47: 69–&.

13

Albumin Antioxidant Response to Stress in Diabetic Nephropathy Progression

Rafael Medina-Navarro[1]*, Itzia Corona-Candelas[2], Saúl Barajas-González[2], Margarita Díaz-Flores[3], Genoveva Durán-Reyes[3]

1 Department of Experimental Metabolism, Center for Biomedical Research of Michoacán (CIBIMI-IMSS), Morelia, Michoacán, México, **2** Department of Nephrology, General Regional Hospital N° 1, IMSS, Morelia, Michoacán, Mexico, **3** Biochemistry Medical Research Unit, National Medical Center, IMSS, México City, México

Abstract

Background: A new component of the protein antioxidant capacity, designated Response Surplus (RS), was recently described. A major feature of this component is the close relationship between protein antioxidant capacity and molecular structure. Oxidative stress is associated with renal dysfunction in patients with renal failure, and plasma albumin is the target of massive oxidation in nephrotic syndrome and diabetic nephropathy. The aim of the present study was to explore the albumin redox state and the RS component of human albumin isolated from diabetic patients with progressive renal damage.

Methods/Principal Findings: Serum aliquots were collected and albumin isolated from 125 diabetic patients divided into 5 groups according to their estimated glomerular filtration rate (GFR). In addition to clinical and biochemical variables, the albumin redox state, including antioxidant capacity, thiol group content, and RS component, were evaluated. The albumin antioxidant capacity and thiol group content were reciprocally related to the RS component in association with GFR reduction. The GFR decline and RS component were significantly negatively correlated ($R = -0.83$, $p < 0.0001$). Age, creatinine, thiol groups, and antioxidant capacity were also significantly related to the GFR decline ($R = -0.47$, $p < 0.001$; $R = -0.68$, $p < 0.0001$; $R = 0.44$, $p < 0.001$; and $R = 0.72$, $p < 0.0001$).

Conclusion/Significance: The response of human albumin to stress in relation to the progression of diabetic renal disease was evaluated. The findings confirm that the albumin molecular structure is closely related to its redox state, and is a key factor in the progression of diabetes nephropathy.

Editor: Maria P. Rastaldi, Fondazione IRCCS Ospedale Maggiore Policlinico & Fondazione D'Amico per la Ricerca sulle Malattie Renali, Italy

Funding: This study was supported by a grant from Consejo Nacional de Ciencia y Tecnología (CONACYT) SALUD-2010-01-141937/CONACYT- FIS/IMSS/PROT/ 896 and a grant from the Fondo de Investigación en Salud FIS/IMSS/PROT/G10/854. The funders had no role in study design, data collection and analysis, decision to publish, or preparation of the manuscript.

Competing Interests: The authors have declared that no competing interests exist.

* Email: cibimirmn@gmail.com

Introduction

Increasing evidence supports an association between the serum albumin concentration and mortality from several diseases [1–4]. The integrity of the albumin molecule might be a key determinant of its biologic activity [5,6]. Structural stress induced by nonenzymatic glycation or reactive oxygen species impairs the antioxidant capabilities of albumin, a factor associated with the development of diabetes complications [7]. Evidence of protein stress (carbonyl content, dityrosine content) is detected in patients with diabetic nephropathy (DN) [8,9]. How this stress affects the protein antioxidant response and the molecular structure of albumin during the progression of DN, however, is not clear. Albumin has antioxidant properties [10] and is the major antioxidant in plasma, which is continuously exposed to oxidative stress [11].

A new component of the protein antioxidant capacity (AC) was recently described [12]. The intrinsic component, designated

Response Surplus (RS), represents the antioxidant response that occurs when proteins undergo a structural perturbation by a stressor, such as temperature, short-wave ultraviolet (UV) light, and reactive oxygen species. The change in the AC of proteins is closely related to their molecular structure [12].

Previous studies demonstrated the impaired antioxidant capacity of albumin as a consequence of various structural modifications [13,14]. The changes were clearly linked to the passive redox state of the thiol groups and particularly to the albumin redox active thiol group (Cys-34). Therefore, the AC of any biologic system is very complex, and the albumin Cys-34-dependent AC of albumin represents a passive component, whereas the RS represents an active component linked to changes in the molecular structure [12].

The model proposed to explain the RS is presented in Figure 1. Using this model, we aim to explain the mechanism underlying the albumin antioxidant response to stress that can only be partially

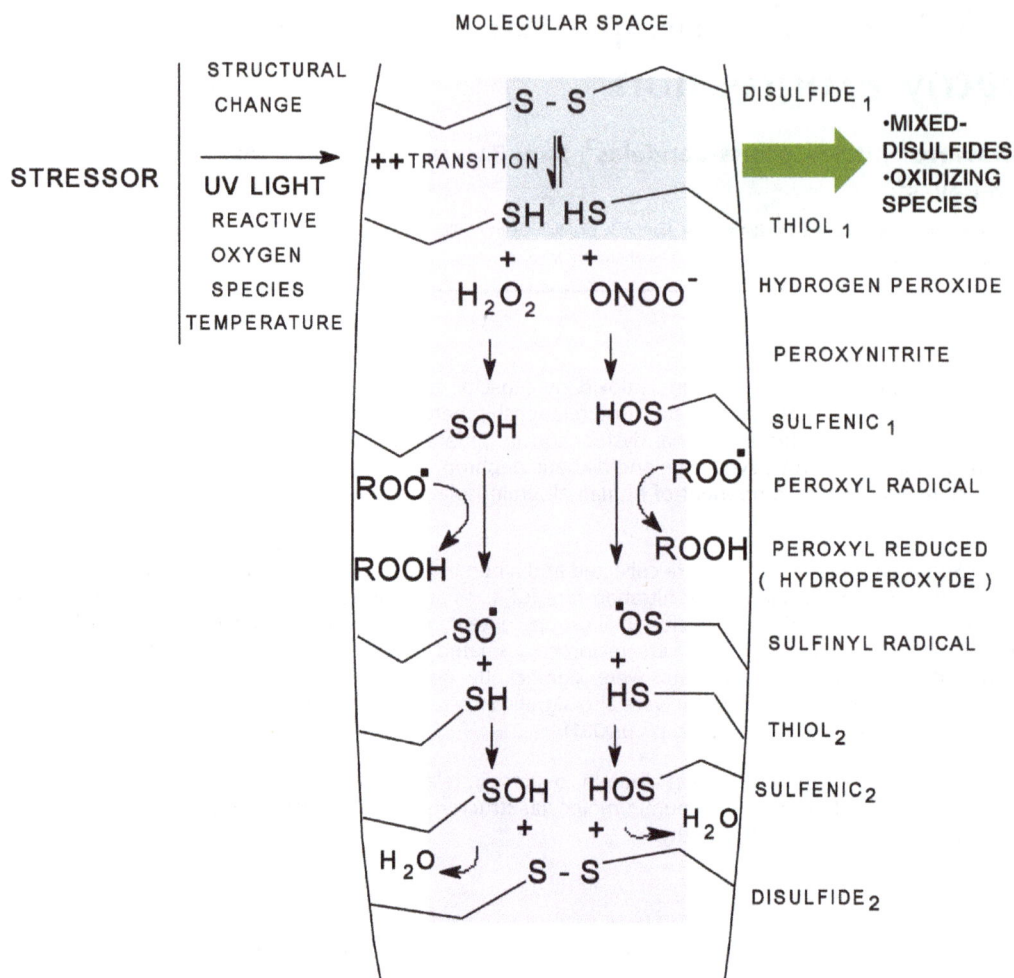

Figure 1. In accordance with the model proposed to explain the Response Surplus (RS), a stressor introduces instability into the core of a protein, which promotes the transition (structural perturbation) between the disulfide and thiol (SH) groups. The SH groups then interact with reactive oxygen species and form mixed disulfides. Although it is a primer, it is part of the passive antioxidant capacity of a protein and is Cys34-dependent. In contrast, in a second dynamic phase, a sulfenic acid derivative forms from the reaction between reactive oxygen species and SH groups can reduce free radicals such as peroxyl radicals (ROO) with high efficiency, as previously reported [15], generating sulfinyl radical derivatives (RSO) as the product of the reaction. Then, a second stock of sulfenic acid can be recycled from RSO if enough SHs are available, which will be converted to a new S-S group. From this model, we can conclude that the Cys34 chemistry provides only a limited explanation and cannot account for how albumin functions as an antioxidant in an oxidative milieu; the RS component represents a possible explanation.

accounted for by the Cys34 chemistry. The RS component represents a measure of the response to this stress.

In the present work, we determined for the first time the RS component of human serum albumin (HSA) isolated from diabetic patients with progressive renal damage. In addition, the AC and evidence of structural alterations of the albumin molecule were determined using fluorescence spectroscopy to measure hydrophobicity and intrinsic fluorescence and quenching. The results demonstrated a reciprocal progressive reduction of the AC, progression of the RS component, and evidence of structural changes, all of which correlated with the progressive stages of DN.

Results and Discussion

Patient characteristics, biochemistry, and redox variables

The principal clinical and biochemical differences between patients in Stages 1–5 DN compared with the Stage 0 control group are presented in Table 1. As expected, together with the estimated glomerular filtration rate (GFR), the glycated hemoglo-

bin (HbA1c), blood creatinine levels, and age differed between all stages and the control group; differences were greater in patients in advanced stages of DN. Age was not significantly different between Stage 1 and Stage 0. Mean body mass index (BMI) in patients from Stages 1–5 was 27.8 Kg m^2, lower than the mean value of the control group; however, the BMI of the Stage 1 group was significantly higher than that in all other groups, including the control group (33.6±6.1). Mean HbA1c was 7.1% and significantly differed between Stages 1–5 and the control group; however, there were no differences among Stage 1–5 groups. The corresponding BMI indicated that all groups except Stage 5 were overweight; female sex predominated in Stages 1–3, and male sex predominated in Stage 5. Additional information regarding the medications prescribed for patients included in the study, some of which are associated with comorbidities, is included in (Table S1). Additional information regarding variables Stages 1–5 and Control in (File S1).

Table 2 summarizes the biochemical results with respect to the redox state and RS of albumin between groups. A progressive

Table 1. Characteristics and differences between the patients grouped by [a]GFR (Stages 0–5).

Characteristics	CONTROL (Stage 0)	Stage 1	Stage 2	Stage 3	Stage 4	Stage 5
		GFR≥90	GFR 60–89	GFR 30–59	GFR 15–29	GFR<15
n (male:female)	20(7:13)	25(7:18)	25(6:19)	25(8:17)	25(12:13)	25(16:9)
Age (years)	49±8	48±8	58±6**	62±6***	60±9**	61±7***
[b]BMI (kg m^{-2})	28.3±11	33.6±6.1*	28.2±4.9	27.3±5.1	26.2±6.1	23.7±9.3
[c]HbA1c (%)	6.2±0.60	7.34±1.3*	7.1±1.3*	7.1±0.9*	7.1±0.2*	7.2±0.9*
Blood Creatinine (mg dl^{-1})	0.75±0.18	0.79±0.09**	0.98±0.11**	1.02±0.11***	3.94±0.94***	8.9±2.46***
Serum Albumin (mg/dl)	5.7±1.8	4.7±2.1	4.0±0.4***	4.0±0.5***	3.7±0.5***	3.6±0.5***
GFR (mL min^{-1} 1.73 m^{-2})	122.2±46	101.0±9.4*	74.0±6.0***	50.2±8.8***	18.53±4.3***	5.12±1.9***

[a]glomerular filtration rate (mL min^{-1} 1.73 m^{-2});
[b]body mass index;
[c]glycated hemoglobin.
TEU, Trolox equivalent units; RS, response surplus.
*P<0.05 vs. control; **P<0.01 vs. control; ***p<0.001 vs. Control.

increase in the RS% component correlated with the DN stage; from Stage 2, the differences were statistically significant (p< 0.001). The differences between the Stage 1 group and the Stage 0 group were not significant. A significant difference in the concentration of thiol groups was identified in the albumin of patients from Stages 4 and 5 compared with the Stage 0 group (p< 0.05), and the AC tended to decrease with significant differences between Stages 3 and 4 compared to the Stage 0 group (p<0.05) and Stage 5 with respect to the Stage 0 group (p<0.01).

The relationship between the GFR and clinical variables is shown in Table 3. Age was significantly negatively correlated with GFR (R = –0.47, p<0.001). Creatinine was significantly correlated with the GFR (R = –0.68, p<0.0001); similarly, although with less significance, serum albumin and sex (female) correlated with the GFR (R = 0.147, p<0.043 and R = 0.195, p<0.011 respectively).

The relationship between the GFR and the redox variables and RS of albumin is shown in Table 4. Between the redox variables, the thiol groups (SHs) were significantly positively correlated with GFR (R = 0.44, p<0.001). The AC was also significantly positively correlated with GFR (R = 0.72, p<0.0001). Finally, the response

to stress or RS was very strongly significantly negatively correlated with GFR (R = –0.83, p<0.0001).

The factors associated with renal function decline that strongly correlated with GFR were included in a multivariate regression analyses to test the association of the variable with renal function decline (Table 5). Sex, serum albumin, HbA1c, AC, and thiol groups were excluded from the final model. RS, creatinine, and variables that significantly correlated with GFR in the bivariate analyses were tested in stepwise linear regression analyses. GFR was inversely associated with RS, creatinine, and age, and positively associated with BMI, all variables accounting for 74% of the variability in the GFR (Table 5). These findings indicate that the redox state of albumin is highly related with DN and the decline of renal function. Although the antioxidant properties of HSA are largely attributed to the Cys34 chemistry and the ratio of the albumin forms mercaptoalbumin and non-mercaptolbumin [16], the results obtained here and previously [12] suggest that additional mechanisms tightly linked with the molecular structure are involved in the redox balance and albumin antioxidant properties. Interestingly, even with a high correlation, the redox

Table 2. Redox state and RS of purified albumin between the patients grouped by [a]GFR (Stages 0–5).

Characteristics	CONTROL (Stage 0)	Stage 1	Stage 2	Stage 3	Stage 4	Stage 5
		GFR≥90	GFR 60–89	GFR 30–59	GFR 15–29	GFR<15
Thiol groups (SHs) (μmol g^{-1} protein)	4.71±1.5	5.18±1.77	4.92±1.39	4.66±1.69	3.63±0.87*	3.28±1.21*
Antioxidant Capacity ([b]TEU nM)	7.72±1.99	7.94±0.29	6.94±0.58	6.48±0.62*	6.34±0.50*	6.04±0.54**
RS[c] (%)	11.0±8.3	14.5±5.4	30.0±8.3***	51.9±10.8***	60.5±10.2***	74.7±11.2***

*P<0.05 vs. Control;
**P<0.01 vs. Control;
***p<0.001 vs. Control.
[a]glomerular filtration rate (mL min^{-1} 1.73 m^{-2});
[b]TEU, Trolox equivalent units;
[c]RS, response surplus.

Table 3. Correlation between glomerular filtration rate (GFR) with principal clinical variables.

	GFR	Age	BMI[a]	HbA1c[b]	Creatinine	Serum albumin	Sex
	(ml min⁻¹ 1.73 m⁻²)	(years)	(kg m⁻²)	(%)	(mg/dl)	(mg/dl)	(0:M, 1:F)
Range	3.2–134	28–75	15.5–52.8	5.0–8.9	0.7–17.0	2.0–9.71	
Pearson R	1.00	-0.47	0.371	0.028	-0.68	0.147	0.195
p value	<0.0001	<0.001	<0.001	0.376	<0.0001	0.043	0.011

[a]body mass index; [b]glycated hemoglobin.

components AC and thiol groups were excluded from the final regression analyses model in contrast with the strong association of RS, which suggests that RS is related with the GFR through pathways other than oxidative stress and possibly related with albumin structural changes.

The native albumin response to short-wave UV light as a stressor

The antioxidant response of proteins to several stressors has been described [12]. In the present work, short-wave UV light (245 nm) was used as the stressor, because the results are particularly clean and reproducible. The energy of the light used in the experiments was calculated at 10 mW cm^{-2} and the samples were exposed to UV radiation as described in Materials and Methods. The change in the antioxidant potential in relation to the protein concentration is shown in Figure 2, and corresponds to commercial delipidated albumin with and without exposure to UV light. Under normal conditions and without a stress pulse, the AC of native albumin is solely protein-concentration dependent, and a linear relationship is established between AC and the protein concentration (Figure 2, without UV light). In contrast, in response to a stressor, the albumin redox behavior is quite different and it is biphasic in relation with the protein amount (Figure 2, with UV light). With a low concentration of protein, the AC was lower than that of albumin without a stressor (1.56 μg is the minimal amount of albumin assayed). From this point, the AC of the albumin is increased in relation to the concentration, remaining lower than albumin without UV light up to 12.5 μg protein, a point at which it is not possible to observe differences in AC with or without a stressor. From here, the AC increases progressively to 1-, 2-, and 3-fold the basal amount in the stressed albumin in a concentration-dependent manner (1.76, 2.61, and 3.0 equivalents of Trolox [nmol] with 25, 50, and 100 μg of protein, respectively [$p < 0.001$]).

The information gained from these experiments provides important clues to understanding the nature of the RS mechanism. In the experimental condition assays, with a low protein concentration, the oxidative stress produced by UV light oxidized all of the Cys34 SH groups and then the AC decreased. With a specific protein concentration (12.5 μg), a balance can be observed between albumin oxidation and thiolation (reversible states of oxidation) with and without stressor effects. Under stressed conditions, beginning at 12.5 μg the AC observed is not produced exclusively from the Cys34 SH and intermediates, and the loss of linearity becomes evident; a positive response to the stress becomes evident as well (Figure 2).

Under physiologic conditions and in the extracellular environment, HSA exists in two different forms, human mercaptoalbumin (HSA-reduced) with the thiol group in a reduced form, and human reversible non-mercaptoalbumin (HSA-oxidized), which presents as several kinds of mixed disulfides, with cysteine, cysteinylglycine, glutathione, homocysteine, and γ-glutamylcysteine [17,18]. Additionally, HSA can comprise varying amounts (1%–2%) of strongly oxidized human non-mercaptoalbumin, and sulfinic and sulfonic acid (HSA-SO$_2$H and HSA-SO$_3$H). The reduced and oxidized forms of HSA are in dynamic equilibrium, depending on the redox state of Cys34 and the thiol disulfide exchange, patient age, and disease state [16].

The differences in AC between the stressed and non-stressed albumin presented here cannot be explained by the thiol disulfide exchange (Figure 2), particularly in a system without low molecular-weight thiols. Based on the experiments described, we confirmed the existence of an active component that was determined by changes in the tertiary structure of the protein.

Table 4. Correlation between GFR[a] with redox variables and RS.

	GFR ($mL\ min^{-1}\ 1.73\ m^{-2}$)	Albumin Thiol Groups ($\mu mol\ g^{-1}$ protein)	Albumin AC[b] (TEU nM)	Albumin RS[c] (%)
Range	3.2–134	1.42–7.78	5.51–8.41	1.27–16.02
Pearson R	1.00	0.44	0.72	−0.83
p value	<0.000	<0.001	<0.0001	<0.0001

[a]glomerular filtration rate ($mL\ min^{-1}\ 1.73\ m^{-2}$).
[b]TEU, Trolox equivalent units.
[c]RS, response surplus.

We isolated and studied albumin from diabetic patients at different stages of DN, looking for evidence of changes in AC and the RS component that correlated with the conformational changes in albumin and the progression of the disease.

Protein antioxidant capacity, response surplus, and the progression of DN

In the present study, we established a positive correlation between the DN stage and the antioxidant response to stress (RS; $R^2 = 0.7927$) (Figure 3). The results demonstrated that albumin isolated and purified from diabetic patients with advanced stages of renal damage and lower GFR has a higher RS value than albumin from normal patients or patients in earlier stages of DN. In diabetes, the structural changes induced by nonenzymatic glycation or by reactive oxygen species impair the antioxidant capabilities of albumin; the stress imposed by the former changes

represent the possible stressors associated with the induction of an increased response to stress (RS). In this respect, the progressive reduction in the albumin AC observed in the present study (Figure 4) accompanied by the progressive reduction in the GFR greatly support the notion that changes in the native albumin structure are associated with the loss of protein function. The postulated tight relationship between the AC and the protein conformation [12] should explain the higher response to stress as a function of the higher instability of the albumin molecule in the advanced stages of renal damage. The RS represents an indirect measure of these structural changes.

Fluorescence spectroscopy and quenching

The intrinsic fluorescence of the HSA tryptophan residue was used to monitor changes in its molecular conformation. Fluorescence of albumin samples was measured using a LS 45

Table 5. Stepwise linear regression models for parameters associated with GFR.

Independent variable	Regression Coefficient (95% CI)	Standarized Regression Coefficient	P
Model 1			
RS	−1.537 (−1.709, −1.364)	0.834	<0.0001
Model 2			
RS	−1.317 (−1.524, −1.111)	−0.715	<0.0001
Creatinine	−2.666 (−4.159, −1.173)	−0.200	0.001
Model 3			
RS	−1.168 (−1.387, −0.949)	−0.634	<0.0001
Creatinine	−2.825 (−4.271, −1.379)	−0.212	<0.0001
Age	−0.773 (−1.244, −0.303)	−0.161	0.001
Model 4			
RS	−1.126 (−1.344, −0.907)	−0.611	<0.0001
Creatinine	−2.649 (−4.080, −1.219)	−0.199	<0.0001
Age	−0.754 (−1.218, −0.291)	−0.157	0.002
BMI	0.625 (0.96, 1.158)	0.107	0.021
Excluded			
Sex	−0.028		0.534
Serum albumin	0.032		0.462
AC	−0.002		0.973
Thiol groups	−0.001		0.988

RS, response surplus; BMI, body mass index; AC, antioxidant capacity; GFR, glomerular filtration rate. Only variables that significantly correlated with the dependent variable GFR in the bivariate analyses were tested in stepwise linear regression analyses. GFR was inversely associated with RS, creatinine, and age, and positively associated with BMI; all variables accounting for 74% of the variability in the GFR.

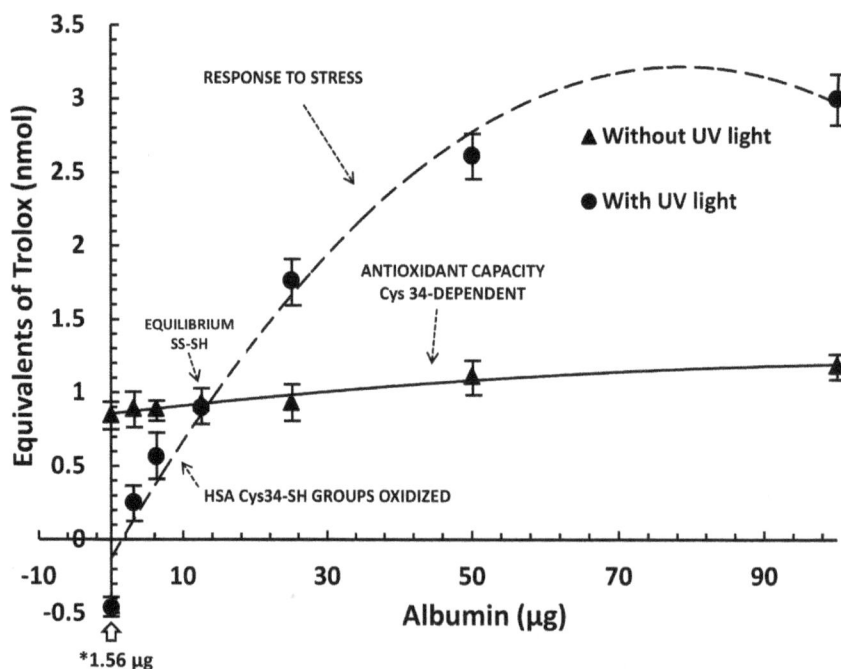

Figure 2. Native albumin antioxidant response to stress produced by short-wave UV light. Triangles (p) represent the antioxidant capacity (AC) of native serum albumin in relation to protein concentration. Circles (~) represent the antioxidant capacity of the increased amounts of native albumin previously exposed to UV light (254 nm) in the conditions described in the Materials and Methods. Results are expressed as mean ± SD. The light energy was calculated at 10 mW cm^{-2}; * 1.56 µg corresponds to the minimal amount of protein. The AC of non-stressed albumin has a concentration–dependent linear relationship. In stressed albumin, however, the AC has a biphasic in relation to the albumin concentration. Beginning at a concentration of 12.5 µg (SS-SH equilibrium), the AC progressively increases to 1-, 2-, and 3-fold the basal amount in stressed albumin in a concentration-dependent manner (1.76, 2.61, and 3.0 equivalents of Trolox (nmol) with 25, 50, and 100 µg of protein, respectively. At a low concentration, the human serum albumin (HSA) Cys34-SH groups are oxidized and the AC decreases. At a specific protein concentration (12.5 µg), a balance can be observed between albumin oxidation and thiolation (reversible states of oxidation). In stressed conditions at above12.5 µg, the AC could not be produced exclusively from the Cys34 SH and intermediates, and there is a loss of linearity; the point at which the Response Surplus is produced.

fluorescence spectrophotometer (Perkin-Elmer, Llantrisant, UK) with excitation and emission wavelengths of 280 and 340 nm, respectively, as described in the Materials and Methods. The differences in the fluorescence between albumin samples from controls and patients at different stages of renal disease are shown in Figure 5. The results indicate a progressive decrease in tryptophan fluorescence with the change in the estimated GFR from Stages 1–5; significant differences were achieved in Stages 2–5 compared to the Stage 0 group (p<0.001). The gradual decrease in the fluorescence emission intensity of HSA suggests conformational changes attributable to the glycation of protein that affect the environment of the residue [5,19]. The fluorescence of HSA is dominated by tryptophan emission, and the emission spectrum of HSA is mainly due to a single residue, Trp214, in subdomain IIA [19].

The emission intensity of tryptophan on the HSA is decreased in the presence of a water-soluble quencher. The quenching by acrylamide of the fluorescence of the tryptophan residue of HSA was measured. The acrylamide Stern-Volmer plots for HSA from patients in the progressive stages of DN are shown in Figure 6. The results indicated that the tryptophan residue of the HSA was more accessible to quenching by acylamide in the incipient stages of DN than in the advanced stages of the disease, where the tryptophan residue is inaccessible or only slightly accessible to acrylamide. The Stern-Volmer curves and the progressive reduction in the calculated Stern-Volmer constants representing the slopes of the plots suggest that conformational changes correlate with the progression of the DN. Changes in accessibility

due to conformational changes have been reported [20,21], and structural changes induced by glucose or free radicals have been demonstrated [5,13]. Although we observed differences in the tryptophan accessibility to the quencher in DN stages, significant differences were observed only when comparing Stages 3, 4, or 5 with the Stage 0 control group (p = 0.043, p = 0.047, and p = 0.0035, respectively; Figure 6).

Protein surface hydrophobicity

Structural alterations of proteins are generally accompanied by changes in surface hydrophobicity. In the present study, we used an adapted UV photolabeling method with the fluorescent probe 4,4'-dianilino-1, 1'- binaphthyl -5,5'-disulfonic acid (BisANS) [22] to monitor changes in the surface hydrophobic domains and changes in protein unfolding of albumin from patients in different stages of diabetic renal disease, such as described in the Materials and Methods. Incorporation of BisANS was decreased in DN patients from incipient Stages 1–3 and advanced Stages 4–5 compared with the Stage 0 control group (p<0.001; Figure 7). In fact, albumin from all of the groups, including Stage 0, showed a lower incorporation of the fluorescent probe than native commercial albumin (p<0.01). The results are consistent with our previous results [12] where commercial albumin showed higher levels of hydrophobicity and albumin modified with acrolein and partially glycated albumin quenched the level of fluorescence. A marked loss of BisANS incorporation by other proteins has also been demonstrated, particularly when exposed

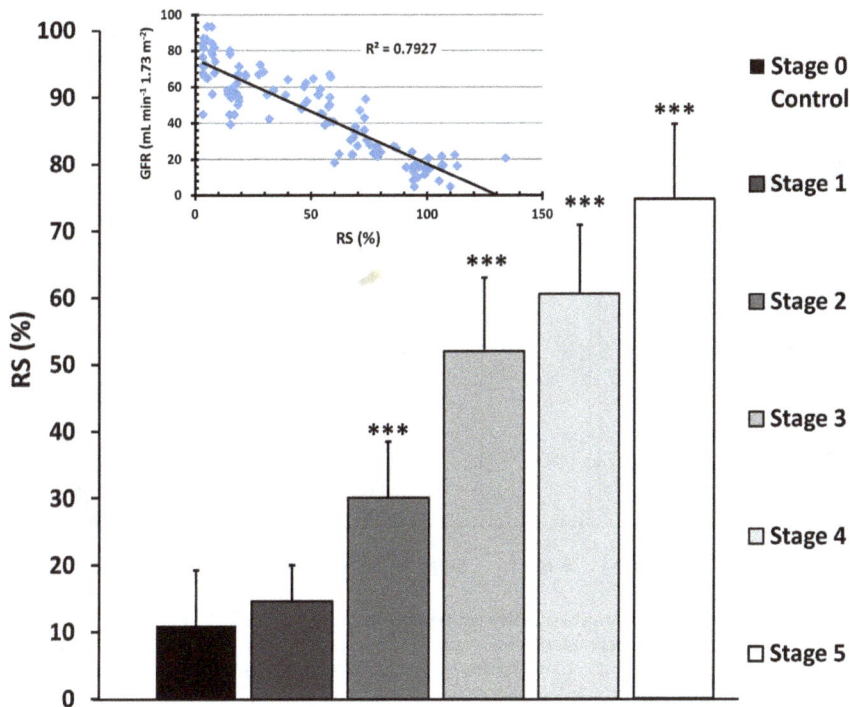

Figure 3. Albumin Response Surplus (RS) in the diabetic nephropathy progression. Albumin from the Stage 0 control group and patients with different stages of renal disease were isolated and the response to stress was measured as described in the Materials and Methods. The glomerular filtration rate (GFR) was highly correlated with the response surplus (RS) (internal graph, $R^2 = 0.7927$). The patients were grouped as Stage 0 (control) and Stages 1–5 depending of their GFR; with the exception of Stage 1, patients in all stages showed a progressive increase in RS with a high level of significance ($p<0.001$). Results are expressed as mean + SD. ***$P<0.001$ vs. Control group.

in vitro and in vivo to oxidative stress: a loss in the surface hydrophobic domain in creatine kinase and glyceraldehydes-3-phosphate dehydrogenase was demonstrated by the BisANS photo-incorporation before in vitro metal-catalyzed oxidation and in vivo from oxidative stress generated by rat muscle denervation [22]; the loss of hydrophobicity was accompanied by a marked reduction in activity. The changes observed in surface hydrophobicity in albumin from patients with DN can be interpreted as a disturbance in the protein structure produced by continuous exposure to the oxidative stress accompanying diabetes. An alternative or complementary explanation for this result is that hydrophobic patches in the surface of HSA were

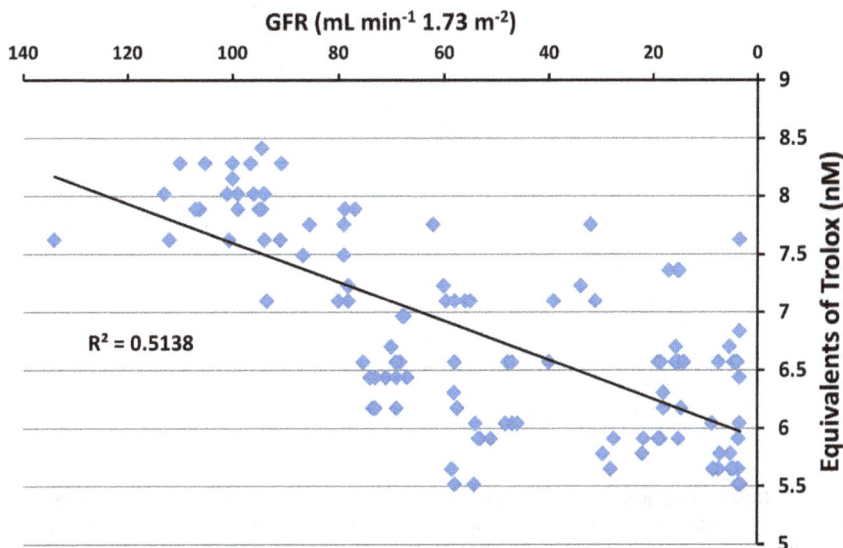

Figure 4. Antioxidant capacity (AC) measurement in relation to the glomerular filtration rate (GFR) of patients at different stages of diabetic nephropathy. The AC was calculated in Trolox Equivalent Units (nM) as described in the Materials and Methods section. A progressive and statistically significant correlation was observed between AC and the decline of renal function. R^2 = coefficient of determination.

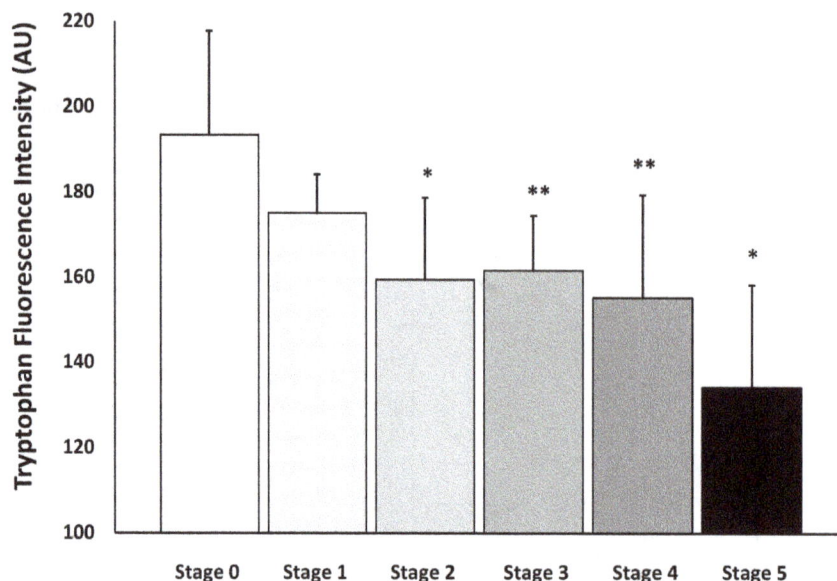

Figure 5. Effect of the progression of diabetic nephropathy on tryptophan fluorescence of albumin samples isolated from diabetic patients. The stages were determined from the estimated glomerular filtration rate as described in the Materials and Methods. A progressive reduction of the fluorescence intensity was observed compared to the Stage 0 control group. Values shown are mean + SD of the fluorescence of the total samples of each group. *$p < 0.05$ vs. Stage 0 group; **$p < .01$ vs. Stage 0 group.

modified in advanced stages of DN as a consequence of glycation. Analyses of native and modified bovine serum albumin revealed that protein surface hydrophobicity decreases upon both albumin glycation and drug binding [23].

Implications

The changes observed in the present work included: the RS component in albumin from DN patients increased proportionally with a reduction of the GFR; the AC of albumin from DN patients

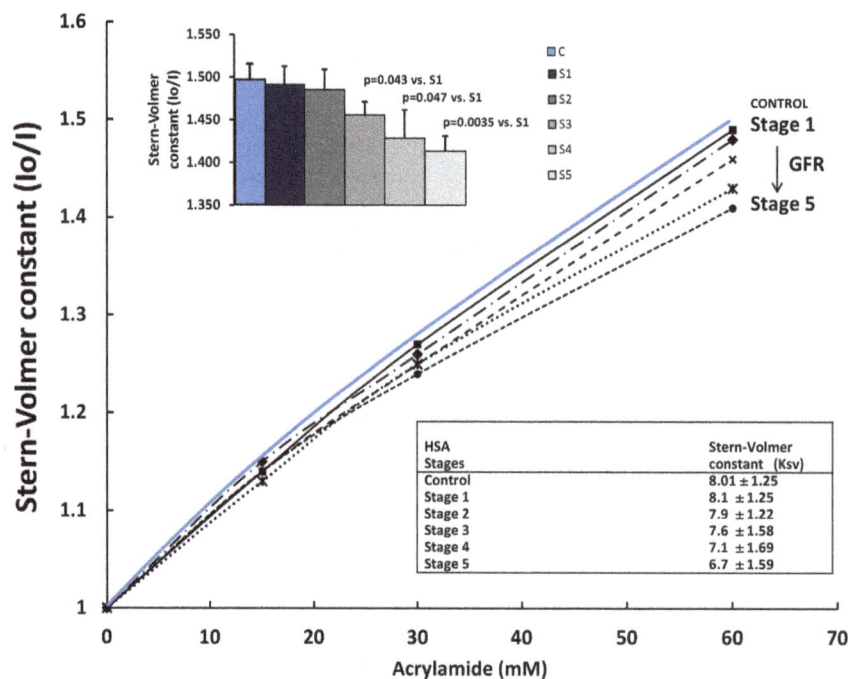

Figure 6. Tryptophan accessibility to the quencher in relation to the progression of diabetic nephropathy (DN). Stern-Volmer curves representing tryptophan accessibility to acrylamide reveal the existence of albumin structural changes between the advanced stages of renal disease compared with the Stage 0 control group without evidence of renal damage and a normal filtration rate. In the upper graph and table, the Stern-Volmer constants were calculated from the plot with an acrylamide concentration of 60 mM. Tryptophan fluorescence was assessed with an excitation and emission wavelengths of 280 and 340 nm, respectively. Results are mean + SD of 3 experiments with different samples from each group. Io and I correspond to the fluorescence intensities without and with the quencher, respectively. The Stern-Volmer formula for constant calculation is presented in the Materials and Methods.

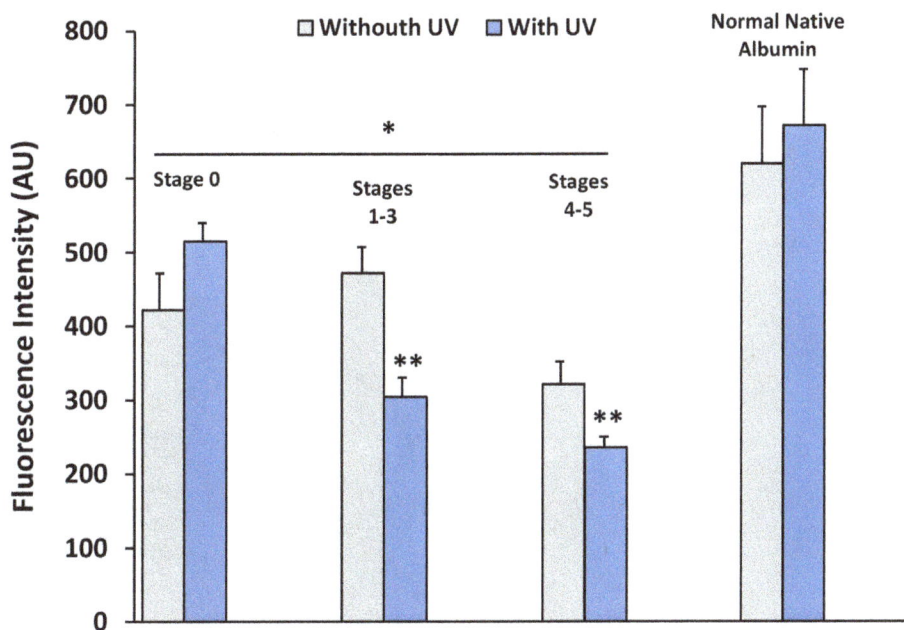

Figure 7. Changes in surface hydrophobicity of albumin isolated from control subjects and diabetic patients with diabetic nephropathy (DN). Changes in the surface hydrophobicity were monitored using the fluorescent probe 4,4'-dianilino-1,1'- binaphthyl - 5,5'disulfonic acid (BisANS) in response to urea (3 M)-induced protein unfolding. Incorporation of the BisANS is decreased in DN patients in Stages 1–3 and decreases further in Stages 4–5 compared to the Stage 0 control group. Similarly, albumin from all groups, including the Stage 0 control group, showed lower incorporation of the fluorescent probe than native commercial albumin. Values are mean + SD of three experiments with different samples from each group. *p<0.05 vs. native albumin without UV; **p<.001 vs. Stage 0 group and native albumin with UV.

and the thiol group content tended to diminish in the advanced stages of renal disease; and a reduction in the intrinsic fluorescence and quenching of acrylamide and a reduced surface hydrophobicity, evidence of albumin conformational changes, correlated with the advanced stages of DN.

Oxidative stress is correlated with renal dysfunction in patients with renal failure [24] and plasma albumin undergoes massive oxidation in primary nephrotic syndrome [25]. Diabetic patients with DN are frequently in a state of increased oxidative stress, because free radical production is increased in diabetes. Glycation and oxidative damage cause albumin modifications, impairing its antioxidant properties [26], and producing main chain fragmentation and loss of both secondary and tertiary structure [27].

Albumin intramolecular disulfide bonds formed by the presence of Cys-Cys residues suggest that an active mechanism is linked with the sulfhydryl-disulfide turnover. The normal sulfhydryl-disulfide interchange reaction that occurs within the protein molecule leading to reorganization of the position of the disulfide-bridges was postulated several years ago [28,29]. The Cys34 chemistry and the oxidation states, that is, reversible sulfenic (–SOH) and sulfinic (–SO$_2$H), and irreversible sulfonic (–SO$_3$H) acid derivatives, continue to be the prevailing explanation for the bulk of the redox potential of albumin. We postulated that an intramolecular sulfhydryl-disulfide interchange positively explains, at least partially, the RS component and the increased antioxidant potential of albumin when it undergoes structural stress. Although it has been suggested that the formation of albumin isomers in the aging process is due to thiol disulfide interchanges catalyzed by the thiol group of cysteine [30], the results presented here demonstrate an active interchange without the intervention of low molecular-weight thiols and independent of Cys34. In this respect, the possibility of the formation of multiple isomeric forms of albumin must be shown and, in fact, the formation of albumin isomers, the

trapping of the intermediates free –SH groups from thiolate anions and the isolation of the isomeric forms have been reported [31]. Similarly, albumin isomerization must accompany structural alterations and aging.

In this regard, the modifications introduced by isomerization, including structural changes, decreased fluorescence, and aging, have consequences such as decreased resistance to catabolism and the formation of isomers with very fast catabolic rates [29,32]. Some of these changes correlate with the results presented here and are also compatible with the mechanism proposed to explain the RS component (Figure 1). A complementary scheme represents the hypothesis of an antioxidant response to stress depending on the molecular structure, and explains the results presented here (Figure 8).

Based on our results and the scheme presented, we conclude the following:

1. The albumin AC decreases with a reduction in the filtration rate and advancing stages of DN as a consequence of the oxidation of thiol groups, derived essentially from Cys34. Under oxidative stress, the free thiol groups react, resulting in the formation of progressively more oxidized species. The formation of reversible sulfenic and sulfinic acid (HSA-SOH and HSA-SO$_2$H, respectively) eventually maintains the redox state with the plasma environment at moderate, or at least not extended, exposure to oxidative stress. Evidence indicates that the HSA redox system in plasma can be activated and recover after a single bout of exercise [33]. With higher levels or extended exposure to oxidative stress, the formation of sulfonic acid (HSA-SO$_3$H), the irreversible or endproduct of Cys34 SH oxidation occurs [34]. In the present work, a clear trend to a reduction in the albumin AC and reduced thiol group content together represent evidence of a progressive oxidized condition

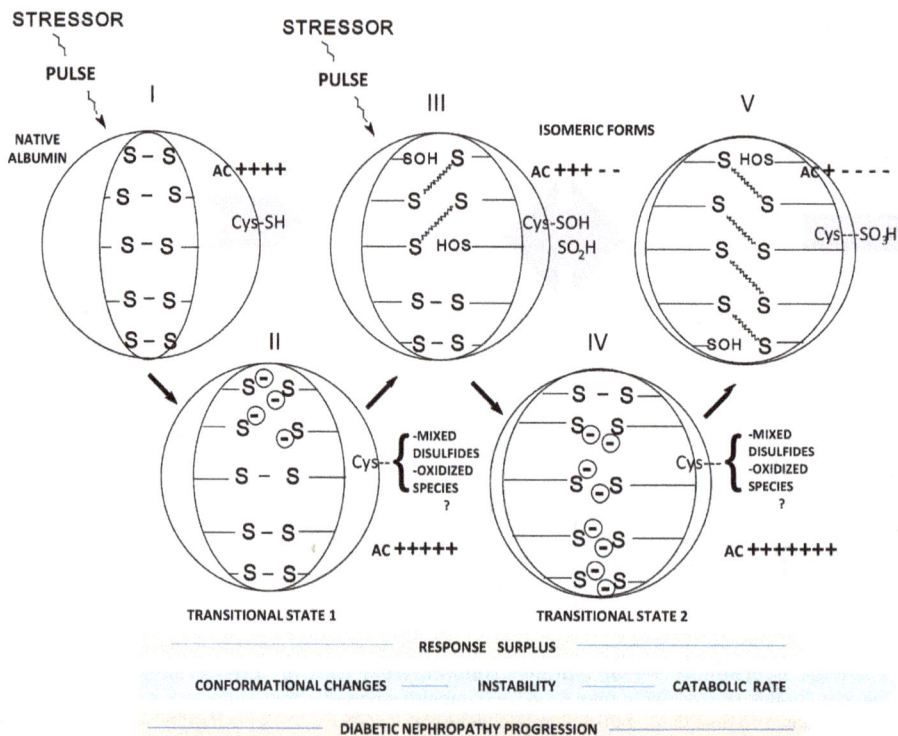

Figure 8. Schematic representation of the possible mechanism underlying the albumin response to stress in the progression of diabetic nephropathy. The internal region of albumin with a section of the disulfide pairing occurring between α-helices in native albumin (I) is shown. In the native state, the albumin antioxidant capacity is Cys34-dependent (violet figures). When a stressor compromises the molecular structure, cleavage of the disulfide linkage generates transition stages and the antioxidant capacity is thiolate anion-dependent (II). Transition state 1 eventually becomes an isomeric form of albumin (III). A new pulse of stress on a structurally weak isomeric form generates a second-order transition state (IV) that eventually becomes a second albumin isomeric form (V). A repetitive sequence of events results in a high response (increased Response Surplus [RS]), more and deeper structural changes, instability, and an accelerated catabolic rate (pale green, blue, and orange arrows; structural changes are represented by the shape of the internal semicircles). In general terms, isomeric forms III and V should provide low AC and the transitional states II and IV high AC. Pale violet = chemical intermediates and final product (irreversible) deriving from Cys34: thiol, sulfenic acid, sulfinic acid, and sulfonic acid (Cys-SH, Cys-SOH, Cys SO$_2$H, and Cys SO$_3$H, respectively). The chemical intermediates (mixed disulfides, oxidized species) of Cys34 in transition states II and IV are unknown and may depend on the specific albumin milieu. The stressor could be, for example, UV radiation, high levels or long-standing oxidative stress, high temperature, glycation, etc. AC = Antioxidant Capacity; RS = Response Surplus; Cys = Cysteine; S- = Thiolate anion.

in advanced stages of DN, possibly a consequence of the continuous exposure to oxidative stress during diabetes.

2. In extreme or extended oxidative stress levels where the albumin molecular structure is compromised (structural stress), we propose transient states of transition between HSA-Cys-Cys sequences and thiolate anion (HSAS$^-$) and probably the reduced moiety (HSA-SH), from the core of the protein (Figure 8). These transient intermediates with a higher AC eventually form albumin isomers. The isomeric forms of albumin could be isolated, but the transition intermediates are spontaneously transformed in a short period of time to the corresponding isomers. The albumin structure possesses an unusual disulfide-bonding pattern characterized by the eight Cys-Cys sequences (31). When albumin is stressed by short-wave UV light irradiation, the AC is increased as a response to the stress, and the cleavage of some HSA-SS groups leads to the formation of reactive thiolate anion groups (HSA-S$^-$) (formation of intermediates II and IV, Figure 8). The use of a stressor *in vitro* reproduces in short time a phenomenon that is possibly extended *in vivo* with longer exposure to the stressor.

3. Transition stages are characterized by an increase in AC, but eventually all transition states result in unstable isomeric forms.

4. The increased stress response (RS) of albumin in progressively advanced stages of renal disease is likely an indirect reflection of the structural changes and instability produced by extended exposure to oxidative stress. We cannot confirm whether RS constitutes a compensatory response to oxidative threats; however, some of the albumin structural and functional consequences should include changes in the redox balance, reduced resistance to catabolism, and binding property effects of endogenous and exogenous compounds.

5. Although previous studies suggest that in extracellular fluids the albumin antioxidant activity mostly relies on the free thiol group of Cys34, the results obtained here suggest that the entire albumin molecular structure is highly involved in the antioxidant response of the protein and that the structural stress that accompanies degenerative pathologies constitutes a cumulative and measurable event.

6. Based on the present study, the RS represents an independent variable with the potential to be functional in diagnostic or prognostic possibilities for DN. For now, the changes described in point 4 above provide indirect evidence of cumulative protein instability or liability. The specific participation of the albumin conformational and functional changes in the pathogenesis of DN must be established in future studies. A

specific example is the possible participation of these changes in the induction of endoplasmic reticulum stress produced in the renal proximal tubular cells by the accumulation of misfolded proteins [35,36]. The strong association of RS with albumin structure and the analyses of the data obtained here suggest that renal function decline is related through pathways other than oxidative stress and possibly related more with albumin structural changes.

Materials and Methods

Chemicals

Blue sepharose was obtained from GE-HealthCare. All solutions were prepared with chemicals of pure analytical grade and filtered with a 0.45-μm filter (Millipore, Bedford, MA, USA). Horseradish peroxidase Type I (E.C. 1.11.1.7); hydrogen peroxide 30% (w/w); sodium HEPES (N-(2-hydroxyethyl)piperazine-N'- (2-ethanesulfonic acid) 99.5%; p-iodophenol; luminol (5-amino-2,3-dihydro-1,4-phtalazinedione); albumin from human serum (lyophilized powder, 99%, essential fatty acid free, prepared from essentially globulin free albumin; Trolox (6-hydroxy-2, 5, 7, 8-tetramethylchroman-2- carboxylic acid) were obtained from Sigma-Aldrich Inc., St. Louis, MO, USA. The centrifuge (Ultra-free-MC Centrifugal Filter device, with a 5000 and 10,000 NMWL cut-off) was obtained from AmiconMillipore Corporation.

Patients and samples

Serum aliquots were collected from 145 institutionalized diabetic patients from the Nephrology Unit at Regional General Hospital #1, Morelia, Michoacán, México, from the hospital system of the Mexican Social Security Institute. Patients with diabetes, as assessed by American Diabetes Association criteria (fasting glucose level ≥ 126 mg dl^{-1} [7.0 mmol l^{-1}] or 2-h glucose level ≥ 200 mg/dl [11.1 mmol l^{-1}]) [37] were included. The exclusion criteria included a diagnosis of nondiabetic kidney disease, cancer, hypertension, supplement consumption, or medication changes during the 12-week study period.

All of the patients were informed of the study purposes and procedures and all individuals provided written informed consent prior to participation. The protocol for the research project #2010-785-061 was approved by the Ethics and Investigation National Committee (CNIC), Mexican Social Security Institute (IMSS), México. The investigation was performed in accordance with the principles of the Declaration of Helsinki and Good Clinical Practice.

The patients were divided into five groups according to their estimated GFR [38] and an additional control group (Stage 0) of early-diagnosed patients with no evidence of structural or functional renal decline. The Stage 1 group comprised patients with a normal or high GFR (between 90 and 130 mL min^{-1} 1.73 m^{-2}) with history of urine or blood test abnormalities, or imaging or pathology abnormalities; Stage 2 group comprised patients with a GFR between 60 and 89 mL min^{-1} 1.73 m^{-2}; Stage 3 comprised patients with a GFR between 30 and 59 mL min^{-1} 1.73 m^{-2}; Stage 4 comprised patients with a GFR between 15 and 29 mL min^{-1} 1.73 m^{-2}, and Stage 5 comprised patients with end-stage chronic renal failure and a GFR≤ 15 mL min^{-1} 1.73 m^{-2}.

After an initial consultation and examination, the patients were instructed to visit the hospital Nephrology department after an 8- to 12-h overnight fast for clinical examination and sample collection; fasting blood was drawn, and serum and plasma aliquots were transferred to plastic tubes and were either used

fresh or stored at −80°C for 2 to 4 weeks until analysis. Fresh samples were used for glycated hemoglobin (HbA1c), creatinine, and sulfhydryl group measurements, and albumin isolation was performed during the next 3 months in groups of six samples corresponding to each stage of DN.

Isolation of human serum albumin

The albumin was separated and purified from 0.5-mL aliquots of human serum. The protein was purified by affinity chromatography using Blue Sepharose, packed in columns (2 mL) designed especially for this study. Before equilibration, the columns were washed with 5 bed volumes of starting buffer (50 mM citric acid, 100 mM Na$_2$HPO$_4$, pH 3.0). Each diluted sample was filtered with a 0.45-μm Millipore membrane, passed through the column and washed with 10 additional bed volumes of starting buffer. The albumin bound to the gel was eluted with the corresponding elution buffer (50 mM KH$_2$PO$_4$ and 1.5 M KCl, pH 7.0), and the albumin fraction was dialyzed in 10 mM phosphate buffer (pH 7.4), using a 3000-Dalton cut-off membrane. The protein concentration was determined and each eluted fraction was examined in 10% sodium dodecyl sulfide-polyacrylamide gels to verify protein integrity.

Antioxidant capacity system

The AC was measured using an enhanced chemiluminescence-based assay. Horseradish peroxidase is an enzyme that catalyzes the decomposition of peroxides and forms free radicals. Peroxidase/H$_2$O$_2$ mixtures are used to generate free radicals [39] from several substrates. Hydrogen peroxide removes two electrons of the enzyme, and each of these is replaced in two one-electron steps, in each of which a substrate molecule forms a radical [40]. Oxygen consumption has been documented for the sodium horseradish peroxidase/hydrogen peroxide system and free radical generation [41]. Radicals formed and peroxides and oxygen oxidized luminol, producing light that could be detected and measured. Given that a constant rate of radical generation can be achieved, light emission depends on the constant production of free radicals. Enhanced chemiluminescence is used to measure the AC in biologic fluid [42]. The system used in the present work contains horseradish peroxidase (5 μU l^{-1}), hydrogen peroxide 30% w/w (3.0 μM), substrate p-iodophenol (25 μM), and luminol (300 μM from a stock solution of 10 mM in DMSO). A reaction mixture containing perborate base buffer prepared with sodium tetraborate (100 mM) and sodium carbonate (100 mM) was used. Intra-assay variation for 15 samples analyzed in triplicate was 0.29%–5.2%, with a mean of 4.05%.

Protein antioxidant capacity and protein response surplus

RS is defined as the increase in AC of a protein when it is exposed to a structural stressor [12]. To obtain RS measurements, the AC of a protein sample was determined before (ACb) and after (ACa) an oxidative or structural challenge; in the present work, UV light (254 nm) was used as the stressor. The results obtained in Trolox concentration (Trolox Equivalent Units, nM TEU) with the use of standard curves were transformed to %, considering ACb to be 100%. In this way, the values reported as RS represent the percentage of AC surplus above the mean AC of proteins before treatment. In the same way, the percent accumulated AC (%AAC) represents the sum of AC plus the RS value in percentage. To obtain RS from the results, 100 was subtracted from %AAC.

1. AC = Antioxidant Capacity (Trolox Equivalent Units, TEU)

2. %AAC, Percent Accumulated Antioxidant Capacity = (ACa)
   (100) ÷ ACb

3. RS%, Response Surplus % = (%AAC) − 100

Thiol group measurement

Total serum sulfhydryl groups were measured using Ellman's reagent (5,5′-dithio-bis [2-nitrobenzoic acid)] or DTNB. The DTNB reacts with a sulfhydryl group to yield a mixed disulfide and 2-nitro-5-thiobenzoic acid, with high molar extinction coefficient at 412 nm. Samples containing 0.5 mg of purified albumin in 0.1 M sodium phosphate, pH 8.0, and 3.0 mM of EDTA were incubated for 15 min in the dark at 20°C with a DTNB solution (2.5 mM); absorbance was recorded at 412 nm, but to obtain the final value, the baseline absorbance was subtracted. Thiol values are expressed as $\mu mol \; g^{-1}$ of albumin using a molar absorption coefficient of $14,130 \; M^{-1}cm^{-1}$. A calibration curve was constructed based on the sequential dilution of 1 mmol freshly prepared cysteine stock solution at the time [43].

Antioxidant standard preparation

Standard curves were achieved using increasing amounts of the water soluble tocopherol analog Trolox (6-hydroxy-2,5,7,8-tetramethylchroman 2-carboxylic acid). Trolox standards were prepared in milli-Q water (Millipore Corporation, Bedford, MA, USA) before initiating each experiment and starting with a stock solution of 80 µM. The AC through the experiments was expressed as nanomolar Trolox (Trolox Equivalent Units or TEU nM). Approximately 320 nM Trolox is enough to completely suppress the chemiluminescent signal and corresponds to the maximal AC achieved. One standard curve was prepared before each experiment.

Treatment with UV light

UV light was used in the present work as a structural stressor of proteins. Isolated albumin was illuminated by a 3 UV™-36; 6 Watt, 0.16 A lamp (UVP, Upland, CA, USA) with a wavelength adjusted at 254 nm. The energy of the light used in the experiments was calculated to be $10 \; mW \; cm^{-2}$, monitored using a radiometer UVP model UVX-25 (UVP). The temperature for the experiments was maintained at 25±0.1°C.

Albumin fluorescence and quenching

The presence of two tryptophan residues in albumin that generate autofluorescence allows for changes in the protein conformation to be monitored. Fluorescence measurements were performed using an LS 45 fluorescence spectrophotometer (Perkin-Elmer, Llantrisant, UK) at excitation and emission wavelengths of 280 and 340 nm, respectively. The slit width was set to 10 nm for excitation and emission. The decrease in fluorescence was analyzed according to the Stern-Volmer equation [44]:

$$Io/I = 1 + Ks_{sv}[Q]$$

Where Io and I represent the fluorescence intensities at a wavelength in the absence and presence of the quencher, respectively, Ksv is the Stern-Volmer constant for the collisional quenching process, and [Q] is the concentration of the quencher. The quencher used in the present work was acrylamide. In the acrylamide-quenching experiments, the quencher agent was used at a final concentration of up to 60 mM, and the native albumin and isolated albumin from Stage 1–5 groups were used at a concentration of 2 g L^{-1} in phosphate-buffered saline.

Protein hydrophobicity

Albumin hydrophobicity was measured by the incorporation of BisANS by 3 M urea treatment followed by long-wave UV light irradiation, as described previously [45] with modifications. Briefly, purified albumin samples (1 mg mL^{-1}) from Stage 0 (control) and Stages 1–5 were diluted in a 3 M urea solution prepared in labeling buffer containing 50 mM Tris-HCL, 10 mM $MgSO_4$ at pH 7.4. Samples were treated with 3 M urea for 2 h at room temperature. At the end, 100 mM BisANS was added and the samples agitated thoroughly with vortex. Aliquots (350 µL) were placed in a 96-well plate and incubated on ice for 1 h under exposure to long-wave UV light (365 nm) for BisANS incorporation. The samples then were placed in microcentifuge filters (10-kDa cutoff, Amicon Ultra, Millipore, Tullagreen, Carrigtwohill CO, Ireland) and washed two times with labeling buffer. Samples were re-dissolved in the same buffer and the measurement of fluorescence was performed using an LS 45 fluorescence spectrophotometer (Perkin-Elmer) at excitation and emission wavelengths of 355 and 460 nm, respectively.

Protein sample conditioning

Before the incubation period, protein samples were placed in microcentrifuge filters (10 kDa cutoff, Amicon Ultra, Millipore) and centrifuged at 5000 g for 1 h. The filter residues were washed and dialyzed two times and dissolved in phosphate buffer (pH 7.4); the protein concentration was determined using the Bradford method [46]. Aliquots containing 100 µg of protein were irradiated with UV light for varying periods of time for each experiment and added to the reaction mixture in the luminometer cuvette after the signal was stable. The residual solvent remaining after microfiltration was collected and reacted with Folin-Ciocalteu reagent to monitor protein fragmentation. In addition, light scattering was measured at 400 nm in a spectrophotometer to check for the aggregation of irradiated proteins. The short period of albumin exposure to the UV light produced reproducible results without inducing aggregation or fragmentation. In contrast, some studies have reported structural changes and aggregation produced by UV light on lysozyme and albumin, but with very long exposure to 285-nm UV light [47].

Statistical Analysis

Analysis of the results was performed using the statistical software SPSS 10.0 (IBM, Chicago, IL, USA). Values are expressed as mean ± SD. Groups were compared using the one-way ANOVA test followed by Dunnett's test. Results were considered to be statistically significant when p<0.05. Pearson's correlation coefficient (R) was used to determine the correlation between variables. The correlation was determined to be significant when the p value was less than 0.05 (5%) with Fisher's z transformation. Multivariate models were used to assess the association between variables and the GFR decline.

Supporting Information

Table S1 Medications prescribed for patients included in the study from stages 0 to 5.

File S1 Database with the results of the principal variables from stages 1 to 5 and control group. Sex, age, estimated glomerular filtration rate (GFR), glycated hemoglobin

(HbA1c), blood creatinine (CREAT), antioxidant capacity (AC), Response Surplus (RS), accumulated antioxidant capacity (ACA%), thiol groups (SHs), weight, height and body mass index (BMI).

Acknowledgments

We are grateful to Dr. Roman Acosta Rosales and the administrative authorities of IMSS for the assistance during the realization of the present work.

Author Contributions

Conceived and designed the experiments: RMN. Performed the experiments: RMN. Analyzed the data: RMN MDF GDR ICC SBG. Contributed reagents/materials/analysis tools: RMN MDF GDR ICC SBG. Contributed to the writing of the manuscript: RMN.

References

1. Phillips A, Shaper AG, Whincup PH (1989) Association between serum albumin and mortality from cardiovascular disease, cancer and other causes. Lancet 16: 1434–1436.
2. Goldwasser P, Feldman J (1997) Association of serum albumin and mortality risk. J Clin Epidemiol 50: 693–703.
3. Owen WF Jr, Lew NL, Liu Y, Lowrie EG, Lazarus JM (1993) The urea reduction ratio and serum albumin concentration as predictors of mortality in patients undergoing hemodialysis. N Engl J Med 329: 1001–1006.
4. Iseki K, Kawazoe N, Fukiyama K (1993) Serum albumin is a strong predictor of death in chronic dialysis patients. Kidney Int 44: 115–119.
5. Bourdon E, Loreau N, Blache D (1999) Glucose and free radicals impair the antioxidant properties of serum albumin. FASEB J 13: 233–244.
6. Roche M, Rondeau P, Singh NR, Tarnus E, Bourdon E (2008) The antioxidant properties of serum albumin. FEBS Lett 582: 1783–1787.
7. Lim PS, Cheng YM, Yang SM (2007) Impairments of the biological properties of serum albumin in patients on haemodialysis. Nephrology (Carlton) 12: 18–24.
8. Yamada N, Nakayama A, Kubota K, Kawakami A, Suzuki E (2008) Structure and function changes of oxidized human serum albumin: physiological significance of the biomarker and importance of sampling conditions for accurate measurement. Rinsho Byori 56: 409–415.
9. Barzegar A, Moosavi-Movahedi AA, Sattarahmady N, Hosseinpour-Faizi MA, Aminbakhsh M, et al. (2007) Spectroscopic studies of the effects of glycation of human serum albumin on L-Trp binding. Prot Pept Lett 14: 13–18.
10. Kouoh F, Gressier B, Luyckx M, Brunet C, Dine T, et al. (1999) Antioxidant properties of albumin: effect on oxidative metabolism of human neutrophil granulocytes. Farmaco 54: 695–699.
11. Soriani M, Pietraforte D, Minetti M (1994) Antioxidant potential of anaerobic human plasma: role of serum albumin and thiols as scavengers of carbon radicals. Arch Biochem Biophys 312: 180–188.
12. Medina-Navarro R, Durán-Reyes G, Díaz-Flores M, Vilar-Rojas C (2010) Protein antioxidant response to the stress and the relationship between molecular structure and antioxidant function. PLoS One 5: e8971. doi: 10.1371/journal.pone.0008971
13. Faure P, Troncy L, Lecomte M, Wiernsperger N, Lagarde M, et al (2005) Albumin antioxidant capability is modified by methylglyoxal. Diabetes Metab 31: 169–177.
14. Kawakami A, Kubota K, Yamada N, Tagami U, Takehana K, et al. (2006) Identification and characterization of oxidized human serum albumin. A slight structural change impairs its ligand-binding and antioxidant functions. FEBS J 273: 3346–3357.
15. Vaidya V, Ingold KU, Pratt DA (2009) Garlic: source of the ultimate antioxidants–sulfenic acids. Angew Chem Int Ed Engl 48: 157–160.
16. Anraku M, Chuang VT, Maruyama T, Otagiri M (2013) Redox properties of serum albumin. Biochim Biophys Acta 1830: 5465–5472.
17. Turell L, Radi R, Alvarez B (2013) The thiol pool in human plasma: The central contribution of albumin to redox processes. Free Radic Biol Med 65: 244–53.
18. Candiano G, Petretto A, Bruschi M, Santucci L, Dimuccio V, et al. (2009) The oxido-redox potential of albumin methodological approach and relevance to human diseases. J Proteomics 73: 188–195.
19. Shaw AK, Pal SK (2008) Spectroscopic studies on the effect of temperature on pH-induced folded states of human serum albumin. J Photochem Photobiol B 90: 69–77.
20. Wells TA, Nakazawa M, Manabe K, Song PS (1994) A conformational change associated with the phototransformation of Pisum phytochrome A as probed by fluorescence quenching. Biochemistry 33: 708–712.
21. Weber J, Senior AE (2000) Features of F(1)-ATPase catalytic and noncatalytic sites revealed by fluorescence lifetimes and acrylamide quenching of specifically inserted tryptophan residues. Biochemistry 39: 5287–5294.
22. Pierce A, deWaal E, Van Remmen H, Richardson A, Chaudhuri A (2006) A novel approach for screening the proteome for changes in protein conformation. Biochemistry 45: 3077–3085.
23. Khodarahmi R, Karimi SA, Ashrafi Kooshk MR, Ghadami SA, Ghobadi S, et al. (2012) Comparative spectroscopic studies on drug binding characteristics and protein surface hydrophobicity of native and modified forms of bovine serum albumin: possible relevance to change in protein structure/function upon non-enzymatic glycation. Spectrochim Acta A Mol Biomol Spectrosc 89: 177–186.
24. Terawaki H, Yoshimura K, Hasegawa T, Matsuyama Y, Negawa T, et al. (2004) Oxidative stress is enhanced in correlation with renal dysfunction: examination with the redox state of albumin. Kidney Int 66: 1988–1993.
25. Musante L, Bruschi M, Candiano G, Petretto A, Dimasi N, et al. (2006) Characterization of oxidation end product of plasma albumin 'in vivo'. Biochem Biophys Res Commun 349: 668–673.
26. Otagiri M, Chuang VT (2009) Pharmaceutically important pre- and posttranslational modifications on human serum albumin. Biol Pharm Bull 32: 527–534.
27. Coussons PJ, Jacoby J, McKay A, Kelly SM, Price NC, et al. (1997) Glucose modification of human serum albumin: a structural study. Free Radic Biol Med 22: 1217–1227.
28. Smithies O (1965) Disulfide-bond cleavage and formation in proteins. Science 150: 1595–1598.
29. Wallevik K (1976) Spontaneous in vivo isomerization of bovine serum albumin as a determinant of its normal catabolism. J Clin Invest 57: 398–407.
30. Wallevik K (1976) SS-interchanged and oxidized isomers of bovine serum albumin separated by isoelectric focusing. Biochim Biophys Acta 420: 42–56.
31. Gabaldon M (2009) Thiol dependent isomerization of bovine albumin. Int J Biol Macromol 44: 43–50.
32. Gabaldón M (2002) Preparation and characterization of bovine albumin isoforms. Int J Biol Macromol 30: 259–267.
33. Lamprecht M, Greilberger JF, Schwaberger G, Hofmann P, Oettl K (2018) Single bouts of exercise affect albumin redox state and carbonyl groups on plasma protein of trained men in a workload-dependent manner. J Appl Physiol 104: 1611–1617.
34. Turell L, Botti H, Carballal S, Ferrer-Sueta G, Souza JM, et al. (2008) Reactivity of sulfenic acid in human serum albumin. Biochemistry 47: 358–367.
35. Wu X, He Y, Jing Y, Li K, Zhang J (2010) Albumin overload induces apoptosis in renal tubular epithelial cells through a CHOP-dependent pathway. OMICS 14: 61–73.
36. Takeda N, Kume S, Tanaka Y, Morita Y, Chin-Kanasaki M, et al. (2013) Altered unfolded protein response is implicated in the age-related exacerbation of proteinuria-induced proximal tubular cell damage. Am J Pathol 183: 774–85.
37. ADA, American Diabetes Association (2013) Standards of medical care in diabetes–2013. Diabetes Care 36: S12–S21.
38. Cockcroft DW, Gault MH (1976) Prediction of creatinine clearance from serum creatinine. Nephron 16: 31–41.
39. Moreno SN, Stolze K, Janzen EG, Mason RP (1988) Oxidation of cyanide to the cyanyl radical by peroxidase/H_2O_2 systems as determined by spin trapping. Arch Biochem Biophys 265: 267–271.
40. Halliwell B, Gutteridge JM (1999) Antioxidant defenses. In: Free Radicals in Biology and Medicine. 3th Edn. New York: Oxford University Press.
41. Kalyanaraman B, Janzen EG, Mason RP (1985) Spin trapping of the azidyl radical in azide/catalase/H_2O_2 and various azide/peroxidase/H_2O_2 peroxidizing systems. J Biol Chem 260: 4003–4006.
42. Whitehead TP, Thorpe GHG, Maxwell SRL (1992) Enhanced chemiluminescent assay for antioxidant capability in biological fluids. Analytica Chimica Acta 266: 265–277.
43. Riener CK, Kada G, Gruber HJ (2002). Quick measurement of protein sulfhydryls with Ellman's reagent and with 4, 4′-dithiodipyridine. Anal Bioanal Chem 373: 266–276.
44. Peterman BF, Laidler KJ (1980) Study of reactivity of tryptophan residues in serum albumins and lysozyme by N-bromosuccinamide fluorescence quenching. Arch Biochem Biophys 199: 158–164.
45. Eftink MR, Ghiron CA (1981) Fluorescence quenching studies with proteins. Anal Biochem 114: 199–227.
46. Bradford MM (1976) A rapid and sensitive method for the quantification of microgram quantities of protein utilizing the principle of protein-dye binding. Anal Biochem 72: 248–254.
47. Xie J, Qin M, Cao Y, Wang W (2011) Mechanistic insight of photo-induced aggregation of chicken egg white lysozyme: the interplay between hydrophobic interactions and formation of intermolecular disulfide bonds. Proteins 79: 2505–2516.

14

Toxic Effects of Maternal Zearalenone Exposure on Intestinal Oxidative Stress, Barrier Function, Immunological and Morphological Changes in Rats

Min Liu, Rui Gao, Qingwei Meng, Yuanyuan Zhang, Chongpeng Bi, Anshan Shan*

Institute of Animal Nutrition, Northeast Agricultural University, Harbin, P. R. China

Abstract

The present study was conducted to investigate the effects of maternal zearalenone (ZEN) exposure on the intestine of pregnant Sprague-Dawley (SD) rats and its offspring. Ninety-six pregnant SD rats were randomly divided into four groups and were fed with diets containing ZEN at concentrations of 0.3 mg/kg, 48.5 mg/kg, 97.6 mg/kg or 146.0 mg/kg from gestation days (GD) 1 to 7. All rats were fed with mycotoxin-free diet until their offspring were weaned at three weeks of age. The small intestinal fragments from pregnant rats at GD8, weaned dams and pups were collected and studied for toxic effects of ZEN on antioxidant status, immune response, expression of junction proteins, and morphology. The results showed that ZEN induced oxidative stress, affected the villous structure and reduced the expression of junction proteins claudin-4, occludin and connexin43 (Cx43) in a dose-dependent manner in pregnant rats. Different effects on the expression of cytokines were also observed both in mRNA and protein levels in these pregnant groups. Ingestion of high levels of ZEN caused irreversible damage in weaned dams, such as oxidative stress, decreased villi hight and low expression of junction proteins and cytokines. Decreased expression of jejunal interleukin-8 (IL-8) and increased expression of gastrointestinal glutathione peroxidase (GPx2) mRNA were detected in weaned offspring, indicating long-term damage caused by maternal ZEN. We also found that the Nrf2 expression both in mRNA and protein levels were up-regulated in the ZEN-treated groups of pregnant dams and the high-dose of ZEN group of weaned dams. The data indicate that modulation of Nrf2-mediated pathway is one of mechanism via which ZEN affects gut wall antioxidant and inflammatory responses.

Editor: Kartik Shankar, University of Arkansas for Medical Sciences, United States of America

Funding: This work was supported by National Basic Research Program (2012CB124703); the China Agriculture Research System (CARS-36); and Program for Innovative Research Team of Universities in Heilongjiang Province (2012TD003). The funders had no role in study design, data collection and analysis, decision to publish, or preparation of the manuscript.

Competing Interests: The authors have declared that no competing interests exist.

* Email: asshan@neau.edu.cn

Introduction

The global occurrence of mycotoxins constitutes a major risk factor for human and animal health, and an estimated 25% of the world's crop production is contaminated [1,2]. Zearalenone (ZEN) is a non-steroidal estrogenic mycotoxin biosynthesized through a polyketide pathway by a variety of *Fusarium* fungi that are commonly found in feed and foodstuffs [3,4]. It is frequently implicated in reproductive disorders of farm animals and occasionally in hyperoestrogenic syndromes of humans [5]. The adverse effects of ZEN may be even more pronounced during pregnancy, as the fetuses are susceptible to toxins due to their fragile developmental state and inadequate defense mechanism. Many studies have shown that ZEN can change the intrauterine environment during early gestation by affecting the secretory mechanism of the endometrium [6,7]. Indeed, exposure of fetuses to ZEN led to impaired development and decreased litter size [8,9]. It was demonstrated by Zhang, et al. that exposure to ZEN during early gestation affected maternal reproductive capability and delayed fetal development [10]. Likewise, many studies have shown that ZEN affects fecundity, and the effects could placental transfer to fetuses [11–13].

The intestinal tract is the first physical barrier against ingested food contaminants [14]. After ingestion of ZEN-contaminated food, enterocytes may be exposed to high concentrations of the toxin [15]. It has been demonstrated that ZEN easily crossed the intestinal barrier, and rapidly absorbed by enterocytes [16,17]. Although xenobiotic biotransformation reactions occur mainly in the liver, the intestine may also contribute to overall biotransformation. At present, available data support that ZEN has hepatotoxic, haematotoxic, cytotoxic and genotoxic activities, which are not related to its binding affinity to estrogen receptor sites [18–21]. Furthermore, evidence of its effects on oxidative stress has emerged from several studies, which demonstrate that ZEN induces lipid peroxidation in the liver, spleen, kidney, and testis [22–25]. ZEN has also been shown to have immunotoxicity, which results in several alterations of immunological parameters [26–28]. Several studies *in vitro* have reported that ZEN induces cytotoxicity and oxidative damage, and inhibits protein and DNA syntheses in the human Caco-2 cell line [29,30] or in the swine jejunal epithelial cells [31,32]. However, to the best of our knowledge, the information *in vivo* that supports these observations is limited.

The epithelial surface consists of a simple columnar epithelium, which is increased by the presence of villi [33]. Intestinal mucosal permeability is closely related to the integrity of intestinal barrier. The function of the intestinal barrier is affected by its morphology and cellular junctions, including tight junctions and gap junctions [34]. The gut barrier is formed to a large extent by tight junctions containing occludin and one or more claudin isoforms [35]. Tight junctions seal the luminal end of intercellular space and limit transport via this paracellular route to relatively small hydrophilic molecules. In addition, gap junctions are a kind of structures that localized at the plasma membranes, and allow the exchange of ions, nucleotides, metabolites and other small molecules including second messengers between adjacent cells, thereby facilitating electrical and metabolic coupling [36,37]. Gap junctions are clusters of transmembrane channels composed of connexin (Cx) dodecamers, of which the most widely expressed isoform is Cx43 [38,39]. Based on its function as a physical barrier, intestine is also an active component of the immune system and creates a kind of barrier against invading pathogens [40,41]. However, the increased intestinal barrier permeability may lead to intestinal inflammation [42].

The purpose of the present study was to investigate the effects of early pregnancy dietary exposure to different doses of ZEN on the intestine of the pregnant rats. We investigated the effects of ZEN on intestinal antioxidant status, cytokines expression, barrier function, and morphology in maternal rats. These indicators were also examined in weaned pups to test toxic effects of maternal ZEN exposure on intestinal development of offspring.

Materials and Methods

Ethics statement

This study was performed in strict accordance with the recommendations of the National Research Council Guide, and all of the animal experimental procedures were approved by the Ethical and Animal Welfare Committee of Heilongjiang Province, China. Rats were housed in a temperature-controlled room with proper darkness-light cycles, fed with a regular diet, and maintained under the care of the Laboratory Animal Unit, Northeast Agricultural University, China. All of the surgeries were performed through the ether anesthetization, and every effort was made to minimize suffering.

Animals and feed treatments

Sprague-Dawley (SD) rats and basal diet used in this study were purchased from Jilin University Laboratory Animal Centre (Changchun, China). Ninety-six female rats weighing between 190 and 210 g and 24 male rats weighing between 300 and 325 g were prepared for the experiment. Male rats were used only as sires and were not subjected to any treatments. Rats were housed in multiple mouse racks and were acclimated for one week in groups of five per cage with free access to mycotoxin-free diet and tap water. The animal holding rooms were maintained on a 12-h light/12-h dark cycle at 24.5±0.5°C with 55±5% relative humidity.

After acclimation, the females were mated through naturally breeding (at a ratio of one male to two females). Every morning, each female was examined for the presence of sperm in the vaginal lavage. Each pregnant rat was maintained individually in a polycarbonate metabolic cage. The pregnant rats were randomized into four groups (ZEN0, ZEN50, ZEN100, and ZEN150), and were fed with diets containing ZEN at concentrations of 0.3 mg/kg, 48.5 mg/kg, 97.6 mg/kg or 146.0 mg/kg (equal to 0, 4.5, 9, and 13.5 mg/kg bw/d) respectively from gestation days (GD) 1 to 7. ZEN was purchased as pure crystals from Fermentek Ltd. (Jerusalem, Israel) and diluted in acetonitrile. The substance

has been proven to be stable for at least eight months at room temperature. The doses applied in this study were selected on the basis of literatures reported by Ruddick et al. (1976), Collins et al. (2006) and Arora et al. (1981) [43–45].

Sample collection

Twelve pregnant rats from each group reflecting the average body weight were selected and sacrificed by ether anesthetization on GD8. The remaining pregnant rats were fed with the standard mycotoxin-free diet until the pups were weaned at three weeks of age. Remaining twelve weaned dams and twelve pups with similar body weight selected randomly from each group were sacrificed by ether anesthetization. After slaughter, the jejunal sections were immediately excised and rinsed in normal saline. Afterwards, the samples were quickly dissected, frozen in the liquid nitrogen, and stored at $-80°C$ until subjected to RNA extraction and Western blotting. Other sections of the jejunum were separated into two parts. One part was fixed in 10% formalin for morphological analysis and immunohistochemistry. The other part was sealed into pockets and preserved at $-20°C$ until used for the evaluation of ZEN residues, antioxidant status, and cytokine synthesis.

Detection of ZEN residues in the intestine

The jejunal tissues were analyzed for ZEN residues using ELISA kit. The ELISA method has been used for ZEN determination by Bennett et al. (1994) and Nuryono et al. (2005) [46,47]. Jejunal samples of 0.5 g were homogenized with 2.5 ml of diluteapp: addword: dilute methanol (methanol: water = 6:4) and then extracted by shaking the mixture for 15 min. The sample solution was filtered through a folded paper filter, and the filtrate was diluted with the buffer solution provided in the kit. The sample solution was then treated according to the ZEN Kit, and absorbance was measured at 450 nm using an ultraviolet spectrophotometer.

Assessment of lipid peroxidation

Jejunal tissues homogenates (10%) were prepared in chilled normal saline. The suspension was centrifuged at 3500 rpm for 10 min. Lipid peroxidation was determined by measuring the amounts of malondialdehyde (MDA) through the thiobarbituric acid method described by Bloom and Westerfe (1971) using a commercial MDA kit [48]. The absorbance was measured at 532 nm. Total proteins were quantified through the classical Bradford method with Coomassie Brilliant Blue G-250. Concentration of MDA was expressed as nmol/mg of protein.

Evaluation of antioxidant enzyme activity

The activities of total superoxide dismutase (SOD), glutathione peroxidase (GPx) and catalase (CAT) were analyzed using commercial reagent kits [49]. Analysis of the SOD activity was based on the SOD-mediated inhibition of the formation of nitrite from hydroxylammonium in the presence of O^{2-} generators (xanthine/xanthine oxidase) [50]. Activity of SOD was expressed as units/mg of protein and determined by measuring the reduction of in the optical density of the reaction solution at 550 nm. GPx is an enzyme that catalyzes glutathione oxidation by oxidizing the reduced tripeptide glutathione (GSH) into oxidized glutathione [51]. GPx activity was expressed as units/mg of protein, and one unit was defined as the amount required to decrease the GSH by 1 mM/min after subtracting the decrease in GSH per minute obtained with the nonenzymatic reaction. CAT activity was assayed by the method developed by Aebi (1984) [52], and calculated as nM H_2O_2 consumed/min/mg of tissue protein.

Cytokine synthesis

Jejunal tissue homogenates (10%) were analyzed for cytokines content by sandwich ELISA. Commercial reagent kits were used to detected the interleukin (IL)-1α, IL-1β, IL-6, IL-8, and tumor necrosis factor (TNF)-α (Shanghai, China). A purified fraction of anti-rat cytokines were used as the capture antibodies in conjunction with biotinylated anti-rat cytokines. Streptavidin-HRP and tetramethylbenzidine were used for the detection. The absorbance was read at 450 nm using a microplate reader. Recombinant rat IL-1α, IL-1β, IL-6, IL-8, and TNF-α were used as standards, and results are expressed as nanograms of cytokine/mL. All of the tests were performed using four independent replicates.

Real time-PCR (RT-PCR)

Total RNA was extracted from approximately 100 mg of frozen jejunal tissues using the reagent box of Total RNA Kit, according to the manufacturer's instructions. The concentration of RNA was measured by using a spectrophotometer, and the purity was ascertained by the *A260/A280* ratio. Total RNA from each sample was converted into cDNA according to the manufacturer's instructions and used for RT-PCR.

SYBR Green I RT-PCR Kit was used to measure mRNA expression of antioxidant genes (Nrf2 and gastrointestinal GPx (GPx2)), tight junctions (occludin and claudin-4) and inflammatory cytokines (IL-1α, IL-1β, IL-6, IL-8, and TNF-α) expressed relative to the quantity of the β-actin endogenous control. Rat-specific primers were designed from published GenBank sequences and were synthesized by Sangon (Table 1).

For analyses on an ABI PRISM 7500 SDS thermal cycler, PCR reactions were performed with 2.0 μl of first-strand cDNA and 0.4 μl of sense and anti-sense primers in a final volume of 20 μl. Samples were centrifuged briefly and run on the PCR machine using the default fast program (1 cycle at 95°C for 30 s, 40 cycles of 95°C for 5 s and 60°C for 34 s). All of the PCR reactions were performed in triplicate. The relative gene expression levels were determined using the $2^{-\Delta\Delta Ct}$ method [53].

Immunohistochemistry

Cx43 expression was analyzed on formalin-fixed, paraffin-embedded intestinal sections to evaluate intestinal-cell gap junctions. Tissue sections were deparaffinised with xylene and dehydrated through a graded ethanol series. Heat-mediated antigen retrieval was done by heating the sections (immersed in EDTA buffer, pH 8.0) in a microwave oven (750 W) for 15 min. Endogenous peroxidase activity was blocked by incubation in methanol-H_2O_2 solution. The tissue sections were quenched in normal goat serum for 30 min and incubated overnight at 4°C with the primary antibody (diluted 1:75. Secondary antibody was applied followed by the streptavidin conjugated to horseradish peroxides. Finally, 3,3′-diaminobenzidine was used for color development, and hematoxylin was used for counter staining. The proportion of the intestinal section expressing Cx43 was estimated. Each sample was assessed as showing either normal or reduced staining. Normal staining was considered when homogeneous and strong basolateral membrane staining of the enterocytes were detected. Heterogeneous and weak staining were considered to indicate reduced expression.

Western blotting

Total proteins in the intestinal tissues were extracted and the protein lysate were added. After schizolysis for 1 h on ice, the extracts were centrifugalization for 20 min in the speed of 12000 r/min at 4°C. Then the supernatant was taken, and Coomassie Brilliant Blue protein assay was used to detect the concentration of Nrf2 protein. Proteins (50 μg per lane) were

Table 1. Nucleotide sequences of primers for RT-PCR.

Gene	Forward primer(from 5′ to 3′) Reverse primer(from 5′ to 3′)	Fragment length (bp)	Genbank no.
Nrf2	CTGCTGCCATTAGTCAGTCG	101	NM_031789.2
	GCCTTCAGTGTGCTTCTGGT		
GPx2	CCGTGCTGATTGAGAATGTG	113	NM_183403.2
	AGGGAAGCCGAGAACCACTA		
occludin	CTCCAACGGCAAAGTGAATG	104	NM_031329.2
	CGGACAAGGTCAGAGGAATC		
claudin-4	CTGCCTGGAGTCTTGGTGTC	122	NM_001012022.1
	GAGGGTAGGTGGGTGGGTAA		
IL-1β	GCCAACAAGTGGTATTCTCCA	120	NM-031512.2
	TGCCGTCTTTCATCACACAG		
IL-1α	GAGTCGGCAAAGAAATCAAGA	112	NM_017019.1
	TTCAGAGACAGATGGTCAATGG		
IL-6	AGTTGCCTTCTTGGGACTGA	102	NM_012589.2
	ACTGGTCTGTTGTGGGTGGT		
IL-8	AAGAGGGCTGAGAACCAAGA	124	XM_004833923.1
	CCCACACAATACACAAAGAACTG		
TNF-α	TTCCGTCCCTCTCATACACTG	149	NM_012675.3
	AGACACCGCCTGGAGTTCT		
β-actin	ACCCGCGAGTACAACCTTC	207	NM_031144
	CCCATACCCACCATCACACC		

transblotted to polyvinylidene difluoride membranes in standard Tris-glycine transfer buffer, pH 8.3, containing 0.5% SDS. After transfer, membranes were blocked for 1 h at room temperature in TBST (10 mmol/L Tris-HCl, pH 8.0, 150 mmol/L NaCl, 0.2% Tween-20) containing 5% non-fat milk powder, and incubated overnight at 4°C with diluted primary antibody (Nrf2) (1:500 dilution). Membranes were then washed in TBST for three times. The second antibodies (1:2000 dilution) labeled by horseradish peroxydase (Nrf2) were added, with incubation for 2 h at 37°C and washing the membrane for three times. To demonstrate equal loading, membranes were stripped and reprobed with a specific antibody recognizing β-actin. Band densities were obtained by scanning the membranes using the Quantity One software. Density data were standardised within membranes by expressing the density of each band of interest relative to that of β-actin in the same lane.

Morphological analysis

Cross-sectional jejunal samples from the formalin-preserved segments were cut into 2-mm^2 sections and fixed by standard paraffin embedding. Samples were sectioned at 5 μm and stained with hematoxylin eosin. Each slide was divided into three single segments and analyzed the microstructures of jejunum using an optical microscope (Nikon Eclipse E400). The villous heights and crypt depths of 30 randomly chosen villi were measured.

Statistical analysis of data

The indices were analyzed through ANOVA and Duncan's multiple range tests using the SPSS19.0 statistical software. The data are expressed as the means ± standard error of the mean (mean ±SEM). The level of significance was accepted as P<0.05.

Results

ZEN induces oxidative stress in the jejunum

ZEN induced a significant increase of MDA formation in the ZEN-treated groups of pregnant dams and the ZEN150 group of weaned dams (P<0.05). The MDA concentrations in the offspring exhibited an increasing tendency, but this difference was not significant (P>0.05). Dietary ZEN reduced the activity of SOD in the jejunum of pregnant dams, and the significantly decreased activity of SOD was also observed in the ZEN150 group of weaned dams (P<0.05). We also found that the MDA concentrations and SOD activity were not different between the ZEN100 and ZEN150 group of pregnant dams. The activity of CAT in the pregnant dams decreased in a dose-dependent manner, but this difference was not significant (P>0.05). Significant increases both in GPx activity and GPx2 mRNA were seen in all ZEN-treated groups of pregnant dams (P<0.05). GPx2 mRNA expression was also up-regulated in the ZEN150 group of weaned dams and in the ZEN100 and ZEN150 groups of weaned pups (P<0.05). The jejunal extracts of pups showed normal levels in CAT, SOD and GPx activity compared with the ZEN0 group (Figure 1).

ZEN alters cytokines both in protein and mRNA levels of the jejunum

We investigated the effects of different doses of ZEN on cytokine secretion involved in the inflammatory response in the intestinal samples. Pregnant dams exposed to ZEN50 exhibited increased IL-1α and decreased TNF-α synthesis (P<0.05). Both the IL-1β and TNF-α synthesis were down-regulated in the ZEN150 group of pregnant dams (P<0.05). Although IL-6 and IL-8 synthesis had a decreased tendency in the ZEN-treated groups, the differences were not statistically significant (P>0.05). No negative effects were

observed on the cytokines concentrations in the weaned dams. However, the concentration of IL-8 in the offspring exhibited a significant decrease in the ZEN100 and ZEN150 groups (P<0.05) (Table 2).

We further quantified the expression of genes coding for cytokines, using RT-PCR. Significantly increased IL-1α mRNA and decreased IL-1β, IL-6, IL-8 and TNF-α mRNA were observed in the ZEN-treated groups of pregnant dams with a dose-dependent manner (P<0.05). For the weaned dams, the IL-6 (ZEN150), IL-8 (ZEN100, ZEN150) and TNF-α (ZEN150) mRNA were down-regulated (P<0.05). For the weaned pups, the IL-8 mRNA was also down-regulated in the ZEN100 and ZEN150 groups (P<0.05) (Figure 2).

ZEN up-regulates the expression of Nrf2 in the jejunum

The results of RT-PCR and Western blotting showed that the expression of Nrf2 was increased in pregnant dams in a dose-dependent manner, and the changes were significant in ZEN100 and ZEN150 groups (P<0.05). In weaned dams, a significant up-regulation of Nrf2 mRNA was observed in the ZEN150 group, while the protein levels were increased in ZEN100 and ZEN150 groups (P<0.05). Nrf2 expression was not affected in other groups (Figure 3).

ZEN reduces mRNA expression of tight junction proteins

Effects of ZEN on the expression of the tight junction proteins, including occludin and claudin-4, were assessed using RT-PCR. Both the occludin and claudin-4 mRNA expression levels in pregnant dams showed a declining trend with increase of ZEN. Expression of the occludin mRNA in the ZEN100, ZEN150 groups and the claudin-4 mRNA in the ZEN150 group were significantly down-regulated in pregnant dams (P<0.05). The ZEN150 group still exhibited low occludin and claudin-4 mRNA expression levels in weaned dams (P<0.05). ZEN had no significant effects on the mRNA expression of the occludin and claudin-4 in the offspring (P>0.05) (Figure 4).

ZEN decreases the expression of the gap-junction Cx43 in the jejunum

Cx43 immunoreactivity appeared along the smooth muscle surface both in the outer circular layer and innermost circular layer of the jejunum in dams. Strongly positive immunoreactivity results were also obtained in the intestinal glands (arrow). Significant and dose-related decreases in Cx43 immunoreactivity were seen in all of the ZEN-treated groups of pregnant dams. In the ZEN150 group, low Cx43 immunoreactivity was also observed in weaned dams. There was less obvious Cx43 immunoreactivity in the jejunum of the weaned pups, and the difference was not significant from the ZEN0 group (Figure 5).

ZEN alters villous structure in the jejunum

ZEN damaged the villous structure in a dose-dependent manner in the jejunum of pregnant dams. The ZEN0 group showed normal intestinal morphology. Mild local disruptions of villus tips were observed in the ZEN50 group. Most regions of the jejunum showed denudated villi with part digestion in the ZEN100 group. In the group of ZEN150, villi loss, disruption in integrity of villi were commonly observed. The intestinal mucosal structure could be partially recovered in the ZEN50, ZEN100 groups of the dams at weaning. However, in the ZEN150 group, jejunal sections still showed mild disruption of villi. Histological examination of the pups groups revealed normal mucosal structure and exhibited structural integrity (Figure 6).

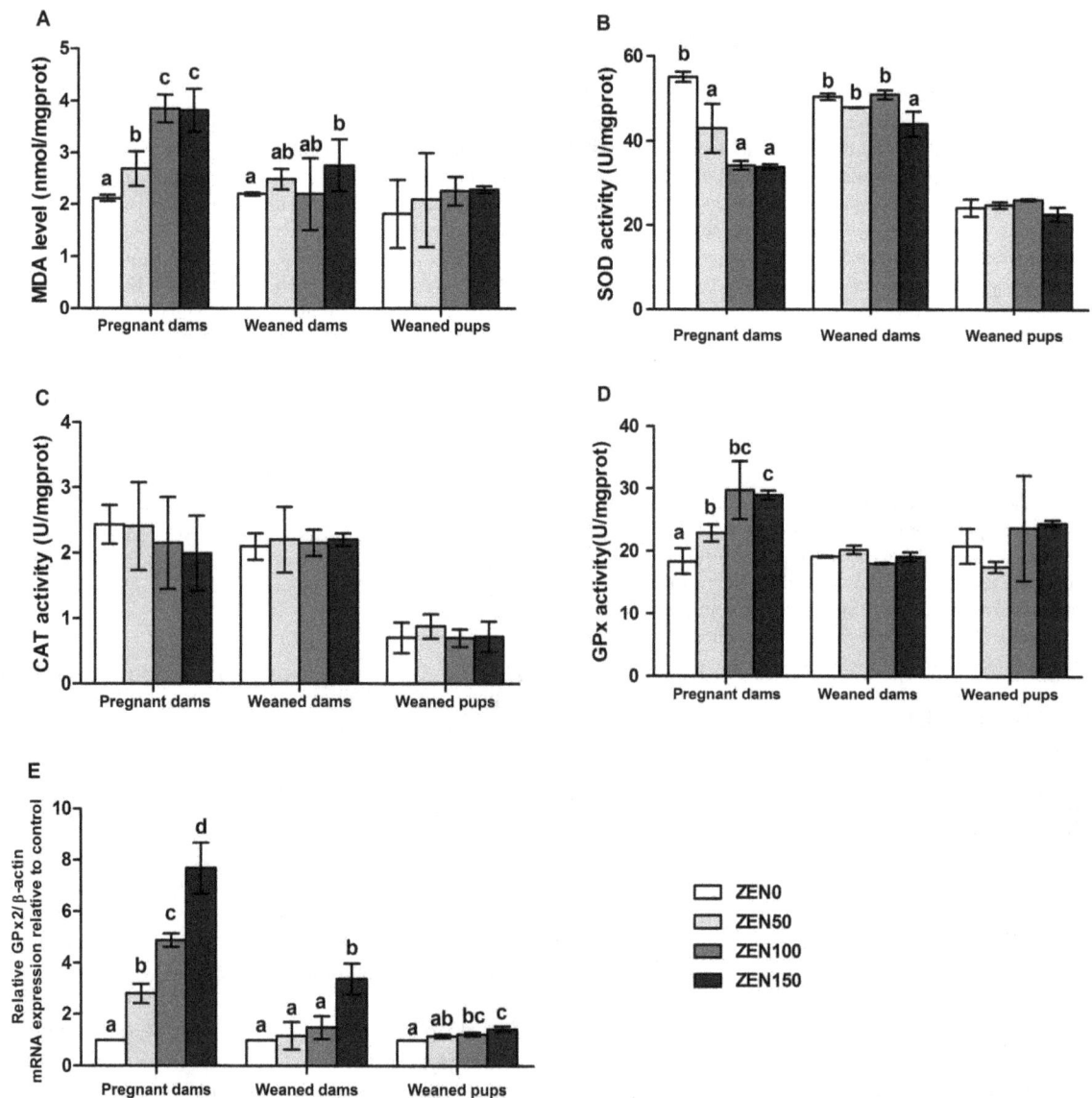

Figure 1. Effects of zearalenone (ZEN) on oxidative stress in the jejunum of rats. (A) malondialdehyde (MDA) level; (B) total superoxide dismutase (SOD) activity; (C) catalase (CAT) activity; (D) glutathione peroxidase (GPx) activity; (E) intestinal glutathione peroxidase (GPx2) mRNA expression. The relative gene expression for GPx2 was calculated relative to the gene expression of the ZEN0 group. β-actin was chosen as a reference gene having the same relative expression mean in the ZEN0 group as in the treated groups. ZEN 0, ZEN 50, ZEN 100, and ZEN 150 means a daily feeding with ZEN at doses of 0.3, 48.5, 97.6, and 146.0 mg/kg respectively, at gestation days 1 to 7. All of the values are expressed as the means ±SEM (n = 12). a, b, c means within a row with no common superscripts differ significantly (P<0.05).

By examining the microstructure of the intestinal mucosa, we found that ZEN affected villus and crypt structures. In the jejunum of pregnant rats, villus height was dose-dependently decreased in the ZEN-treated groups, and the ZEN150 group exhibited a significantly lower villus height compared with the ZEN0 group (P<0.05). The jejunal crypt depth was increased in the ZEN-treated groups, but the changes were only significant in the ZEN50 group (P<0.05). The villus height in the jejunum of weaned dams were significantly decreased in the ZEN150 group compared with other groups (P<0.05), whereas no differences were observed in the crypt depth. For the weaned pups, neither the villus height nor the crypt depth was affected by the treatment of female rats with ZEN (Table 3).

Discussion

Early pregnancy is a sensitive period that can be influenced by ZEN exposure which affects early pregnancy events, including fertilization, embryo development, embryo transport, and embryo implantation [54]. Important developmental changes take place among early gestational days, when the primitive yolk sac placenta is replaced by the chorioallantoic placenta [55]. The chorioallantoic placenta is vascularized by allantoic vessels and becomes more permeable [56]. Time that ZEN exists in the body could be extended due to the extensive enterohepatic cycling [6,57]. Thus, a high dose of the toxin could transfer to the embryos.

An important function of intestinal epithelia is to provide a barrier against the penetration of food contaminants and pathogens present in the intestinal lumen. The disruption of the

Table 2. Effect of zearalenone (ZEN) on the cytokine synthesis on intestine in pregnant dams at gestation day 8, and weaned dams and pups at three weeks after born (ng/L).

	ZEN0	ZEN 50	ZEN 100	ZEN 150
Pregnant dams				
IL-1β	30.51±0.01[b]	29.88±0.10[b]	30.00±0.30[b]	28.68±0.66[a]
IL-1α	33.467±0.03[b]	34.50±0.36[a]	33.81±0.12[b]	33.26±0.39[b]
IL-6	87.46±0.08	87.32±0.85	87.33±1.99	87.00±1.80
IL-8	315.35±4.79	314.80±6.84	312.43±6.84	312.42±0.01
TNF-α	259.65±2.50[c]	253.28±1.05[b]	257.83±1.58[c]	245.87±1.50[a]
Weaned dams				
IL-1β	26.99±0.03	26.83±0.82	27.16±0.88	27.22±0.51
IL-1α	33.94±0.48	34.08±0.04	33.12±0.39	33.41±0.77
IL-6	84.53±0.11	81.08±1.23	81.74±1.04	81.32±0.66
IL-8	276.85±0.99	288.71±2.74	288.12±5.82	291.38±6.29
TNF-α	247.81±0.31	238.71±6.31	237.52±0.41	241.25±5.27
Weaned pups				
IL-1β	4.73±0.09	4.78±0.09	4.66±0.07	4.69±0.05
IL-1α	27.52±0.29	27.54±0.65	26.83±0.51	27.99±0.41
IL-6	65.74±1.31	65.17±1.38	65.99±2.36	65.26±2.29
IL-8	253.74±5.12[bc]	257.76±2.53[c]	245.05±4.03[b]	230.62±0.67[a]
TNF-α	198.44±2.51	200.11±3.19	196.42±4.22	193.62±2.20

All of the values are expressed as the means ±SEM (n = 12).
ZEN 0, ZEN 50, ZEN 100, and ZEN 150 means a daily feeding with ZEN at doses of 0.3, 48.5, 97.6, and 146.0 mg/kg respectively, at gestation days 1 to 7.
[a, b, c]means within a row with no common superscripts differ significantly (P<0.05).

intestinal barrier induced increased penetration of the normally excluded luminal substances that could promote intestinal disorders [41,58]. Studies about the effects of ZEN on the intestine have focused on combinational action with other *Fusarium* mycotoxins, such as deoxynivalenol, T-2, or fumonisin [15,59–61], while few studies have investigated the *in vivo* effects of ZEN on the intestine. The present investigation was conducted to study the toxic effects of maternal ZEN exposure on antioxidant status, immune response, expression of junction proteins, and villous structure in the intestinal samples of pregnant rats. These indexes were also examined in the jejunal samples of weaned offspring to investigated the effects of ZEN on intestinal development of the pups during the fetal period.

Following ingestion of food or feed, ZEN is absorbed from the upper part of intestinal tract. ZEN is extensively absorbed after its oral administration in rats, rabbits and humans [17]. We detected ZEN residues in all of the jejunal samples, and found no significant amounts of ZEN, indicating the rapid absorption of ZEN in jejunum.

Several studies both *in vitro* and *in vivo* reported that ZEN enhanced the formation of reactive oxygen species (ROS) and caused oxidative damage [22–25,62]. Oxidative stress results in damage to cellular structures and has been linked to many diseases [63]. MDA is the end product of lipoperoxidation and is considered as an excellent index of lipid peroxidation [64,65]. As showed by previous research, the concentrations of MDA were significantly increased in the liver, kidney, and testis of Balb/c mice treated with ZEN 40 mg/kg bw [25,66]. Our results showed that the MDA concentrations were significantly increased in the ZEN-treated group of the pregnant dams and in the ZEN150 group of weaned dams, indicating the presence of oxidative stress in the jejunum. The variational trend of SOD activity was opposite

to MDA with the increase of ZEN in the pregnant and weaned dams. The activity of SOD is known to serve protective function for the elimination of reactive free radicals and thus it represents an important antioxidant defense in nearly all cells exposed to oxygen [67,68]. The finding in our result may be due to the increase of ROS in the jejunal tissue. CAT as an early marker of oxidative stress exhibited a decreasing tendency of activity in pregnant dams, which supports the hypothesis.

GPx can modify the poisonous peroxide to a non-toxic hydroxyl compound in order to protect the membrane structure and function. The increased GPx activity was observed in all of the ZEN-treated groups of pregnant dams, which may be due to the defending function of the organism itself against the ROS generation induced by ZEN. The mRNA expression of GPx2 was increased in the ZEN-treated groups of pregnant dams and the ZEN150 group of weaned dams. In addition, GPx2 mRNA was increased in the ZEN100 and ZEN150 groups of weaned pups. The up-regulated expression of GPx2 has been reported as an intestinal defending function against hydroperoxide absorption [69], which may explain the similar results of the MDA, SOD, GPx between the ZEN100 and ZEN150 groups of the pregnant dams. Results in weaned pups indicated the existence of oxidative stress in the jejunum. Other studies have demonstrated that ZEN induced increased activity of GPx in the duodenum mucosa of chicken [70] and increased mRNA expression of GPx in the liver of piglets [23] which was similar to our results. Antioxidant enzymes are considered to be the first line of cellular defense against oxidative damage. Based on the tendency of the antioxidant enzyme activity exhibited in the study, the oxidative stress damage caused by ZEN in the pregnant dams was further confirmed and in a dose-dependent manner. We have observed that the oxidative stress caused by ZEN during early pregnancy in

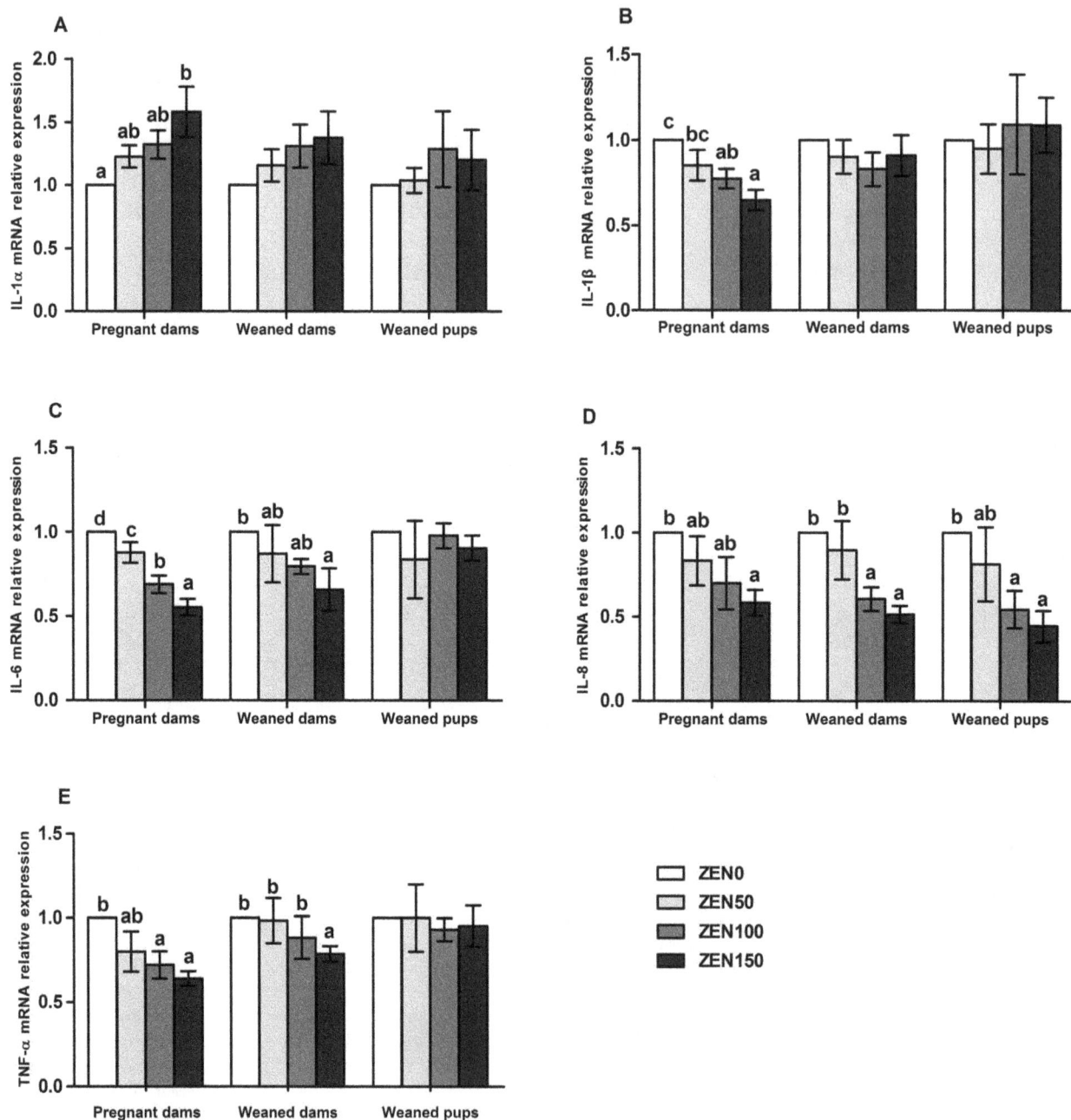

Figure 2. Effects of zearalenone (ZEN) on the expression of immune genes in the jejunum of rats. (A) Interleukin (IL)-1α; (B) IL-1β; (C) IL-6; (D) IL-8; (E) Tumor necrosis factor (TNF)-α. The relative gene expression for each gene was calculated relative to the gene expression of the ZEN0 group. β-actin was chosen as a reference gene having the same relative expression mean in the ZEN0 group as in the treated groups. ZEN 0, ZEN 50, ZEN 100, and ZEN 150 means a daily feeding with ZEN at doses of 0.3, 48.5, 97.6, and 146.0 mg/kg respectively, at gestation days 1 to 7. All of the values are expressed as the means ±SEM (n=12). a, b, c means within a row with no common superscripts differ significantly (P<0.05).

the ZEN50 and ZEN100 groups of weaned dams could be recovered. However, the up-regulated MDA level and GPx activity and down-regulated SOD activity in ZEN150 group of weaned dams suggested the long-term oxidative stress caused by the high dose of ZEN, and the oxidative stress was unrecoverable even through the defending function of the intestine.

Oxidative stress and inflammation are tightly correlated. Lipid peroxidation may bring about protein damage either through direct attack by free radicals or through chemical modification by its end products, e.g. MDA [71]. Pathways that generate

mediators of inflammation (e.g., adhesion molecules, and interleukins) are induced by oxidative stress [72,73]. It is generally accepted that intestinal epithelium, as the interface between the highly antigenic luminal environment and the mucosal immune system, plays an active role in the immune responsiveness of intestinal mucosa. Immune responses in the intestinal mucosa are partly controlled by cytokines release in response to environmental stimuli. In the present study, we investigated the effects of different doses of ZEN on the production of pro-inflammatory cytokines (IL-1α, IL-1β, IL-6, IL-8 and TNF-α) in the jejunum, which are

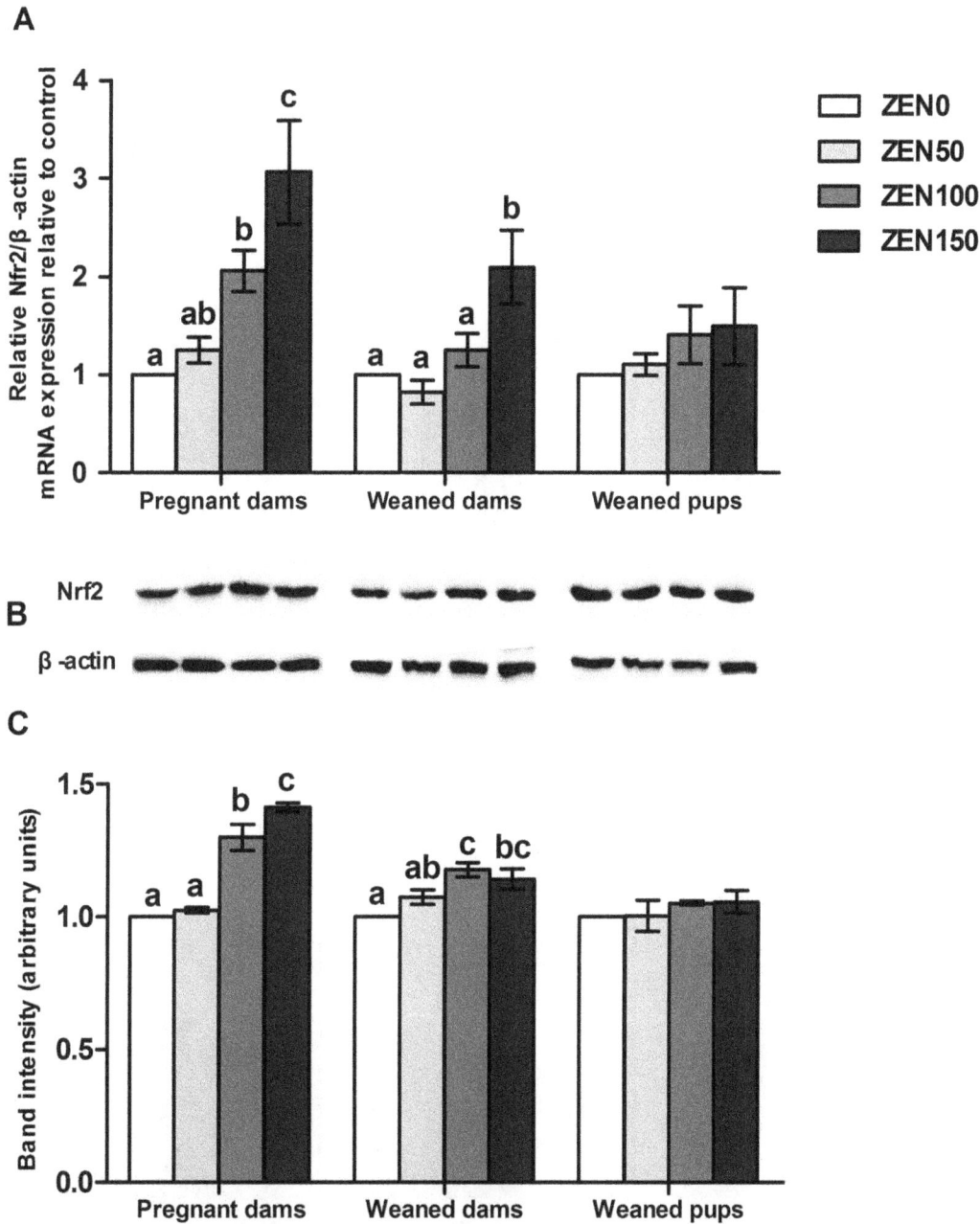

Figure 3. Effects of zearalenone (ZEN) on the expression of Nrf2 in jejunum of rats. (A) Nrf2 mRNA expression; (B) The immunoblot; (C) The expression of the Nrf2 protein estimated by densitometric analyses after normalisation with the β-actin signal. ZEN 0, ZEN 50, ZEN 100, and ZEN 150 means a daily feeding with ZEN at doses of 0.3, 48.5, 97.6, and 146.0 mg/kg respectively, at gestation days 1 to 7. All values are expressed as mean ±SEM (n = 6). [a, b, c] means within a row with no common superscripts differ significantly (P<0.05).

considered to play a key role in the regulation of the immune and inflammatory responses. ZEN has been described as either an inductor or a suppressor of pro-inflammatory cytokines [26,74,75]. In our result, the expression of IL-1β, IL-6, IL-8 and TNF-α in gene level exhibited a decreased tendency in pregnant dams. However, the expression of IL-1α mRNA was up-regulated in pregnant dams. The concentrations of cytokines were consistent with the changes in gene levels to a large extent, but not dose-related. Cytokine assays are not sufficiently sensitive to detect minute amounts of cytokines secreted in potential target tissues *in vivo*, and the adsorption and uptake of cytokines may impair the accurate quantitation of secreted cytokines [76], which may

explain the different response of cytokines in protein and gene levels. The results of the inhibition of cytokine secretions were similar to those of Marin et al., who demonstrated that ZEN depressed the inflammatory cytokine secretions (TNF-a and IL-1β) both in the PBMCs at 5 and 10 μM of ZEN and in the liver of weanling piglets treated with 250ppb of ZEN [23,75]. On the other hand, it has been shown that 40 μM ZEN could significantly increase the mRNA expression of IL-1α in porcine jejunal epithelial cell line [61].

As one of the pro-inflammatory cytokines produced in intestinal mucosal cells, IL-8 could enhance cell proliferation and control the repair processes during injury of the intestinal mucosa or cytotoxic

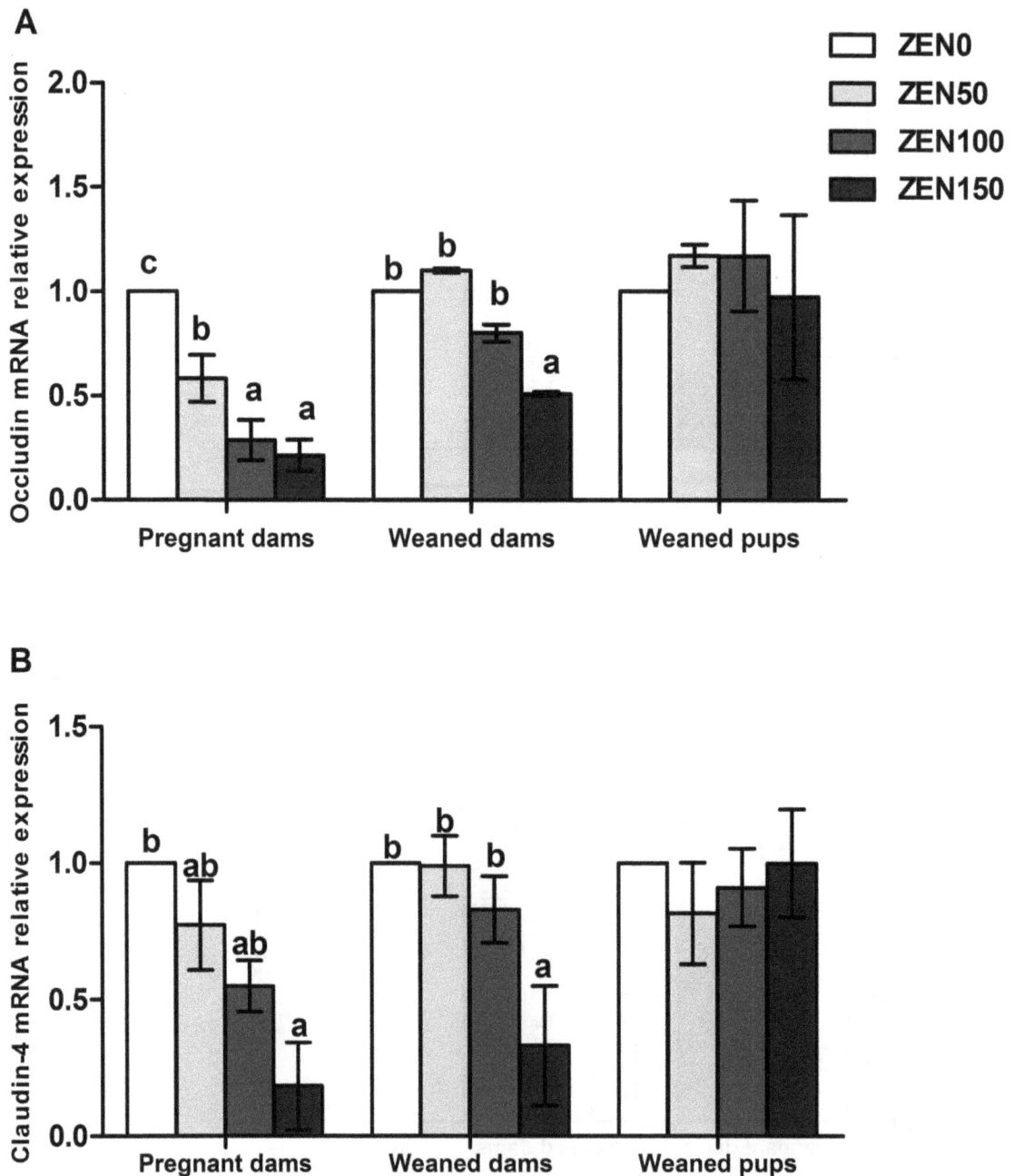

Figure 4. Effects of zearalenone (ZEN) on the mRNA expression of tight junctions in the jejunum of rats. (A) occludin; (B) claudin-4. The relative gene expression for each gene was calculated relative to the gene expression of the ZEN0 group. β-actin was chosen as a reference gene having the same relative expression mean in the ZEN0 group as in the treated groups. ZEN 0, ZEN 50, ZEN 100, and ZEN 150 means a daily feeding with ZEN at doses of 0.3, 48.5, 97.6, and 146.0 mg/kg respectively, at gestation days 1 to 7. All values are expressed as mean ±SEM (n = 12). a, b, c means within a row with no common superscripts differ significantly (P<0.05).

stress [77,78]. Because of its impact on the constitutive synthesis of IL-8, ZEN could perturb the maintenance of the steady homeostasis and the healing properties of the intestine. The weaned pups in the ZEN100 and ZEN150 groups exhibited a significant decrease in IL-8 expression, indicating the unrecoverable immune disorder caused by maternal ZEN. It has been established that the intestine has its own immune network, which causes localized induction of various cytokines and chemokines [79]. In our study, the divergent changes in inflammatory response indicate that ZEN induce immunotoxic effects in the jejunum, which impacted the capacity of the organism both to eradicate injurious stimuli and to initiate healing process.

Nrf2 is reported to be the key transcription factor that regulates cellular antioxidant response and counteracts inflammation [80–82]. The Nrf2-induced mechanism is activated by the stimulation of invading pathogens, and then Nrf2 was translocated from the cytoplasm to the nucleus, through its nuclear localization sequence. Up-regulated expression of the Nrf2 could activate the expression of downstream antioxidant genes (such as GPx2), reduce the inflammatory response and ameliorate the intestine

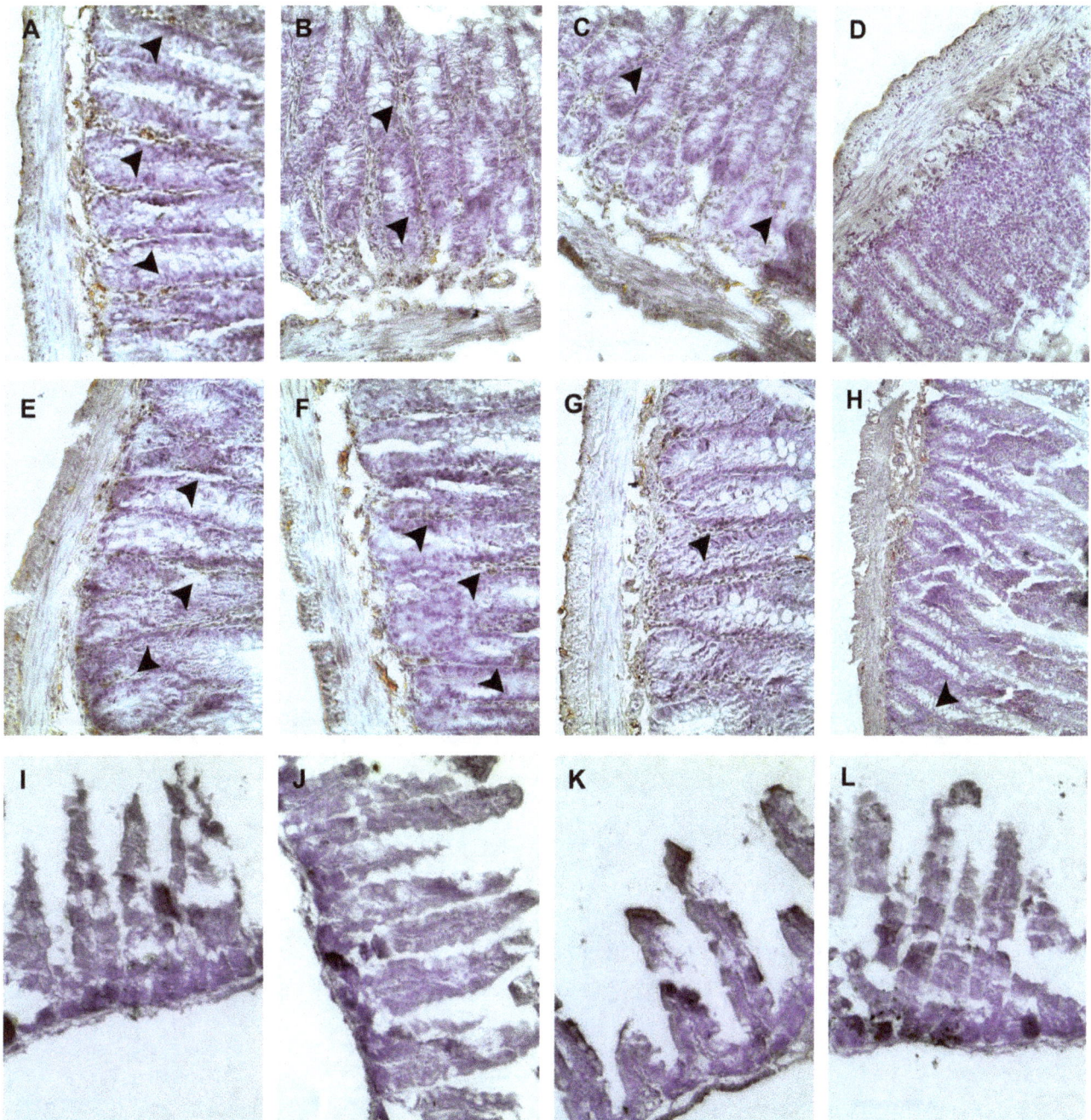

Figure 5. Effect of zearalenone (ZEN) on the expression of gap-junction Cx43 in rats. Samples were collected from the pregnant dams (A–D) at gestation day 8, and weaned dams (E–H) and pups (I–L) at three weeks after born. The results were showed from left to right-ZEN0, ZEN50, ZEN100, ZEN150 in each period of the rats. (A–H) Cx43 immunoreactivity appeared along the smooth muscle surface both in the outer circular layer and innermost circular layer of the jejunal samples. (A–D) The Cx43 immunoreactivity observed in the intestinal glands was down-regulated in the ZEN-treated groups of pregnant dams (arrow). (E–H) the weak immunoreactivity of Cx43 was also observed in the ZEN150 group. (I–L) less Cx43 immunoreactivity was observed. 400×. ZEN 0, ZEN 50, ZEN 100, and ZEN 150 means a daily feeding with ZEN at doses of 0.3, 48.5, 97.6, and 146.0 mg/kg respectively, at gestation days 1 to 7. [a, b, c] means within a row with no common superscripts differ significantly (P<0.05).

damage [83]. In the present study, dose-related increase of Nrf2 expression was observed in pregnant dams. The results suggested that the expression of Nrf2 was activated with the ZEN ingestion in the jejunum. We also observed the increased expression of Nrf2 in the ZEN-treated groups of weaned dams. Considering the results of the normal level of cytokines and antioxidant capacity in the ZEN50 and ZEN100 groups of weaned dams, we believe that

the increased Nrf2 is one of the ways to ameliorate the intestine damage through regulating the cellular antioxidant response and counteracting inflammation. However, in the ZEN150 group of weaned dams, the damage caused by ZEN was irreversible.

Intestinal barrier function is affected by the expression of the tight-junction and gap-junction proteins. In our study, the mRNA expression of tight junction proteins showed a strong correlation

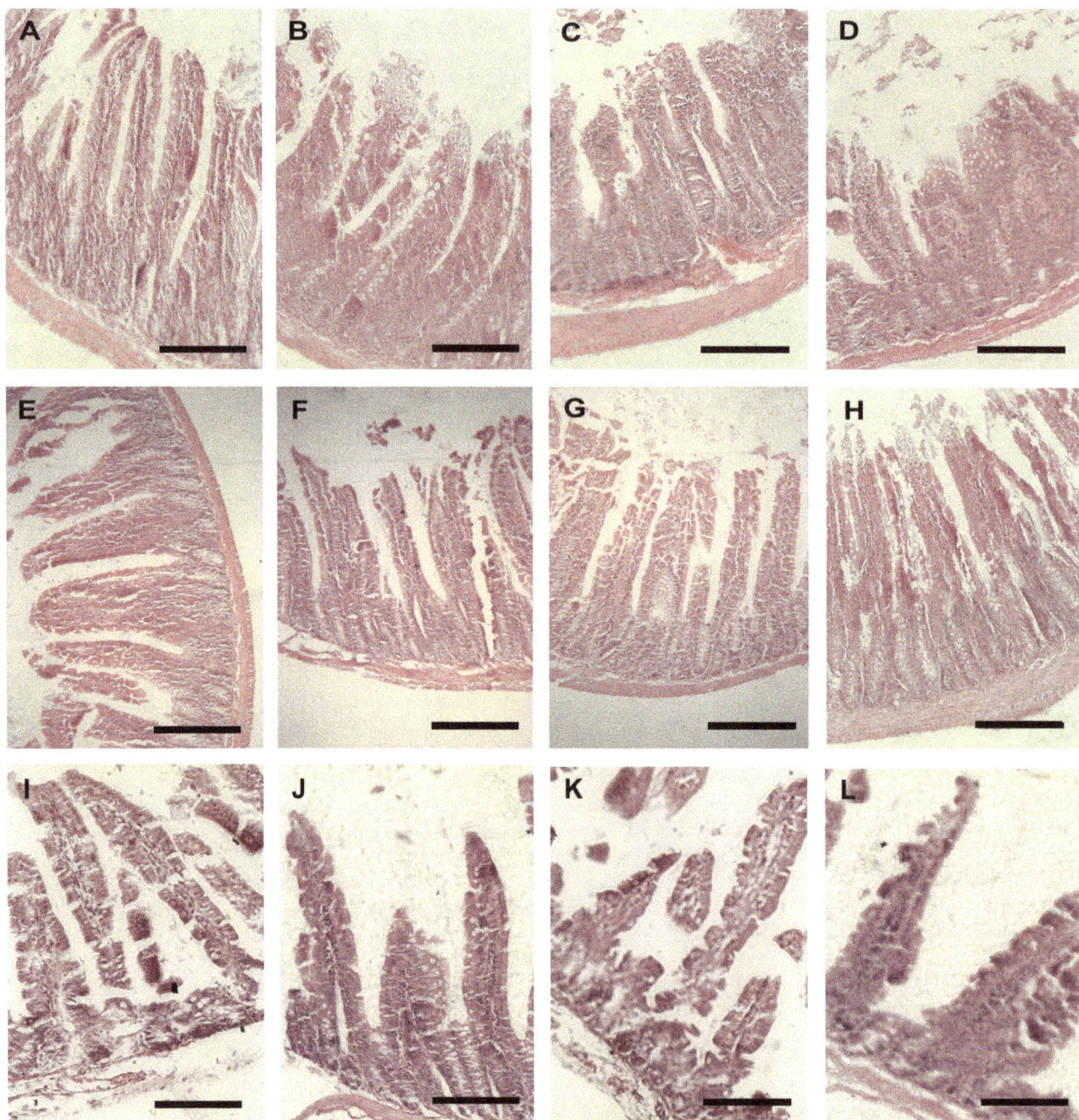

Figure 6. Effect of zearalenone (ZEN) on the jejunum histology in rats. Samples were collected from the pregnant dams (A–D) at gestation day 8, and weaned dams (E–H) and pups (I–L) at three weeks after born. The results were showed from left to right-ZEN0, ZEN50, ZEN100, ZEN150 in each period of the rats. (A)(E, F): jejunal histology demonstrated intact intestinal villi structure. (B) (G): mild local disruption of villus tips. (C) (H): denudated villi with partial digestion with moderate to massive epithelial lifting. (D): villi loss and disruption in integrity of villi. (A–H): HE, 200×; (I–L) HE, 400×. ZEN 0, ZEN 50, ZEN 100, and ZEN 150 means a daily feeding with ZEN at doses of 0.3, 48.5, 97.6, and 146.0 mg/kg respectively, at gestation days 1 to 7.

with the concentrations of ZEN in pregnant dams. To the best of our knowledge, this study provided the first demonstration that the ZEN-contaminated diet reduced mRNA expression of occludin and claudin-4. The claudin-4 and occludin are important for the formation of the actual tight junction seal in intestinal epithelium, which is involved in gut barrier function. The reduction of claudin-4 and occludin suggests defective adhesive properties in enterocytes that would correlate with an increased intestinal translocation of toxic luminal antigens, promoting intestinal inflammation [84]. Except the tight junctions, gap junctions exist between adjacent cells, allowing the transfer of small molecules (under 1000 daltons). The intestinal epithelial cells are physically and functionally interconnected via membrane channels that are composed of the gap junction Cx43 regulating their migration [85]. ZEN was found to inhibit gap junction intercellular communications [86]. As the results showed, the expression of

Table 3. Effect of zearalenone (ZEN) on villus hight and crypt depth of the jejunum in pregnant dams at gestation day 8, and weaned dams and pups at three weeks after born (μm).

	ZEN0	ZEN50	ZEN100	ZEN150
Pregnant dams				
Villus	245.80±9.82[a]	224.00±5.31[ab]	209.74±10.92[ab]	197.33±13.75[b]
Crypt	136.79±12.86[a]	167.01±4.72[b]	154.28±5.21[ab]	141.56±9.59[ab]
Weaned dams				
Villus	233.89±12.48[b]	233.16±6.17[b]	231.68±5.63[b]	218.19±7.86[a]
Crypt	131.816±11.48	138.08±4.71	131.21±7.83	139.42±6.76
Weaned pups				
Villus	97.24±4.01	96.548±2.04	97.79±3.97	97.37±4.05
Crypt	55.80±3.26	56.238±1.22	57.04±4.03	55.67±1.14

All of the values are expressed as the means ±SEM (n = 12).
ZEN 0, ZEN 50, ZEN 100, and ZEN 150 means a daily feeding with ZEN at doses of 0.3, 48.5, 97.6, and 146.0 mg/kg respectively, at gestation days 1 to 7.
[a, b, c]means within a row with no common superscripts differ significantly (P<0.05).

Cx43 was decreased in a dose-dependent manner in pregnant dams. Changes were not recovered in the weaned dams of the ZEN150 group. Previous studies have shown that Cx43 plays an important role in innate immune control of commensal-mediated intestinal epithelial wound repair [87]. So the data are also confirmed by the results that ZEN induced oxidative stress and caused inflammation in the pregnant dams. Decreased levels of TNF-a and IL-1β observed in the jejunum could also contribute to tight-junction barrier defects [88,89]. The down-regulated expression of junction proteins could increase intestinal barrier permeability, resulting in damage of the intestinal barrier function in rat.

The decreased expression of junction proteins induced by other toxins is always followed by the changes of intestinal villous structure [90,91]. In addition, hypersensitivity to antigens in diet could induce morphological changes in the intestine [92]. However, there are few studies about the effects of ZEN on the intestinal morphology which is a commonly indicator of intestinal health in the studies focusing on the influence of food-derived antigens. The main histological changes observed in pregnant dams included the decreased villus height (ZEN150), the increased crypt depth (ZEN50). We also observed that pregnant rats exhibited mild to moderate intestinal lesions, including the detachment of the intestinal epithelial cells and the denudation of villi with part digestion. Damage was not recovered in the high-dose ZEN groups of weaned dams. Intestinal epithelial cells act as a barrier against the penetration of microbial pathogens, cytotoxic agents, and other intestinal contents [93]. Once the jejunum intestinal epithelial cells shed and microvillus height became shorter and sparser, the permeability increased, which then augmented inflammatory and amplified disturbances in gut motor [94]. The histopathological changes observed in the present study indicate that the intestine is inflamed or damaged as a result of ZEN. Possible explanation for these histological changes can be a direct irritant effect of ZEN or suppression of mitosis or protein synthesis [22,95]. However, the normal mucosal appearance and junction proteins in the pups indicated the intestinal structure was not affected by maternal ZEN treatment.

As one of the natural estrogen-like molecules present in our daily environment, ZEN can represent an important exposure source at critical times, such as early gestation. The early events of pregnancy are associated with rapid changes in the expression of genes required for nutrient transport, cellular remodeling, angiogenesis, and relaxation of vascular tissues, as well as cell proliferation and migration [96]. Maternal effects can impact the morphology, physiology and behavior in future generations and even last a lifetime. Previous work have proposed that ZEN exposure during early gestation induced teratogenesis in the fetuses [10]. In the present study, we have observed the down-regulated expression of pro-inflammatory IL-8 and the up-regulated GPx2 mRNA in the jejunum of weaned pups. We considered that these deleterious effects in pups caused by maternal ZEN exposure, lasted from the stages of pregnancy to weaning.

Conclusions

The toxic effects of the ZEN exposure on the intestinal function of pregnant dams are summarized in four aspects: causing oxidative stress, inducing inflammation, altering villous structure, and impairing intestinal barrier function. These changes are in a dose-dependent manner in pregnant dams and are unrecovered to some extent in the ZEN100 and ZEN150 groups of weaned dams. The expression of IL-8 and GPx2 in the pups were affected by the maternal exposure to ZEN100 and ZEN150, indicating the toxic effects of high-dose ZEN on the intestinal development during early pregnancy. We also found that the activation of Nrf2 was related to the jejunal antioxidant status and the inflammatory response, and the Nrf2-mediated signal pathway was one of mechanism that ZEN mediated toxicity in the jejunum.

Author Contributions

Conceived and designed the experiments: YZ AS. Performed the experiments: ML RG QM YZ CB. Analyzed the data: ML. Contributed reagents/materials/analysis tools: ML RG QM YZ CB AS. Wrote the paper: ML.

References

1. Fink-Gremmels J (1999) Mycotoxins: their implications for human and animal health. Vet Quart 21: 115–120.
2. Oswald IP, Marin DE, Bouhet S, Pinton P, Taranu I, Accensi F (2005) Immunotoxicological risk of mycotoxins for domestic animals. Food Addit Contam 22: 354–360.
3. Richard JL (2007) Some major mycotoxins and their mycotoxicoses-an overview. Int J Food Microbiol 119: 3–10.
4. Tabuc C, Marin D, Guerre P, Sesan T, Bailly JD (2009) Molds and mycotoxin content of cereals in southeastern. Romania J Food Prot 72: 662–665.
5. Etienne M, Dourmad JY (1994) Effects of zearalenone or glucosinolates in the diet on reproduction in sows: a review. Livest Prod Sci: 99–113.
6. Appelgren LE, Arora RG, Larsson P (1982) Autoradiographic studies of ^{3}H zearalenone in mice. Toxicology 25: 243–253.
7. Etienne M, Jemmali M (1982) Effects of zearalenone (F2) on estrous activity and reproduction in gilts. J Anim Sci 55: 1–10.
8. Diekman MA, Long GG (1989) Blastocyst development on days 10 or 14 after consumption of zearalenone by sows on days 7 to 10 after breeding. Am J Vet Res 50: 1224–1227.
9. Young LG, Ping H, King GJ (1990) Effects of feeding zearalenone to sows on rebreeding and pregnancy. J Anim Sci 68: 15–20.
10. Zhang Y, Jia Z, Yin S, Shan A, Gao R, et al. (2013) Toxic effects of maternal zearalenone exposure on uterine capacity and fetal development in gestation rats. Reprod Sci 1933719113512533.
11. Dänicke S, Brüssow KP, Goyarts T, Valenta H, Ueberschär KH, et al. (2007) On the transfer of the Fusarium toxins deoxynivalenol (DON) and zearalenone (ZON) from the sow to the full-term piglet during the last third of gestation. Food Chem Toxicol 45: 1565–1574.
12. Schnurrbusch U, Heinze A (2002) Achtung Mykotoxine. Tierhaltung 10: 112–117.
13. Tiemann U, Dänicke S (2006) In vivo and in vitro effects of the mycotoxins zearalenone and deoxynivalenol on different non-reproductive and reproductive organs in female piglets. Food Addit Contam 24: 306–314.
14. Soderholm J, Perdue MH (2001) Stress and gastrointestinal tract. II. Stress and intestinal barrier function. Am J Physiol Gastrointest Liver Physiol 280: G7–13.
15. Bouhet S, Oswald IP (2005) The effects of mycotoxins fungal food contaminants on the intestinal epithelial cell-derived innate immune response. Vet Immunol Immunop 108: 199–209.
16. Videmann B, Mazallon M, Tep J, Lecoeur S (2008) Metabolism and transfer of the mycotoxin zearalenone in human intestinal Caco-2 cells. Food Chem Toxicol 46: 3279–3286.
17. Kuiper-Goodman T, Scott PM, Watanabe H (1987) Risk assessment of the mycotoxin zearalenone. Regul Toxicol Pharmacol 7: 253–306.
18. Maaroufi K, Chekir L, Creppy EE, Ellouz F, Bacha H (1996) Zearalenone induces modifications of haematological and biochemical parameters in rats. Toxicon 34: 535–540.
19. Zinedine A, Soriano JM, Molto JC, Manes J (2007) Review on the toxicity occurrence metabolism detoxification regulations and intake of zearalenone: an oestrogenic mycotoxin. Food Chem Toxicol 45: 1–18.
20. Kim IH, Son HY, Cho SW, Ha CS, Kang BH (2003) Zearalenone induces male germ cell apoptosis in rats. Toxicology Letters 138: 185–192.
21. Pfohl-Leszkowicz A, Chekir-Ghedira L, Bacha H (1995) Genotoxicity of zearalenone an oestrogenic mycotoxin: DNA adductsformation in female mouse tissues. Carcinogenesis 16: 2315–2320.
22. Abid-Essefi S, Ouanes Z, Hassen W, Baudrimont I, Creppy EE, et al. (2004) Cytotoxicity inhibition of DNA and protein syntheses and oxidative damage in cultured cells exposed to zearalenone. Toxicol In Vitro 18: 467–474.
23. Marin DE, Pistol GC, Neagoe IV, Calin L, Taranu I (2013) Effects of zearalenone on oxidative stress and inflammation in weanling piglets. Food Chem Toxicol 58: 408–415.
24. Jia Z, Liu M, Qu Z, Zhang Y, Yin S, Shan A (2014) Toxic Effects of zearalenone on oxidative stress, inflammatory cytokines, biochemical and pathological changes induced by this toxin in the kidney of pregnant rats. Environ Toxicol Phar 37: 580–591.
25. Salah-Abbès JB, Abbès S, Abdel-Wahhab MA, Oueslati R (2009) Raphanus sativus extract protects against zearalenone induced reproductive toxicity oxidative stress and mutagenic alterations in male Balb/c mice. Toxicon 53: 525–533.
26. Salah-Abbès JB, Abbès S, Houas Z, Abdel-Wahhab MA, Oueslati R (2008) Zearalenone induces immunotoxicity in mice: Possible protective effects of radish extract (Raphanus sativus). J Pharm Pharmacol 60: 761–770.
27. Pestka JJ, Tai JH, Witt MF, Dixon DE, Forsell JH (1987) Suppression of immune response in the B6C3F1 mouse after dietary exposure to the Fusarium mycotoxins deoxynivalenol (vomitoxin) and zearalenone. Food Chem Toxicol 25: 297–304.
28. Luongo D, De Luna R, Russo R, Severino L (2008) Effects of four Fusarium toxins (fumonisin B(1) alpha-zearalenol nivalenol and deoxynivalenol) on porcine whole-blood cellular proliferation. Toxicon 52: 156–162.
29. Abid-Essefi S, Baudrimont I, Hassen W, Ouanes Z, Mobio TA, et al. (2003) DNA fragmentation apoptosis and cell cycle arrest induced by zearalenone in cultured DOK Vero and Caco-2 cells: prevention by Vitamin E. Toxicology 192: 237–248.
30. Kouadio JH, Mobio TA, Baudrimont I, Moukha S, Dano SD, et al. (2005) Comparative study of cytotoxicity and oxidative stress induced by deoxynivalenol zearalenone or fumonisin B1 in human intestinal cell line Caco-2. Toxicology 213: 56–65.
31. Marin DE, Taranu I, Pistol G, Stancu M (2013) Effects of zearalenone and its metabolites on the swine epithelial intestinal cell line: IPEC 1. P Nutr Soc 72: E40.
32. Wan LYM, Turner PC, El-Nezami H (2013) Individual and combined cytotoxic effects of Fusarium toxins (deoxynivalenol, nivalenol, zearalenone and fumonisins B1) on swine jejunal epithelial cells. Food Chem Toxicol 57: 276–283.
33. DeSesso JM, Jacobson CF (2001) Anatomical and physiological parameters affecting gastrointestinal absorption in humans and rats. Food Chem Toxicol 39: 209–228.
34. Ma Y, Semba S, Khan RI, Bochimoto H, Watanabe T, et al. (2013) Focal adhesion kinase regulates intestinal epithelial barrier function via redistribution of tight junction. Biochim Biophys Acta. 1832: 151–159.
35. Harhaj NS, Antonetti DA (2004) Regulation of tight junctions and loss of barrier function in pathophysiology. Int J Biochem Cell Biol 36: 1206–1237.
36. Bruzzone R, White TW, Paul DL (1996) Connections with connexins: the molecular basis of direct intercellular signalling. Eur J Biochem 238: 1–27.
37. Kumar NM, Gilula NB (1996) The gap junction communication channel. Cell 84: 381–388.
38. Goodenough DA (1975) The structure of cell membranes involved in intercellular communication. Am J Clin Pathol 63, 636–645.
39. Li Z, Zhou Z, Daniel EE (1993) Expression of gap junction connexin 43 and connexin 43 mRNA in different regional tissues of intestine in dog. Am J Physiol 265: G911–G916.
40. Brandtzaeg P (1996) History of oral tolerance and mucosal immunitya. Ann NY Acad Sci 778: 1–27.
41. Oswald IP (2006) Role of intestinal epithelial cells in the innate immune defence of the pig intestine. Vet Res 37: 359–368.
42. Osselaere A, Santos R, Hautekiet V, De Backer P., Chiers K, et al. (2013) Deoxynivalenol impairs hepatic and intestinal gene expression of selected oxidative Stress, tight junction and inflammation proteins in broiler chickens, but addition of an adsorbing agent shifts the effects to the distal parts of the small intestine. PloS one 8: e69014.
43. Ruddick JA, Scott PM, Harwig J (1976) Teratological evaluation of zearalenone administered orally to the rat. B Environ Contam Tox 15: 678–681.
44. Collins TF, Sprando RL, Black TN, Olejnik N, Eppley RM, et al. (2006) Effects of zearalenone on in utero development in rats. Food Chem Toxicol 44: 1455–1465.
45. Arora RG, Frolen H, Nilsson A (1981) Interference of mycotoxins with prenatal development in the mouse I Influence of aflatoxin B1 ochratoxin A and zearalenone. Acta Vet Scand 22: 524–534.
46. Bennett GA, Nelsen TC, Miller BM (1994) Enzyme-linked immunosorbent assay for detection of zearalenone in maize wheat and pig feed: Collaborative study. J AOAC Int 77: 1500–1509.
47. Nuryono N, Noviandi CT, Böhm J, Razzazi-Fazeli E (2005) A limited survey of zearalenone in Indonesian maize-based food and feed by ELISA and high performance liquid chromatography. Food control 16: 65–71.
48. Bloom RJ, Westerfe WW (1971) The thiobarbituric acid reaction in relation to fatty livers. Arch Biochem Biophys 145: 669–675.
49. Rongzhu L, Suhua W, Guangwei X, Chunlan R, Fangan H, et al. (2009) Effects of acrylonitrile on antioxidant status of different brain regions in rats. Neurochem Int 55: 52–557.
50. Elstner EF, Heupel A (1976) Inhibition of nitrite formation from hydroxylam-monium-chloride: a simple assay for superoxide dismutase. Anal Biochem 70: 616–620.
51. Sedlak J, Lindsay RH (1968) Estimation of total protein-bound and nonprotein sulfhydryl groups in tissue with Ellman's reagent. Anal Biochem 25: 192–205.
52. Aebi H (1984) Catalase in vitro. Meth Enzymol 105: 121–126.
53. Livak KJ, Schmittgen TD (2001) Analysis of relative gene expression data using real-time quantitative PCR and the $2^{-\Delta\Delta Ct}$ method. Methods 25: 402–408.
54. Zhao F, Li R, Xiao S, Diao H, Viveiros MM, et al. (2013) Postweaning exposure to dietary zearalenone a mycotoxin promotes premature onset of puberty and disrupts early pregnancy events in female mice. Toxicol Sci 132: 431–442.
55. Beaudoin AR (1980) Embryology and teratology In: Baker HJ Lindsey JR Weisbroth SH editors. The laboratory rat Volume II Research applications New York: Academic Press 75–101.
56. DeSesso JM (1997) Comparative Embryology In: Hood RD editors Handbook of Developmental Toxicology London: CRC Press 111–174.
57. Biehl ML, Prelusky DB, Koritz GD, Hartin KE, Buck WB, et al. (1993) Biliary excretion and enterohepatic cycling of zearalenone in immature pigs. Toxicol Appl Pharm 121: 152–159.
58. Arrieta MC, Bistritz L, Meddings JB (2006) Alterations in intestinal permeability. Gut 55: 1512–1520.
59. Obremski K, Zielonka L, Gajecka M, Jakimiuk E, Bakuła T, et al. (2007) Histological estimation of the small intestine wall after administration of feed containing deoxynivalenol, T-2 toxin and zearalenone in the pig. Pol J Vet Sci 11: 339–345.

60. Kouadio JH, Dano SD, Moukha S, Mobio TA, Creppy EE (2007) Effects of combinations of Fusarium mycotoxins on the inhibition of macromolecular synthesis, malondialdehyde levels, DNA methylation and fragmentation, and viability in Caco-2 cells. Toxicon 49: 306–317.

61. Wan MLY, Woo CSJ, Turner PC, Wan JMF, Hani EN (2013) Individual and combined effects of Fusarium toxins on the mRNA expression of pro-inflammatory cytokines in swine jejunal epithelial cells. Toxicology letters, 200, 238–246.

62. Yu JY, Zheng ZH, Son YO, Shi X, Jang YO, et al. (2011) Mycotoxin zearalenone induces AIF- and ROS-mediated cell death through p53- and MAPK-dependent signaling pathways in RAW2647 macrophages. Toxicol In Vitro 25: 1654–1663.

63. Marin DE, Taranu I (2012) Overview on aflatoxins and oxidative stress. Toxin Rev 31: 32–43.

64. Bird RP, Draper HH (1984) Comparative studies on different methods of malonaldehyde determination. Method Enzymol 105: 299.

65. Tomita M, Okuyama T, Kawai S (1990) Determination of malonaldehyde in oxidized biological materials by high-performance liquid chromatography. J Chromatogr A 515: 391–397.

66. Zourgui L, Golli EE, Bouaziz C, Bacha H, Hassen W (2008) Cactus (Opuntia ficus-indica) cladodes prevent oxidative damage induced by the mycotoxin zearalenone in Balb/C mice. J Food Chem Toxicol 46: 1817–1824.

67. Liska DJ (1998) The detoxification enzyme systems. Altern Med Rev 3: 187–198.

68. Cheung CC, Zheng GJ, Richardson BJ, Lam PKS (2001) Relationship between tissue concentrations of polycyclic aromatic hydrocarbons and antioxidative responses of marine mussels. Aquat Toxicol 52: 189–203.

69. Brigelius-Flohé R (1999) Tissue-specific functions of individual glutathione peroxidases. Free Radic. Biol Med 27: 951–965.

70. Grešáková Ľ, Bořutová R, Faix Š, Plachá I, Čobanová K, et al. (2012) Effect of lignin on oxidative stress in chickens fed a diet contaminated with zearalenone. Acta Vet Hung 60: 103–114.

71. Halliwell B, Gutteridge JMC (1999) Free radicals in biology and medicine (3rd ed). Oxford: Clarendon Press.

72. Jenny NS (2012) Inflammation in aging: cause effect or both? Discov Med 13: 451–460.

73. Kim YJ, Kim EH, Hahm KB (2012) Oxidative stress in inflammation-based gastrointestinal tract diseases: challenges and opportunities. J Gastroenterol Hepatol 27: 1004–1010.

74. Ruh MF, Bi Y, Cox L, Berk D, Howlett AC, et al. (1998) Effect of environmental estrogens on IL-1b promoter activity in a macrophage cell line. Endocrine 9: 207–211.

75. Marin DE, Taranu I, Burlacu R, Manda G, Motiu M, et al. (2011) Effects of zearalenone and its derivatives on porcine immune response. Toxicol In Vitro 25: 1981–1988.

76. Azconaolivera JI, Ouyang Y, Murtha J, Chu FS, Pestka JJ (1995) Induction of cytokine mRNAs in mice after oral exposure to the trichothecene vomitoxin (deoxynivalenol): relationship to toxin distribution and protein synthesis inhibition. Toxicol Appl Pharm 133: 109–120.

77. Maheshwari A, Lacson A, Lu W, Fox SE, Barleycorn AA, et al. (2004) Interleukin-8/CXCL8 forms an autocrine loop in fetal intestinal mucosa. Pediatr Res 56: 240–249.

78. Zachrisson K, Neopikhanov V, Wretlind B, Uribe A (2001) Mitogenic action of tumour necrosis factor-alpha and interleukin-8 on explants of human duodenal mucosa. Cytokine 15: 148–155.

79. Stadnyk AW (2002) Intestinal epithelial cells as a source of inflammatory cytokines and chemokines. Can J Gastroenterol 16: 241–246.

80. Rangasamy T, Guo J, Mitzner WA, Roman J, Singh A, et al. (2005) Disruption of Nrf2 enhances susceptibility to severe airway inflammation and asthma in mice. J Exp Med 202: 47–59.

81. Pan H, Wang H, Zhu L, Mao L, Qiao L, et al. (2011) Depletion of Nrf2 enhances inflammation induced by oxyhemoglobin in cultured mice astrocytes. Neurochem Res 36: 2434–2441.

82. Thimmulappa RK, Lee H, Rangasamy T, Reddy SP, Yamamoto M, Kensler TW, et al. (2006) Nrf2 is a critical regulator of the innate immune response and survival during experimental sepsis. J Clin Invest 116: 984–995.

83. Banning A, Deubel S, Kluth D, Zhou Z, Brigelius-Flohé R (2005) The GI-GPx gene is a target for Nrf2. Mol Cell Biol 25: 4914–4923.

84. Maresca M, Fantini J (2010) Some food-associated mycotoxins as potential risk factors in humans predisposed to chronic intestinal inflammatory diseases. Toxicon 56: 282–294.

85. Leaphart CL, Qureshi F, Cetin S, Li J, Dubowski T, et al. (2007) Interferon-gamma inhibits intestinal restitution by preventing gap junction communication between enterocytes. Gastroenterology 132, 2395–2411.

86. Ouanes-Ben Othmen Z, Essefi SA, Bacha H (2008) Mutagenic and epigenetic mechanisms of zearalenone: prevention by Vitamin E. World Mycotoxin J 1: 369–374.

87. Ey B, Eyking A, Gerken G, Podolsky DK, Cario E (2009) TLR2 mediates gap junctional intercellular communication through connexin-43 in intestinal epithelial barrier injury. J Biol Chem 284: 22332–22343.

88. Ye D, Ma I, Ma TY (2006) Molecular mechanism of tumor necrosis factor-a modulation of intestinal epithelial tight junction barrier. Am J Physiol Gastrointest Liver Physiol 290: 496–504.

89. Al-Sadi R, Ye D, Dokladny K, Ma TY (2008) Mechanism of IL-1beta-induced increase in intestinal epithelial tight junction permeability. J Immunol 180: 5653–5661.

90. McLaughlin J, Padfield PJ, Burt JP, O'Neill CA (2004) Ochratoxin A increases permeability through tight junctions by removal of specific claudin isoforms. Am J Physiol-Cell Ph 287: C1412–1417.

91. Pinton P, Nougayrède JP, Del Rio JC, Moreno C, Marin DE, et al. (2009) The food contaminant deoxynivalenol decreases intestinal barrier permeability and reduces claudin expression. Toxicol Appl Pharm 237: 41–48.

92. Li DF, Nelssen JL, Reddy PG, Blecha F, Hancock JD, et al. (1990) Transient hypersensitivity to soybean meal in the early-weaned pig. J Anim Sci 68: 1790–1799.

93. Yu LCH (2005) SGLT-1-mediated glucose uptake protects intestinal epithelial cells against LPS-induced apoptosis and barrier defects: a novel cellular rescue mechanism? FASEB J 19: 1822–1835.

94. Collins SM (2001) Stress and the Gastrointestinal Tract IV. Modulation of intestinal inflammation by stress: basic mechanisms and clinical relevance. American Journal of Physiology. Gastrointest Liver Physiol 280: G315–G318.

95. Aida EM, Hassanane MS, Alla ESAA (2001) Genotoxic evaluation for the estrogenic mycotoxin zearalenone. Reprod Nutr Dev 41: 79–89.

96. Bazer FW, Spencer TE, Johnson GA, Burghardt RC, Wu G (2009) Comparative aspects of implantation. Reproduction 138: 195–209.

15

Mangiferin Attenuates Diabetic Nephropathy by Inhibiting Oxidative Stress Mediated Signaling Cascade, TNFα Related and Mitochondrial Dependent Apoptotic Pathways in Streptozotocin-Induced Diabetic Rats

Pabitra Bikash Pal, Krishnendu Sinha, Parames C. Sil*

Division of Molecular Medicine, Bose Institute, Kolkata, India

Abstract

Oxidative stress plays a crucial role in the progression of diabetic nephropathy in hyperglycemic conditions. It has already been reported that mangiferin, a natural C-glucosyl xanthone and polyhydroxy polyphenol compound protects kidneys from diabetic nephropathy. However, little is known about the mechanism of its beneficial action in this pathophysiology. The present study, therefore, examines the detailed mechanism of the beneficial action of mangiferin on STZ-induced diabetic nephropathy in Wister rats as the working model. A significant increase in plasma glucose level, kidney to body weight ratio, glomerular hypertrophy and hydropic changes as well as enhanced nephrotoxicity related markers (BUN, plasma creatinine, uric acid and urinary albumin) were observed in the experimental animals. Furthermore, increased oxidative stress related parameters, increased ROS production and decreased the intracellular antioxidant defenses were detected in the kidney. Studies on the oxidative stress mediated signaling cascades in diabetic nephropathy demonstrated that PKC isoforms (PKCα, PKCβ and PKCε), MAPKs (p38, JNK and ERK1/2), transcription factor (NF-κB) and TGF-β1 pathways were involved in this pathophysiology. Besides, TNFα was released in this hyperglycemic condition, which in turn activated caspase 8, cleaved Bid to tBid and finally the mitochorndia-dependent apoptotic pathway. In addition, oxidative stress also disturbed the proapoptotic-antiapoptotic (Bax and Bcl-2) balance and activated mitochorndia-dependent apoptosis via caspase 9, caspase 3 and PARP cleavage. Mangiferin treatment, post to hyperglycemia, successfully inhibited all of these changes and protected the cells from apoptotic death.

Editor: Srinivasa M. Srinivasula, IISER-TVM, India

Funding: The authors have no support or funding to report.

Competing Interests: The authors have declared that no competing interests exist.

* Email: parames@bosemain.boseinst.ac.in

Introduction

Diabetic mellitus is one of the most recognizable endocrine metabolic disorders fundamentally characterized by hyperglycemia via disruption of carbohydrate, fat and protein metabolism from insufficiency of secretion or action of endogenous insulin (WHO). Hyperglycemia is a well distinguished pathogenic factor of chronic complications in diabetic mellitus and not only generates excessive free radicals (reactive oxygen species; ROS) but also attenuates antioxidative machineries through glycation of the antioxidant enzymes [1]. Streptozotocin (STZ) is commonly used as a diabetic inducer in experimental animals and its toxicity is generated by nitric oxide (NO) on pancreatic β-cells. The cellular toxicity of STZ is linked with the ROS formation resulting oxidative damage of various organ tissues [2]. Hence, oxidative stress has been considered to be a general pathogenic factor of diabetic complications including nephropathy [3]. Diabetic nephropathy is the most serious micro vascular complication of diabetes mellitus and the most common cause of the end stage renal disease (ESRD). This is usually found in both type 1 or type 2 diabetes worldwide [4]. It is caused by the damage to small blood vessels in the kidneys that in turn become less efficient or ultimately fail to function [5]. Diabetic nephropathy has been characterized by glomerular hypertrophy, glomerular hyperfiltration, increased urinary albumin secretion, increased basement membrane thickness and mesangial expansion with the accumulation of extracellular matrix proteins (ECM) [6,7].

Hyperglycemia is strongly associated with increased production of reactive oxygen species (ROS). The plausible major sources of ROS in the diabetic nephropathy are: the activation of polyol pathways, advanced glycation end products (AGEs), autoxidation of glucose, xanthine oxidase activity, mitochondrial respiratory chain deficiencies, NAD(P)H oxidase and nitric oxide synthase (NOS) [8,3]. Therefore, a molecule possessing both hypoglycemic and antioxidant properties might be considered a protective agent against diabetic nephropathy [9,10].

Herbal medicinal plants are of importance to the health of the individual as well as for the communities. The beneficial effects of these medicinal plants are typically due to the occurrence of several chemically active materials that generate a specific physiological action in the human body. Numerous herbal medicinal plants like *Terminalia arjuna* [11–14], *Phyllanthus* niruri [15–19], *Pithecellobium dulce* [20], *Cajanus indicus* [21–25], etc. are natural sources of antioxidants in India and worldwide. These bioactive antioxidants have been used for the treatment of various organ dysfunctions. In addition, fruits, vegetables, tea and wine are the major sources of flavonoids, xanthanoids or polyphenolic compounds and are typically well known to have strong antioxidant activity [26]. Mangiferin (1,3,6,7-Tetrahydroxyxanthone C2-β-D-glucoside; $C_{19}H_{18}O_{11}$; M.W.- 422,34 g/mol), a natural C-glucosyl xanthone and polyhydroxy polyphenol compound, can be found in various plant species, such as the bark of mango tree (*Mangifera indica*) [27,28]. Various earlier reports showed that mangiferin contains antidiabetic [29], antidiarrhea [30], anticancer [31], antiallergic [32], anti-HIV [33] antibacterial [34] and antioxidant [35] properties. In our previous studies we have also established that mangiferin act as a very good protective agent against $Pb(NO_3)_2$ induced oxidative stress in hepatic and cardiac pathophysiology by enhancing antioxidant defense and acting through apoptotic pathways [36,37]. A number of literature reported the beneficial role of mangiferin in STZ induced type 1 diabetic nephropathy via oxidative stress, but the details of the mechanisms is still unknown.

In the present study we have, therefore, designed experiments to investigate the mechanisms underlying the protective action of mangiferin in renal oxidative damage induced by streptozotocin (STZ) type1 (insulin dependent) diabetes using rats as the working model.

Materials and Methods

Chemicals

STZ (Streptozotocin), bovine serum albumin (BSA), DHE (dihydroethidium) and Bradford reagent were purchased from Sigma-Aldrich Chemical Company (St. Louis, MO, USA). Antibodies were purchased from cell signaling technology, Inc. (CST) and Abcam. Kits for measurement of blood glucose, blood uria nitrogen (BUN), creatinine, uric acids were purchased from Span Diagnostic Ltd. India and kit for urinary albumin was purchased from Abcam. All other chemicals were bought from Sisco Research Laboratory, India.

Animals

Adult male Wister rats weighing approximately 160–180 g were purchased from M/S Gosh Enterprises, Kolkata, India. Animals were acclimatized under laboratory conditions for 2 weeks prior to experiments. They were maintained under standard conditions of temperature ($23\pm2°C$) and humidity ($50\pm10\%$) with an alternating 12 h light/dark cycles. The animals had free access to tap water and were fed a standard pellet diet (Agro Corporation Private Ltd., Bangalore, India). All the experiments with animals were carried out according to the guidelines of the Institutional Animal Ethical Committee (IAEC), Bose Institute, Kolkata (the permit number is IAEC/BI/3(I) cert./2010) and full details of the study was approved by both the IAEC and Committee for the Purpose of Control and Supervision on Experiments on Animals (CPCSEA), Ministry of Environment & Forests, New Delhi, India (the permit number is 95/99/CPCSEA).

Extraction and isolation of mangiferin

Mangiferin was extracted in our laboratory following the method as described by Ghosh et al. [38]. Briefly, crudely powdered bark of *Mangifera indica* was thoroughly extracted thoroughly with ethanol (95%) in a Soxhlet apparatus for 56 h. The combined alcohol extracts was concentrated under reduced pressure until a yellow amorphous powder was obtained. The dried alcoholic extract was adsorbed on silica gel (60–120 mesh) and chromatographed in silica gel column packed in petroleum ether (60–80°). The column was eluted with chloroform: methanol (1:1) which yielded mangiferin as a pale yellow amorphous powder that crystallized with the ethanol to form pale yellow needle shaped mangiferin crystals. Purity of the product has been checked by the HPLC, HRMS (ESI) analysis and NMR (^{1}H, ^{13}C) spectroscopy (data not shown).

Determination of dose dependent activity of mangiferin by BUN level

For this study, rats were randomly distributed into seven groups each consisting of six animals. First two groups were served as normal control (received only water as vehicle) and STZ-treated (STZ single dose, 65 mg/kg mg/kg body weight, inject intraperitonally) respectively. Remaining five groups of animals were treated with five different doses of mangiferin (10 mg, 20 mg, 40 mg, 60 mg and 80 mg/kg body weight for 30 days, orally) after STZ administration.

In vivo experimental design

Group 1 (Normal Group: abbreviated as Cont): Rats received neither STZ nor mangiferin, but received vehicle only.

Group 2 (Mangiferin treated Group: abbreviated as Mang): Rats received only an oral sample of mangiferin at a dose of 40 mg/kg body weight.

Group 3 (Diabetic Group: abbreviated as STZ): Overnight fasted rats received a single dose of streptozotocin (STZ, 65 mg/kg mg/kg body weight, inject intraperitonally) in 0.1 M cold citrate buffer (pH 4.5). The blood glucose levels above 250 mg/dl on the 3rd day after STZ injection were considered as diabetic.

Group 4 (STZ+Mangiferin: abbreviated as STZ+Mang): Post treatment Group: Rats orally received mangiferin at a dose of 40 mg/kg body weight/day in water for 30 days orally from the 4th day after STZ injection.

Collection of blood, urine and kidney from experimental animals

The animals were sacrificed [anesthetizing with ketamin (IM) and thiobutabarbital (IP) at a dose of 30 and 50 mg/kg body wt respectively] after the experimental period and the kidneys were collected and stored at −80°C until further analysis. The body weight and kidney weight were measured and evaluated between groups. Blood samples were withdrawn from the caudal vena cava, collected in test tubes having heparin solution and centrifuged at 1,500 g for 10 min to obtain plasma. The plasma was instantly stored at −80°C until use. For albumin measurements urine was collected from the bladder and immediately stored at −80°C.

Preparation of mitochondrial, cytosolic and microsomal fractions

The kidneys were minced, rinsed with PBS and homogenized in a Dounce glass homogenizer in homogenizing buffer (50 mM phosphate buffer, pH 7.5, containing 1 mM EDTA, 1.5 mM $MgCl_2$, 10 mM KCl and supplemented with protease and phosphatase inhibitors). The homogenates were centrifuged for 10 min at 2,000 g at 4°C. The pellet was thrown away and the supernatant was re-centrifuged at 12,000 g for 10 min at 4°C. The pellet was then re-suspended in 200 mM mannitol, 50 mM sucrose, 10 mmol/L Hepes–KOH (pH 7.4) and stored as a mitochondrial fraction at −80°C. The final supernatant was further centrifuged at 105,000 g for 60 min at 4°C. The resulting microsomal pellets were then suspended in a 0.25 mM sucrose solution containing 1 mM EDTA and stored at −80°C until use. The supernatant was received and used as cytosolic fraction and stored at 4°C.

Measurement of protein content

The protein content of the experimental samples was measured by the method of Bradford (1976) [39] using crystalline BSA as standard.

Measurement of plasma glucose levels and kidney dysfunction markers

Plasma glucose levels and specific markers related to kidney dysfunction such as BUN, creatinine, uric acid in the plasma and urinary albumin were estimated using standard kits.

Histological studies

For histological assessments, small segments of kidneys from the normal and experimental rats were fixed in 10% buffered formalin and were processed for paraffin sectioning. Sections of about 5 μm width were stained with hematoxylin and eosin (H&E) for assessment under light microscope.

Measurement of lipid peroxidation and protein carbonyl content

The lipid peroxidation in terms of malondialdehyde (MDA) formation in kidney tissue homogenate (containing 1 mg of protein) was measured following the method of Esterbauer and Cheeseman [40].

Protein carbonyl contents were measured according to the methods Uchida and Stadtman [41].

Measurement of intracellular ROS production

Intracellular ROS production was measured by using 2,7-dichlorofluorescein diacetate (DCFDA) as a probe according to the method of LeBel and Bondy [42] followed by some modifications introduced by Kim et al. [43]. The formation of DCF was assessed in a fluorescence spectrometer (HITACHI, Model No F4500) equipped with a FITC filter at the excitation wavelength of 488 nm and emission wavelength of 510 nm for 10 minutes.

The oxidative fluorescent dye dihydroethidium (DHE) was used to detect superoxide ($O_2^{·-}$) production in kidney from normal and experimental rats [44]. Cryosections (10 μm) from kidney tissue, were stained with the dye DHE (10 μmol/L) in a light-protected and humidified chamber for 30 min at 37°C. Images for each section were analyzed with a fluorescent microscope.

Measurement of the activity of antioxidant enzymes

Activities of antioxidant enzymes such as superoxide dismutase (SOD), catalase (CAT), glutathione peroxidase (GP_X) and glutathione reductase (GR) in the kidney tissue were measured following the methods as described elsewhere [45].

Measurement of the levels of cellular metabolites

Cellular GSH levels were measured by using Ellman's reagent (DTNB; 5,5-dithiobis-2-nitrobenzoic acid) [46]. Oxidized glutathione (GSSG) contents in the samples were measured following the method of Hissin and Hilf [47].

Measurement of plasma AGEs level by ELISA

The levels of plasma AGEs was measured by ELISA kits according to the manufacturer's instructions (Abcam, UK).

Measurement of xanthine oxidase activity

The activity of xanthine oxidase was assessed by measuring the enzymatic oxidation of xanthine. The reactive mixture contained 1.9 mL of 50 mM potassium phosphate buffer, pH 7.5 and 1 mL of 0.15 mM xanthine. The reaction was started by adding 100 μL of kidney tissue extract and the increase in absorbance was measured at 290 nm for 4 min.

Measurement of renal hydroxyproline level

The kidney hydroxyproline levels were measured according to the method of Woessner (1961) [48].

Determination of mitochondrial membrane potential

The mitochondrial membrane potential from isolated mitochondrial fraction of kidney tissue was carried out by using a FACScan flow cytometer with an argon laser excitation at 488 nm and a 525 nm band-pass filter. Mitochondrial membrane potential ($\Delta\Psi_m$) has been estimated on the basis of cell preservation of the fluorescent cationic probe rhodamine 123.

Agarose gel electrophoresis for DNA fragmentation

The DNA fragmentation assay was performed by using electrophoresing genomic DNA samples, isolated from normal as well as experimental kidney, on agarose/EtBr gel by the procedure described by Sellins and Cohen [49].

TUNEL assay for DNA fragmentation

Paraffin embedded renal tissue sections (5 μm) was warmed for 30 min (64°C), deparaffinized and rehydrated. Terminal transferase mediated dUTP nick end-labeling of nuclei has been performed by using APO-BrdU TUNEL Assay kit (A-23210; Molecular Probes, Eugene, OR) following the manufacturer's protocol.

Immunoblotting

An equal amount of protein (50 μg) from each sample was resolved by 10% SDS-PAGE and transferred to PVDF membrane. Membranes were blocked at room temperature for 2 h in blocking buffer containing 5% non-fat dry milk to prevent nonspecific binding. The membranes were then incubated with each of these anti-PKCα (1:250 dilution), anti-PKCβ (1:250 dilution), anti-PKCε (1:250 dilution), anti-ERK (1:1,000 dilution), anti-JNK (1:1,000 dilution), anti-p38 (1:1,000 dilution), anti-NF-κB(1:1,000 dilution), anti-TGF-β1(1:1,000 dilution), anti-TNFα (1:1,000 dilution), anti-caspase-8 (1:1,000 dilution), anti-t-Bid (1:1,000 dilution), anti-Bid (1:1,000 dilution), anti-Bcl-2 (1:1,000 dilution), anti-Bax (1:1,000 dilution), anti-cytochrome C (1:1,000 dilution), anti-caspase-3 (1:1,000 dilution), anti-caspase-9 (1:1,000 dilution) and anti-PARP (1:1000) primary antibodies separately at 4°C overnight. The membranes were washed in TBST (50 mmol/L Tris–HCl, pH 7.6, 150 mmol/L NaCl, 0.1% Tween 20) for 30 min and incubated with appropriate HRP-conjugated secondary antibody (1:2,000 dilution) for 2 h at room temperature and developed by the HRP substrate, 3,30 diaminobenzidine tetrahydrochloride (DAB) system (Bangalore Genei, India).

Statistical analysis

All the experimental values were expressed as mean ± SEM (n = 6). Significant differences between the groups were determined with SPSS 10.0 software (SPSS Inc., Chicago, IL, USA) for Windows using one-way analysis of variance (ANOVA) and the group means were compared by Duncan's Multiple Range Test (DMRT). A difference was considered significant at the $p < 0.05$ level.

Results

Dose dependent effect of mangiferin by BUN level

BUN assay was used to determine the optimum dose of mangiferin for its protective action of kidney tissue against STZ-induced oxidative damage. Experimental results suggest that STZ-induced increased BUN level and that could be prevented by mangiferin treatment linearly up to a dose of 40 mg/kg body weight for 30 days (Figure 1A). This dose of mangiferin has, therefore, been used for the subsequent experiments.

Effects of mangiferin on the body weight, kidney weight and plasma glucose of STZ induced diabetic rats

Approximately 90% of rats injected with STZ developed type 1 diabetes characterized by the significant increase in plasma glucose level (395 mg/dL) (Figure 1E). Physical inertia was detected in the STZ-induced diabetic rats; animals gained kidney weight and the kidney to body weight ratio (a marker for the development of diabetic nephropathy) was also increased (Figure 1B, C, D) compared to normal rats. Post-treatment with mangiferin for 30 days after STZ exposure, however, decreased the plasma glucose levels (214 mg/dL) and reduced the growth accelerating activities suggesting its role as a good anti-hyperglycemic and growth-inhibiting agent in STZ induced diabetic animals.

Effects of mangiferin on STZ-induced nephrotoxicity

STZ-induced diabetic rats exhibited significant alterations in the markers of nephrotoxicity. Plasma BUN, plasma creatinine, plasma uric acid and urinary albumin showed significant elevation (Figure 2A, B, C, D). However, mangiferin treatment efficiently reduced the alterations of these parameters and appeared to act as a nephroprotective agent in diabetes.

Mangiferin protects from STZ-induced renal injury

Histological studies (H&E stained) on STZ-induced diabetic kidney showed increased glomerular size and significant hydropic changes in the proximal convoluted tubules (Figure 3). These alterations were effectively decreased on post treatment with mangiferin for 30 days. These results again suggest the protective action of mangiferin in diabetic renal injury.

Effects of mangiferin on STZ-induced oxidative stress related parameters

In our present studies, induction of diabetes results in increased lipid peroxidation, protein carbonylation and oxidized glutathione (GSSG) content in association with decreased reduced glutathione (GSH) as well as GSH to GSSG ratio in kidney the kidney tissue (Figure 4A, B, C, D, E). However, post-treatment with mangiferin for 30 days effectively decreased the alterations in these oxidative stress related parameters suggesting it to be a good antioxidant agent that protects rat kidney from diabetes-induced oxidative damage.

Effects of mangiferin on STZ -induced ROS levels

ROS mediated oxidative stress due to hyperglycemia plays an important role in diabetic nephropathy. In the present study, STZ-induced diabetic animals showed increased production of intra-cellular ROS in the kidney tissue (Figure 4F). Mangiferin, on the other hand, reduced the ROS production nearly normal level in the kidney tissue. Therefore, mangiferin in this situation acts as a powerful ROS scavenger.

We have also performed dihydroethidium (DHE) staining to assess superoxide (O_2·$^-$) production in the kidney tissue. Figure 4G clearly showed significant increase in superoxide production in the STZ-induced diabetic kidneys and its reduction by mangiferin treatment.

Effects of mangiferin on STZ-induced changes in cellular antioxidant enzymes

Antioxidant enzymes act as the first line of cellular defense molecules in ROS mediated oxidative damage. Effect of mangiferin on the activities of the antioxidant enzymes (CAT, SOD, GP_X and GR) in STZ-induced kidney was shown in Figure 5 (Figure 5A, B, C, D). In the STZ-induced diabetic kidney, the activities of these antioxidant enzymes are significantly lower. However, treatment with mangiferin for 30 days restored the activities of these antioxidant enzymes in STZ-induced diabetic kidney.

Effects of mangiferin on STZ-induced AGE and xanthine oxidase levels

Advance glycation end product (AGE) and xanthine oxidase are important ROS inducer under diabetic condition. In the present studies, STZ-induced diabetic rats increased the plasma AGE formation and increased the activities of xanthon oxidase (in kidney tissue) (Figure 6A, B). Treatment with mangiferin for 30 days, however, suppressed AGE formation and inhibited the activities of xanthon oxidase suggesting that it could block the activities of these ROS inducers in diabetic condition.

Effects of mangiferin on STZ-induced renal hydroxyproline levels

Collagen is normally considered as a reliable marker of fibrosis and is frequently determined by measuring the hydroxyproline content in the sample. The renal hydroxyproline level in STZ-induced diabetic rats was significantly higher (Figure 6C). However, post treatment with mangiferin in STZ-induce diabetic rats effectively reduced that hydroxyproline level. The results suggest that mangiferin could efficiently prevent renal fibrosis in diabetic rats.

Mangiferin inhibits STZ-induced activation of PKCs

Hyperglycemia induced ROS production initiates the activation of PKCs in the development of diabetic nephropathy. Immunoblot analysis showed that STZ induced diabetes was strongly associated with the increased expression of PKCα, PKCβ and PKCε in kidney tissue (Figure 7) and this expression could significantly be reduced by the treatment with mangiferin for 30 days.

Mangiferin inhibits STZ-induced activation of MAPKs

Hyperglycemia mediated oxidative stress is related to the activation of MAPKs family proteins. This family is known to act as the inducers of apoptotic cell death under a variety of pathophysiological circumstances [50]. In our present study, immunoblot analysis shows the stimulated phosphorylation of p38, JNK and ERK1/2 MAPKs in the renal tissue of STZ-induced diabetic rat (Figure 8). On the other hand, mangiferin treatment, post to diabetic induction, significantly reversed the activation of p38, JNK and ERK1/2 MAPKs.

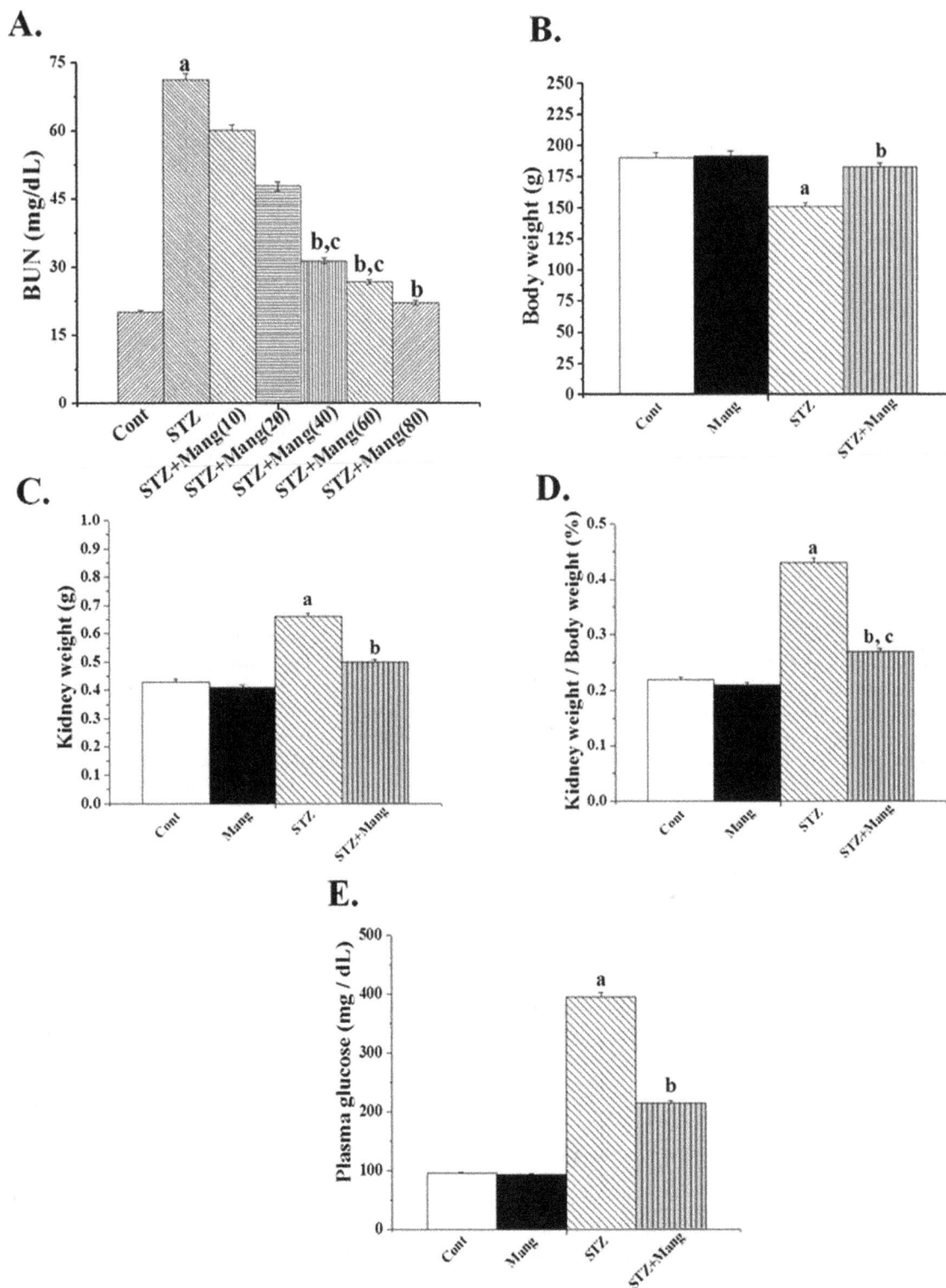

Figure 1. Effects of mangiferin (Mang) on the body weight, kidney weight, plasma glucose and nephrotoxicity of STZ-induced type 1 diabetic rats. Cont: normal control, Mang: treated with mangiferin, STZ: STZ-induced (diabetic), STZ+Mang: Mangiferin treated post to STZ-induced. (A) Dose dependent effect of mangiferin on BUN level against STZ induced toxicity in the kidney tissue of the experimental rats. [Cont: BUN level in normal mice, STZ: BUN level in STZ induced mice, STZ+Mang(10), STZ+Mang(20), STZ+Mang(40), STZ+Mang(60) and STZ+Mang(80): BUN level in mangiferin (Mang) treated mice for 30 days at a dose of 10, 20, 40, 60 and 80 mg/kg body weight respectively post to STZ administration], (B) Body weight, (C) kidney weight, (D) kidney weight to body weight ratio, (E) plasma glucose. Each column represents mean ± SEM, n = 6. "a" indicates the significant difference between the normal control and STZ-induced groups, "b" indicates the significant difference between the STZ-induced and mangiferin treated groups and "c" indicates the significant difference between the STZ+Mangiferin group and normal control group (P^a<0.05, P^b<0.05, P^c<0.05).

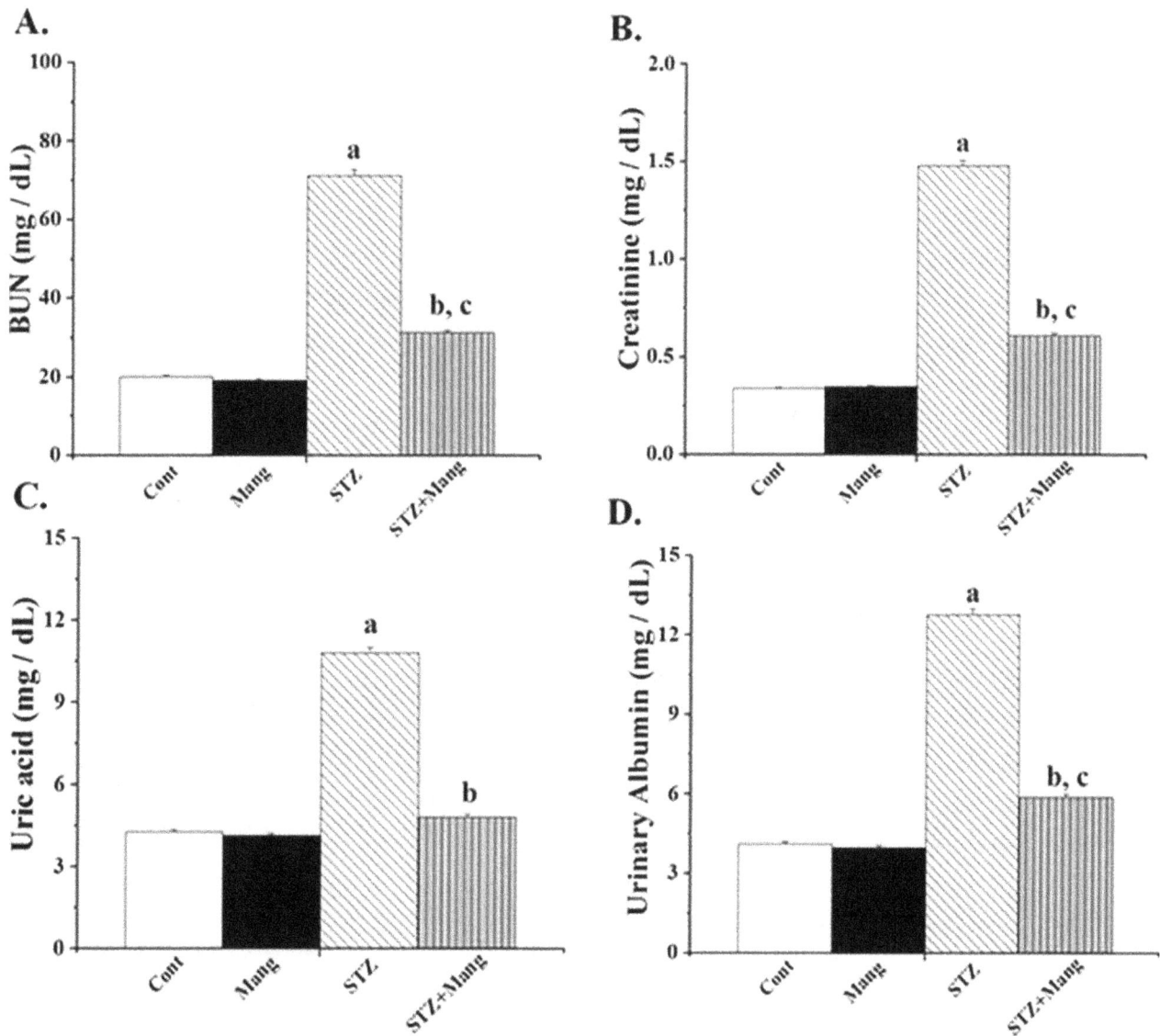

Figure 2. Effects of mangiferin (Mang) on the nephrotoxicity of STZ-induced type 1 diabetic rats. Cont: normal control, Mang: treated with mangiferin, STZ: STZ-induced (diabetic), STZ+Mang: Mangiferin treated post to STZ-induced. (A) BUN, (B) Creatinine, (C) Uric acid and (D) urinary albumin were measured. Each column represents mean ± SEM, n = 6. "a" indicates the significant difference between the normal control and STZ-induced groups, "b" indicates the significant difference between the STZ-induced and mangiferin treated groups and "c" indicates the significant difference between the STZ+Mangiferin group and normal control group ($P^a < 0.05$, $P^b < 0.05$, $P^c < 0.05$).

Mangiferin inhibits STZ-induced activation of NF-κB pathway

NF-κB, an important inflammation transcription factor, plays a critical role for the pathogenesis of diabetic nephropathy. To examine whether the activation of NF-κB pathway has any role in STZ-induced diabetic nephropathy and whether mangiferin can inhibit this occurrence, we performed an immunoblot analysis. Our result shows that, in STZ-induced diabetic kidney tissue, expression of NF-κB and the phosphorylation of IKKα increased and expression of IκBα decreased (Figure 9A). Mangiferin treatment could, however, effectively decrease the expression of NF-κB and IKKα but increased the expression IκBα. So, we say that mangiferin could inhibit diabetes induced NF-κB pathway.

Mangiferin inhibits STZ-induced overexpression of TGF-β1

Transforming growth factor-bita 1 (TGF-β1) plays a critical role in the development of diabetic nephropathy. Immunoblot analysis shows that the expression of TGF-β1 was significantly increased in the in kidney tissue of STZ-induced diabetic rats and was efficiently inhibited by the post treatment with mangiferin (Figure 9B).

Effects of mangiferin on STZ-induced TNFα releases

In the present study, we have investigated the effect of mangiferin on the expression of TNFα in kidney tissues of normal and experimental rats. The expressions of TNFα were increased significantly in STZ-induced diabetic rats (Figure 10A). On the other hand, treatment with mangiferin effectively normalized the

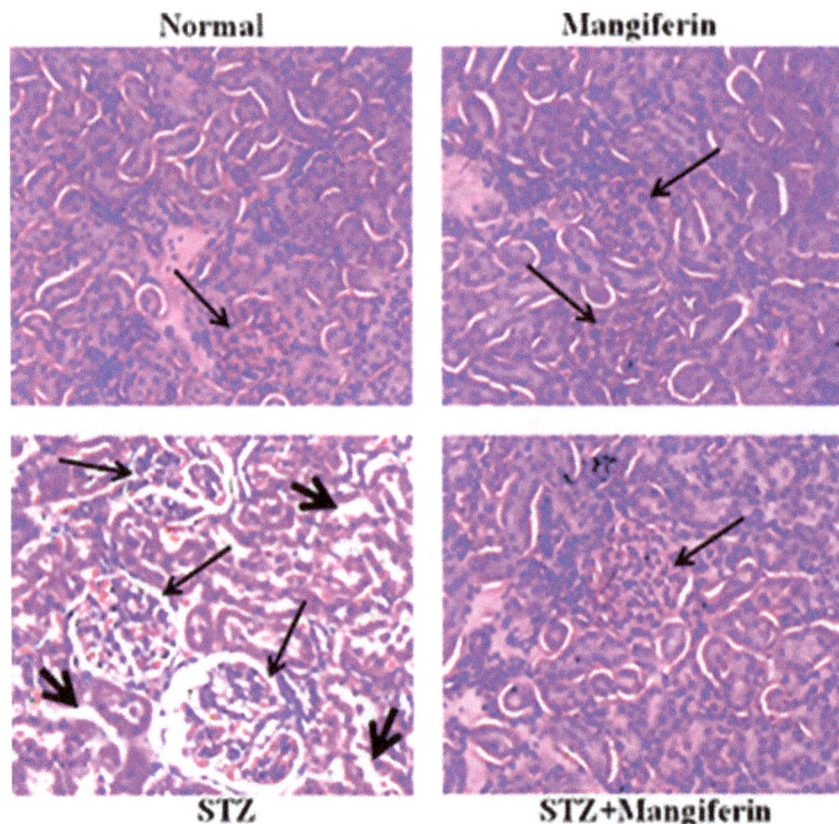

Figure 3. Effects of mangiferin on STZ-induced diabetic nephropathy in rats. Histological examination (H&E stained) (200x) in rat kidney of normal group, Mangiferin (Only) group, STZ-induced diabetic group, and STZ + Mangiferin group. Large arrows represent the size of glomerulus and small arrows represent the hydropic changes of the proximal convoluted tubules.

expression of TNFα. So mangiferin could successfully inhibit the TNFα in diabetic kidney.

Mangiferin inhibits STZ-induced activation of caspase 8 and t-Bid

Caspase-8 can function independently by initiating TNFα related apoptosis in various pathophysiological conditions. Therefore, to investigate whether the activation of TNFα related caspase 8 and t-Bid (cleaved form of Bid) has any role in STZ-induced diabetic nephropathy and whether mangiferin can inhibit these activations; immunoblot analysis has been carried out using appropriate antibodies. Our results reveal that the expression of caspase 8 and t-Bid was significantly increased although the expression of Bid in STZ-induced diabetic kidney tissue of experimental rats was decreased. Mangiferin post-treatment effectively reversed these expressions (Figure 10B). So these results suggest that mangiferin could efficiently suppress the TNFα related apoptotic pathway.

Anti-apoptotic effects of mangiferin on STZ-induced Bcl-2 family proteins

In hyperglycemia mediated oxidative stress, proapoptotic (Bax, Bad etc.) and antiapoptotic (Bcl-2, Bcl-xl) Bcl-2 family proteins are involved in response to apoptosis. Immunoblot analysis shows that the expression of Bcl-2 (antiapoptotic) was downregulated and that of Bax (proapoptotic) was upregulated in STZ-induced diabetic renal tissue (Figure 11A). However, treatment with mangiferin significantly altered the expression of these proteins. So,

mangiferin could act as an effective anti-apoptotic agent by increasing the expression of anti-apoptotic Bcl-2 family proteins in the mitochondria of diabetic kidney.

Anti-apoptotic effects of mangiferin on STZ-mediated mitochondrial dependent apoptotic pathways

Translocation of Bax (pro-apoptotic Bcl-2 family protein) from cytosol to mitochondria in hyperglycemia induces oxidative stress and causes damages to the outer mitochondrial membrane; which in turn, initiates the mitochondrial dependent apoptotic pathway [51]. To establish whether mangiferin applies its anti-apoptotic activities in the STZ-induced mitochondrial dependent apoptotic pathway, we measured the mitochondrial membrane potential in the kidney tissue (by flow cytometry) as well as the expression cytosolic cytochrome C, caspase 9, caspase 3 and PARP cleavage in kidney tissue (by immunoblot analysis). Results showed that STZ-induced diabetic condition significantly reduced the mitochondrial membrane potential (MMP) (Figure 11B), increased the expression of cytosolic cytochrome C, caspase 3 as well as caspase 9 and PARP cleavage (Figure 12). However, treatment with mangiferin, post to STZ exposure, effectively inhibited these parameters suggesting that mangiferin has a potential anti-apoptotic effect in diabetes-mediated mitochondrial dependent apoptotic pathways in kidney.

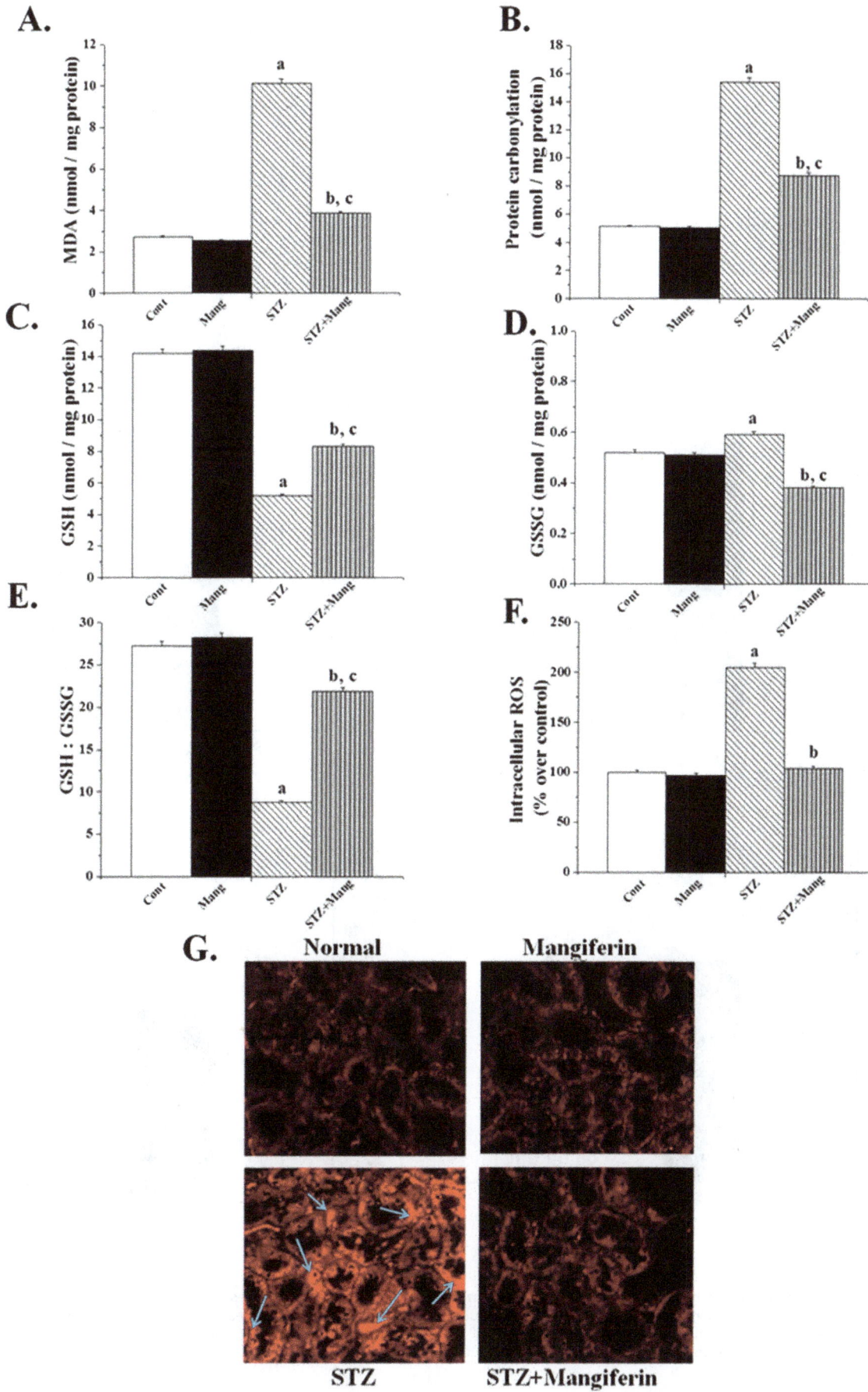

Figure 4. Effects of mangiferin (Mang) on STZ-induced oxidative stress related parameters and ROS levels. (A) MDA, (B) Protein carbonylation, (C) GSH (reduced glutathione), (D) GSSG (oxidized glutathione), (E) GSH to GSSG ratio, (F) Effect of mangiferin on STZ-induced intracellular ROS production in kidney and (G) Effect of mangiferin on superoxide ($O_2{\cdot}^-$) production in kidney tissue of STZ-induce diabetic rats. Superoxide was detected by dihydroethidium (DHE) staining. Each column represents mean $\pm$ SEM, n = 6. "a" indicates the significant difference between the normal control and STZ-induced groups, "b" indicates the significant difference between the STZ-induced and mangiferin treated groups and "c" indicates the significant difference between the STZ+Mangiferin group and normal control group ($P^a < 0.05$, $P^b < 0.05$, $P^c < 0.05$).

Mangiferin protects from STZ-induced apoptosis in kidney

STZ-induced apoptosis has also been evident from the ladder pattern (hallmark of apoptosis) in DNA gel electrophoresis (Figure 13A) and TUNEL assay (Figure 13B). In the figure, TUNEL positive nuclear staining has been shown in STZ-induced diabetic rat kidney representing the apoptosis of kidney cells in this pathophysiology. On the other hand, mangiferin significantly reduced the STZ-induced disturbances in the number of TUNEL positive nuclei and protected DNA in the tissue. Data from the DNA fragmentation assay and the TUNEL assay confirm the anti-apoptotic role of mangiferin in diabetic nephropathy.

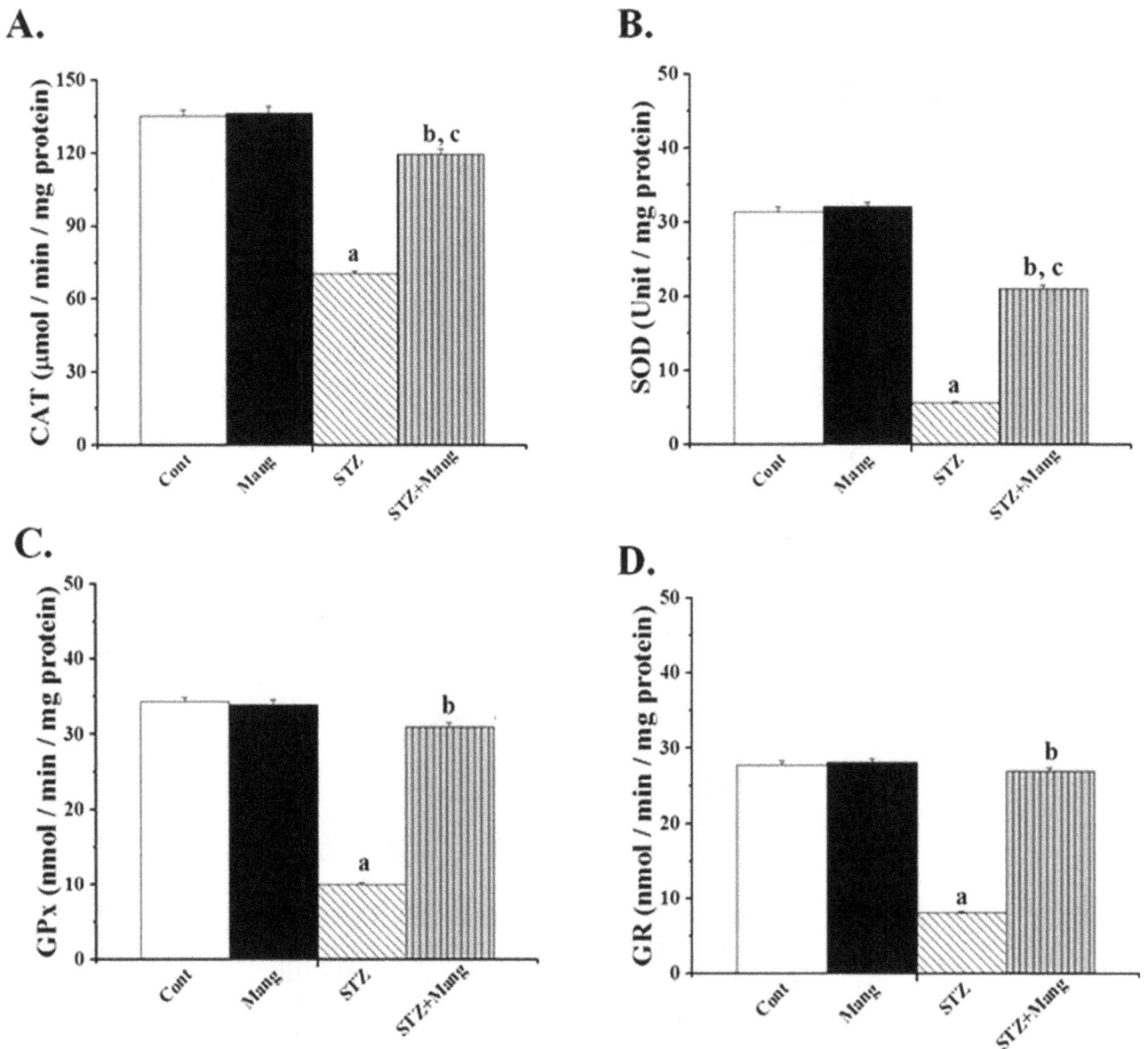

Figure 5. Effects of mangiferin on STZ-induced changes in cellular antioxidant enzymes. (A) CAT (Catalase), (B) SOD (Superoxide dismutase), (C) GPx (glutathione peroxidase), (D) GR (glutathione reductase). Each column represents mean $\pm$ SEM, n = 6. "a" indicates the significant difference between the normal control and STZ-induced groups, "b" indicates the significant difference between the STZ-induced and mangiferin treated groups and "c" indicates the significant difference between the STZ+Mangiferin group and normal control group ($P^a < 0.05$, $P^b < 0.05$, $P^c < 0.05$).

A.

B.

C.

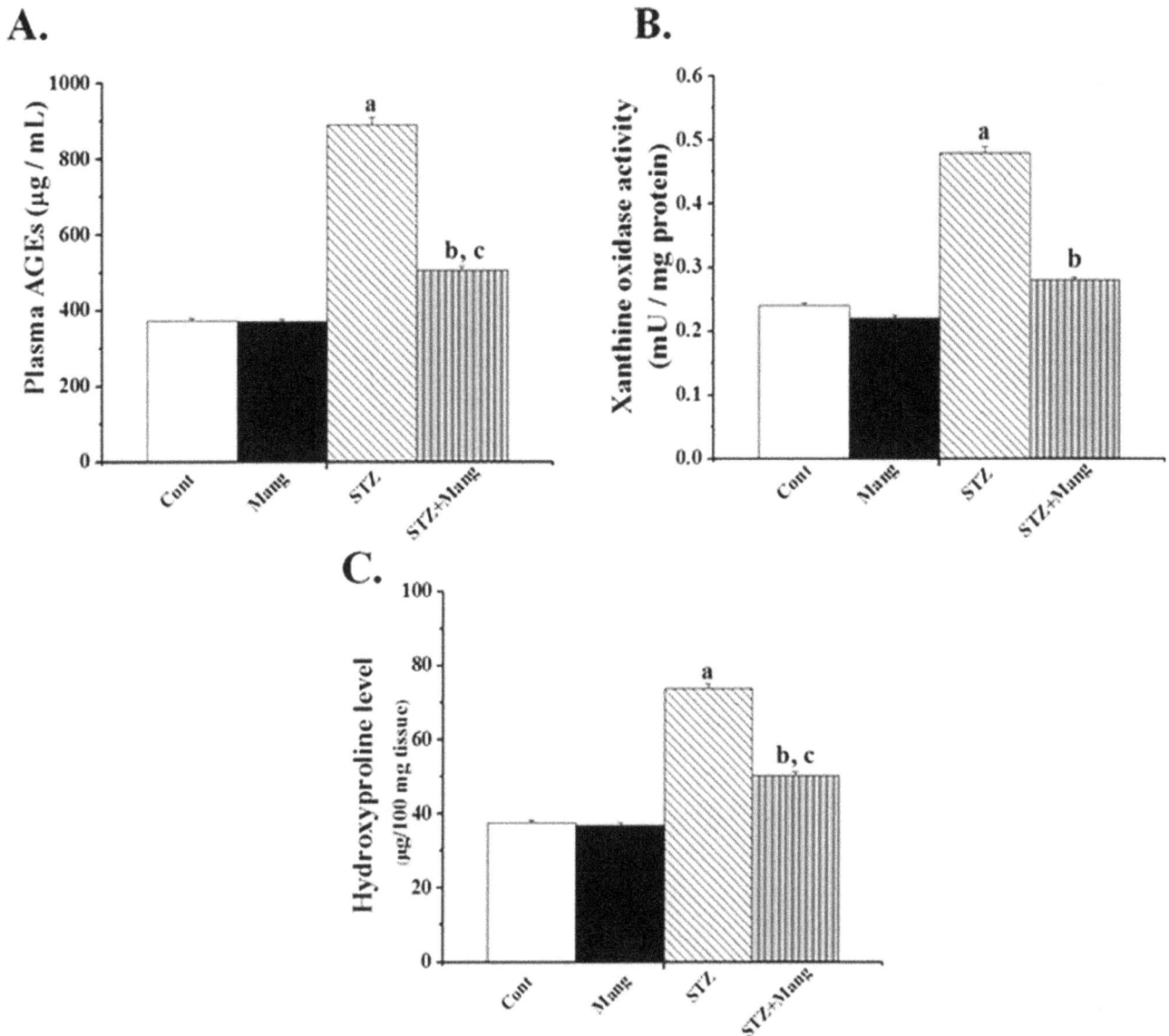

Figure 6. Effects of mangiferin on STZ-induced AGE, xanthine oxidase and renal hydroxyproline levels. (A) AGE (advance glycation end product), (B) xanthine oxidase and (C) renal hydroxyproline were measured. Each column represents mean ± SEM, n = 6. "a" indicates the significant difference between the normal control and STZ-induced groups, "b" indicates the significant difference between the STZ-induced and mangiferin treated groups and "c" indicates the significant difference between the STZ+Mangiferin group and normal control group ($P^a < 0.05$, $P^b < 0.05$, $P^c < 0.05$).

Discussion

Various experimental and clinical reports suggest that oxidative stress plays a major role in the pathogenesis of diabetic nephropathy in both type 1 and type 2 diabetes mellitus [52,53]. Our present study established that mangiferin could provide protection against diabetic nephropathy (STZ; 65 mg/kg b.w. single dose) [54] via the reversal of the activation of PKCs, MAPKs, NF-κB as well as TGF-β1 and inhibiting both the extrinsic (mitochondrial independent) and intrinsic (mitochondrial dependent) apoptotic cell death involved in this pathophysiology. A number of recent studies showed that mangiferin could protect various organs including the kidney, probably because of its strong anti hyperglycemic effect and free radical (ROS) scavenging activity [55–57] in STZ-induce diabetic condition. The free radical scavenging property of mangiferin is probably understood

from the structural point of view as it contains four phenolic H-atoms, two of which could easily be abstracted by suitable free radicals (e.g., ROS) to form two phenoxyl radicals that are stabilized by resonance (Figure 14) [58,36]. Furthermore, mangiferin could decrease the xanthon oxidase activation and advance glycation end product (AGE) formation; these two phenomena are considered to be the crucial mechanism for increased production of ROS. In this study, we characterized diabetes mellitus by increased plasma glucose level along with decreased kidney to body weight ratio. A significant increase in plasma BUN, creatinine, uric acid and urinary albumin also indicated the progressive nephrotoxicity in the animals. Mangiferin, on the other hand, effectively reversed this pathophysiology by lowering plasma BUN, creatinine, uric acid and urinary albumin. Mangiferin also reduced the plasma glucose level, restored kidney to body weight ratio and altered diabetes-induced oxidative stress

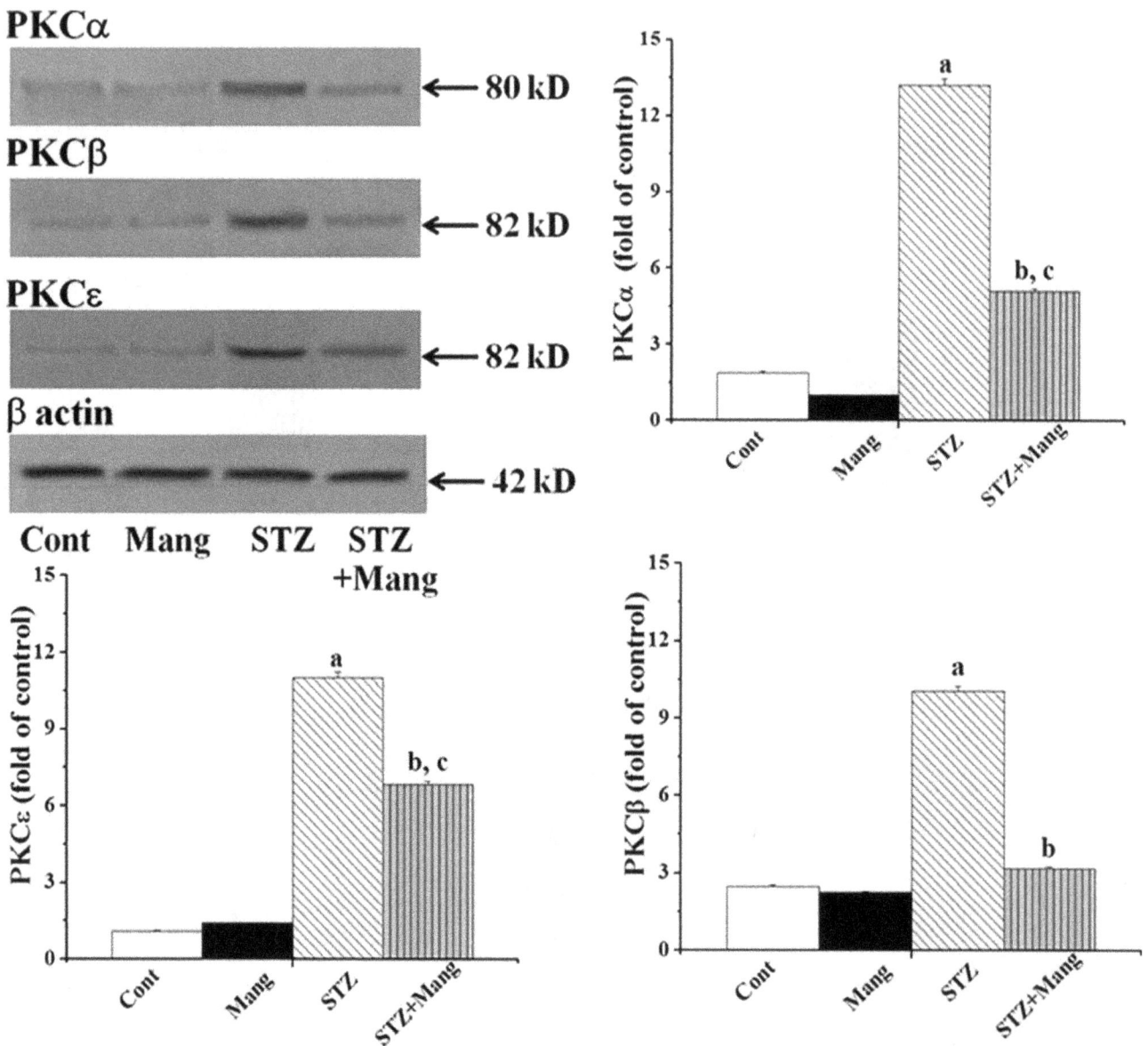

Figure 7. Mangiferin inhibits STZ-induced activation of PKCs (Immunoblot analysis). β actin was used as an internal control. Cont: normal control, Mang: treated with mangiferin, STZ: STZ-induced (diabetic), STZ+Mang: Mangiferin treated post to STZ-induced. PKCs (PKCα, PKCβ and PKCε expressions. Each column represents mean ± SEM, n = 6. "a" indicates the significant difference between the normal control and STZ-induced groups, "b" indicates the significant difference between the STZ-induced and mangiferin treated groups and "c" indicates the significant difference between the STZ+Mangiferin group and normal control group (P^a<0.05, P^b<0.05, P^c<0.05).

related parameters such as MDA content, protein carbonylation, ROS production, GSH and GSSG levels as well as GSH:GSSG ratio in the kidney tissue.

Besides dietary antioxidants, cells also develop numerous antioxidant defense systems against free radical assaults. GSH plays a major role in protecting cells against oxidative stress and it is decreased in the kidney due to STZ-mediated oxidative stress. GR reduces GSSG to GSH, thereby supporting the antioxidant defense system. GR has a disulfide bond in its active site, but ROS interferes with the disulfide bond and inhibits the enzyme. SOD and CAT mutually plays a vital role in keeping out ROS. GPx requires GSH in their course of reactions for removal of excess free radicals from the system. Decreased GSH content occurs because of the reduced functions of GPx, making cells more susceptible to oxidative damage. In the present study, we found that the

mangiferin effectively attenuated the decreased activities of antioxidant enzymes such as CAT, SOD, GP_X and GR in STZ-induced diabetic kidney.

Kidney tubule fibrosis plays an important role in the development of diabetic nephropathy [59]. In our present study, we observed a significant increase in the levels of renal hydroxyproline content in STZ-induce diabetic rats; this enhancement in turn increases the severity of the kidney lesions and fibrosis in experimental animals. Mangiferin, on the other hand, effectively diminished the hydroxyproline level, suggesting its anti-fibrotic efficacy in diabetic condition.

Hyperglycemia contributes to the formation of AGEs in diabetic condition and AGEs has been considered as an important biological source of ROS. Increased amounts of AGEs have been reported to be present in diabetic renal glomeruli [60]. AGEs can

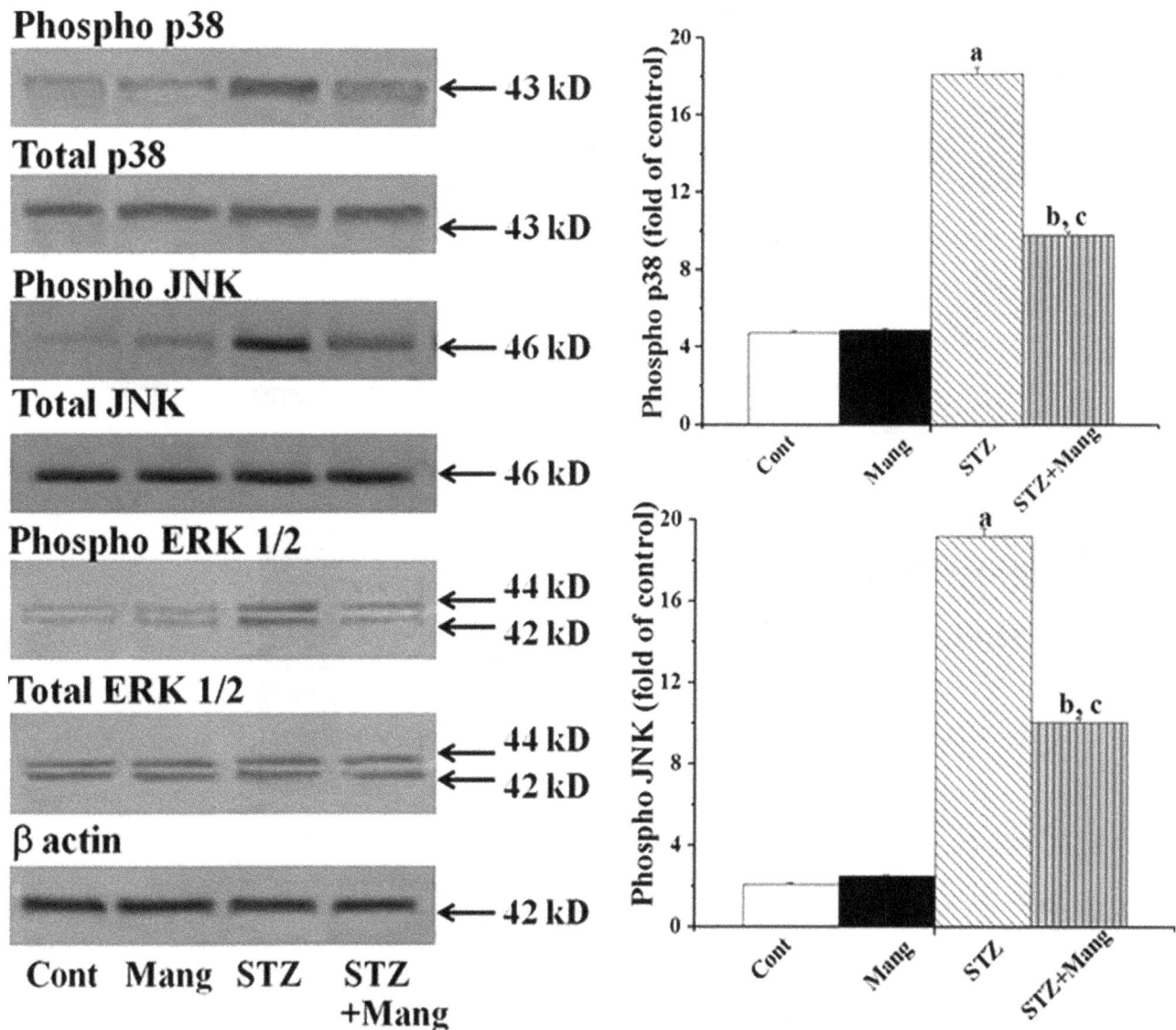

Figure 8. Mangiferin inhibits STZ-induced activation of MAPKs (Immunoblot analysis). β actin was used as an internal control. Cont: normal control, Mang: treated with mangiferin, STZ: STZ-induced (diabetic), STZ+Mang: Mangiferin treated post to STZ-induced. MAPKs (p38, JNK and ERK1/2) expressions. Each column represents mean ± SEM, n = 6. "a" indicates the significant difference between the normal control and STZ-induced groups, "b" indicates the significant difference between the STZ-induced and mangiferin treated groups and "c" indicates the significant difference between the STZ+Mangiferin group and normal control group (P^a<0.05, P^b<0.05, P^c<0.05).

form from intracellular auto-oxidation of glucose to glyoxal (activation of polyol pathway also occur in hyperglycemic condition), breakdown of the Amadori product (1-amino-1-deoxyfructose lysine adducts) to 3-deoxyglucosone and fragmentation of glyceraldehyde-3-phosphate and dihydroxyacetone phosphate to methylglyoxal. These reactive intracellular dicarbonyls (such as glyoxal, methylglyoxal and 3-deoxyglucosone) react with amino groups of intracellular and extracellular proteins to form AGEs [8]. STZ also considerably increased the activity of xanthine oxidase considerably, which is another important biological source of ROS in diabetic conditions [3]. However, our present study showed that mangiferin could suppress the AGEs formation and inhibit the activities of xanthon oxidase in STZ-induced diabetic renal tissue.

Investigation of the ROS activated signal transduction pathways in STZ-induced diabetic renal dysfunction showed that ROS could activate PKCs, MAPKs and transcription factor (NF-κB); and could also upregulate TGF-β1. Various reports suggest that hyperglycemia mediated and ROS-induced activation of PKCs and MAPKs play a significant role in the development and progress of diabetic nephropathy [61,62]. Lee et al. (2003) [61] also reported that the activation PKCs is also responsible for the generation of ROS in STZ-induced diabetic kidney. In our present study, we observed the increased expression of PKCs (PKCα, PKCβ and PKCε) and MAPKs (phospho- p38, phospho-JNK and phospho-ERK1/2) in STZ-induced diabetic kidney. However, treatment with mangiferin, inhibits the activations of ROS-mediated PKCs as well as MAPKs suggesting that STZ-induced diabetic nephropathy is mediated via the activation of PKCs MAPKs family proteins. A previous report suggests that hyperglycemia mediated ROS induces the activation of the transcription factor, NF-κB [63]. Due to the activation of MAPKs

A.

B.

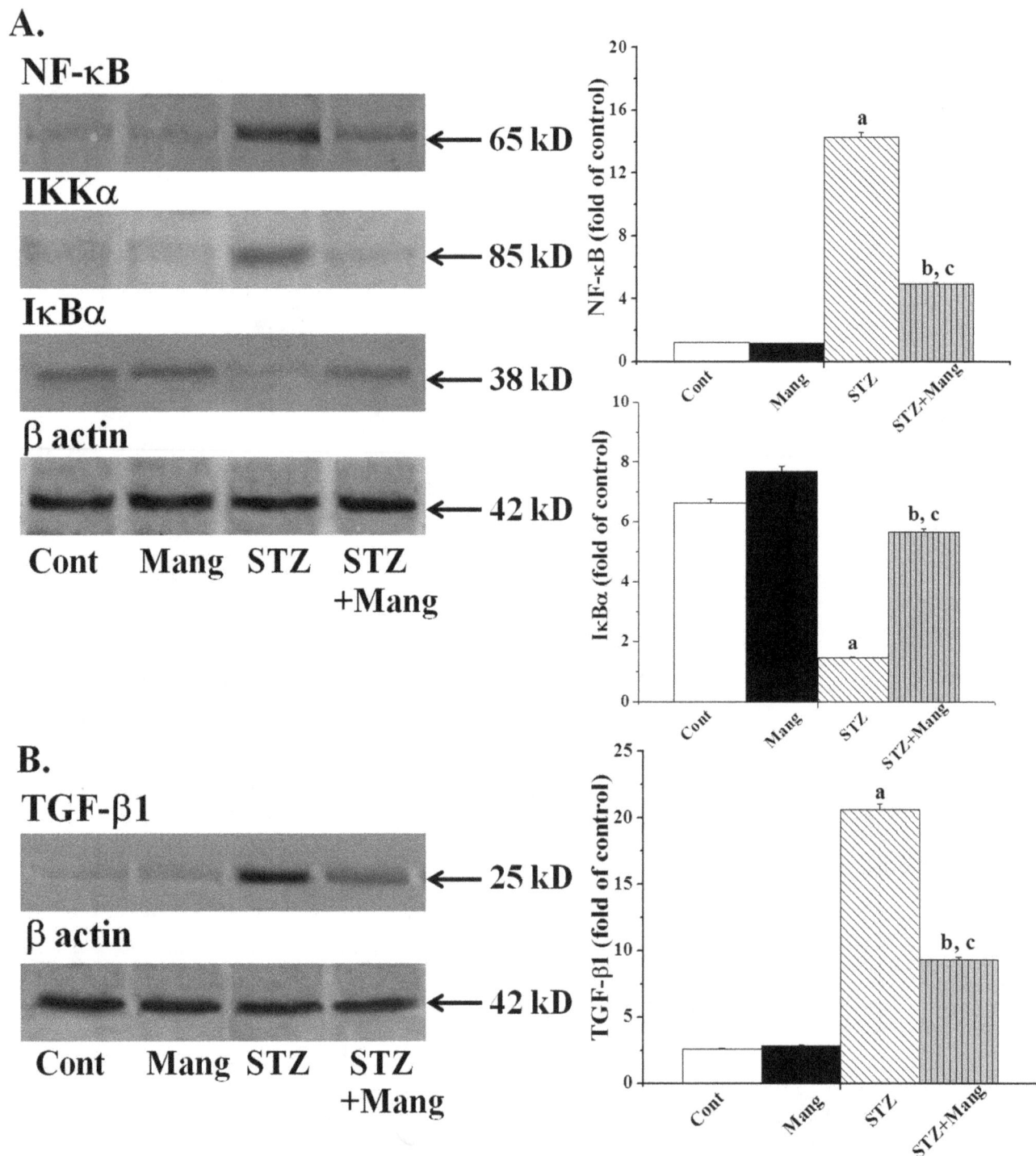

Figure 9. Mangiferin (Mang) inhibits STZ-induced activation of NF-κB pathway and overexpression of TGF-β1 (Immunoblot analysis). (A) NF-κB (NF-κB, IKKα and IκBα expressions and (B) TGF-β1 expressions. Each column represents mean ± SEM, n = 6. "a" indicates the significant difference between the normal control and STZ-induced groups, "b" indicates the significant difference between the STZ-induced and mangiferin treated groups and "c" indicates the significant difference between the STZ+Mangiferin group and normal control group ($P^a < 0.05$, $P^b < 0.05$, $P^c < 0.05$).

family proteins in our present study in STZ-induced diabetic kidney, the transcription factor, NF-κB, was also activated. Activation of this transcription factor could be regulated by the phosphorylation of its p65 subunits and degradation of its inhibitor-κB (IκB) via phosphorylation of IKKα/β resulting its

translocation into the nucleus [64]. Then, transcription of DNA to mRNA occurs followed by its translation into protein, helping cellular apoptosis. NF-κB also up-regulates TGF-β1 [65] which plays an important role in the development of renal hypertrophy and the accumulation of extracellular matrix (ECM) resulting in

A.

B.

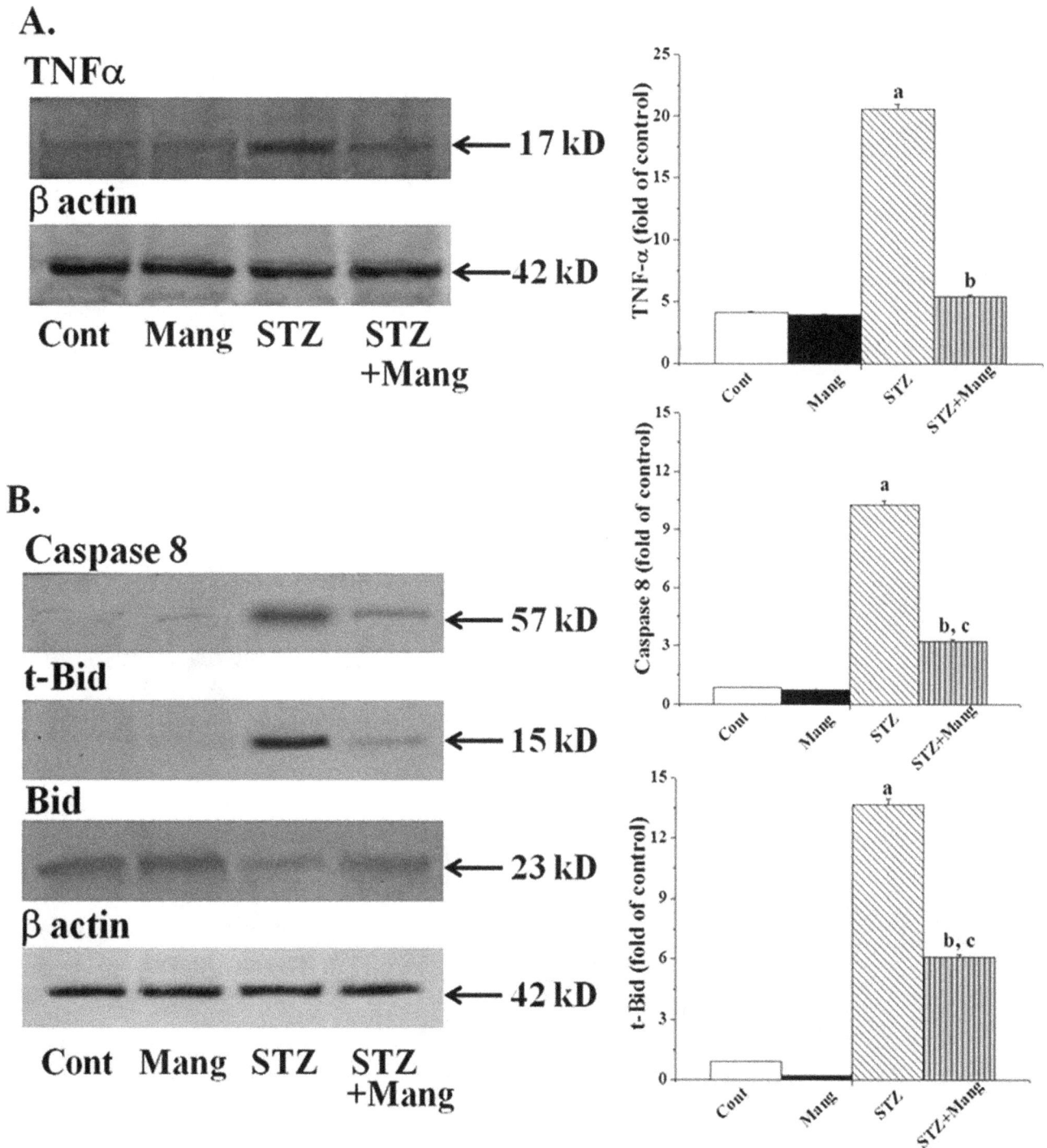

Figure 10. Mangiferin (Mang) inhibits STZ-induced TNFα releases and the activation of caspase 8 and t-Bid (Immunoblot analysis). (A) TNFα expression levels and (B) caspase 8, t-Bid and Bid expressions. Each column represents mean ± SEM, n=6. "a" indicates the significant difference between the normal control and STZ-induced groups, "b" indicates the significant difference between the STZ-induced and mangiferin treated groups and "c" indicates the significant difference between the STZ+Mangiferin group and normal control group (Pᵃ<0.05, Pᵇ<0.05, Pᶜ< 0.05).

the development of diabetic nephropathy in experimental animals [66,67]. In our present study we also observed that STZ up-regulated the TGF-β1 in the kidney. Mangiferin, could, however, inhibit the activation of NF-κB and attenuate the up-regulated TGF-β1 in this pathophysiological condition. Finally we suggest that mangiferin could inhibit hyperglycemia-mediated and ROS-

induced activation of signal transduction cascade such as PKCs, MAPKs and transcription factor (NF-κB) as well as up-regulation of TGF-β1 in diabetic kidney.

Our recent interest has also been focused to determine the role of the most important cytokine, TNFα, released in the inflammatory process which can activate signaling pathways related to cell

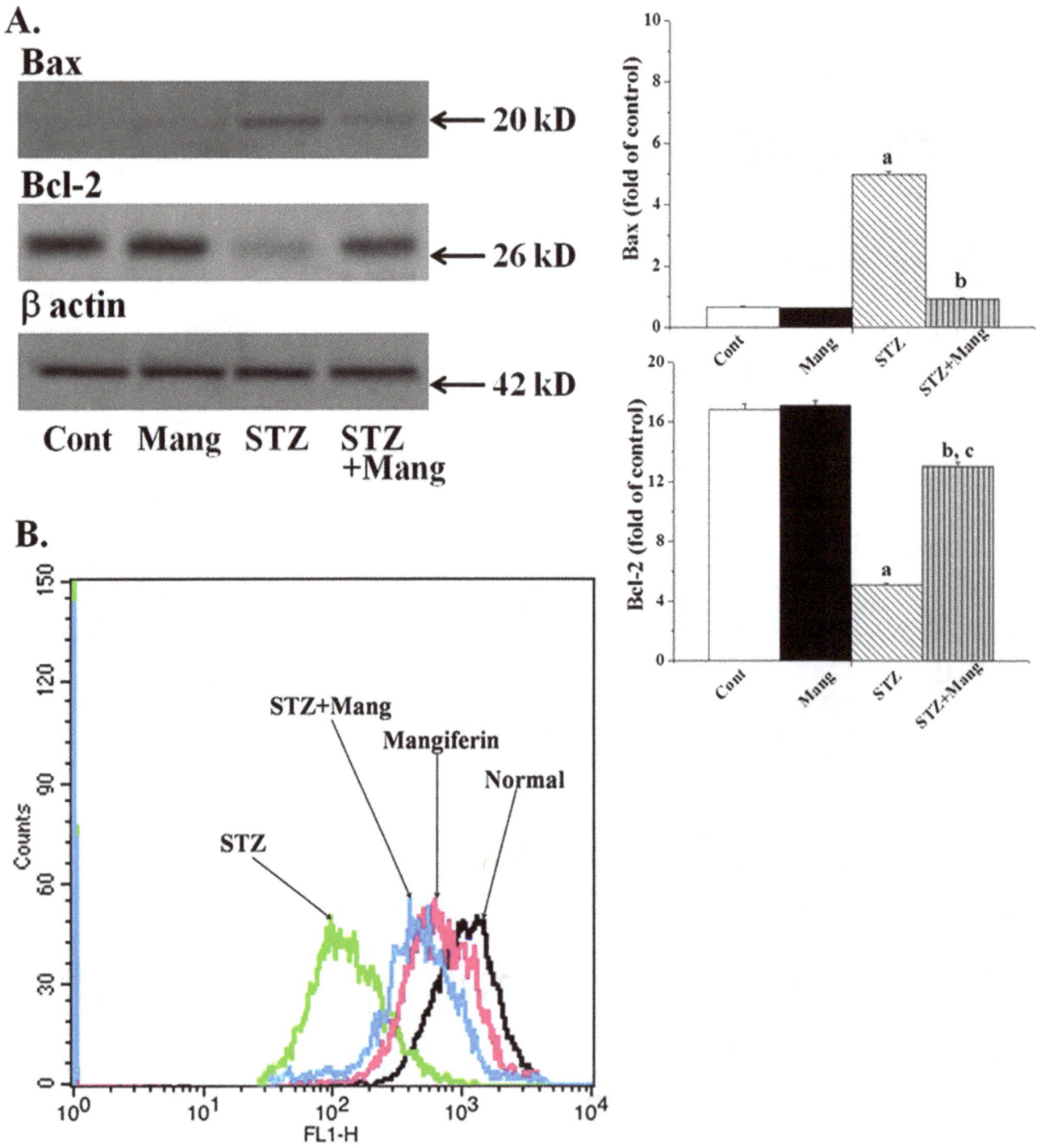

Figure 11. Mangiferin ameliorates Bcl-2 family proteins (Immunoblot analysis) and inhibits the mitochondrial dependent apoptotic pathways in STZ-induced diabetic kidney of rats (Immunoblot analysis). (A) Expressions of pro-apoptotic (Bax) and anti-apoptotic (Bcl-2) Bcl-2 family proteins, (B) mitochondrial membrane potential (flow cytometry). Each column represents mean ± SEM, n = 6. "a" indicates the significant difference between the normal control and STZ-induced groups, "b" indicates the significant difference between the STZ-induced and mangiferin treated groups and "c" indicates the significant difference between the STZ+Mangiferin group and normal control group (P[a]<0.05, P[b]< 0.05, P[c]<0.05).

survival and apoptosis (extrinsic apoptosis) in diabetic circumstances. TNFα initiates the activation of caspase 8 in the cytosol via its binding to the death receptor, TNF-R1. Caspase 8, in this scenario, plays the most important role in the implementation of programmed cell death [68] via two different pathways, either type I or type II. In type I pathway, the initiator caspase, caspase-8, directly activates the downstream effector caspase, caspase 3 is subsequently cleaved resulting in apoptosis [69,70]. In type II pathway, caspase 8 does not directly activate the caspase 3 to execute cell death. In this process, caspase 8 executes cell death via

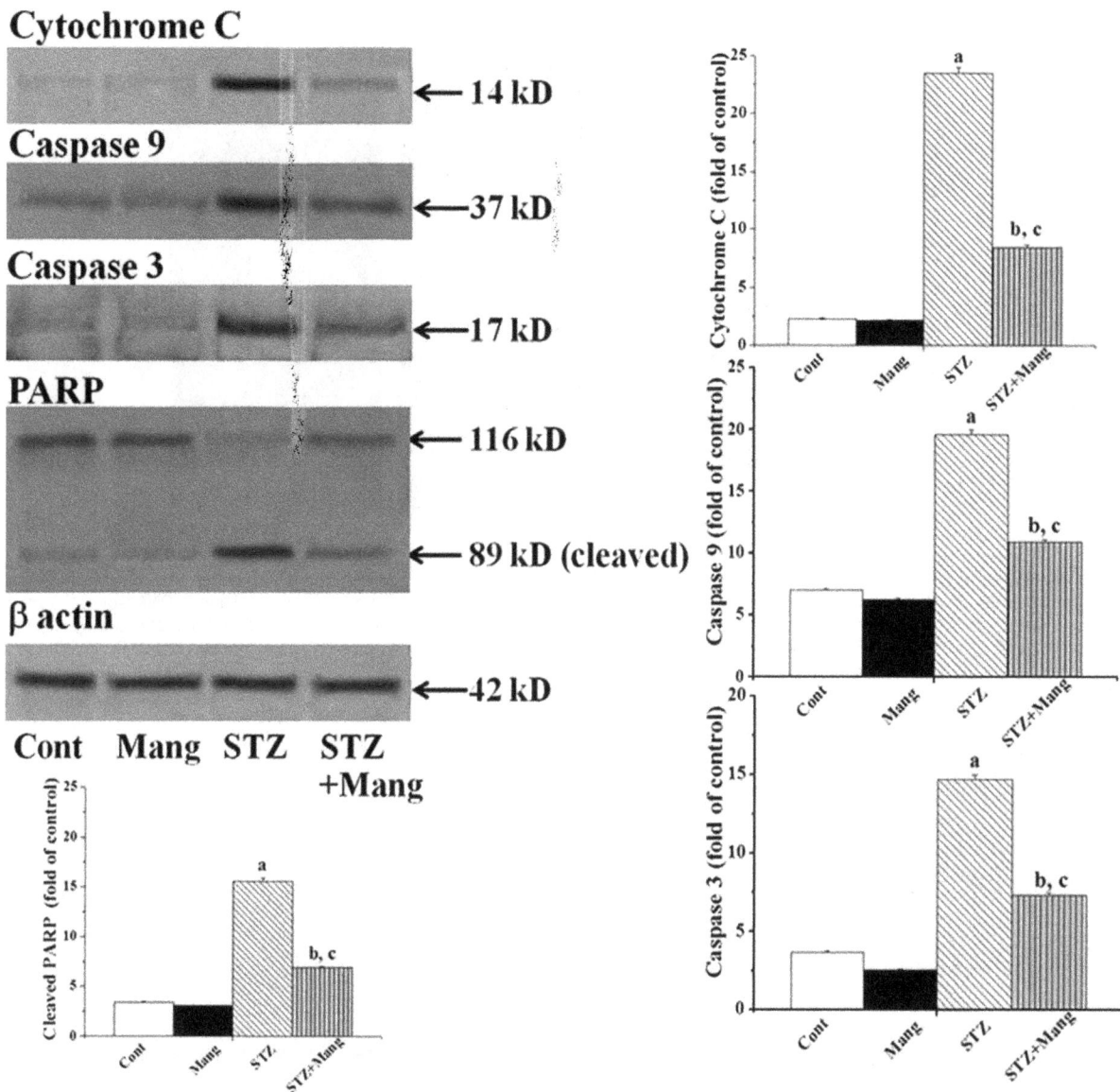

Figure 12. Mangiferin inhibits the mitochondrial dependent apoptotic pathways in STZ-induced diabetic kidney of rats (Immunoblot analysis). Cytochrome C, Caspase 9, Caspase 3 and PARP expressions. Each column represents mean ± SEM, n = 6. "a" indicates the significant difference between the normal control and STZ-induced groups, "b" indicates the significant difference between the STZ-induced and mangiferin treated groups and "c" indicates the significant difference between the STZ+Mangiferin group and normal control group ($P^a < 0.05$, $P^b < 0.05$, $P^c < 0.05$).

mitochondria-dependent apoptotic pathways and is going through mitochondrial Bcl-2 family protein, Bid. Bid is cleaved to 15 kD truncated form (t-Bid) by caspase-8 which in turn translocates to the mitochondria and promotes mitochondria-dependent apoptotic pathways [71]. In the present study, we found that STZ-induced diabetic kidney could release TNFα which in turn activates caspase 8 followed by the activation of mitochondrial dependent cell death pathways (type II) via the cleavage of Bid to t-Bid in the cytosol. Mangiferin, on the other hand, inhibits these TNFα mediated apoptotic events via decreased level of released TNFα as well as decreased expression of caspase 8, t-Bid and increased expression of Bid in the cytosol.

Bcl-2 family proteins are the family of proteins involved in programmed cell death due to oxidative stress and mitochondria and play a vital role in the regulation of this process. ROS

mediated oxidative stress disturbs the balance between pro-apoptotic (such as bcl-2 and bcl-XL) and anti-apoptotic (such as Bad, Bax or Bid) Bcl-2 family proteins, resulting an excess of pro-apoptotic proteins in the cells which are more susceptible to apoptosis. An excess of pro-apoptotic Bcl-2 proteins at the surface of the mitochondria is considered to be an important factor for the development of the Permeability Transition (PT) pore due to the loss of mitochondrial membrane potential and released cytochrome C in cytosol. Cytochrome C efficiently to interacts with Apaf-1 and leads to the recruitment of pro-caspase 9 to form a multiple-protein complex called the apoptosome. Apoptosome formation leads to the activation of caspase 9 as well as caspase 3. Caspase 3 plays a very important role in executing apoptosis by activating DNases and inhibiting the important DNA repair enzyme poly (ADP-ribose) polymerase (PARP). The function of

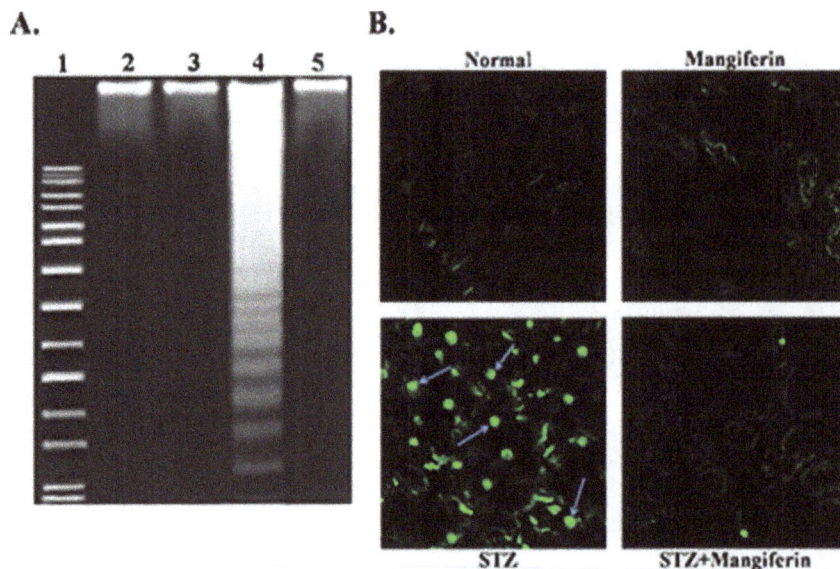

Figure 13. Mangiferin protects from STZ-induced apoptosis in kidney. (A) DNA fragmentation on agarose/ethydium bromide gel. DNA isolated from experimental rat kidney was loaded onto 1% (w/v) agarose gels. Lane 1: Marker (1 kb DNA ladder); Lane 2: DNA isolated from normal kidney tissue; Lane 3: DNA isolated from mangiferin treated kidney tissue; Lane 4: DNA isolated from STZ-induced diabetic kidney; Lane 5: DNA isolated from Mangiferin treated in STZ-induced diabetic kidney. DNA ladder formation in STZ-induced diabetic kidney. (B) TUNEL staining in kidney tissue sections (10x) in experimental rats. Arrows indicate TUNEL positive nucleus.

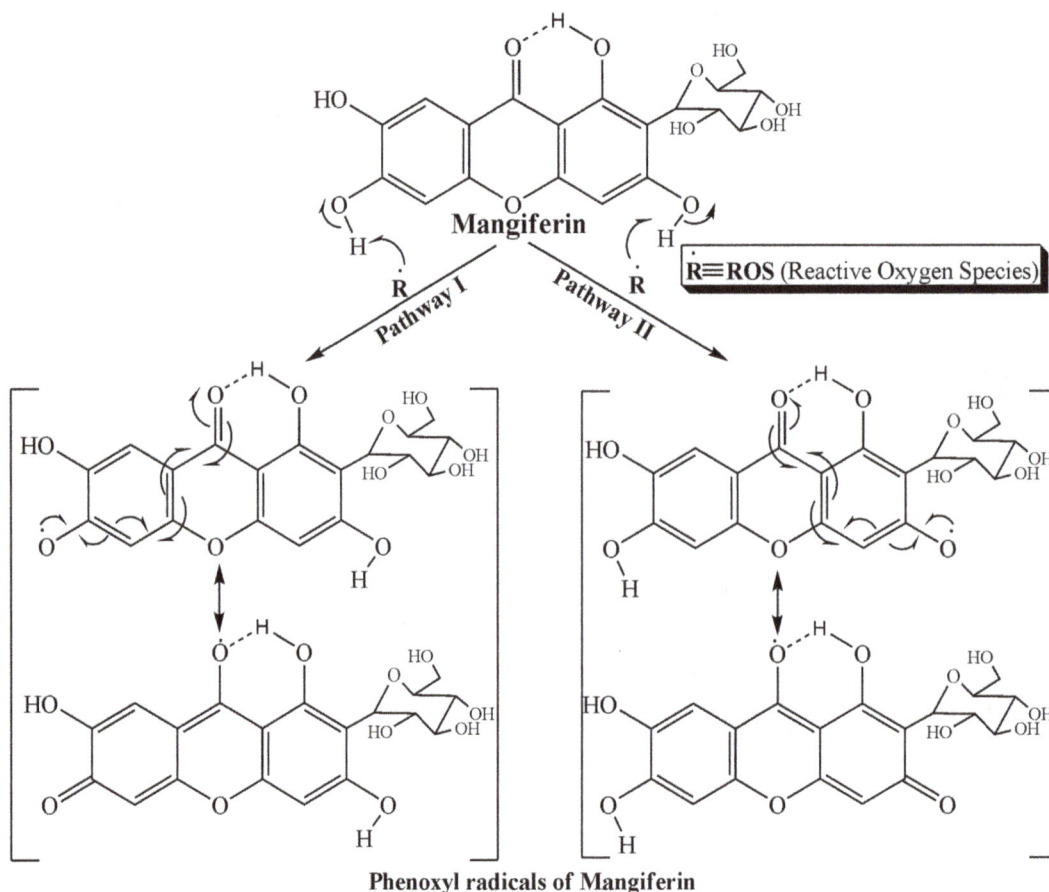

Figure 14. Plausible molecular mechanisms of mangiferin how it scavenge reactive free radicals (ROS).

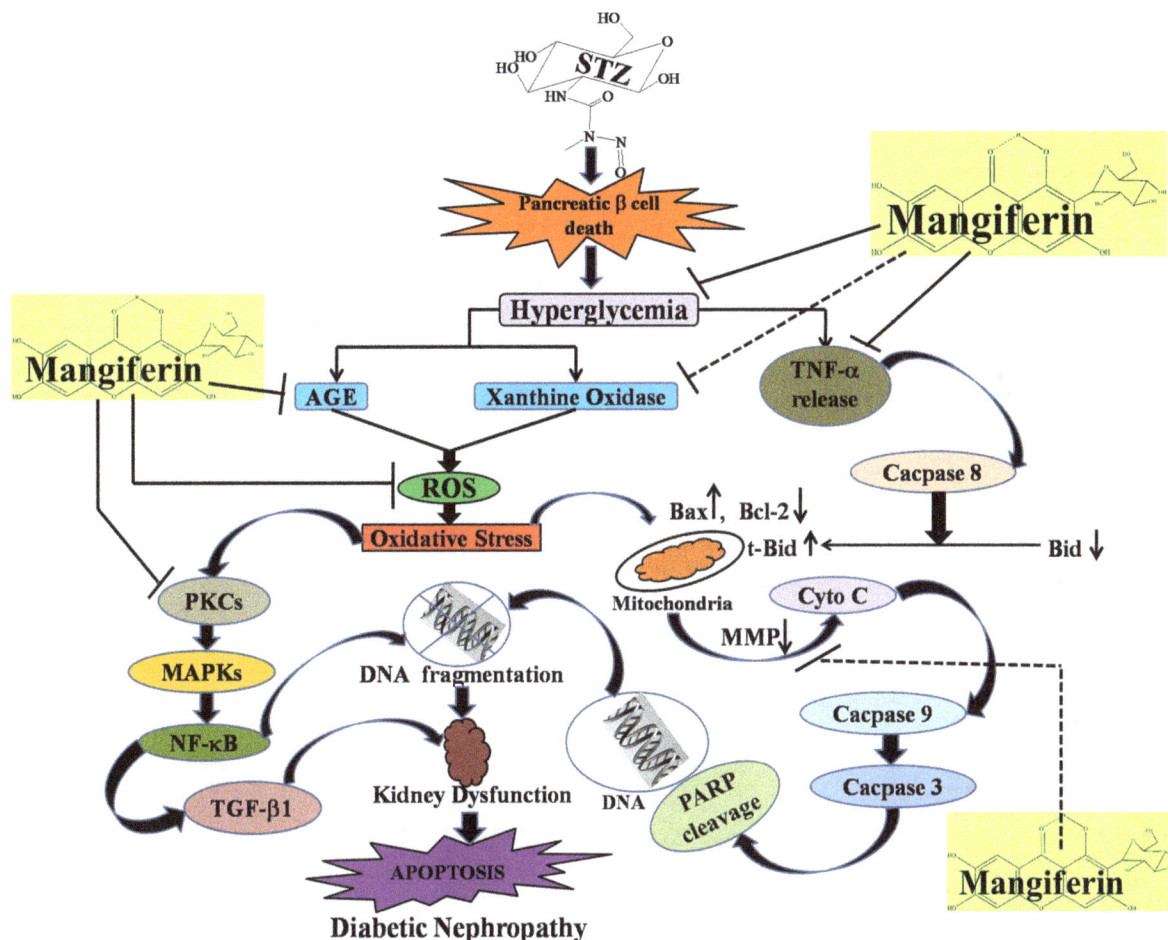

Figure 15. Schematic representation of Pb(II) induced hepatotoxicity and its protection by mangiferin.

PARP (repair DNA damage) is prevented by caspase 3 via the cleavage of PARP. On the other hand, the fragmentation of DNA is due to the enzyme caspase activated DNase (CAD), which exists as ICAD (inhibitor of CAD; an inactive complex) in normal condition. When caspase 3 is activated in stressed condition, ICAD is cleaved to CAD by caspase 3 as well as fragmentation of chromosomal DNA into nucleosomal units and ultimately apoptotic cell death occurs [72]. In our studies, we found the increased expression of Bax (pro-apoptotic), decreased expression of Bcl-2 (anti-apoptotic) as well as activated mitochondrial dependent pathways via reduced mitochondrial membrane potential, enhanced cytochrome C, increased expression of caspase 9 and caspase 3 in the cytosol and cleavage of PARP in STZ-induced diabetic nephropathy. However, treatment with mangiferin, effectively ameliorates the changes in Bcl-2 family proteins and inhibits the mitochondrial dependent apoptotic pathways in this pathophysiology. STZ-induced renal apoptotic cell death and the protective action of mangiferin was also confirmed from DNA fragmentation analysis (ladder formation) and TUNEL assays (showing TUNNEL positive nuclear staining).

In conclusion, the findings of our present study established, for the first time, that mangiferin treatment could provide effective protection against oxidative injury in the renal tissue of STZ-induced type 1 diabetic rats via ROS-induced, PKCs, MAPKs, NF-κB and TGF-β1 mediated, TNFα related mitochondrial dependent apoptotic pathways (Figure 15).

Acknowledgments

The authors acknowledge to the Council of Scientific & Industrial Research (CSIR, Human Resource Development Group), India.

Author Contributions

Conceived and designed the experiments: PBP KS PCS. Performed the experiments: PBP KS PCS. Analyzed the data: PBP KS PCS. Contributed reagents/materials/analysis tools: PBP KS PCS. Contributed to the writing of the manuscript: PBP KS PCS.

References

1. Bonnefont-Rousselot D, Beaudeux JL, Therond P, Peynet J, Legrand A, et al. (2004) Diabetes mellitus, oxidative stress and advanced glycation endproducts. Ann Pharm Fr 62: 147–157.

2. Szkudelski T (2001) The mechanism of alloxan and streptozotocin action in B cells of the rat pancreas. Physiol Res 50: 537–546.

3. Forbes JM, Coughlan MT, Cooper ME (2008) Oxidative Stress as a Major Culprit in Kidney Disease in Diabetes. Diabetes 57: 1446–1454.

4. Yan HD, Li XZ, Xie JM, Li M (2007) Effects of advanced glycation end products on renal fibrosis and oxidative stress in cultured NRK-49F cells. Chin Med J 120: 787–793.

5. Ibrahim HN, Hostetter TH (1997) Diabetic nephropathy. J Am Soc Nephrol 8: 487–493.

6. Zelmanovitz T, Gerchman F, Balthazar AP, Thomazelli FC, Matos JD, et al. (2009) Diabetic nephropathy. Diabetol Metab Syndr 21: 1–10.

7. Soetikno V, Watanabe K, Sari FR, Harima M, Thandavarayan RA, et al. (2011) Curcumin attenuates diabetic nephropathy by inhibiting PKC-α and PKC-β1 activity in streptozotocin-induced type I diabetic rats. Mol Nutr Food Res 55: 1655–1665.

8. Brownlee M (2001) Biochemistry and molecular cell biology of diabetic complications. Nature 414: 813–820.

9. Palsamy P, Subramanian S (2011) Resveratrol protects diabetic kidney by attenuating hyperglycemia-mediated oxidative stress and renal inflammatory cytokines via Nrf2-Keap1 signaling. Biochim Biophys Acta 1812: 719–731.

10. Das J, Sil PC (2012) Taurine ameliorates alloxan-induced diabetic renal injury, oxidative stress-related signaling pathways and apoptosis in rats. Amino Acids 43: 1509–1523.

11. Ghosh J, Das J, Manna P, Sil PC (2009) Arjunolic acid, a triterpenoid saponin, prevents acetaminophen (APAP)-induced liver and hepatocyte injury via the inhibition of APAP bioactivation and JNK-mediated mitochondrial protection. Free Radical Biol Med 48: 535–553.

12. Manna P, Ghosh J, Das J, Sil PC (2010) Contribution of type 1 diabetes to rat liver dysfunction and cellular damage via activation of NOS, PARP, IκBα/NF-κB, MAPKs and mitochondria dependent pathways: Prophylactic role of arjunolic acid. Free Radical Biol Med 48: 1465–1484.

13. Ghosh J, Das J, Manna P, Sil PC (2010) Protective effect of the fruits of Terminalia arjuna against cadmium-induced oxidant stress and hepatic cell injury via MAPK activation and mitochondria dependent pathway. Food Chem 123: 1062–1075.

14. Pal S, Pal PB, Das J, Sil PC (2011) Involvement of both intrinsic and extrinsic pathways in hepatoprotection of arjunolic acid against cadmium induced acute damage in vitro. Toxycology 283: 129–139.

15. Chatterjee M, Sil PC (2006) Hepatoprotective effect of aqueous extract of Phyllanthus niruri on nimesulide-induced oxidative stress in vivo. Indian J Biochem Biophy 43: 299–305.

16. Bhattacharjee R, Sil PC (2007) Protein isolate from the herb, Phyllanthus niruri L. (Euphorbiaceae), plays hepatoprotective role against carbon tetrachloride induced liver damage via its antioxidant properties. Food Chem Toxicol 45: 817–826.

17. Sarkar MK, Kinter M, Mazumder B, Sil PC (2009) Purification and characterization of a novel antioxidant protein molecule from Phyllanthus niruri. Food Chem 111: 1405–1412.

18. Bhattacharyya S, Pal PB, Sil PC (2013) A 35 kD Phyllanthus niruri protein modulates iron mediated oxidative impairment to hepatocytes via the inhibition of ERKs, p38 MAPKs and activation of PI3k/Akt pathway. Food Chem Toxicol 56: 119–30.

19. Bhattacharjee S, Ghosh S, Sil PC (2014) Amelioration of aspirin induced oxidative impairment and apoptotic cell death by a novel antioxidant protein molecule isolated from the herb Phyllanthus niruri. PLoS One 9: e89026.

20. Pal PB, Pal S, Manna P, Sil PC (2012) Traditional extract of Pithecellobium dulce fruits protects mice against CCl₄ induced renal oxidative impairments and necrotic cell death. Pathophysiology 19: 101–114.

21. Sarkar K, Ghosh A, Kinter M, Mazumder B, Sil PC (2006) Purification and characterization of a 43 kD hepatoprotective protein from the herb Cajanus indicus L. Protein J 25: 411–421.

22. Manna P, Sinha M, Sil PC (2007) A 43 kD protein isolated from the herb Cajanus indicus L attenuates sodium fluoride-induced hepatic and renal disorders in vivo. J Biochem Mol Biol 40: 382–395.

23. Ghosh A, Sil PC (2009) Protection of acetaminophen induced mitochondrial dysfunctions and hepatic necrosis via Akt-NF-κB pathway: Role of a novel plant protein. Chem Biol Interact 177: 96–106.

24. Pal S, Sil PC (2012) A 43 kD protein from the leaves of the herb Cajanus indicus L. modulates doxorubicin induced nephrotoxicity via MAPKs and both mitochondria dependent and independent pathways. Biochimie 94: 1356–67.

25. Pal S, Ahir M, Sil PC (2012) Doxorubicin-induced neurotoxicity is attenuated by a 43-kD protein from the leaves of Cajanus indicus L. via NF-κB and mitochondria dependent pathways. Free Radic Res 46: 785–98.

26. Rice-Evans CA, Miller NJ, Bolwell PG, Bramley PM, Pridham JB (1995) The relative antioxidant activities of plant derived polyphenolic flavonoids. Free Radic Res 22: 375–383.

27. Matkowski A, Kuś P, Góralska E, Woźniak D (2013) Mangiferin- a bioactive xanthonoid, not only from mango and not just antioxidant. Mini Rev Med Chem 13: 439–55.

28. Aritomi M, Kawasaki T (1969) A new xanthone C-glucoside, position isomer of mangiferin, from Anemarrhena asphodeloides Bunge. Tetrahedron Lett 12: 941–944.

29. Aderibigbe AO, Emudianughe TS, Lawal BA (1999) Antihyperglycaemic effect of Mangifera indica in rat. Phytother Res 13: 504–507.

30. Sairam K, Hemalatha S, Kumar A, Srinivasan T, Ganesh J, et al. (2003) Evaluation of anti-diarrhoeal activity in seed extracts of Mangifera indica. J Ethnopharmacol 84: 11–15.

31. Yoshimi N, Matsunaga K, Katayama M, Yamada Y, Kuno T, et al. (2001) The inhibitory effects of mangiferin, a naturally occurring glucosylxanthone, in bowel carcinogenesis of male F344 rats. Cancer Lett 163: 163–170.

32. Garcia D, Escalante M, Delgado R, Ubeira FM, Leiro J (2003) Anthelminthic and antiallergic activities of Mangifera indica L. stem bark components Vimang and mangiferin. Phytother Res 17: 1203–1208.

33. Guha S, Ghosal S, Chattopadhyay U (1996) Antitumor, immunomodulatory and anti-HIV effect of mangiferin, a naturally occurring glucosylxanthone. Chemotherapy 42: 443–451.

34. Bairy I, Reeja S, Siddharth RPS, Bhat M, Shivananda PG (2002) Evaluation of antibacterial activity of Mangifera indica on anaerobic dental microflora based on in vivo studies. Indian J Pathol Microbiol 45: 307–310.

35. Das J, Ghosh J, Roy A, Sil PC (2012) Mangiferin exerts hepatoprotective activity against D-galactosamine induced acute toxicity and oxidative/nitrosative stress via Nrf2-NFκB pathways. Toxicol Appl Pharmacol 260: 35–47.

36. Pal PB, Sinha K, Sil PC (2013) Mangiferin, a natural xanthone, protects murine liver in Pb(II) induced hepatic damage and cell death via MAP kinase, NF-κB and mitochondria dependent pathways. PLoS One 8: e56894.

37. Sinha K, Pal PB, Sil PC (2013) Mangiferin, a naturally occurring xanthone C-glycoside, ameliorates lead (Pb)-induced murine cardiac injury via mitochondria-dependent apoptotic pathways. Signpost Open Access J Org Biomol Chem 1: 47–63.

38. Ghosh M, Das J, Sil PC (2012) D(+) galactosamine induced oxidative and nitrosative stress-mediated renal damage in rats via NF-κB and inducible nitric oxide synthase (iNOS) pathways is ameliorated by a ployphenol xanthone, mangiferin. Free Rad Res 46: 116–32.

39. Bradford MM (1976) A rapid and sensitive method for the quantitation of microgram quantities of protein utilizing the principle of protein-dye binding. Anal Biochem 72: 248–254.

40. Esterbauer H, Cheeseman KH (1990) Determination of aldehydic lipid peroxidation products: Malonaldehyde and 4-hydroxynonenal. Methods Enzymol 186: 407–421.

41. Uchida K, Stadtman ER (1993) Covalent attachment of 4-hydroxynonenal to glyceraldehydes-3-phosphate dehydrogenase. J Biol Chem 268: 6388–6393.

42. LeBel CP, Bondy SC (1990) Sensitive and rapid quantitation of oxygen reactive species formation in rat synaptosomes. Neurochem Int 17: 435–440.

43. Kim JD, McCarter RJM, Yu BP (1996) Influence of age, exercise and dietary restriction on oxidative stress in rats. Aging Clin Exp Res 8: 123–129.

44. Lara LS, McCormac M, Semprum-Prieto LC, Shenouda S, Majid DS, Kobori H, Navar LG, Prieto MC (2012) AT1 receptor mediated augmentation of angiotensiongen, oxidative stress, and inflammation in ANGII-salt hypertension. Am J Physiol Renal Physiol 302(1): F85–94.

45. Pal PB, Pal S, Das J, Sil PC (2012) Modulation of mercury-induced mitochondria-dependent apoptosis by glycine in hepatocytes. Amino Acids 42: 1669–83.

46. Ellman GL (1959) Tissue sulphydryl group. Arch Biochem Biophys 82: 70–77.

47. Hissin PJ, Hilf RA (1976) A fluorometric method for determination of oxidized and reduced glutathione in tissues. Anal Biochem 74: 214–226.

48. Woessner JF (1961) The determination of hydroxyproline in tissueand protein samples containing small proportions of this iminoacid. Arch Biochem Biophys 93: 440–447.

49. Sellins KS, Cohen JJ (1987) Gene induction by gamma-irradiation leads to DNA fragmentation in lymphocytes. J Immunol 139: 3199–3206.

50. Das J, Sil PC (2012) Taurine ameliorates alloxan-induced diabetic renal injury, oxidative stressrelated signaling pathways and apoptosis in rats. Amino Acids 43: 1509–23.

51. Das J, Vasan V, Sil PC (2012) Taurine exerts hypoglycemic effect in alloxan-induced diabetic rats, improves insulin-mediated glucose transport signaling pathway in heart and ameliorates cardiac oxidative stress and apoptosis. Toxicol Appl Pharm 258: 296–308.

52. Baynes JW, Thorpe SR (1999) Role of oxidative stress in diabetic complications: A new perspective on an old paradigm. Diabetes 48: 1–9.

53. Chen HC, Guh JY, Chang JM, Hsieh MC, Shin SJ, et al. (2005) Role of lipid control in diabetic nephropathy. Kidney Int Suppl 94: S60–2.

54. Lee WC, Chen HC, Wang CY, Lin PY, Ou TT, et al. (2010) Cilostazol ameliorates nephropathy in type 1 diabetic rats involving improvement in oxidative stress and regulation of TGF-β and NF-κB. Biosci Biotechnol Biochem 74: 1355–1361.

55. Li X, Cui X, Sun X, Li X, Zhu Q, et al. (2010) Mangiferin Prevents Diabetic Nephropathy Progression in Streptozotocin-Induced Diabetic Rats. Phytother Res 24: 893–899.

56. Miura l T, Ichiki H, Hashimoto l I, Iwamoto l N, Kato l M, et al. (2001) Antidiabetic activity of a xanthone compound, mangiferin. Phytomedicine 8: 85–87.

57. Muruganandan S, Gupta S, Kataria M, Lal J, Gupta PK (2002) Mangiferin protects the streptozotocin-induced oxidative damage to cardiac and renal tissues in rats. Toxicology 176: 165–173.

58. Mishra B, Priyadarsini IK, Sudheerkumar M, Unnikrishnan MK, Mohan H (2006) Pulse radiolysis studies of mangiferin: A C-glycosyl xanthone isolated from Mangifera indica. Rad Phys Chem 75: 70–77.

59. Thomson SC, Deng A, Bao D, Satriano J, Blantz RC, Vallon V (2001) Ornithine decarboxylase, kidney size, and the tubular hypothesis of glomerular hyperfiltration in experimental diabetes. J Clin Invest 107: 217–224.

60. Horie K, Miyata T, Maeda K, Miyata S, Sugiyama S, et al. (1997) Immunohistochemical colocalization of glycoxidation products and lipid peroxidation products in diabetic renal glomerular lesions. Implication for glycoxidative stress in the pathogenesis of diabetic nephropathy. J Clin Invest 100: 2995–2999.

61. Dunlop ME, Muggli EE (2000) Small heat shock protein alteration provide a mechanism to reduce mesangial cell contractility in diabetes and oxidative stress. Kidney Int 57: 464–475.

62. Lee HB, Yu MR, Yang Y, Jiang Z, Ha H (2003) Reactive oxygen species-regulated signaling pathways in diabetic nephropathy. J Am Soc Nephrol 14: S241-S245.

63. Ha H, Yu MR, Choi YJ, Kitamura M, Lee HB (2002) Role of high glucoseinduced nuclear factor-KB activation in monocyte chemoattractant protein-I expression by mesangial cells. J Am Soc Nephrol 13: 894–902.

64. Sakurai H, Suzuki S, Kawasaki N, Nakano H, Okazaki T, et al. (2003) Tumor necrosis factor-alpha-induced IKK phosphorylation of NF-kappaB p65 on serine 536 is mediated through the TRAF2, TRAF5, and TAK1 signaling pathway. J Biol Chem 278: 36916–36923.

65. Shah SV, Baliga R, Rajapurkar M, Fonseca VA (2007) Oxidants in chronic kidney disease. J Am Soc Nephrol 18: 16–28.

66. Wolf G, Ziyadeh FN (1999) Molecular mechanisms of diabetic renal hypertrophy. Kidney Int 56: 393–405.

67. Park IS, Kiyomoto H, Abboud SL Abboud, HE (1997) Expression of transforming growth factor-β and type IV collagen in early streptozotocin-induced diabetes. Diabetes 46: 473–480.

68. Budihardjo I, Oliver H, Lutter M, Luo X, Wang X (1999) Biochemical pathways of caspase activation during apoptosis. Annu Rev Cell Dev Biol 15: 269–290.

69. Scaffidi C, Fulda S, Srinivasan A, Friesen C, Li F, et al. (1998) Two CD95 (APO-1/Fas) signaling pathways. Embo J 17: 1675–87.

70. Sinha K, Pal PB, Sil PC (2014) Cadmium (Cd2+) exposure differentially elicits both cell proliferation and cell death related responses in SK-RC-45. Toxicol in Vitro 28: 307–318.

71. Luo X, Budihardjo I, Zou H, Slaughter C, Wang X (1998) Bid, a Bcl2 interacting protein, mediates cytochrome c release from mitochondria in response to activation of cell surface death receptors. Cell 94: 481–90.

72. Lawen A (2003) Apoptosis- an introduction. BioEssays 25: 888–896.

16

Behavior of the Edible Seaweed *Sargassum fusiforme* to Copper Pollution: Short-Term Acclimation and Long-Term Adaptation

Hui-Xi Zou[1,9], Qiu-Ying Pang[2,9], Li-Dong Lin[2], Ai-Qin Zhang[2], Nan Li[1], Yan-Qing Lin[1], Lu-Min Li[1], Qin-Qin Wu[1], Xiu-Feng Yan[1]*

1 Zhejiang Provincial Key Lab for Subtropical Water Environment and Marine Biological Resources Protection, College of Life and Environmental Science, Wenzhou University, Wenzhou, People's Republic of China, 2 Key Laboratory of Saline-Alkali Vegetation Ecology Restoration in Oil Field, Northeast Forest University, Harbin, People's Republic of China

Abstract

Aquatic agriculture in heavy-metal-polluted coastal areas faces major problems due to heavy metal transfer into aquatic organisms, leading to various unexpected changes in nutrition and primary and/or secondary metabolism. In the present study, the dual role of heavy metal copper (Cu) played in the metabolism of photosynthetic organism, the edible seaweed *Sargassum fusiforme*, was evaluated by characterization of biochemical and metabolic responses using both ^{1}H NMR and GC-MS techniques under acute (47 µM, 1 day) and chronic stress (8 µM, 7 days). Consequently, photosynthesis may be seriously inhibited by acute Cu exposure, resulting in decreasing levels of carbohydrates, e.g., mannitol, the main products of photosynthesis. Ascorbate may play important roles in the antioxidant system, whose content was much more seriously decreased under acute than that under chronic Cu stress. Overall, these results showed differential toxicological responses on metabolite profiles of *S. fusiforme* subjected to acute and chronic Cu exposures that allowed assessment of impact of Cu on marine organisms.

Editor: Fanis Missirlis, CINVESTAV-IPN, Mexico

Funding: This material is based upon work funded by Zhejiang Provincial Natural Science Foundation of China (LQ13C030005, http://www.zjnsf.gov.cn), Natural Science Foundation of China (31270541, http://www.nsfc.gov.cn), Plan for Qianjiang Talent of Zhejiang (QJD1202014), and Program for Wenzhou Science & Technology Innovative Research Team of China (C20120007-08). The funders had no role in study design, data collection and analysis, decision to publish, or preparation of the manuscript.

Competing Interests: The authors have declared that no competing interests exist.

* Email: yanxiufeng@wzu.edu.cn

9 These authors contributed equally to this work.

Introduction

Over the last few decades, heavy metal pollution has become a global problem posing threat on both soil and marine ecosystems, as a result of the mass industrialization and various agricultural activities such as the intensive use of chemical fertilizers, wastewater and biosolids [1]. Many heavy metals accumulate in marine organisms, which may be subsequently transferred to human body via the food chain [2].

Marine algae, particularly seaweeds, are a food source for marine animals such as sea urchins and fishes, and are the base of many marine food webs. For several centuries, there has been a traditional use of seaweeds as food in East-Asian countries, like China, Japan and the Republic of Korea. *Sargassum fusiforme* (Sargassaceae, Phaephyceae), an endemic brown algae from the western coast of the North Pacific, is widely consumed in Japan and Korea. This alga, in great demand, is also cultivated in East-Asian counties, especially in China, where the cultivation area was 2.6% (2,482 ha) of the entire coastal area for commercial cultivation of seaweeds with a total production reached 32,000 tonnes per year (freshweight) [3].

It is reported that concentration of copper (Cu) found in standard reference material (oyster tissue) from China was above $100 \ \mu g \ g^{-1}$ [4], indicating a serious Cu pollution in these coastal areas. Although Cu is an essential micronutrient, excessive amount can be extremely harmful to algae [5]. Seaweeds are often exposed to low concentrations of metals including Cu for long periods. In the cases of ocean outfall, they may even abruptly exposed to high levels of metals. In the study of short- and long-term response of the marine green macroalga *Ulva fasciata* to Cu excess, regulation of mRNA expression involved in redox homeostasis and antioxidant defense were different [6]. In another study, distinct changes in the antioxidant responses to acute or chronic treatment with Cu were observed in the unicellular alga *Gonyaulax polyedra*, suggesting a different oxidative status of these two types of metal stresses [7]. Thus, it seems that both micro- and macroalgae have different responsive mechanisms to short- and long-term exposures of Cu.

Chemical analysis alone is not able to provide a satisfying assessment of the environmental quality of an ecosystem due to the biotransforming of an individual pollutant by living organisms [8,9]. To gain more information regarding the health state of a particular ecosystem, it is important to monitor the response of biota to the pollutants as well. Metabolomics characterizes and quantifies end products-the metabolites that produced by living organismsunder a given set of conditions. Metabolomics has shown considerable potential as a tool for environmental toxicology [10–14]. Both GC-MS and NMR techniques have been widely used in metabolomics and metabolite profiling [15–18]. GC-MS is particularly effective in the analysis of primary metabolites, while NMR, inherently quantitative, provides universal detection for organic components without coupling to a separation technique [19,20]. Because of the complementary analytical features of NMR and MS, opportunities for leveraging both methods are being considered which will create a more comprehensive metabolic profiling [19,21–22].

It is now well known that synthesis of antioxidant and metal-chelating components and activation of antioxidant enzymes are key factors for tolerance to heavy metals and other abiotic stress in plants [23]. The toxic effect of heavy metals appears to be related to production of reactive oxygen species (ROS), which usually leads to lipid peroxidation and oxidation of some enzymes and a massive protein. To better understand oxidative stress under acute and chronic conditions, the content of malondialdehyde (MDA), which represents the level of lipid peroxidation, was measured, as well as the activities of antioxidant enzymes superoxide dismutase (SOD), catalase (CAT) and peroxidase (POD). Additionally, activity of nitrate reductase (NR) that primarily involved in maintenance of a favorable cellular oxidation/reduction potential was also determined. In this study, we characterized the impact of Cu on the marine brown algae S. fusiforme using both NMR- and GC-MS-based metabolomics, which allowed identifying more analytes and created an opportunity to expand the scope of metabolomics research.

Materials and Methods

Algal material and culture conditions

S. fusiforme samples were collected from the Northeastern coast of Wenzhou, China (28.0°N, 121.2°E) in September 2012. This location is not privately-owned or protected in any way, thus no specific permissions were required, and the field studies did not involve endangered or protected species. After collection, algae were immediately transported to the laboratory in a cooler (4°C) within 2 h. Fronds were then washed with filtered natural seawater (with salinity of 27 ‰) and maintained in high-density polypro-pylene containers for 2 days at 20°C before Cu treatment, using a photoperiod of 12:12 h and a photon flux density of 100 μmol $m^{-2} s^{-1}$. Culture medium was aerated and changed daily.

Two experiments were carried out to evaluate the metabolic differences between responses of S. fusiforme to acute and chronic Cu exposures. For short-term treatment (1 day), thalli were cultivated in filtered natural seawater containing $CuCl_2$ (Sigma, USA) in the final concentration of 47 μM. For long-term treatment (7 days), culture medium containing $CuCl_2$ was prepared in the same manner as for short-term treatment but with final concentration of 8 μM. After treatment, thalli were immediately frozen in liquid nitrogen and stored at −80°C for further analysis.

In addition, the term "acute exposure" was employed for conditions of exposure to high Cu concentrations (47 μM) after 1 day in this study. On the other hand, "chronic exposure" defines the exposure to lower sub-lethal Cu concentrations (8 μM) after 7 days.

Measurement of enzyme activities and MDA content

Approximately 0.1 g algal samples (fresh weight, $n = 5$) were homogenized in liquid nitrogen and extracted with 1 mL of 0.05 M potassium phosphate buffer (pH 7.0) containing 0.25% (v/v) Triton X-100 and 1% (w/v) polyvinylpolypyrrolidone (PVPP). The enzyme activities in the algal tissues were detected using commercial kits (Nanjing Jiancheng Biotech., China) according to the manufacturer's instructions. In this study, the antioxidant enzymes included SOD (EC 1.15.1.1), CAT (EC 1.11.1.6) and POD (EC 1.11.1.7). In addition, activity of nitrate NR (EC 1.7.99.4) was also determined.

Protein concentration was determined according to the method of Bradford [24] with bovine serum albumin as standard. The unit of each enzyme was defined as the activity of an enzyme per milligram of total protein (expressed in $\mu mol\ min^{-1}$ per mg protein, or U per mg protein).

As a measure of lipid peroxidation, MDA levels in algal tissue were estimated by measuring thiobarbituric acid reactive substances following the standard protocol using MDA detection kit (Nanjing Jiancheng Biotech., China) and were expressed as nmol per mg of protein.

NMR analysis and data processing

Polar metabolites were extracted from the algal tissues using the solvent system of methanol/water (1/1) as described previously [25,26]. NMR spectra were acquired using Bruker AV-500 spectrometer (Bruker Bio Spin, Canada), with ^1H observation frequencies of 500.18 MHz, spectral width 6,009.6 Hz, mixing time 0.1 s, and relaxation delay 0.3 s as described previously [27].

All the NMR spectra were converted to a format for pattern recognition analysis using custom-written ProMetab software based on the Matlab software package (version 7.0; The Math-Works, Natick, MA, USA) [28]. NMR spectral peaks were identified following tabulated chemical shifts [29] and using the software, Chenomx (Evaluation Version, Chenomx Inc., Canada).

GC-TOF MS analysis and data processing

Extraction and fractionation of metabolites for GC-TOF MS analysis were performed as described [30] and about 100 mg of each tissue sample was weighed accurately. After derivatization, the metabolites were analyzed by GC-TOF MS analysis that was performed using an Agilent 7890 gas chromatograph system (Agilent, CA, USA) coupled with a Pegasus 4D time-of-flight mass spectrometer (LECO Corp., MI, USA). The system utilized a DB-5MS capillary column coated with 5% diphenyl cross-linked with 95% dimethylpolysiloxane (30 m×250 μm inner diameter, 0.25-μm film thickness; J&W Scientific, Folsom, CA, USA). A 1 μL aliquot of the analyte was injected in splitless mode. Helium was used as the carrier gas, the front inlet purge flow was 15 mL min^{-1}, and the gas flow rate through the column was 1 mL min^{-1}. The initial temperature was kept at 80°C for 0.2 min, then raised to 190°C at a rate of 10°C min^{-1}, then to 220°C at a rate of 3°C min^{-1} and finally to 280°C at a rate of 20°C min^{-1} for 16.8 min. The injection, transfer line, and ion source temperatures were 280, 270, and 220°C, respectively. The energy was −70 eV in electron impact mode. The mass spectrometry data were acquired in full-scan mode with the m/z range of 20–600 at a rate of 10 spectra per second after a solvent delay of 480 s.

A total of 288 peaks were found in the 24 samples. Data processing was performed as following: after the missing values were imputed using k nearest neighbor method of Bioconductor (www.bioconductor.org) impute package, data was filtered in order to eliminate the noise using interquantile range, resulting in 268 peaks. The filtered data were subsequently normalized where adonitol, pentakis (trimethylsilyl) ether were used as the internal references.

Statistical analysis

Data were expressed as the mean ± standard deviation (SD). Statistical analysis included one-way analysis of variance (AN-OVA). Principal component analysis (PCA) was used to reduce the dimensionality of data and summarize the similarities and differences between multiple NMR spectra. The principal component score plots were used to visualize general clusters between various groups of samples. ANOVA was conducted on PC scores from each group to test statistical significance ($P<0.05$) of separations.

Results and Discussion

In total, 25 metabolites were exclusively quantified by NMR (Table 1 and 2), while this number for MS was much more (288 peaks detected). Various metabolite classes were identified in NMR spectra, including amino acids, carbohydrates, and intermediates in the tricarboxylic acid (TCA) cycle. Fig. 1A shows the score plots for the acute Cu stress experiment, where PC1 and PC2 represent 76.26% and 7.03%, respectively. The acute Cu stress experiment score plots contain six distinct groups that represent the control samples and the individual treatments. The separations between the control (inverted red triangles) and exposed (green cycles) were obviously observed from the PC scores plots ($P<0.05$), while no separation for the chronic stress (Fig. 1B).

GC-TOF MS is also a powerful analytical tool employed in metabolomics studies, which provides the detection of different metabolite peaks. The components measured by GC-TOF MS that Fig. 2 summarizes are mainly amino acids, amines, organic acids, polyols and sugars. It has been proposed that decrease in utilization of carbohydrates for growth produced by heavy metals is more pronounced than the decrease in CO_2 fixation resulting in an increased accumulation of carbohydrates [31]. However, as shown in Fig. 2, the levels of most carbohydrates, e. g., mannose, fructose, hexitol and manitol, were decreased under acute stress, so were those under chronic Cu stress, though to a lesser extent and some even with increasing levels.

Amino acid concentrations were mainly affected

The summarized results for the amino acids profiling from NMR are shown in Table 1. Generally, more than half of the amino acids exhibit different profiles in response to the acute and chronic Cu stress. Concentrations of alanine and glutamine increased (by 30% and 9%, respectively), while those of aspartate and glutamate significantly decreased (by nearly 30%), resulting in a general decrease in total amount of the predominant amino acids (alanine, glutamate, glutamine and aspartate) under acute stress. It is particularly interesting as all the predominant amino acids content increased, especially glutamine (by 338%), under chronic stress.

Branched chain and aromatic amino acids were present only in small quantities in *S. fusiforme*. Interestingly, aromatic amino acids content increased (by 23%, Table 1) under acute stress, in contrast, which decreased by 22% under chronic stress. In more detail, the content of tyrosine and tryptophan both increased by approximately 50% except that of phenylalanine, decreased by 14% under acute stress, while minor changes or a slight decrease were observed for aromatic acids under chronic stress, indicating a quite different responding pattern to these two Cu stress in *S. fusiforme*.

Aromatic amino acids are synthesized from the common precursor metabolite chorismate, which originates from the shikimate pathway as shown in Fig. 3 [32], whose importance was demonstrated by the fact that 20% of the carbon fixed by plants flows through it under normal growth conditions [33]. Presence of the shikimate pathway in macroalgae has been experimentally verified only in green and red algae, as well as in the diatom *Thalassiosira pseudonana* [34]. However, the reduction in phlorotannin content and mortality in *Fucus vesiculosus* caused by glyphosate indicated the existence of the shikimate pathway in brown algae [35]. Moreover, little is known about the influence of heavy metal stress on enzymes involved in the shikimate pathway [33], especially in algae. Surprisingly, our results indicated an enhancement of the shikimate pathway under acute but not chronic Cu stress. In pepper, due to the much more accumulated Cu in the roots than in the aerial parts, the induction of shikimate dehydrogenase (SKDH), which is the enzyme that catalyses the fourth step in the shikimate pathway, only existed in the hypocotyl [33]. It seems that the amount we used were both not high enough to inhibit the activity of this enzyme. However, influences of different concentrations of Cu on other enzymes involved in the shikimate pathway are also unclear. One hypothesis could be that more other enzymes were induced under acute than chronic Cu stress. Moreover, phenylalanine is required for the synthesis of various phenolic compounds that play important roles in non-enzymatic antioxidant defense processes. This may explain the decrease in the content of phenylalanine under both Cu stress conditions.

Different nitrogen assimilation patterns

Nitrate assimilation is an apparently simple process in photosynthetic eukaryotes, involving two transports and two reduction steps to produce ammonium in the chloroplast [36], within which NR is a key enzyme that catalyzes the first, also the rate-limiting step in the reduction of nitrate to ammonium [37]. After a short exposure to Cu (1 day), a significant decrease in the NR activation state was observed (Fig. 4I). However, after 7 days of Cu treatment, the NR activation state in thalli was found to be approximate to that in untreated thalli (Fig. 4J). The strong inhibition of NR activity by acute Cu exposure in *S. fusiforme* was entirely in agreement with results obtained with other organisms [38,39]. Furthermore, excessive Cu causes a drastic change in nitrogen metabolism affecting also other enzymes involved in nitrate reduction and amino acid metabolism and leading to diminution of total nitrogen [40]. As a result, the levels of primary amino acid products of nitrogen assimilation (glutamine and glutamate) were reduced.

During chronic Cu exposure, the total content of glutamate and glutamine, especially glutamine in *S. fusiforme* even increased comparing to the control (Table 1), meanwhile the activity of NR was little reduced (Fig. 4). Though excessive cadmium could significantly inhibit activities of glutamate dehydrogenase (GDH), glutamine synthetase (GS) and glutamine oxoglutarate aminotransferase (GOGAT) [41], through which ammonium is further incorporated into the amino acids [42,43], data concerning the inhibitory effect of excess Cu are scarce [44]. It is reported that GS appeared highly increased in the model brown alga *Ectocarpus siliculosus* exposed to 50 mg L^{-1} of Cu after 10 days [45]. Consequently, it may suggest that GS be strongly activated in *S. fusiforme*, though whose activity was not determined in the present study, resulting in the dramatic increase in glutamine content. The ratio of glutamine to glutamate has been proposed as an indicator of nitrogen status [46], which was raised in thalli under both acute and chronic Cu stress, by approximately 55% and 200% (Table 1), respectively, showing great variations where the latter was almost 4 times as the former.

Physiological responses

The algae were cultivated with increasing concentrations of Cu, which would trigger the synthesis of ROS that led to lipids peroxidation [47]. In general, increases in peroxidase activity are regarded as a reliable indicator of stress or potential phytotoxicity of heavy metals, the increase in peroxidase activity being a response to an increase in peroxides [48]. As a product of lipid peroxidation that accumulates greatly following heavy metal exposure, MDA is an indicator of lipid peroxidation. The algae responded similarly in the content of MDA to both acute and chronic Cu stress, resulted in 254% and 193% increase, respectively (Fig. 4A and B).

SOD has been called the cell's first line of defense against ROS that catalyzes the disproportionation of O_2^- to O_2 and H_2O_2 [49]. Generally speaking, peroxidative stress triggers higher level of MDA or lower level of SOD or both. In *S. fusiforme*, SOD activity was activated, a little increased by 34% and 28% (Fig. 4C and D), respectively, under acute and chronic Cu exposure, consistent with the result reported [47], which was also coinciding with increases in MDA content (Fig. 4A and B).

CAT is another important ROS-scavenging enzyme associated with antioxidant stress in algae, which catalyzes the production of H_2O and O_2 from the degradation of H_2O_2 in cytosol and persoxisomes [50]. It is reported that at high concentrations (above 20 μM), Cu might be responsible for the inhibition of CAT, resulting in an insufficient ROS detoxification with enhanced H_2O_2 accumulation and lipid peroxidation [51]. Furthermore, inhibition of CAT activity may be caused by increased levels of O_2^- [52]. However, the activity of CAT declined significantly by 74% in thalli under chronic stress (Fig. 4F), while surprisingly only by 30% in that under acute stress (Fig. 4E). In addition, proper levels of Cu might lead to an increase in activity of CAT to cope with Cu stress [53]. As expected, CAT activity was activated and significantly increased in thalli under 8 μM Cu treatment after 1 day in our study (data not shown). Thus, it is plausible that the only explanation might be that O_2^- accumulated under chronic Cu exposure was more than that under acute treatment. In other word, it is the long-term exposure, rather than the high levels, as the main reason that led to the more accumulated O_2^-. In another aspect, the results above also suggested CAT as a more sensitive antioxidant enzyme than SOD, in agreement with other studies [47].

Besides SOD, activity of POD was also up-regulated in our experiment by both treatments, where chronic stress induced more activation of POD activity than acute stress (by 118% and 47%, respectively, Fig. 4H and G).

Mannitol as the main product of photosynthesis

ROS were formed either after acute or chronic heavy metal exposure, where the former abruptly generated into high levels that exceeded the ability of the antioxidant system to cope with them, while the latter increased steadily, resulting in different levels of damage to cellular compounds [7].

Mannitol is almost universally present in brown algae, being the main product of photosynthesis instead of sucrose [54], which may also function as carbohydrate storage, translocatable assimilate, source of reducing power, osmoregulator and/or antioxidant [55]. Changes in the monnitol content of marine brown algae have been reported in many field-based studies except that of heavy metal [56]. Based on the visual inspection, mannitol is the most abundant metabolite in the NMR spectrum from tissues of *S.*

fusiforme (data not shown). The concentration of mannitol decreased strongly by 72% (Table 2) in algae under acute stress, while only by 14% (Table 2) in that under chronic stress.

A mannitol cycle has been proposed in a number of organisms, including micro and macroalgae, where the latter is essentially the same as the fungal cycle [55]. In some yeasts, Cu^{2+} supplementation activates mannitol dehydrogenase involved in the biosynthesis of mannitol, resulting in an increased mannitol production. However, little is known about the affection of Cu on these enzymes involved in the metabolism of mannitol, especially in brown algae.

At the cell membrane, Cu may interfere with cell permeability [57,58]. In the present study exces Cu treatment caused much mannitol lost in the cell of *S. fusiforme*, indicating an enormous increase in permeability to it. In another aspect, as a compatible solute, mannitol is frequently used as a scavenger of hydroxyl radicals *in vitro* [59] and *in vivo* [60]. It may be involved in the cellular ROS-scavenging system to detoxify the oxidative stress. Therefore, it seems that in the long-term adaptation to low-concentration Cu stress of this algae, no significant differences were observed in the cell permeability, resulting in a little reduction in the content of mannitol.

Malate and aspartate may play important roles

Similar to C_4 plants, malate and aspartate were accumulated as candidates for the organic store in *Fucus* spp. that has a quite close phylogenetic relationship with *Sargassum* spp., both of which belong to Fucales [61]. Additionaly, C_3- and CAM-like photosynthesis were also observed in this species [61]. This coexistence of different photosynthetic pathways may be normal in aquatic environment [62], e.g., the both C_3 and C_4 photosynthetic pathways involved in the green-tide-forming alga, *Ulva prolifera* [63–65]. Anyway, malate and aspartate may play important roles in photosynthesis. As a potent inhibitor of photosynthesis, Cu dramatically reduced the levels of malate and aspartate in acute Cu treated *S. fusiforme*, where the former was even more strongly reduced (Table 1 and 2). However, content of aspartate was surprisingly increased by nearly 30% when under chronic Cu stress (Table 1). Moreover, malate can function as a vacuolar osmolyte and may also serve as an additional sink for carbon assimilation and reducing equivalents [66].

Aspartate was found to be the most abundant amino acid, which is in line with the results previously described [67]. More than half of the content was D-aspartate, whose cellular localization was also confirmed in *S. fusiforme* [68]. It was proposed that D-aspartate may play an important role in the growth of *S. fusiforme*, as well as in both germination and growth of higher plants [68,69]. However, the two isomers of aspartate were not elucidated by the methods of our study. To gain deeper insight into the biological role, especially under Cu stress, further studies, for example, the influences of Cu on the levels of both isomers and genes involved in their metabolism, will be required.

Metabolites involved in choline metabolism

Though the quaternary ammonium compound choline, which is the major component of membrane lipids in eukaryotes [70], was not detected in the present study, its precursor o-phosphocholine was found to be decreasing in its content under Cu stress. The metabolic pathway related to choline is summarized in Fig. 5. Significant reductions to a similar extent were observed for phosphocholine, dimethylglycine and glycine in *S. fusiforme* under acute Cu stress, while a quite different metabolic pattern was observed for chronic stress, where trimethylamine showed the opposite behavior and increased 2.3-fold over the control. The

Table 1. Concentrations of amino acids and related components.

µmol g^{-1}	Acute stress			Chronic stress		
	Control	Treated	Percentage	Control	Treated	Percentage
Alanine	0.2743	0.3557±0.0427*	130%±16%	0.2096	0.2978±0.0428*	142%±20%
Glutamate	0.6674	0.4692±0.0735*	70%±11%	0.4509	0.4907±0.0888	109%±20%
Glutamine	0.1216	0.1324±0.0412	109%±34%	0.0177	0.0598±0.0240*	338%±136%
Aspartate	0.8634	0.5878±0.0763*	68%±9%	0.5140	0.6744±0.0851*	131%±17%
Tryptophan	0.0237	0.0335±0.0024*	141%±10%	0.0351	0.0369±0.0079	105%±23%
Tyrosine	0.0382	0.0626±0.0082*	164%±21%	0.0332	0.0310±0.0126	93%±38%
Phenylalanine	0.0556	0.0480±0.0072	86%±13%	0.0835	0.0655±0.0142	78%±17%
Glycine	0.2329	0.1175±0.0380*	50%±16%	0.2029	0.1211±0.0141*	60%±7%
Isoleucine	0.0325	0.0200±0.0073	62%±22%	0.0266	0.0205±0.0023*	77%±9%
Leucine	0.0238	0.0138±0.0048*	58%±20%	0.0068	0.0095±0.0041	140%±60%
Valine	0.0439	0.0458±0.0061	104%±14%	0.0241	0.0212±0.0022*	88%±9%
N,N-Dimethylglycine	0.0444	0.0226±0.0040*	51%±9%	0.0469	0.0080±0.0019*	17%±4%

The absolute concentration of components in the control condition is shown the first column of each treatment, followed by the absolute concentration of each treatment and the relative changes compared to each control. Asterisk indicates significant differences between the stress and its control condition ($P<0.05$).

function of trimethylamine in various maritime plants may be related to the common saline habitat, possibly in osmotic regulation or in the transport of ions across membranes [71], i.e. a kind of osmolytes, which principally are sugars, polyhydric alcohols, amino acids and their derivatives, and methylamines, and all are known to be protein stabilizers. This observation suggests trimethylamine be as the preferred osmoprotectant in *S. fusiforme* under chronic Cu stress, rather than dimethylglycine.

Another most common and widely distributed compatible osmolyte proline was only detected by GC-TOF MS (Fig. 2), but not NMR. In plants, proline is synthesized mainly from glutamate [72]. They showed similar behavior that without significant accumulations in the level to chronic Cu stress (Table 1 and Fig. 2).

However, significant reductions were observed in *S. fusiforme* under acute Cu stress (Table 1 and Fig. 2). As intracellular proline levels are determined by biosynthesis, catabolism and transport between cells and different cellular compartments, we hypothesized that, dissimilar to other organisms, *S. fusiforme* may not use proline first as an osmoprotectant under Cu stress.

Ascorbate: an important antioxidant component

The concentrations of all the other detected polyols, organic acids and sugars were decreased, ranging from 20% for myo-inositol to 49% for citrate (Table 2), especially ascorbate and lactate (90% and 79%, respectively), where the former is considered as the main antioxidants in many plants. Protective

Table 2. Concentrations of organic acids, sugars, polyols and related components.

µmol g^{-1}	Acute stress			Chronic stress		
	Control	Treated	Percentage	Control	Treated	Percentage
Mannitol	19.7659	5.4364±1.2094*	28%±6%	19.9864	17.2164±1.1147*	86%±6%
Malate	0.1094	0.0275±0.0105*	25%±10%	0.0434	0.0261±0.0053*	60%±12%
myo-Inositol	0.1238	0.0986±0.0154	80%±12%	0.0803	0.0447±0.0110*	56%±14%
Citrate	0.4010	0.2056±0.0144*	51%±4%	0.2269	0.1915±0.0316	84%±14%
Xylose	0.1538	0.0997±0.0298	65%±19%	0.2121	0.1254±0.0293*	59%±14%
Succinate	0.0571	0.0302±0.0062*	53%±11%	0.0897	0.0520±0.0032*	58%±4%
Ascorbate	0.1087	0.0110±0.0061*	10%±6%	0.0720	0.0300±0.0064*	42%±9%
Lactate	0.3113	0.0652±0.0130*	21%±4%	0.0314	0.0357±0.0038	114%±12%
Trimethylamine N-oxide	0.2342	0.1628±0.0603*	70%±26%	0.2293	0.2147±0.0313	94%±14%
Trimethylamine	0.0070	0.0052±0.0011	74%±16%	0.0030	0.0069±0.0013*	230%±43%
Carnitine	0.0385	0.0297±0.0078*	77%±20%	0.0335	0.0318±0.0049	95%±15%
Acetate	0.0336	0.0215±0.0018*	64%±5%	0.0296	0.0269±0.0037	91%±13%
O-Phosphocholine	0.0347	0.0184±0.0055*	53%±16%	0.0363	0.0324±0.0051	89%±14%

The absolute concentration of components in the control condition is shown the first column of each treatment, followed by the absolute concentration of each treatment and the relative changes compared to each control. Asterisk indicates significant differences between the stress and its control condition ($P<0.05$).

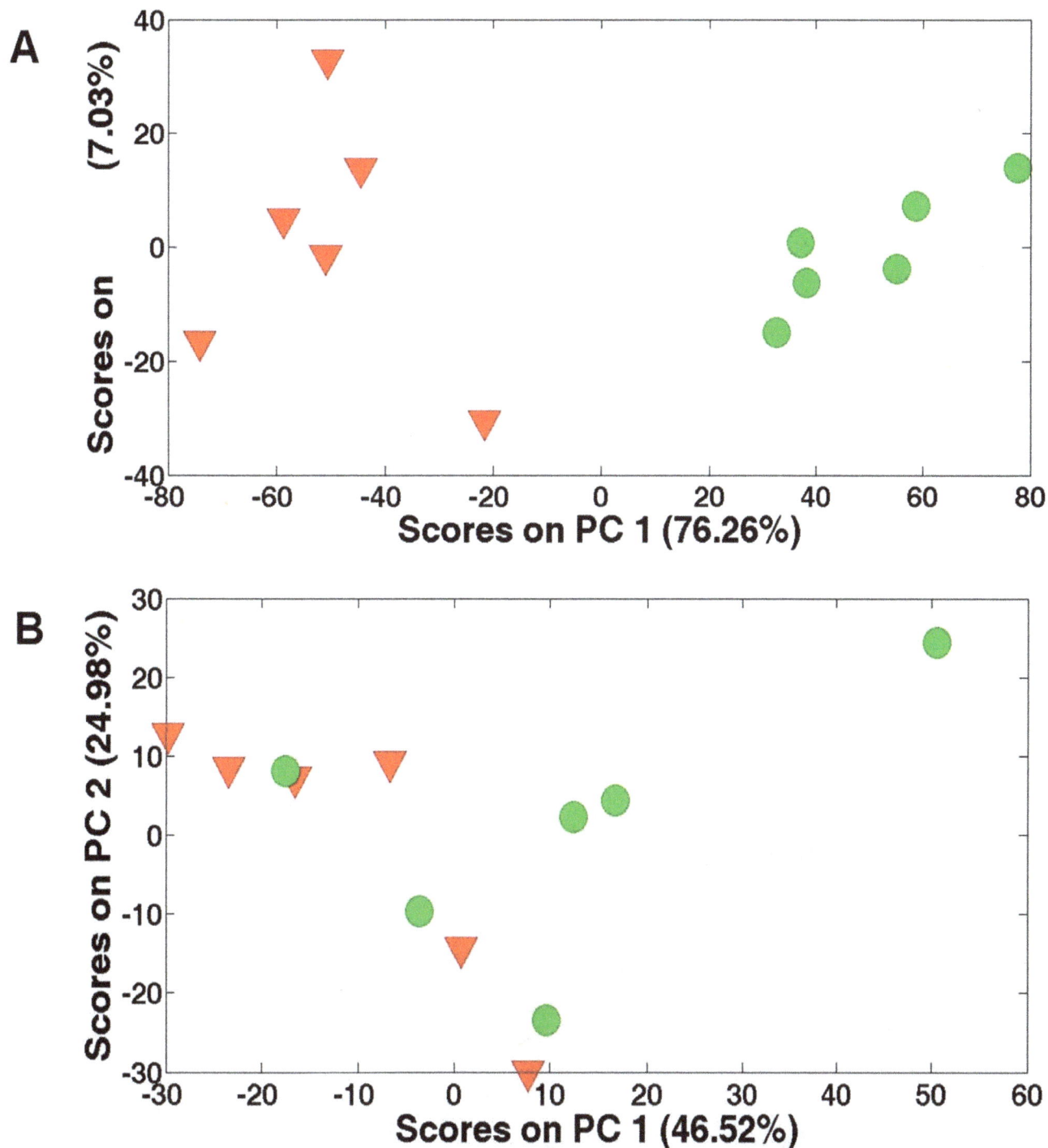

Figure 1. PCA score plots. (A) from the analysis of the 1D ^{1}H NMR spectra of *Sargassum fusiforme* tissue extracts from high concentration copper-treated (47 μM Cu^{2+}) group after exposure for 1 day; (B) from the analysis of the 1D ^{1}H NMR spectra of *Sargassum fusiforme* tissue extracts from low concentration copper-treated (8 μM Cu^{2+}) group after exposure for 7 days.

mechanisms in photosynthetic organisms do not only include ROS enzymes that reduce oxidative stress either with or without the aid of antioxidants but also antioxidants themselves [73].

In the brown algae *Scytosiphon lomentaria*, accumulation of ascorbate was detected in thalli from the Cu-enriched area [74]. In contrast, ascorbate content rapidly decreased and remained low in *Ulva compressa* (Chlorophyta) exposed to excess Cu [75]. Additionally, a low level of ascorbate was also observed in *U.*

compressa collected in Cu-enriched environments, indicating that short-term responses induced by excess Cu were similar to long-term responses occurring in the level of ascorbate [76]. Rapid reduction (almost 90%) of ascorbate was caused by acute Cu exposure in *S. fusiforme* in this study. Though to a relatively smaller extent, the content of ascorbate in thalli decreased by approximately 58% after chronic Cu exposure as compared to the control.

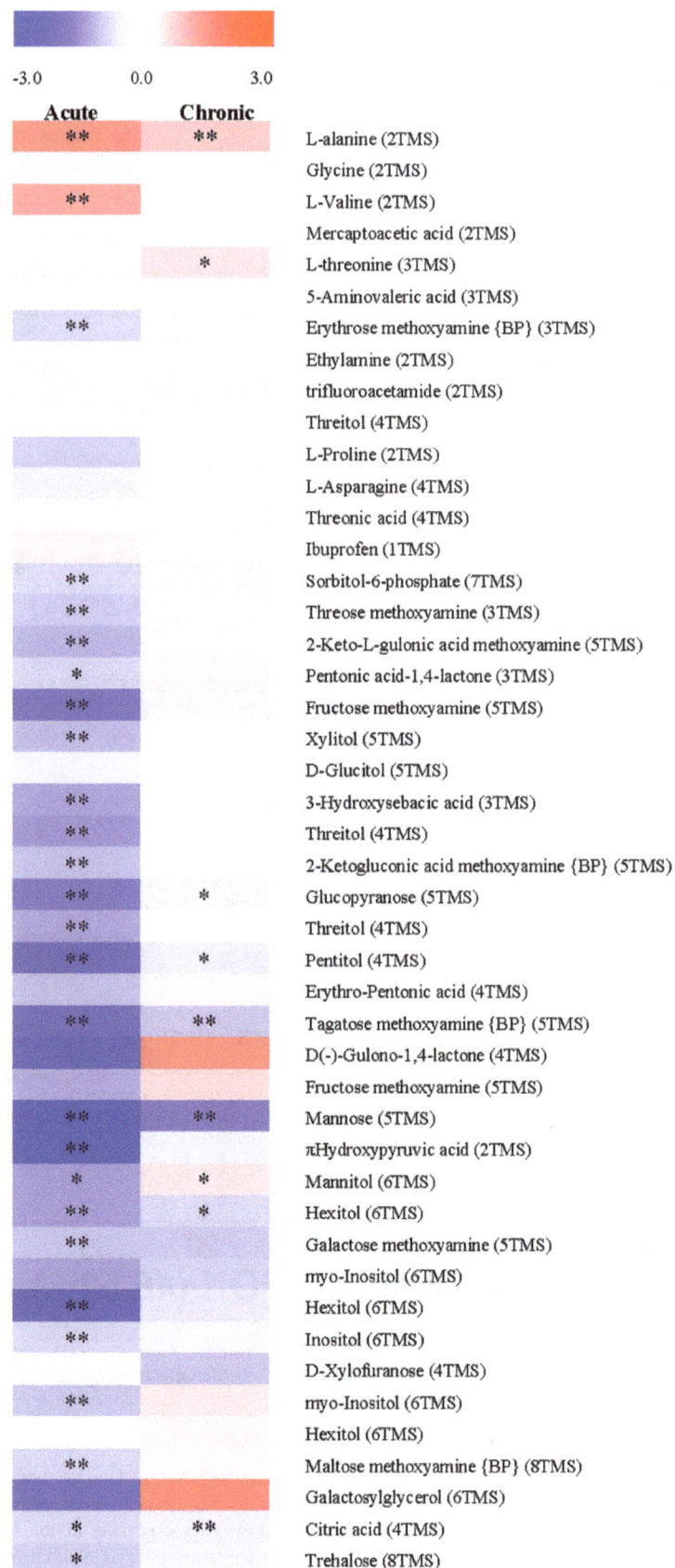

Figure 2. Heat-map of metabolic changes in *Sargassum fusiforme* under acute and chronic Cu stress. Intensity of colors represents log$_2$-transformed ratios of measured means (n = 6) analyte's intensity to its respective mean value in the control conditions. Asterisks mark *t*-test *P*-value, where "**" marks *P*<0.01 and "*" marks *P*<0.05.

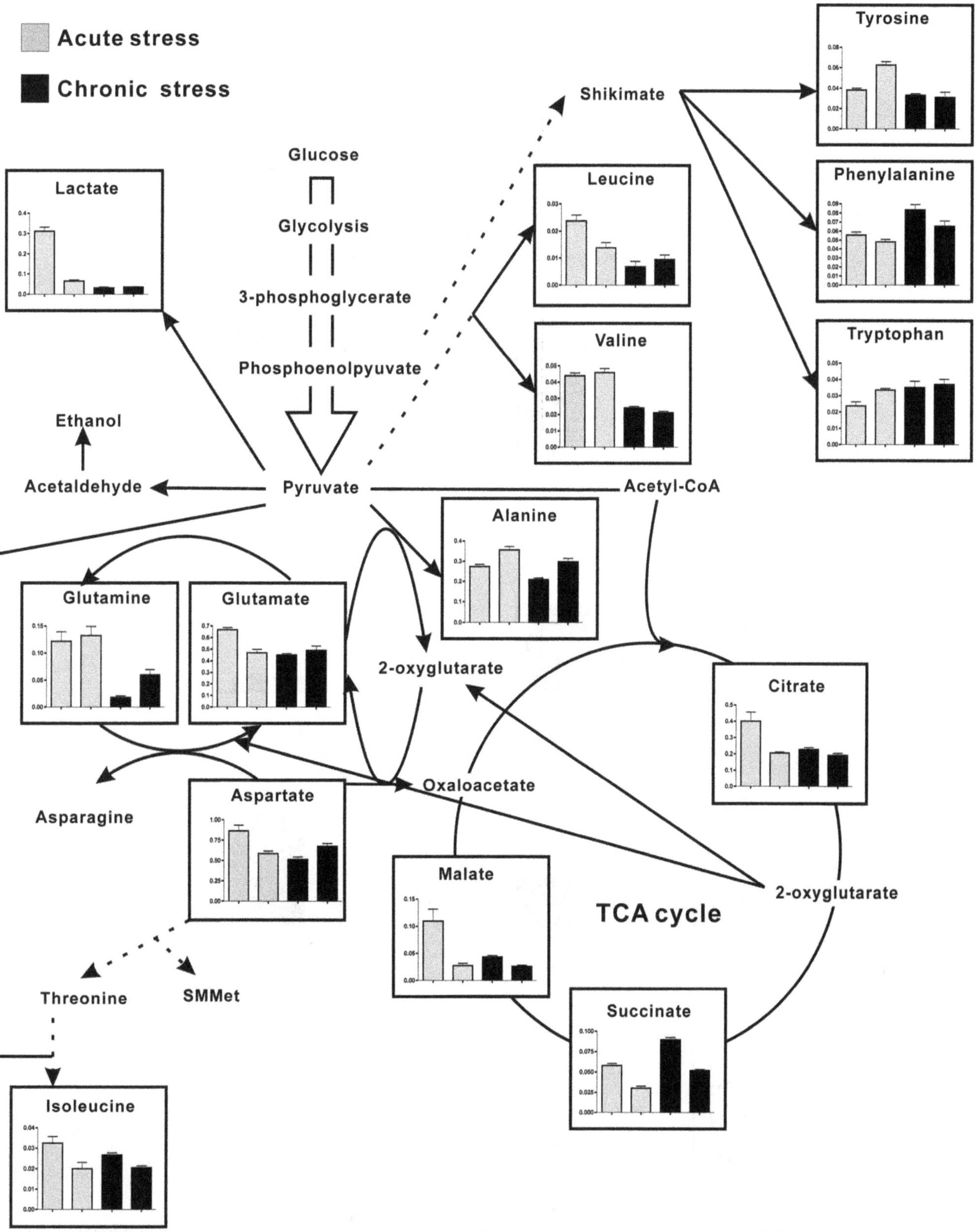

Figure 3. Pathway diagram with bar graphs representing relative metabolite abundance under acute and chronic Cu treatments. The first bar represents the control followed by the Cu treatment of the second bar.

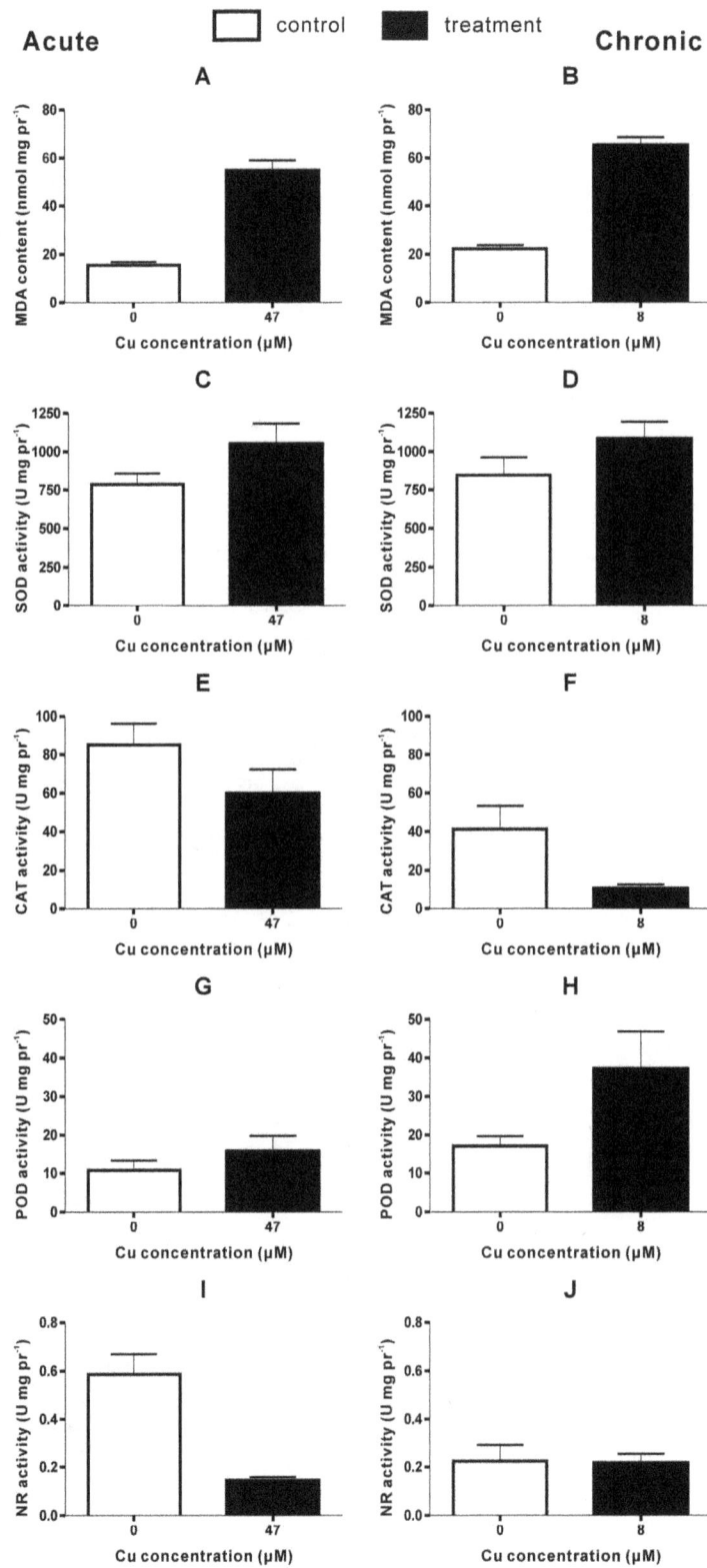

Figure 4. MDA content, antioxidant enzyme activities in *Sargassum fusiforme* under acute and chronic copper stress. Bars represent mean values of independent replicates ±1 SD (n = 4 or 5).

Oxidative stress can be mitigated by the synthesis of antioxidant component ascorbate which directly reduces ROS [23]. It is of particular interest as an electron donor for •OH radicals and also

as a substrate for ascorbate peroxidase (APX). An increase in ascorbate level was absent in *S. fusiforme*, similar to that in *U. compressa* that may be due to a direct oxidation of the newly

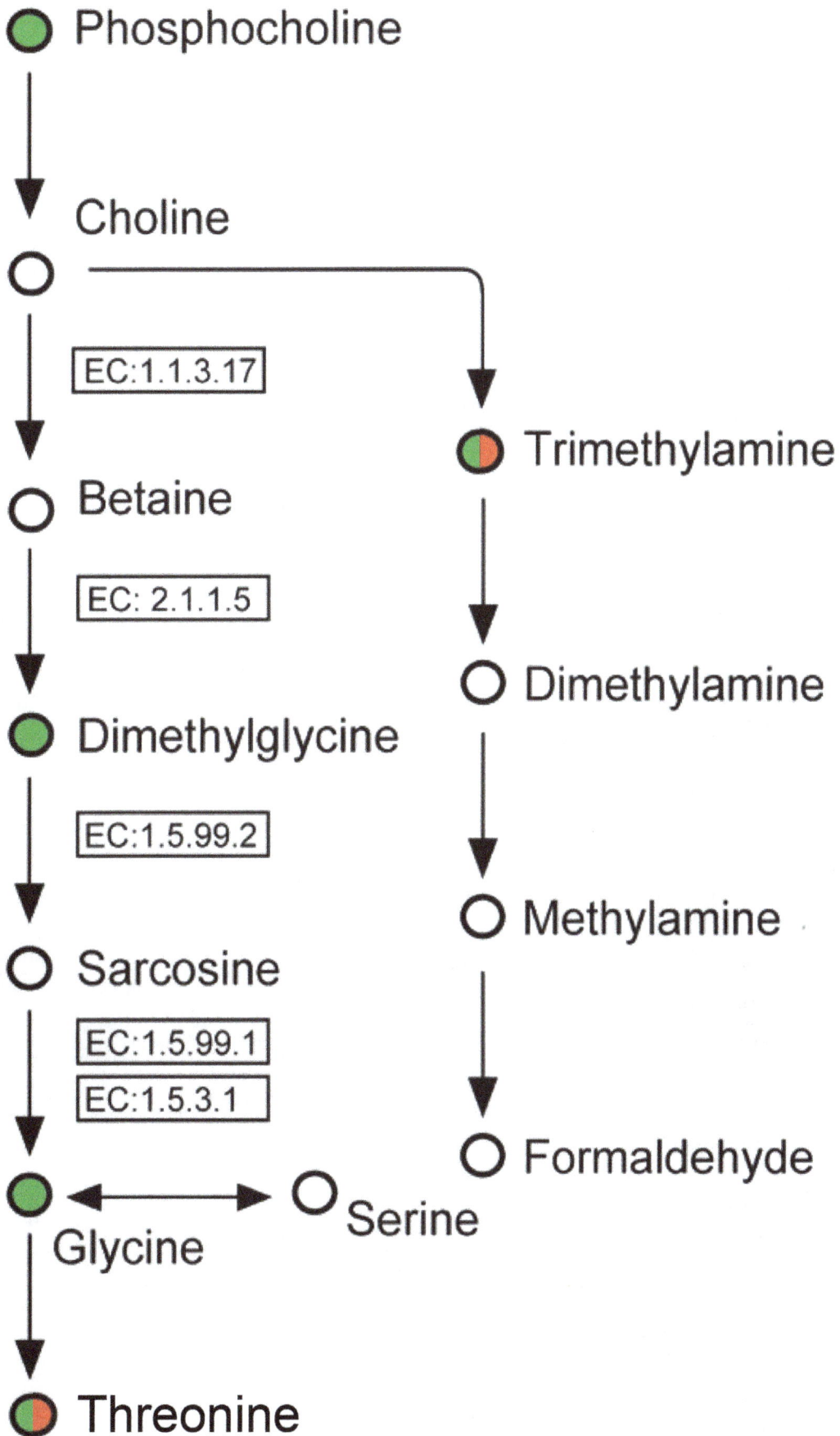

Figure 5. Schematic representation of metabolites related to choline. Identified metabolites from ^{1}H NMR spectroscopy are colored, increased metabolites are shown in red, and decreased metabolites are in green. The left and right half circles represent changes in level of each component under acute and chronic Cu stress, respectively. Key enzymes in choline catabolism include EC 1.1.3.17, choline oxidase; EC 2.1.1.5, betaine-homocysteine methyltransferase; EC 1.5.99.2, demethylglycine dehydrogenase; EC 1.5.3.1, sarcosine oxidase; and EC 1.5.99.1, choline oxidase.

synthesized ascorbate by Cu-induced ROS and/or to the activation of the antioxidant enzyme APX [77]. Cu^{2+} primarily triggers oxidase stress in the chloroplast, in which no active transport of ascorbate has been reported [78]. Therefore, reduced ascorbate was likely to be regenerated by the ascorbate/GSH cycle. As a result, rapid oxidation of ascorbate was provoked by acute Cu stress, which would abruptly generate of high levels of ROS over a short period that usually exceed the total antioxidant capacity of algae. Furthermore, it is reported that the decrease in ascorbate availability would as a result limit not only APX but also all peroxidase activity [78]. This to some extent explained the less activation in POD activity in thalli under acute Cu exposure in this study.

Conclusions

In conclusion, we present metabolic profiles observed for *S. fusiforme* under both acute and chronic Cu exposures. In order to identify as many metabolites as possible, which is also the goal of untargeted metabolomics experiments, ^{1}H NMR and GC-TOF MS were used complementally in the present study. Number of metabolites observed by MS platform was several times as that of NMR in this study, as NMR is generally considered to be less sensitive than GC-MS. These platforms would help expand our understanding of biological mechanisms related to environmental perturbations. Our results demonstrated different patterns of the marine brown algae *S. fusiforme* to acute and chronic Cu exposures in both physiological responses and regulation of metabolic pathways.

Author Contributions

Conceived and designed the experiments: HXZ QYP XFY. Performed the experiments: QYP AQZ LDL YQL LML QQW. Analyzed the data: HXZ QYP NL. Contributed reagents/materials/analysis tools: NL. Contributed to the writing of the manuscript: HXZ XFY.

References

1. Kunhikrishnan A, Bolan NS, Naidu R, Kim WI (2013) Recycled water sources influence the bioavailability of copper to earthworms. J Hazard Mater 261: 784–792.

2. Järup L (2003) Hazards of heavy metal contamination. Br Med Bull 68: 167–182.

3. Pang SJ, Shan TF, Zhang ZH, Sun JZ (2008) Cultivation of the intertidal brown alga *Hizikia fusiformis* (Harvey) Okamura: mass production of zygote-derived seedlings under commercial cultivation conditions, a case study experience. Aquac Res 39: 1408–1415.

4. Fang J, Wang KX, Tang JL, Wang M, Ren SJ, et al. (2004) Copper, Lead, Zinc, Cadmium, Mercury, and Arsenic in Marine Products of Commerce from Zhejiang Coastal Area, China, May 1998. Bull Environ Contam Toxicol 73: 583–590.

5. Contreras L, Mella D, Moenne A, Correa JA (2009) Differential responses to copper-induced oxidative stress in the marine macroalgae *Lessonia nigrescens* and *Scytosiphon lomentaria* (Phaeophyceae). Aquat Toxicol 94: 94–102.

6. Wu TM, Hsu YT, Sung MS, Hsu YT, Lee TM (2009) Expression of genes involved in redox homeostasis and antioxidant defense in a marine macroalga *Ulva fasciata* by excess copper. Aquat Toxicol 94: 275–285.

7. Okamoto OK, Pinto E, Latorre LR, Bechara EJH, Colepicolo P (2001) Antioxidant Modulation in Response to Metal-Induced Oxidative Stress in Algal Chloroplasts. Arch Environ Contam Toxicol 40: 18–24.

8. Dowling VA, Sheehan D (2006) Proteomics as a route to identification of toxicity targets in environmental toxicology. Proteomics 6: 5597–5604.

9. van Lipzig MM, Commandeur JN, de Kanter FJ, Damsten M, Vermeulen NP, et al. (2005) Bioactivation of Dibrominated Biphenyls by Cytochrome P450 Activity to Metabolites with Estrogenic Activity and Estrogen Sulfotransferase Inhibition Capacity. Chem Res Toxicol 18: 1691–1700.

10. Wu H, Wang WX (2011) Tissue-specific toxicological effects of cadmium in green mussel (*Perna viridis*): Nuclear magnetic resonance-based metabolomics study. Environ Toxicol Chem 30: 806–812.

11. Zhang L, Liu X, You L, Zhou D, Wu H, et al. (2011) Metabolic responses in gills of Manila clam *Ruditapes philippinarum* exposed to copper using NMR-based metabolomics. Mar Environ Res 72: 33–39.

12. Wu H, Liu X, Zhao J, Yu J, Pang Q, et al. (2012) Toxicological effects of environmentally relevant lead and zinc in halophyte *Suaeda salsa* by NMR-based metabolomics. Ecotoxicol 21: 2363–2371.

13. Wu H, Liu X, Zhao J, Yu J (2011) NMR-Based metabolomic investigations on the differential responses in adductor muscles from two pedigrees of Manila clam *Ruditapes philippinarum* to cadmium and zinc. Mar Drugs 9: 1566–1579.

14. Wu H, Liu X, Zhao J, Yu J (2012) Toxicological responses in halophyte *Suaeda salsa* to mercury under environmentally relevant salinity. Ecotoxicol Environ Saf 85: 64–71.

15. Liu X, Zhang L, You L, Cong M, Zhao J, et al. (2011) Toxicological responses to acute mercury exposure for three species of Manila clam *Ruditapes philippinarum* by NMR-based metabolomics. Environ Toxicol Pharmacol 31: 323–332.

16. Wu H, Zhang X, Li X, Li Z, Wu Y, et al. (2005) Studies on the acute biochemical effects of La(NO$_3$)$_3$ using ^{1}H NMR spectroscopy of urine combined with pattern recognition. J Inorg Biochem 99: 644–650.

17. Liu X, Yang C, Zhang L, Li L, Liu S, et al. (2011) Metabolic profiling of cadmium-induced effects in one pioneer intertidal halophyte *Suaeda salsa* by NMR-based metabolomics. Ecotoxicol 20: 1422–1432.

18. Farag MA, Porzel A, Wessjohann LA (2012) Comparative metabolite profiling and fingerprinting of medicinal licorice roots using a multiplex approach of GC-MS, LC-MS and 1D NMR techniques. Phytochem 76: 60–72.

19. Barding GA, Béni S, Fukao T, Bailey-Serres J, Larive CK (2012) Comparison of GC-MS and NMR for Metabolite Profiling of Rice Subjected to Submergence Stress. J Proteome Res 12: 898–909.

20. Dunn WB, Erban A, Weber R, Creek D, Brown M, et al. (2013) Mass appeal: metabolite identification in mass spectrometry-focused untargeted metabolomics. Metabolomics 9: 44–66.

21. Pan Z, Raftery D (2007) Comparing and combining NMR spectroscopy and mass spectrometry in metabolomics. Anal Bioanal Chem 387: 525–527.

22. Barding GA, Fukao T, Beni S, Bailey-Serres J, Larive CK (2012) Differential metabolic regulation governed by the rice SUB1A gene during submergence stress and identification of alanylglycine by ^{1}H NMR spectroscopy. J Proteome Res 11: 320–330.

23. Foyer CH, Noctor G (2011) Ascorbate and Glutathione: The Heart of the Redox Hub. Plant Physiol 155: 2–18.

24. Bradford MM (1976) A rapid and sensitive method for the quantitation of microgram quantities of protein utilizing the principle of protein-dye binding. Anal Biochem 72: 248–254.

25. Zhang L, Liu X, You L, Zhou D, Wang Q, et al. (2011) Benzo(a)pyrene-induced metabolic responses in Manila clam *Ruditapes philippinarum* by proton nuclear magnetic resonance (^{1}H NMR) based metabolomics. Environ Toxicol Pharmacol 32: 218–225.

26. Wu H, Zhang X, Wang Q, Li L, Ji C (2013) A metabolomic investigation on arsenic-induced toxicological effects in the clam *Ruditapes philippinarum* under different salinities. Ecotoxicol Environ Safety: 90, 1–6.

27. Wu H, Liu X, Zhao J, Yu J (2013) Regulation of Metabolites, Gene Expression, and Antioxidant Enzymes to Environmentally Relevant Lead and Zinc in the Halophyte *Suaeda salsa*. J Plant Growth Regul 32: 353–361.

28. Parul VP, David MR, Mark RV, David LW (2004) Discrimination Models Using Variance-Stabilizing Transformation of Metabolomic NMR Data. OMICS: J Integr Biol 8: 118–130.

29. Fan JH, Xie GZ, Wen SL (1996) The relativistic beaming model for active galactic nuclei. Astron Astrophys Suppl Ser 116: 409–415.

30. Lisec J, Schauer N, Kopka J, Willmitzer L, Fernie AR (2006) Gas chromatography mass spectrometry-based metabolite profiling in plants. Nat Protoc 1: 387–396.

31. Romanowska E (2002) Gas Exchange Functions in Heavy Metal Stressed Plants. In: Prasad MNV, Strzałka K editor. Physiology and Biochemistry of Metal Toxicity and Tolerance in Plants. Springer Netherlands, Berlin. pp. 257–285.

32. Zeier J (2013) New insights into the regulation of plant immunity by amino acid metabolic pathways. Plant Cell Environ 36: 2085–2103.

33. Díaz J, Bernal A, Pomar F, Merino F (2001) Induction of shikimate dehydrogenase and peroxidase in pepper (*Capsicum annuum* L.) seedlings in response to copper stress and its relation to lignification. Plant Sci 161: 179–188.

34. Richards TA, Dacks JB, Campbell SA, Blanchard JL, Foster PG, et al. (2006) Evolutionary Origins of the Eukaryotic Shikimate Pathway: Gene Fusions, Horizontal Gene Transfer, and Endosymbiotic Replacements. Eukaryot Cell 5: 1517–1531.

35. Pelletreau KN, Targett NM (2008) New Perspectives for Addressing Patterns of Secondary Metabolites in Marine Macroalgae. In: Amsler CD editor. Algal Chemical Ecology. Springer Berlin Heidelberg. pp. 121–146.

36. Fernandez E, Galvan A (2008) Nitrate Assimilation in Chlamydomonas. Eukaryot Cell 7: 555–559.

37. Campbell WH (1999) Nitrate Reductase Structure, Function and Regulation: Bridging the Gap between Biochemistry and Physiology. Annu. Rev. Plant Physiol. Plant Mol Biol 50: 277–303.

38. Harrison WG, Eppley RW, Renger EH (1977) Phytoplankton Nitrogen Metabolism, Nitrogen Budgets, and Observations on Copper Toxicity: Controlled Ecosystem Pollution Experiment. Bull Marine Sci 27: 44–57.

39. Luna CM, Casano LM, Trippi VS (1997) Nitrate reductase is inhibited in leaves of Triticum aestivum treated with high levels of copper. Physiologia Plantarum 101: 103–108.

40. Llorens N, Arola L, Bladé C, Mas A (2000) Effects of copper exposure upon nitrogen metabolism in tissue cultured Vitis vinifera. Plant Sci 160: 159–163.

41. Gouia H, Habib Ghorbal M, Meyer C (2000) Effects of cadmium on activity of nitrate reductase and on other enzymes of the nitrate assimilation pathway in bean. Plant Physiol Biochem 38: 629–638.

42. Fontaine JX, Saladino F, Agrimonti C, Bedu M, Tercé-Laforgue T, et al. (2006) Control of the Synthesis and Subcellular Targeting of the Two GDH Genes Products in Leaves and Stems of Nicotiana plumbaginifolia and Arabidopsis thaliana. Plant Cell Physiol 47: 410–418.

43. Lam HM, Coschigano KT, Oliveira IC, Melo-Oliveira R, Coruzzi GM (1996) The molecular-genetics of nitrogen assimilation into amino acids in higher plants. Annu Rev Plant Physiol Plant Mol Biol 47: 569–593.

44. Burzyński M, Buczek J (1997) The effect of Cu^{2+} on uptake and assimilation of ammonium by cucumber seedlings. Acta Physiol Plantarum 19: 3–8.

45. Ritter A, Ubertini M, Romac S, Gaillard F, Delage L, et al. (2010) Copper stress proteomics highlights local adaptation of two strains of the model brown alga Ectocarpus siliculosus. Proteomics 10: 2074–2088.

46. Flynn KJ, Dickson DMJ, Al-Amoudi OA (1989) The ratio of glutamine:glutamate in microalgae: a biomarker for N-status suitable for use at natural cell densities. J Plankton Res 11: 165–170.

47. Zhu X, Zou D, Du H (2011) Physiological responses of Hizikia fusiformis to copper and cadmium exposure. Botanica Marina 54: 431.

48. MacFarlane GR, Burchett MD (2001) Photosynthetic Pigments and Peroxidase Activity as Indicators of Heavy Metal Stress in the Grey Mangrove, Avicennia marina (Forsk.) Vierh Mar Pollut Bull 42: 233–240.

49. Hassan HM, Scandalios JM (1990) Superoxide dismutases in aerobic organisms. In: Alscher RG, Cumming JR editor. Stress Responses in Plants: Adaptation and Acclimatation Mechanisms. Wiley-Liss, New York. pp. 178–199.

50. Asada K, Takahashi M (1987) Production and scavenging of active oxygen in photosynthesis. In: Kyle DJ, Osmond CB, Arntzen CJ editor. Photoinhibition. Elsevier, Amsterdam. pp. 227–287.

51. Wu TM, Lee TM (2008) Regulation of activity and gene expression of antioxidant enzymes in Ulva fasciata Delile (Ulvales, Chlorophyta) in response to excess copper. Phycologia 47: 346–360.

52. Cakmak I (2000) Possible roles of zinc in protecting plant cells from damage by reactive oxygen species. New Phytol 146: 185–205.

53. Bischof K, Rautenberger R (2012) Seaweed Responses to Environmental Stress: Reactive Oxygen and Antioxidative Strategies. In: Wiencke C, Bischof K editor. Seaweed Biology. Springer Berlin Heidelberg. pp. 109–132.

54. Wickens G (2001) Human and Animal Nutrition. In: Wickens GE editor. Economic Botany. Springer Netherlands. pp. 127–149.

55. Iwamoto K, Shiraiwa Y (2005) Salt-Regulated Mannitol Metabolism in Algae. Mar Biotechnol 7: 407–415.

56. Reed RH, Davison IR, Chudek JA, Foster R (1985) The osmotic role of mannitol in the Phaeophyta: an appraisal. Phycologia 24: 35–47.

57. Overnell J (1975) The effect of heavy metals on photosynthesis and loss of cell potassium in two species of marine algae, Dunaliella tertiolecta and Phaeodactylum tricornutum. Mar Biol 29: 99–103.

58. Sunda WG, Huntsman SA (1983) Effect of competitive interactions between manganese and copper on cellular manganese and growth in estuarine and oceanic species of the diatom Thalassiosira. Limnol Oceanography 28: 924–934.

59. Smirnoff N, Cumbes QJ (1989) Hydroxyl radical scavenging activity of compatible solutes. Phytochemistry 28: 1057–1060.

60. Shen B, Jensen RG, Bohnert HJ (1997) Increased Resistance to Oxidative Stress in Transgenic Plants by Targeting Mannitol Biosynthesis to Chloroplasts. Plant Physiol 113: 1177–1183.

61. Kawamitsu Y, Boyer JS (1999) Photosynthesis and carbon storage between tides in a brown alga, Fucus vesiculosus. Mar Biol 133: 361–369.

62. Xie X, Wang G, Pan G, Sun J, Li J (2014) Development of oogonia of Sargassum horneri (Fucales, Heterokontophyta) and concomitant variations in PSII photosynthetic activities. Phycologia 53: 10–14.

63. Xu J, Fan X, Zhang X, Xu D, Mou S (2012) Evidence of Coexistence of C_3 and C_4 Photosynthetic Pathways in a Green-Tide-Forming Alga, Ulva prolifera. PLoS ONE 7: e37438.

64. Niu J, Hu H, Hu S, Wang G, Peng G, et al. (2010) Analysis of expressed sequence tags from the Ulva prolifera (Chlorophyta). Chinese Journal of Oceanology and Limnology 28: 26–36.

65. Gao S, Chen X, Yi Q, Wang G, Pan G, et al. (2010) A strategy for the proliferation of Ulva prolifera, main causative species of green tides, with formation of sporangia by fragmentation. PLoS One 5: e8571.

66. Doubnerová V, Ryšlavá H (2011) What can enzymes of C_4 photosynthesis do for C_3 plants under stress? Plant Sci 180: 575–583.

67. Nagahisa E, Kan-no N, Sato M, Sato Y (1994) Variations in D-aspartate content with season and part of Hizikia fusiformis. Fisheries Sci 60: 777–779.

68. Yokoyama T, Amano M, Sekine M, Homma H, Tokuda M (2011) Immunohistochemical Localization of Endogenous D-Aspartate in the Marine Brown Alga Sargassum fusiforme. Biosci Biotechnol Biochem 75: 1481–1484.

69. Funakoshi M, Sekine M, Katane M, Furuchi T, Yohda M, et al. (2008) Cloning and functional characterization of Arabidopsis thaliana D-amino acid aminotransferase D-aspartate behavior during germination. FEBS J 275: 1188–1200.

70. Chen C, Li S, McKeever DR, Beattie GA (2013) The widespread plant-colonizing bacterial species Pseudomonas syringae detects and exploits an extracellular pool of choline in hosts. Plant J 75: 891–902.

71. Smith TA (1971) The occurrence, metabolism and functions of amines in plants. Biological Rev 46: 201–241.

72. Szabados L, Savouré A (2010) Proline: a multifunctional amino acid. Trends Plant Sci 15: 89–97.

73. Collén J, Davison IR (1999) Reactive oxygen metabolism in intertidal Fucus spp. (Phaeophyceae). J Phycol 35: 62–69.

74. Contreras L, Moenne A, Correa JA (2005) Antioxidant responses in Scytosiphon lomentaria (Phaeophyceae) inhabiting copper-enriched coastal environments. J Phycol 41: 1184–1195.

75. Mellado M, Contreras RA, González A, Dennett G, Moenne A (2012) Copper-induced synthesis of ascorbate, glutathione and phytochelatins in the marine alga Ulva compressa (Chlorophyta). Plant Physiol Biochem 51: 102–108.

76. Ratkevicius N, Correa JA, Moenne A (2003) Copper accumulation, synthesis of ascorbate and activation of ascorbate peroxidase in Enteromorpha compressa (L.) Grev. (Chlorophyta) from heavy metal-enriched environments in northern Chile. Plant Cell Environ 26: 1599–1608.

77. Gonzalez A, Vera J, Castro J, Dennett G, Mellado M, et al. (2010) Co-occurring increases of calcium and organellar reactive oxygen species determine differential activation of antioxidant and defense enzymes in Ulva compressa (Chlorophyta) exposed to copper excess. Plant Cell Environ 33: 1627–1640.

78. Pinto E, Sigaud-kutner TCS, Leitão MAS, Okamoto OK, Morse D, et al. (2003) Heavy metal-induced oxidative stress in algae. J Phycol 39: 1008–1018.

17

Evaluation of Antioxidant Activities of Ampelopsin and Its Protective Effect in Lipopolysaccharide-Induced Oxidative Stress Piglets

Xiang Hou[1], Jingfei Zhang[1], Hussain Ahmad[1], Hao Zhang[1], Ziwei Xu[1,2]*, Tian Wang[1]

1 College of Animal Science and Technology, Nanjing Agricultural University, Nanjing, Jiangsu, P.R. China, **2** Institute of Animal Husbandry and Veterinary Science, Zhejiang Academy of Agricultural Sciences, Hangzhou, Zhejiang, P.R. China

Abstract

The aim of this study was to investigate the antioxidant potential of ampelopsin (APS) by using various methods *in vitro*, as well as to determine effects of APS on LPS-induced oxidative stress in piglets. The results showed that APS exhibited excellent free radical scavenging by DPPH, ABTS, $O_2{}^{\bullet-}$, H_2O_2 and ferric reducing antioxidant power. Ampelopsin also protected pig erythrocytes against AAPH-induced apoptosis and hemolysis, decreased total superoxide dismutase activity, and increased lipid peroxidation. Furthermore the results demonstrated that APS enhanced the total antioxidant capacity and decreased the malondialdehyde and protein carbonyl contents in LPS-treated piglets. The results of the present investigation suggest that APS possesses a strong antioxidant activity and alleviates LPS-induced oxidative stress, possibly due to its ability to prevent reactive oxygen species.

Editor: Nukhet Aykin-Burns, University of Arkansas for Medical Sciences; College of Pharmacy, United States of America

Funding: This study was funded by the Science Technology Department of Zhejiang Province, and the Modern Agro-industry Technology Research Sytem of China (CARS-36). The funders had no role in study design, data collection and analysis, decision to publish, or preparation of the manuscript.

Competing Interests: The authors have declared that no competing interests exist.

* Email: xzwfyz@sina.com

Introduction

Ampelopsin (3,5,7,3′,4′,5′-hexahydroxy-2,3-dihydroflavonol, APS), is the main bioactive component and most common flavonoid (about 20–30% w/w) in dry tender stems and leaves. In particular, APS comprises more than 40% of the flavonoid content in the cataphyll of the plant *Ampelopsis grossedentata (Hand-Mazz) W.T. Wang*. It is known as vine tea, and has been a health beverage in the south of China for thousands of years. Ampelopsin plays a wide variety of health roles, including anti-inflammatory, anti-hypertension, cough relieving, antioxidant, hepatoprotective, antimicrobial, and anticarcinogenic actions [1,2]. However, to the best of our knowledge, there are so far no systematic reports relating to the antioxidant capacity of APS using both *in vivo* and *in vitro* assays.

Lipopolysaccharides (LPS), a major component of the outer membrane of gram-negative bacteria, are capable of causing a wide variety of pathophysiological effects, such as endotoxic shock and tissue injury, and can be lethal in both humans and animals [3]. Lipopolysaccharides can induce tissue and organ injuries, apart from producing and releasing various pro-inflammatory cytokines and mediators, especially though the increased production of reactive oxygen intermediates such as superoxide radical ($O_2{}^{\bullet-}$), lipid peroxides, and nitric oxide, which cause oxidative stress [4]. Thus, the search for effective, nontoxic, natural agents with antioxidant properties that are efficient in inhibiting the production of reactive oxygen radicals is very important for the prevention and treatment of LPS-induced tissue injuries and sepsis.

Supplementation with dietary antioxidants can be helpful to attenuate the damage to the body induced by oxidative stress due to LPS [5,6], and can be used as potential therapeutic agents or as preventive drugs for the risk of many free radical-mediated diseases [7]. However, there are restrictions on the use of synthetic antioxidants, such as butylated hydroxyanisole (BHA) and butylated hydroxytoluene (BHT), because they may pose a risk to human health [8]. Therefore, efforts to develop alternative antioxidants of natural origin have been reinforced, and numerous studies have been conducted in order to evaluate the antioxidant capacity of certain compounds of plant origin.

The intensity effects of LPS depend on the animal species. Some animals, such as humans, rabbits, and pigs, are very sensitive to low doses of LPS, whereas animals like mice have relatively low sensitivity to LPS [9]. Both mice and rats are commonly used as animal models in clinical studies of oxidative stress in humans. The rodents, however, have limited use as models because their anatomical, physiological, and inflammatory responses, both cellular and humoral, are different from those of humans [10]. Pigs are recognized as an ideal animal model for human health research, based upon their similarity in size, organ development, physiology, disease progression and pathophysiologic responses [11,12]. Therefore, pigs were chosen as an experimental model for functional food researches and this may be more advantageous than using mice. However, the effect of LPS on oxidative stress in pigs and the protective mechanisms involved in flavonoid therapy have remained limited.

The present study was designed to investigate the *in vitro* and *in vivo* antioxidant capacity of APS. In this study, we evaluated the effects of APS as an antioxidant by using several antioxidant assays *in vitro*, including ferric-reducing antioxidant power (FRAP) assay, scavenging activity on 2,2′-azinobis-(3-ethyl-benzthiazoline-6-sulphonate) (ABTS), 1,1-diphenyl-2-picrylhydrazyl (DPPH), hydrogen peroxide (H_2O_2), and $O_2 \cdot^-$. Also, we evaluated the antioxidative effects of APS on the oxidative stress of pig erythrocytes induced with 2,2′-azobis-2-amidinopropane dihydrochloride (AAPH), a free radical generator, to test the markers of oxidative stress, including hemolysis, lipid peroxidation, superoxide dismutase (SOD) enzyme activity, and phosphatidyl-serine exposure on the cell surface, to investigate the protective effect of APS on early cellular apoptotic responses to oxidative stress. Finally, we studied the antioxidant potential of APS *in vivo* by evaluating the activities of antioxidant enzymes and lipid peroxidation levels in the LPS-induced piglets.

Materials and Methods

Chemicals and reagents

APS was purchased from the Aladdin Industrial Corporation (HPLC purity 95%, Shanghai, China). DPPH, ABTS, phenazine methosulfate (PMS), nicotinamide adenine dinucleotide (NADH), nitroblue tetrazolium (NBT), 2,4,6-tri(2-pyridyl)-s-triazine (TPTZ), AAPH, BHA, and LPS (from Escherichia coli O55:B5) were purchased from Sigma-Aldrich (St. Louis, MO). Commercial kits used for the determination of enzymes activities of SOD, glutathione peroxidase (GSH-Px), catalase (CAT), and total protein contents were purchased from the Nanjing Jiancheng Bioengineering Institute (Jiangsu, China). All other chemicals used were obtained from the Shanghai chemical agents company, China and were of analytical grade.

DPPH radical scavenging activity

The DPPH radical scavenging activities of APS and BHA were evaluated by the method of Xu *et al.* [13] with minor modifications. Briefly, 2.0 mL of samples in methanol at different concentrations (2, 4, 6, 8, 10 µg/mL) was mixed with aliquots of 0.1 mM DPPH solution in methanol. The mixture was vortexed for 1 min and incubated at ambient temperature for 30 min in the dark, and then the absorbance at 517 nm was measured using a spectrophotometer. The DPPH radical scavenging activity was calculated by the following equation:

$$\text{DPPH scavenging activity \%}$$
$$= \left(\text{control abs}_{517} - \text{sample abs}_{517}\right) / \text{control abs}_{517} \times 100\%$$

ABTS•+ radical scavenging activity

The ability of antioxidant molecules to quench ABTS radical cation (ABTS•+) was determined according to the method of Okamoto *et al.* [14]. A stable stock solution of ABTS•+ was produced by the reaction of a 7 mM aqueous solution of ABTS with 2.45 mM potassium persulfate (final concentration) and allowing the mixture to stand in the dark at room temperature for 16 h before use. 1 mL of ABTS•+ stock solution was added to the 3 mL of sample solutions at various concentrations (2, 4, 6, 8, 10 µg/mL). The contents were mixed well and incubated at 3°C exactly for 30 min. Then the absorbance was determined at 534 nm. The ABTS•+ radical scavenging activity was calculated as follows:

$$\text{ABTS}^{\bullet+} \text{ scavenging effect (\%)}$$
$$= \left(\text{control abs}_{534} - \text{sample abs}_{534}\right) / \text{control abs}_{534} \times 100\%$$

Ferric - reducing antioxidant power (FRAP) assay

The antioxidant capacity of APS was estimated according to Pulido, *et al.* [15]. 1.5 mL of FRAP reagent, prepared freshly and warmed at 37°C, was mixed with 150 µL of distilled water and 50 µL of test sample solutions at various concentrations (10, 20, 30, 40, 50, 60 µg/mL) or methanol (for the reagent blank). All samples were incubated at 37°C for 30 min in the dark. At the end of incubation, the absorbance readings were taken immediately at 593 nm. The final results were expressed as the concentration of antioxidants having a ferric reducing ability equivalent to that of 1 µM $FeSO_4$. A standard curve ranging from 5 to 150 µM of $FeSO_4 \cdot 7H_2O$ was used for calibration.

H_2O_2 scavenging activity

The ability of scavenging H_2O_2 was determined according to the method reported by Ak & Gulcin [16]. The reaction mixture was composed of 3.4 mL of sample solutions at various concentrations (10, 20, 30, 40, 50, 60 µg/mL) in phosphate buffer (0.1 M, pH = 7.4), 0.6 mL of H_2O_2 solution (0.04 M). Absorbance of reaction mixture at 230 nm was determined after 10 min against a blank contained the phosphate buffer without H_2O_2. The percentage of H_2O_2 scavenging by sample was calculated using the following equation:

$$H_2O_2 \text{ scavenging effect (\%)}$$
$$= \left(\text{control abs}_{230} - \text{sample abs}_{230}\right) / \text{control abs}_{230} \times 100\%$$

$O_2 \cdot^-$ scavenging activity

The $O_2 \cdot^-$ scavenging activity was measured by the method of Yen & Chen [17]. An aliquot of each of the following solutions prepared in 0.1 M phosphate buffer at pH 7.4:150 µM NBT, 60 µM PMS and 468 µM NADH. 1 mL of each of above-mentioned solutions and 1 mL sample solutions at various concentrations (10, 20, 30, 40, 50, 60 µg/mL) were added respectively, to the reaction mixture, and incubated at room temperature for 5 min. The absorbance was measured at 560 nm. The scavenging activity on $O_2 \cdot^-$ was expressed as:

$$O_2 \cdot^- \text{ scavenging effect (\%)}$$
$$= \left(\text{control abs}_{560} - \text{sample abs}_{560}\right) / \text{control abs}_{560} \times 100\%$$

Protective effect of APS on AAPH-induced hemolysis

Swine blood was collected by venipuncture into tubes containing heparin and centrifuged at 1500×g for 10 min at 4°C. After removal of plasma and buffy coat, erythrocytes were washed three times with cold phosphate buffered saline (PBS). 5% suspension of erythrocytes in PBS were preincubated at 37°C for 30 min in the chosen concentrations of APS (2.5, 5, 10 µM), and then incubated with AAPH (final concentration 75 mM) for 5 h with gentle shaking. Aliquots of the reaction mixture were taken out and centrifuged at each hour of incubation, diluted with saline, and centrifuged at 1500×g for 10 min. The percentage of hemolysis was determined by measuring the absorbance of the supernatant (A) at 540 nm and compared with that of complete hemolysis (B) by treating an aliquot with the same volume of the reaction mixture with distilled water. The hemolysis percentage was calculated using the formula: A/B ×100%. Three independent experiments were used for these calculations.

Measurement of erythrocyte SOD activity and lipid peroxidation level

The enzyme activity of SOD was based on its ability to inhibit the oxidation of the sample by the xanthine–xanthine oxidase system. One unit of SOD activity is defined as that amount of enzyme required to inhibit the reduction of SOD by 50% under the specified conditions, and the data were expressed as U/mg Hb. The level of lipid peroxidation was indicated by the content of malondialdehyde (MDA) using a thiobarbituric acid reaction (TBAR) method [18]. MDA content was expressed as nmol/mg Hb.

FACS analysis of annexin V binding

Phosphatidylserine exposure on erythrocytes was performed using the Alexa Fluor 488 Annexin V/Dead cell apoptosis kit (Invitrogen, cat. no V13245) according to the manufacturer's instruction. FACS analysis was performed as described by Agalakova & Gusev [19]. After the preincubation of APS and AAPH, the erythrocytes were washed twice with cold PBS (pH = 7.4, containing 150 mM NaCl, 1.9 mM Na_2HPO_4 and 8.1 mM NaH_2PO_4) and resuspended (5% suspension) in annexin-binding buffer. Then, the cells were stained with FITC- annexin V (1:50) at room temperature in the dark for 15 min. After incubation, samples were finally diluted 1: 5 in the same buffer and measured by flow cytometric analysis on a FACS-Calibur from Becton Dickinson (Heidelberg, Germany). The annexin V fluorescence intensity was measured in FL-1 with 488 nm excitation wavelength and 530 nm emission wavelength.

Assay of antioxidant activity in vivo

Animals and experimental design. Antioxidant enzymes activities, total antioxidant capacity and lipid peroxidation, in vivo, were determined. The animal use protocol for this research was approved by the Animal Care and Use Committee of Nanjing Agricultural University, Nanjing, PR. China. All piglets in this experiment were crossbred (Duroc×Landrace×Yorkshire) with an initial body weight of 7.5–8.3 kg and weaned at 4 weeks old. After weaning, piglets were housed in clean stainless steel metabolic cages in an ambient temperature of 27±1°C with a 12 h light/dark sequence. Each pen had a feeder and a nipple water to allow piglet ad libitum access to the basal diet and water. A total of 16 female piglet were randomly divided into the following four groups: the control group, the animals fed the basal diet for 27 days and receiving intraperitoneal (i.p.) administration of sterile saline on 28th day; LPS group, the animals were administered the basal diet for 27 days and were challenged with LPS (25 µg/kg) i.p. on 28th day; APS+LPS groups, the animals were treated with APS (100 and 400 mg/kg; as-fed basis) p.o. for 27 days, and received LPS (25 µg/kg) on the 28th day.

Measurement of SOD, GSH-Px and CAT activities in lung and liver of piglet

After 6 h of LPS injection, all the animals were sacrificed. Their lung and liver were removed rapidly, weighed and homogenized in ice-cold physiological saline. Then, the homogenate was centrifuged at 3500 g at 4°C for 15 min to remove cellular debris, and the supernatant was collected for analysis. The determination of the activity of SOD, CAT, and GSH-Px were described by Ahmad et al. [20] and measured according to the instructions in the kits. Protein contents were determined by a previously reported method [21].

Measurement of T-AOC in serum, MDA level and carbonyl content in lung and liver of piglet

Blood Samples were collected into plastic tubes (Ganda, Jiangxi China) containing no anticoagulant. Blood samples were allowed to clot at room temperature and were centrifuged at 4000 g for 10 min at 4°C to obtain the required serums. The total antioxidant capability (T-AOC) [20], MDA, and protein carbonyl content were measured according to the instructions in the kits.

Statistical analysis

The experimental results are expressed as mean ± standard error of means (SEM). SPSS (version 16.0) was used for statistical analysis. Data on DPPH, ABTS⁺, $O_2^{\bullet-}$, H_2O_2, and FRAP were analyzed by two-way ANOVA to determine the main effects of antioxidant (AO) and concentration (CT) and their interaction, followed by unpaired t-test for comparisons of two groups at each content point. Data on erythrocyte and piglet were analyzed by one-way ANOVA supplemented with Tukey's HSD post hoc test for multiple comparisons between groups. $P<0.05$ was considered significant.

Results

DPPH radical scavenging activity

The scavenging DPPH radical ability of APS and the synthetic antioxidant BHA is shown in Figure 1A. At concentrations of 2 to 10 µg/mL, the scavenging activities of APS were 66.55% to 96.19%, while the scavenging activities of BHA were 22.96% to 62.18%. The results showed that APS is more efficient ($P<0.05$) in the scavenging DPPH radical ability than the BHA.

ABTS⁺ radical scavenging activity

The ABTS⁺ radical scavenging activities of APS and BHA were positively correlated with their concentrations (Figure 1B). The scavenging activity of APS was significantly higher ($P<0.05$) than that of BHA. At the concentration of 10 µg/mL, the scavenging activities of APS and BHA were 83.13% and 73.02%, respectively.

FRAP assay

The FRAP values of APS were increased in a concentration-dependent manner (Figure 1C). At a concentration of 60 µg/mL, the reducing power of APS and BHA was 55.80 and 22.30 µM $FeSO_4$, respectively. The APS had greater ($P<0.05$) reducing capacity than the reference control of BHA.

H_2O_2 scavenging activity

Figure 1D depicts the scavenging capacity of APS against H_2O_2. The concentration range (10 to 60 µg/mL) of APS was capable of scavenging H_2O_2 in a concentration-dependent manner. Further, when compared with BHA, the APS was more effective ($P<0.05$) for scavenging H_2O_2. The H_2O_2 scavenging

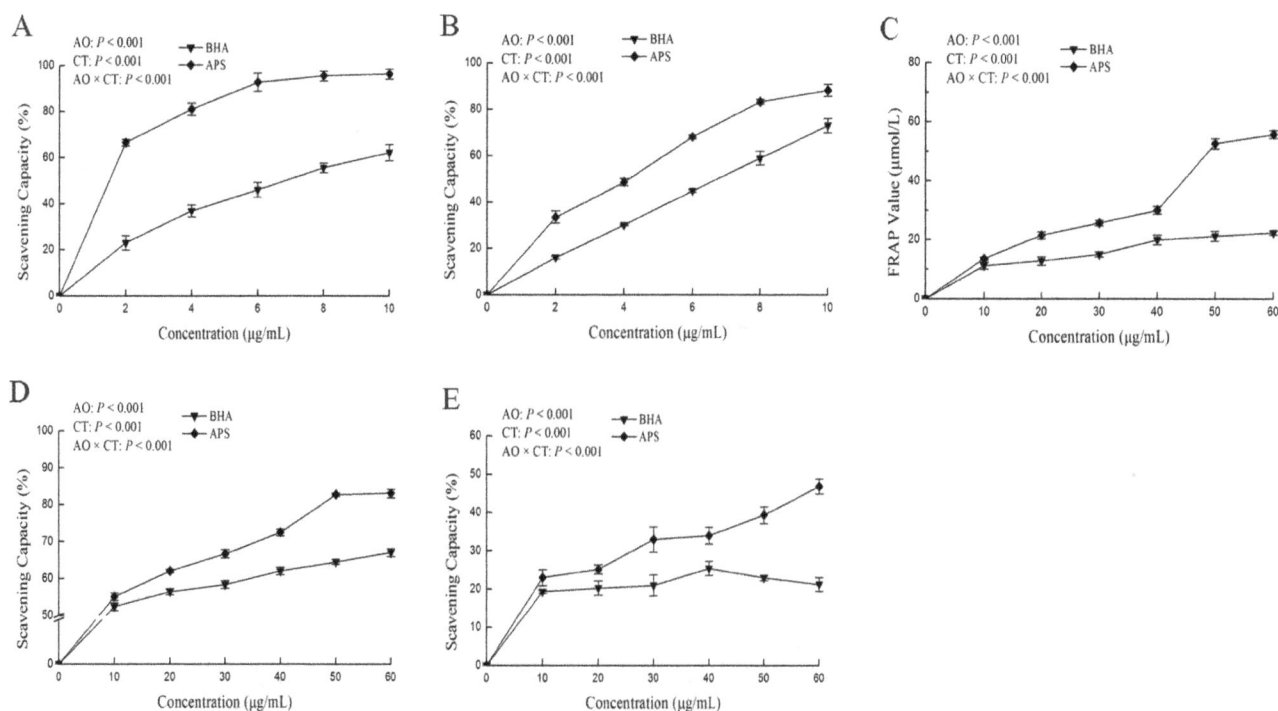

Figure 1. DPPH scavenging activity (A),ABTS scavenging activity (B),FRAP assay (C),H$_2$O$_2$ scavenging activity (D) and O$_2$·$^-$ scavenging activity (E) of APS and BHA. Data are expressed as the mean ± SEM of three parallel experiments. The P value showed the effects of antioxidant (AO) and concentration (CT), and their interaction on radicals scavenging activity.

activity of APS was 83.05% at a concentration of 60 μg/mL, while it was 67.05% for BHA.

O$_2$·$^-$ scavenging activity

Figure 1E depicts the scavenging capacity of APS against O$_2$·$^-$. Increasing the concentrations of APS increased its O$_2$·$^-$ scavenging capacities. However, the O$_2$·$^-$ scavenging capacities of BHA was increased from 10 to 40 μg/mL, and then decreased at higher concentrations. At level of 60 μg/mL, the scavenging activities of APS and BHA were 46.78% and 21.25%, respectively. These results showed that the APS was more efficient ($P<0.05$) than BHA antioxidant.

Effect of APS on AAPH-induced erythrocytes hemolysis

The AAPH-induced hemolysis in pig erythrocytes and the protective effect of the APS at different concentrations (2.5, 5, 10 μg/mL) are shown in Figure 2. Erythrocytes incubated at 37°C as a 5% suspension in PBS remained stable, with little hemolysis within 5 h (3.72% ±0.11). When AAPH was added to the erythrocyte suspension, after a relatively flat growth, hemolysis significantly increased ($P<0.05$) from 18.54±2.87% (1 h) to 87.66±2.14% (2 h), and it also increased in a time-dependent manner. After 5 h of incubation, the hemolytic activities of APS at different concentrations (2.5, 5, 10 μg/mL) against AAPH were 81.12±2.75%, 53.46±1.74%, and 31.38±1.55%, respectively.

Effect of APS on AAPH-induced lipid peroxidation in erythrocytes

The MDA level in the pig erythrocytes of the control group were stable (0.191±0.016 nmol/mg Hb) after 5 h of incubation (Figure 3A). The MDA content was significantly increased ($P<$

0.05) by 443.6%, 774.3%, and 875.4%, at 1, 3, and 5 h after incubation, respectively, with 75 mM AAPH as compared to the control samples. However, pretreatment with the APS concentrations (2.5, 5, 10 μg/mL) significantly decreased ($P<0.05$) the MDA levels compared to the AAPH group. After 5 h of incubation, in AAPH added erythrocytes, the MDA levels were significantly decreased ($P<0.05$) by 16.4%, 35.7% and 40.9% after addition of APS concentrations of 2.5, 5 and 10 μg/mL, respectively.

Effect of APS on AAPH-induced SOD in erythrocytes

Activity of SOD enzyme was significantly decreased ($P<0.05$) in a time-dependent manner after the exposure to AAPH, as compared with control group (Figure 3B). However, after 5 h of incubation, the APS (2.5, 5, 10 μg/mL) pretreated erythrocytes showed a significant ($P<0.05$) protective action against the decrease of SOD enzyme activity induced by AAPH.

Annexin V binding

As shown in the representative dot plot (Figure 4A and 4B), the percentage of annexin-positive erythrocytes from AAPH exposure was significantly higher ($P<0.05$) by 31.8±1.5% than in the APS (2.5 and 10 μg/mL) preincubated groups (18.0±1.7% and 8.8±1.1%), respectively, at 2 h of incubation. After 5 h, the percentage of annexin-binding erythrocytes was significantly increased ($P<0.05$) by 47.2±1.9%, while in the APS (2.5 and 10 μg/mL) pretreated groups, the percentage of annexin-binding erythrocytes was 26.60±1.2% and 15.7±1.1%, respectively.

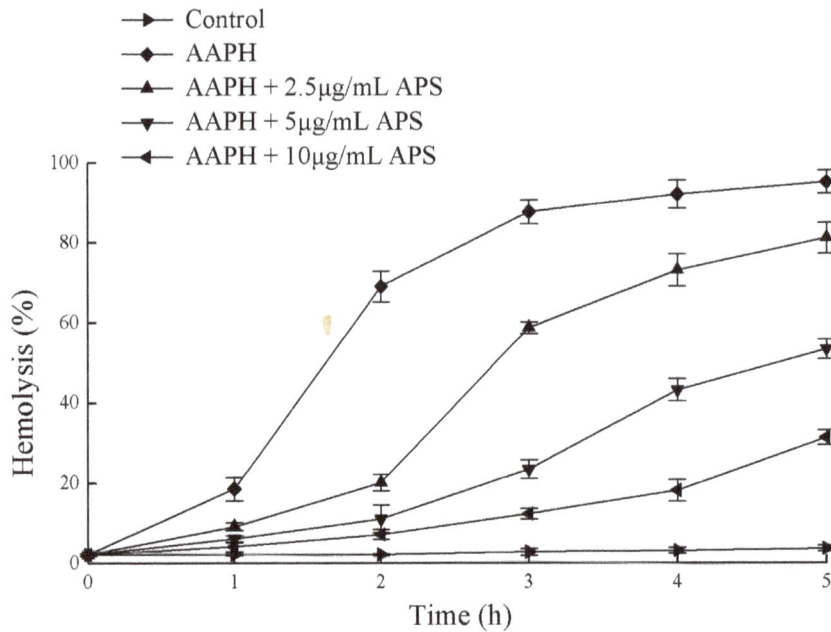

Figure 2. Effect of APS on AAPH-induced erythrocytes hemolysis. The erythrocytes were preincubated with APS at the indicated concentrations for 30 min at 37°C. The cell suspension was then incubated with 75 mM AAPH for 5 h at 37°C. Data are expressed as the mean ± SEM of three parallel experiments.

Effects of APS on serum T-AOC level in LPS-induced piglet

The LPS-treated piglets showed significantly decreased ($P<0.05$) T-AOC levels in the serum as compared with the control piglets (Table 1). The serum T-AOC significantly increased ($P<0.05$) after the supplementation of dietary APS (400 mg/kg) compared with the non-APS-treated LPS injected group.

Effect of APS on antioxidant enzyme activities in LPS-induced piglet

The activities of antioxidant enzymes GSH-Px, SOD, and CAT in tissue homogenates of lung and liver in the present study are shown in Table 1. The LPS-challenged piglets showed a significant decrease ($P<0.05$) in the activities of antioxidant enzymes as compared with the control group pigs. Administration of APS increased LPS-induced depletion of antioxidant enzyme activities both in the liver and lung of piglets. The level of 400 mg/kg APS to LPS-treated piglets increased ($P<0.05$)

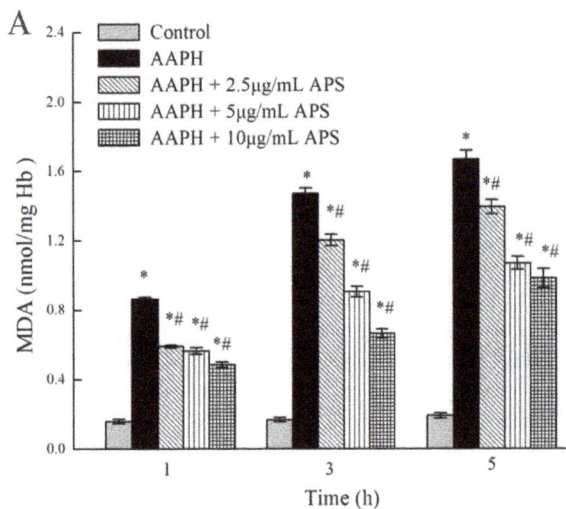

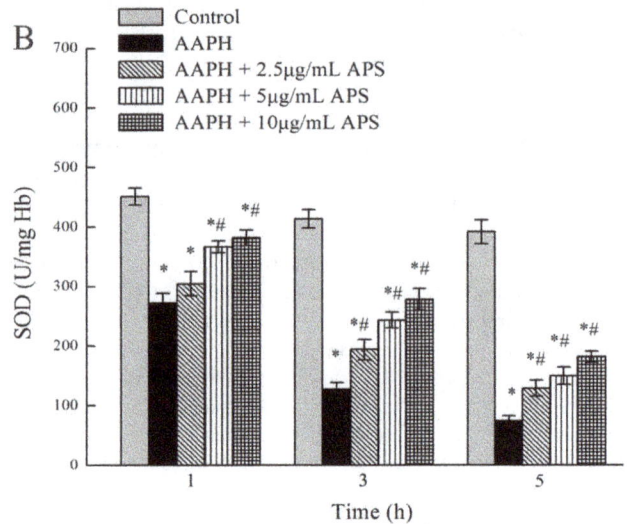

Figure 3. Effect of APS on MDA level (A) and SOD activity (B) in erythrocytes induced by AAPH. The pig erythrocytes were preincubated with APS at the indicated concentrations for 30 min at 37°C. The cell suspension was then incubated with 75 mM AAPH for 1, 3, 5 h at 37°C. Data are expressed as the mean ± SEM of three parallel experiments. *indicates a significant difference from control group ($P<0.05$); # indicates a significant difference from AAPH group ($P<0.05$).

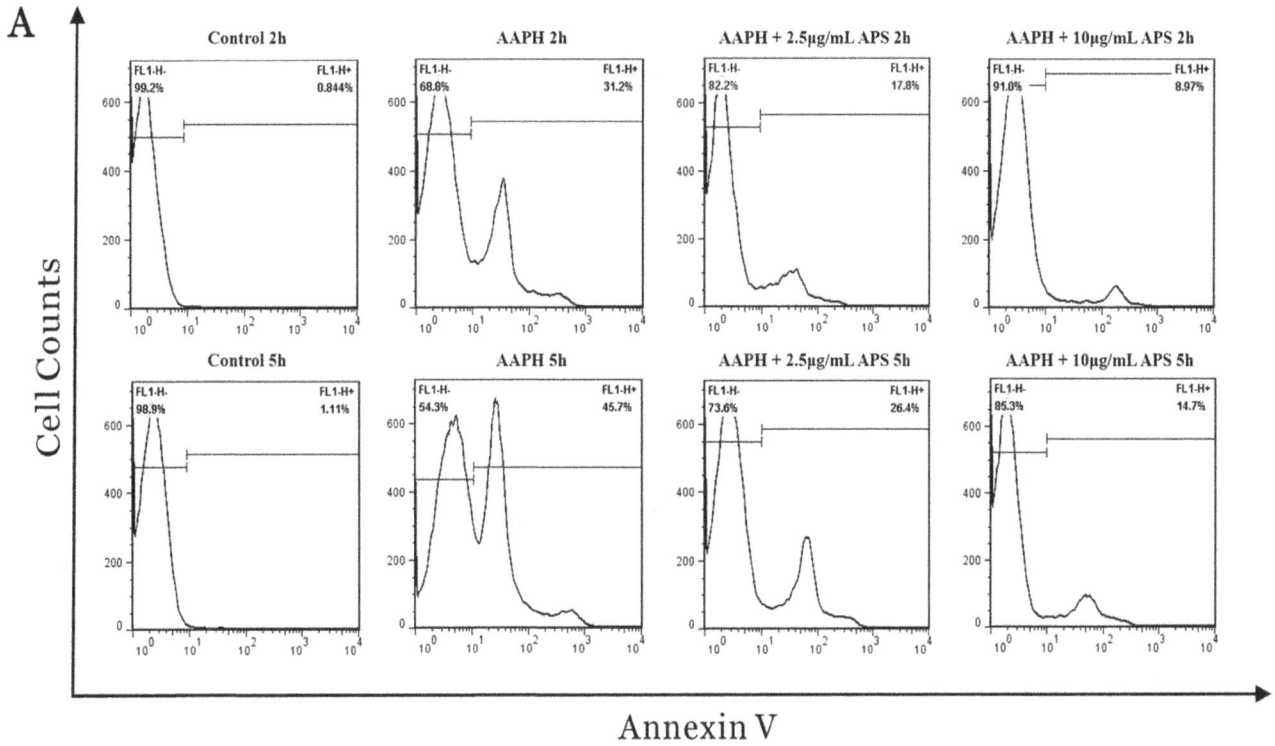

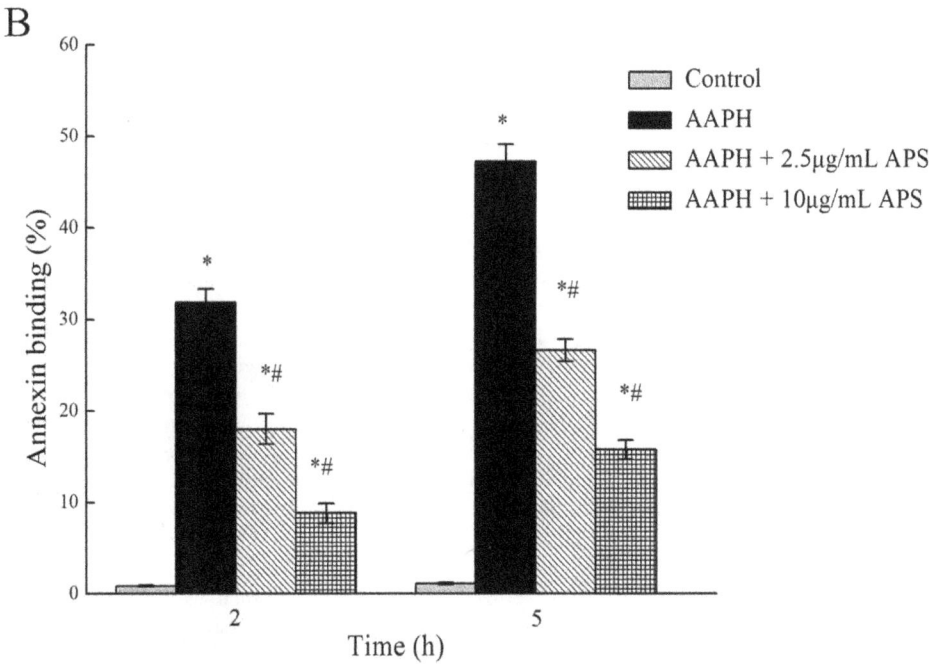

Figure 4. Effect of APS on annexin binding of erythrocytes following AAPH. A Representative original flow cytometry histograms of annexin binding of erythrocytes incubated in APS (2.5 and 10 μg/mL) or 75 mM AAPH for 2, 5 h. B Representative mean ± SEM of erythrocyte annexin binding after a 2, 5 h exposure to AAPH with and without APS. *indicates a significant difference from control group ($P<0.05$); # indicates a significant difference from AAPH group ($P<0.05$).

the activities of hepatic SOD and GSH-Px. However, 100 mg/kg dietary APS to LPS-treated piglets increased the SOD, GSH-Px, and CAT enzyme activities both in the liver and lung, but this was not statistically significant ($P>0.05$) in LPS-stressed piglets.

Effect of APS on lipid peroxidation and protein carbonyl in LPS-induced piglet

The results of present study showed a significant increased ($P< 0.05$) in MDA and protein carbonyl contents in LPS-stressed piglets as compared with the control (Table 1). In piglets, compared with the LPS-injected group, the supplementation of

Table 1. Effect of APS on activities of T-AOC, SOD, GSH-Px and CAT, levels of MDA and contents of protein carbonyl in LPS-induced piglets.

Parameters	control group	LPS group	LPS+APS (100 mg/kg)	LPS+APS (400 mg/kg)
Serum				
T-AOC	33.95±4.58[a]	17.08±2.04[c]	21.76±2.59[bc]	24.12±2.76[b]
Liver				
SOD	353.85±9.63[a]	243.85±8.91[c]	262.67±16.18[c]	295.51±12.61[b]
GSH-Px	38.14±4.34[a]	17.12±7.57[b]	26.64±4.97[a]	29.92±4.76[a]
CAT	27.20±4.61[a]	12.78±2.05[b]	14.36±1.05[b]	15.44±2.78[b]
MDA	2.56±0.31[c]	5.42±0.55[a]	4.58±0.70[ab]	3.70±0.54[bc]
Protein carbonyl	1.12±0.28[b]	3.81±0.84[a]	2.63±1.19[ab]	2.39±0.98[ab]
Lung				
SOD	163.63±31.31[a]	70.39±7.17[b]	100.08±16.8[b]	106.49±19.28[b]
GSH-Px	7.55±1.08[a]	4.37±0.68[b]	5.38±0.94[ab]	6.15±1.64[ab]
CAT	18.36±4.17[a]	11.73±1.61[b]	13.30±1.43[ab]	13.79±1.60[ab]
MDA	1.61±0.15[b]	3.35±0.44[a]	2.88±0.18[a]	2.79±0.12[a]
Protein carbonyl	0.27±0.03[b]	0.46±0.06[a]	0.35±0.05[ab]	0.32±0.07[ab]

Data on T-AOC are expressed as U/mL. Data on SOD, GSH-Px and CAT are expressed as U/mg protein. Values of MDA are expressed in nmol/mg protein. Protein carbonyl content expressed as μmol/mg protein. Data were presented as mean ± SEM (n = 4) and evaluated by one-way ANOVA followed by Tukey's test. Values in a row not sharing same superscript letter are significantly different, $P<0.05$.

dietary APS (100 and 400 mg/kg) decreased the MDA levels by 15.5% ($P>0.05$) and 31.7% ($P<0.05$) respectively in the liver; however, in the lungs the decreased was by 14.0% and 16.7% ($P>0.05$), respectively. The protein carbonyl contents of liver and lung were significantly higher ($P<0.05$) in LPS-injected group than that of the control group, while the dietary APS supplementation (100 and 400 mg/kg) had no significant effect on the protein carbonyl contents of liver and lung in both the control and LPS- injected pigs.

Discussion

In recent years, there has been a great deal of attention towards the field of antioxidants. Food scientists are interested in antioxidants because they can prevent rancidity in diets. Antioxidants have become of interest to biologists and clinicians because they can protect the body from oxidative damage. Ampelopsin is a major flavonoid, which has a close relationship with various kinds of biological activities and antioxidant capacity. In our experiments APS was analyzed using both chemical and biological assays.

DPPH and ABTS[++] radical scavenging activity

The DPPH assay is based on the reduction of DPPH•, a stable free radical that accepts an electron or hydrogen radical to become a stable diamagnetic molecule and is reduced to DPPHH, resulting in decolorization (yellow color) with respect to the number of electrons captured. Thus, it can be widely used as a model to study the reaction rates of antioxidants in order to quantify and compare the scavenging ability of various compounds [22]. Another common organic radical that has been applied to determine the antioxidant activity of both lipophilic and hydrophilic compounds is ABTS radical cation [23]. When estimating the ABTS[++] radical scavenging activities of BHA, similar values were obtained from the DPPH assays, suggesting that these two assays were positively correlated [24]. It was found that APS had a much higher antioxidant capacity in the DPPH assay than in the

ABTS assay, possibly due to the difference in the molecular structures of the two radicals and in the chemical reactions.

FRAP assay

On the basis of previous reports, it is clear that there is an association between the antioxidant capacities of antioxidants and their reducing power [25]. We evaluated the reducing power of APS using FRAP assay. The FRAP method is based on the reducing potential of the antioxidant to convert Fe^{3+} to Fe^{2+} in the presence of TPTZ, producing an intense blue-colored Fe^{2+}–TPTZ complex with an absorption maximum of 593 nm.

The DPPH, ABTS, and FRAP assays belong to an electron transfer mechanism and they are based on measurements of the capacities of antioxidants in the reduction of colored oxidants. The ABTS/DPPH methods measure the active compound capacity against an oxidant, while the FRAP assay directly measures the reducing capacity of substances, which is an important parameter for a compound to be considered a good antioxidant.

H_2O_2 and $O_2•^-$ scavenging activity

Reactive oxygen species (ROS), such as $O_2•^-$, H_2O_2 and hydroxyl radical (OH•), can act as signaling intermediates. However, oxidative stress occurs as a result of higher ROS levels due to an imbalance between ROS production and endogenous antioxidant defense systems [26]. Both DPPH and ABTS represent limited oxidation situations because they do not exist *in vivo*. Therefore, we also investigated the scavenging activity of APS against H_2O_2 and $O_2•^-$.

In contrast, in the more electrophilic forms of ROS, H_2O_2 is a small, stable, uncharged, and freely diffusible molecule that can be rapidly synthesized and destroyed in response to external stimuli [27]. Although H_2O_2 is not very reactive, its high penetrability in cellular membranes leads to OH• formation after its reaction with Fe^{2+} or the $O_2•^-$ in the cells [28,29]. Thus, removing H_2O_2 is very important for living organisms to avoid damage. $O_2•^-$ is a toxic species and forms first in numerous biological reactions among different ROS *in vivo* [30]. Although $O_2\bullet^-$ cannot directly initiate lipid oxidation, it can act as a potential precursor to generate more reactive radical species, such as OH• and peroxynitrite, which interact with lipids, proteins, and DNA molecules, eventually leading to various chronic diseases [31]. Therefore, measuring the potential of compounds to scavenge $O_2\bullet^-$ is one of the most important methods of clarifying the mechanism of antioxidant activity. The results of our present study suggested that the APS could be considered a good scavenger of H_2O_2 and $O_2\bullet^-$.

Effect of APS on AAPH-induced erythrocytes hemolysis

In this study, to confirm the antioxidant capacity of APS, we also evaluated its antioxidant potential by considering its efficacy in inhibiting AAPH-induced oxidative stress in pig erythrocytes. Erythrocytes have a well-established antioxidant defense system; even though, they are more susceptible to oxidation due to the presence of high polyunsaturated fatty acid (PUFA) content in membranes, the O_2 transport associated with redox active hemoglobin molecules, and transition metals such as iron and copper. Hemolysis of pig erythrocytes is a very good model for studying free radical induced oxidative damage to membranes, and for evaluating the antioxidant activity of APS. Erythrocytes are the most abundant cells in humans and animals and can easily become susceptible to oxidative damage with blood flow, usually resulting from exposure to drug side effects, toxic chemicals, and transition metal excesses. Excessive ROS continuously produced in erythrocytes due to these internal and external factors then overwhelmed cellular defenses, leading to protein oxidation and lipid peroxidation, which results in hemolysis and ultimately in cell death.

Erythrocytes have several enzymatic and non-enzymatic cellular antioxidant defenses such as SOD, glutathione (GSH), and ubiquinone. The changes in these biological mechanisms may be responsible for the previous lag phase shown in AAPH-induced hemolysis. Furthermore, APS remarkably delayed AAPH-induced hemolysis and had a dose-dependent correlation. Flavonoids can directly scavenge free radicals, and certain flavonoids also have the ability to incorporate into the hydrophobic core of the bilayer membranes. These changes in their fluidity and stability could strictly limit the diffusion of free radicals, thus decreasing the oxidative process throughout the membrane [32]. Therefore, in the present study, preincubation of the APS before AAPH addition, was considered as the interaction of an antioxidant with the cell membrane of erythrocytes. It was also found that the hemolysis rate inhibited by the APS slowed after the intense attack of peroxyl radicals generated by AAPH. The reason may be that the end products after the oxidation of flavonoids also exhibit antioxidant properties [33]. At the lower APS concentrations, oxidation products can still play the role of antioxidants, thus decreasing the hemolysis rate. All the results of our present study indicated that the APS could efficiently protect erythrocytes against AAPH-induced hemolysis. Our results are also consistent with the findings of previous studies [34–36] and indicated that the different flavonoids exhibit excellent abilities to protect against AAPH-induced hemolysis in erythrocytes [34–36].

Effect of APS on AAPH-induced lipid peroxidation in erythrocytes

In biological systems, lipid peroxidation can reduce membrane fluidity, influence enzymatic activities, inactivate membrane-bound proteins, and break down into cytotoxic aldehydes such as MDA or hydroxynonenal. The MDA, an indicator, is a main product of lipid peroxidation. A higher MDA level suggests that there is more lipid peroxidation and high oxidant stress [37].

Hemolysis caused by radicals can be characterized mainly by two key events: lipid peroxidation and redistribution of oxidized band 3 protein within the membrane [38]. Exposure of erythrocytes to free radicals generates oxidants such as AAPH and PUFA that can be converted into unstable lipid hydroperoxides. Lipid peroxidation continues until the formation of MDA, which can alter the proteins and membrane lipids together, and ultimately leads to the death of cells. The specific protein-lipid interaction provides a mechanism for the formation and stabilization of membrane domains and band 3 protein, a major erythrocyte transmembrane protein that interacts with many phospholipids at its surface [39]. The MDA accumulation can affect the anion transport and functions of the band 3 associated enzymes such as glyceraldehyde-3-phosphate dehydrogenase and phosphofructokinase. This may be the possible reason for the higher MDA content in AAPH-added erythrocyte hemolysate at longer durations.

Effect of APS on AAPH-induced SOD in erythrocytes

The SOD, the first line of defense against oxygen-derived free radicals, catalyze the dismutation of $O_2\bullet^-$ into O_2 and H_2O_2 [20]. Due to the lack of mitochondria, cytoplasmic SOD have key roles in erythrocytes [40]. Our results are in agreement with the previous study by Alvarez-Suarez [41] who reported that the flavonoid quercetin protected against the decrease of SOD activity in erythrocytes induced by AAPH.

Erythrocytes are well protected against ROS by abundant Cu/Zn-SOD, which scavenges free radicals to prevent the formation of methemoglobin. Erythrocyte damage due to oxidative stress is generally thought to be the result of two processes: the oxidation of hemoglobin, followed by the conversion of metHb to hemichromes; and the free radicals attack on membrane components, including the PUFA side chains of the membrane lipids, reduced thiol groups, and other susceptible amino acid chains of membrane proteins [42]. In our present study, the results showed that the APS had a protective action against the formation of MDA. Moreover, the addition of APS in AAPH- induced erythrocytes caused an increase in SOD enzyme activity compared to the AAPH group without APS-treated erythrocytes, possibly linked to the ability of APS to interact with $O_2\bullet^-$ and subsequently scavenge them. Further experiments are required to understand the exact molecular mechanism.

Annexin V binding

We further investigated the protective role of APS in pig erythrocytes using annexin V-FITC for the phosphatidylserine (PS) exposure analysis, one of the markers used for cell death or suicidal erythrocyte death (eryptosis) [43]. Oxidative stress, energy depletion, or lower efficiency of antioxidant defense system [43,44], enhance Ca^{2+} entry via the cation channels, leading to higher intracellular Ca^{2+} concentrations. Erythrocyte PS exposure is accomplished by cell membrane scrambling, which is triggered by an increase in cytosolic Ca^{2+} activity. The exposure of PS at the cell surface favors binding to the respective PS receptors expressed by macrophages [45]; therefore, erythrocytes with PS treatment are more rapidly cleared than non-PS-treated pig erythrocytes. The present study results suggest that the APS prolongs erythrocyte exposure to oxidative stress by increasing the inhibition of apoptotic cellular responses.

Antioxidant effects of APS in LPS-induced piglet

It is well known that the direct toxic effect of LPS in a variety of organs is to increase the formation of reactive oxygen intermediates such as $O_2\bullet^-$, peroxides, and nitric oxide, especially their secondary products, such as MDA, which is an index of oxidative stress, and these increase in a time- and dose-dependent manner [46]. Therefore, the injection of LPS can be a good model for the simulation of oxidative stress in experimental animals. Among the various organs susceptible to LPS, lungs are most commonly affected, which presents as an acute lung injury or, in its extreme manifestation, as acute respiratory distress syndrome. Oxidative stress induced by LPS rapidly affects other organs, such as the liver, which plays a vital role in the detoxification of LPS. Kupffer cells, the resident macrophages of the liver, are the major targets of LPS, and produce excessive amounts of $O_2\bullet^-$ on activation by LPS stress [47]. In addition, LPS stimulation induces polymorphonuclear neutrophil activation and migration into the liver [48], which is another source of free radicals. The *in vitro* antioxidant assays (Figure 1) results showed that the APS could reduce LPS-induced oxidative stress in piglets, especially in organs such as lung and liver.

All organisms have protective systems of antioxidant defense to maintain the cellular redox state, including both antioxidant enzymes and non-enzymatic antioxidants, in order to minimize or even prevent the injuries to organs and body systems caused by ROS. On the other hand, blood is the first stage of LPS detoxification, and measurements of blood biochemical markers have key clinical importance in the diagnosis of malfunctioning of body organs. The serum T-AOC level reflects the total antioxidant status of the body. The results of our present study indicated that APS may increase resistance to oxidative stress by enhancing T-AOC. Our results are in agreement with the previous reports indicating that the different flavonoid-rich extracts have the ability to decrease oxidative stress by promoting T-AOC [49,50].

The three main antioxidant enzymes, GSH-Px, SOD, and CAT, are the major antioxidant enzymes in the antioxidant system, which can help protect the body against oxidative damage [40]. SOD catalyze the production of O_2 and H_2O_2 from $O_2\bullet^-$, which in turn are decomposed to water and oxygen by GSH-Px and CAT enzymes, thus preventing the formation of OH•. Several studies have reported that the antioxidant capacities were decreased in the liver of LPS-induced animal models [46,51]. Previous studies have demonstrated that some natural compounds such as curcumin [46], hesperidin [52] and resveratrol [53]

increased antioxidant enzyme activities under LPS-induced oxidative stress, but no studies have been done using pig models.

Oxidative damage to proteins, as characterized by the formation of carbonyl groups, is also a highly damaging event, and can occur in the absence of lipid peroxidation [54]. We also investigated the oxidative damage in lipids and proteins. *In vitro* experiments (Figure 3) indicated that the APS could reduce MDA levels in oxidative stress in pig erythrocytes. The present results also suggested that the supplementation of dietary APS was effective in preventing the deleterious consequences of oxidative stress. The results of the present study have been confirmed by other researchers, who showed that supplementation with natural antioxidants or plant extracts exerts protective effects against LPS-induced injuries *in vivo* [55,56]. The results suggested that the APS had potent antioxidant activities not only via directly scavenging hydroxyl, superoxide, and metal-induced radicals in various radical-scavenging assays, such as the scavenging activities of free radicals, but also indirectly by enhancing antioxidant enzymes and decreasing lipid peroxidation.

In conclusion, the results of our present study clearly indicated that APS had excellent antioxidant capacity *in vitro*. Dietary APS supplementation can provide protection against LPS-induced oxidative injuries in lung and liver tissues of piglets. These findings reveal a new protective mechanism of APS as a potent antioxidant. The results demonstrated that APS is an accessible source of natural antioxidants and can be used in food and as a therapeutic agent. However, further studies are still required to elucidate the molecular mechanisms underlying the protective effects of APS as an antioxidant.

Acknowledgments

The authors are grateful to thank Zhiyuan Sun for the technical assistance and Xuhui Zhang for her kind help and available discussion.

Author Contributions

Conceived and designed the experiments: XH ZWX. Performed the experiments: XH. Analyzed the data: XH JFZ. Contributed reagents/materials/analysis tools: XH HA HZ TW. Wrote the paper: XH. Designed the software used in analysis: XH JFZ.

References

1. Murakami T, Miyakoshi M, Araho D, Mizutani K, Kambara T, et al. (2004) Hepatoprotective activity of tocha, the stems and leaves of Ampelopsis grossedentata, and ampelopsin. Biofactors 21: 175–178.
2. Ma J, Yang H, Basile MJ, Kennelly EJ (2004) Analysis of polyphenolic antioxidants from the fruits of three Pouteria species by selected ion monitoring liquid chromatography-mass spectrometry. J Agric Food Chem 52: 5873–5878.
3. Westphal M, Stubbe H, Sielenkämper A, Borgulya R, Van Aken H, et al. (2003) Terlipressin dose response in healthy and endotoxemic sheep: impact on cardiopulmonary performance and global oxygen transport. Intens Care Med 29: 301–308.
4. Sugino K, Dohi K, Yamada K, Kawasaki T (1987) The role of lipid peroxidation in endotoxin-induced hepatic damage and the protective effect of antioxidants. Surgery 101: 746.
5. Cadenas S, Rojas C, Barja G (1998) Endotoxin increases oxidative injury to proteins in guinea pig liver: protection by dietary vitamin C. Pharmacol Toxicol 82: 11–18.
6. Ben-Shaul V, Lomnitski L, Nyska A, Zurovsky Y, Bergman M, et al. (2001) The effect of natural antioxidants, NAO and apocynin, on oxidative stress in the rat heart following LPS challenge. Toxicol lett 123: 1–10.
7. Diplock AT, Charleux JL, Crozier-Willi G, Kok FJ, Rice-Evan C, et al. (1998) Functional food science and defence against reactive oxidative species. Brit J Nut 80: 77–112.
8. Yu R, Mandlekar S, Kong ANT (2000) Molecular mechanisms of Butylated hydroxyanisol- induced toxicity: induction of apoptosis through direct release of cytochrome *c*. Am Soci. Pharmacol Exp Ther 58: 431–437.
9. Galanos C, Freudenberg MA (1993) Bacterial endotoxins: biological properties and mechanisms of action. Mediat Inflamm 2: 11–16.
10. Mestas J, Hughes CCW (2004) Of mice and not men: differences between mouse and human immunology. J Immunol 172: 2731–2738.
11. Lunney JK (2007) Advances in swine biomedical model genomics. Int J Biol Sci 3: 179.
12. Nemzek JA, Hugunin KM, Opp MR (2008) Modeling sepsis in the laboratory: merging sound science with animal well-being. Comparative Med 58: 120.
13. Xu R, Shen Q, Ding X, Gao W, Li P (2011) Chemical characterization and antioxidant activity of an exopolysaccharide fraction isolated from Bifidobacterium animalis RH. Eur Food Res Technol 232: 231–240.
14. Okoko T, Ndoni S (2009) The effect of Garcinia kola extract on lipopolysaccharide-induced tissue damage in rats. Trop J Pharm Res 8: 27–31.
15. Pulido R, Bravo L, Saura-Calixto F (2000) Antioxidant activity of dietary polyphenols as determined by a modified ferric reducing/antioxidant power assay. J Agr Food Chem 48: 3396–3402.
16. Ak T, Gulcin I (2008) Antioxidant and radical scavenging properties of curcumin. Chem-Biol Interact 174: 27–37.
17. Yen GC, Chen HY (1995) Antioxidant activity of various tea extracts in relation to their antimutagenicity. J Agr Food Chem 43: 27–32.
18. Ohkawa H, Ohishi N, Yagi K (1979) Assay for lipid peroxides in animal tissues by thiobarbituric acid reaction. Anal Biochem 95: 351–358.
19. Agalakova NI, Gusev GP (2011) Fluoride-induced death of rat erythrocytes in vitro. Toxicol in Vitro 25: 1609–1618.
20. Ahmad H, Tian J, Wang J, Khan MA, Wang Y, et al. (2012) Effects of dietary sodium selenite and selenium yeast on antioxidant enzyme activities and oxidative stability of chicken breast meat. J Agr Food Chem 60: 7111–7120.
21. Lowry OH, Rosebrough NJ, Farr AL, Randall RJ (1951) Protein measurement with the Folin phenol reagent. J Biol Chem 193: 265–275.

22. Bae SH, Suh HJ (2007) Antioxidant activities of five different mulberry cultivars in Korea. LWT-Food Sci Technol 40: 955–962.

23. Miliauskas G, Venskutonis PR, Van Beek TA (2004) Screening of radical scavenging activity of some medicinal and aromatic plant extracts. Food Chem 85: 231–237.

24. Dudonne S, Vitrac X, Coutiere P, Woillez M, Mérillon JM (2009) Comparative study of antioxidant properties and total phenolic content of 30 plant extracts of industrial interest using DPPH, ABTS, FRAP, SOD, and ORAC assays. J Agr Food Chem 57: 1768–1774.

25. Guo C, Yang J, Wei J, Li Y, Xu J, et al. (2003) Antioxidant activities of peel, pulp and seed fractions of common fruits as determined by FRAP assay. Nutr Research 23: 1719–1726.

26. Blokhina O, Virolainen E, Fagerstedt KV (2003) Antioxidants, oxidative damage and oxygen deprivation stress: a review. Ann Bot-London 91: 179–194.

27. Rhee SG, Kang SW, Jeong W, Chang TS, Yang KS, et al. (2005) Intracellular messenger function of hydrogen peroxide and its regulation by peroxiredoxins. Curr Opin Cell Boil 17: 183–189.

28. Rhee SG (2006) Cell signaling. H_2O_2, a necessary evil for cell signaling. Science 312: 1882–1883.

29. Walling C (1975) Fenton's reagent revisited. Accounts Chem Res 8: 125–131.

30. Britigan BE, Pou S, Rosen GM, Lilleg DM, Buettner GR (1990) Hydroxyl radical is not a product of the reaction of xanthine oxidase and xanthine. J Bio Chem 263: 17533–17538.

31. Persinger RL, Poynter ME, Ckless K, Janssen-Heininger YM (2002) Molecular mechanisms of nitrogen dioxide induced epithelial injury in the lung. Mol Cell Biochem 234: 71–80.

32. Arora A, Byrem TM, Nair MG, Strasburg GM (2000) Modulation of liposomal membrane fluidity by flavonoids and isoflavonoids. Arch Biochem Biophys 373: 102–109.

33. Heim KE, Tagliaferro AR, Bobilya DJ (2002) Flavonoid antioxidants: chemistry, metabolism and structure-activity relationships. J Nutr Biochem 13: 572–584.

34. Chen Y, Deuster P (2009) Comparison of quercetin and dihydroquercetin: Antioxidant- independent actions on erythrocyte and platelet membrane. Chem-Bio Interact 182: 7–12.

35. Martínez V, Ugartondo V, Vinardell MP, Torres JL, Mitjans M (2012) Grape epicatechin conjugates prevent erythrocyte membrane protein oxidation. J Agr Food Chem 60: 4090–4095.

36. Dai F, Miao Q, Zhou B, Yang L, Liu ZL (2006) Protective effects of flavonols and their glycosides against free radical-induced oxidative hemolysis of red blood cells. Life Sci 78: 2488–2493.

37. Bagchi D, Bagchi M, Hassoun EA, Stohs SJ (1995) In vitro and in vivo generation of reactive oxygen species, DNA damage and lactate dehydrogenase leakage by selected pesticides. Toxicology 104: 129–140.

38. Sato Y, Kamo S, Takahashi T, Suzuki Y (1995) Mechanism of free radical-induced hemolysis of human erythrocytes: hemolysis by water-soluble radical initiator. Biochemistry 34: 8940–8949.

39. Yeagle PL (1982) 31P nuclear magnetic resonance studies of the phospholipid-protein interface in cell membranes. Biophys J 37: 227–239.

40. Gunduz K, Ozturk G, Sozmen EY (2004) Erythrocyte superoxide dismutase, catalase activities and plasma nitrite and nitrate levels in patients with Behcet disease and recurrent aphthous stomatitis. Clin Exp Dermatol 29: 176–179.

41. Alvarez-Suarez JM, Giampieri F, González-Paramás AM, Damiani E, Astolfi P, et al. (2012) Phenolics from monofloral honeys protect human erythrocyte membranes against oxidative damage. Food Chem Toxicol 50: 1508–1516.

42. López-Revuelta A, Sánchez_Gallego JI, Hernández-Hernández A, Sánchez-Yagüe J, Llanillo M (2006) Membrane cholesterol contents influence the protective effects of quercetin and rutin in erythrocytes damaged by oxidative stress. Chem-Biol Interact 161: 79–91.

43. Lang KS, Duranton C, Poehlmann H, Myssina S, Bauer C, et al. (2003) Cation channels trigger apoptotic death of erythrocytes. Cell Death Differ 10: 249–256.

44. Minetti M, Pietraforte D, Straface E, Metere A, Matarrese P, et al. (2008) Red blood cells as a model to differentiate between direct and indirect oxidation pathways of peroxynitrite. Method Enzymol 440: 253–272.

45. Fadok VA, Bratton DL, Rose DM, Pearson A, Ezekewitz RAB, et al. (2000) A receptor for phosphatidylserine-specific clearance of apoptotic cells. Nature 405: 85–90.

46. Kaur G, Tirkey N, Bharrhan S, Chanana V, Rishi P, et al. (2006) Inhibition of oxidative stress and cytokine activity by curcumin in amelioration of endotoxin-induced experimental hepatoxicity in rodents. Clin Exp Immunol 145: 313–321.

47. Roberts RA, Ganey PE, Ju C, Kamendulis LM, Rusyn I, et al. (2007) Role of the Kupffer cell in mediating hepatic toxicity and carcinogenesis. Toxicol Sci 96: 2–15.

48. Hewett JA, Schultze AE, VanCise S, Roth RA (1992) Neutrophil depletion protects against liver injury from bacterial endotoxin. Lab Invest 66: 347–361.

49. Feng LJ, Yu CH, Ying KJ, Hua J, Dai XY (2011) Hypolipidemic and antioxidant effects of total flavonoids of Perilla Frutescens leaves in hyperlipidemia rats induced by high-fat diet. Food Res Int 44: 404–409.

50. Wang X, Hai CX, Liang X, Yu SX, Zhang W, et al. (2010) The protective effects of Acanthopanax senticosus Harms aqueous extracts against oxidative stress: Role of Nrf2 and antioxidant enzymes. J Ethnopharmacol 127: 424–432.

51. Okoko T, Ndoni S (2009) The effect of Garcinia kola extract on lipopolysaccharide-induced tissue damage in rats. Trop J Pharm Res 8: 27–31.

52. Kaur G, Tirkey N, Chopra K (2006) Beneficial effect of hesperidin on lipopolysaccharide- induced hepatotoxicity. Toxicology 226: 152–160.

53. Dalle-Donne I, Giustarini D, Colombo R, Rossi R, Milzani A (2003) Protein carbonylation in human diseases. Trends in Mol Med 9: 169–176.

54. Dalle-Donne I, Rossi R, Giustarini D, Milzani A, Colombo R (2003) Protein carbonyl groups as biomarkers of oxidative stress. Clin Chim Acta 329: 23–38.

55. Kang KS, Yamabe N, Kim HY, Yokozawa T (2007) Effect of sun ginseng methanol extract on lipopolysaccharide-induced liver injury in rats. Phytomedicine 14: 840–845.

56. Sautebin L, Rossi A, Serraino I, Dugo P, Di Paola R, et al. (2004) Effect of anthocyanins contained in a blackberry extract on the circulatory failure and multiple organ dysfunction caused by endotoxin in the rat. Planta Med 70: 745–752.

Permissions

All chapters in this book were first published in PLOS ONE, by The Public Library of Science; hereby published with permission under the Creative Commons Attribution License or equivalent. Every chapter published in this book has been scrutinized by our experts. Their significance has been extensively debated. The topics covered herein carry significant findings which will fuel the growth of the discipline. They may even be implemented as practical applications or may be referred to as a beginning point for another development.

The contributors of this book come from diverse backgrounds, making this book a truly international effort. This book will bring forth new frontiers with its revolutionizing research information and detailed analysis of the nascent developments around the world.

We would like to thank all the contributing authors for lending their expertise to make the book truly unique. They have played a crucial role in the development of this book. Without their invaluable contributions this book wouldn't have been possible. They have made vital efforts to compile up to date information on the varied aspects of this subject to make this book a valuable addition to the collection of many professionals and students.

This book was conceptualized with the vision of imparting up-to-date information and advanced data in this field. To ensure the same, a matchless editorial board was set up. Every individual on the board went through rigorous rounds of assessment to prove their worth. After which they invested a large part of their time researching and compiling the most relevant data for our readers.

The editorial board has been involved in producing this book since its inception. They have spent rigorous hours researching and exploring the diverse topics which have resulted in the successful publishing of this book. They have passed on their knowledge of decades through this book. To expedite this challenging task, the publisher supported the team at every step. A small team of assistant editors was also appointed to further simplify the editing procedure and attain best results for the readers.

Apart from the editorial board, the designing team has also invested a significant amount of their time in understanding the subject and creating the most relevant covers. They scrutinized every image to scout for the most suitable representation of the subject and create an appropriate cover for the book.

The publishing team has been an ardent support to the editorial, designing and production team. Their endless efforts to recruit the best for this project, has resulted in the accomplishment of this book. They are a veteran in the field of academics and their pool of knowledge is as vast as their experience in printing. Their expertise and guidance has proved useful at every step. Their uncompromising quality standards have made this book an exceptional effort. Their encouragement from time to time has been an inspiration for everyone.

The publisher and the editorial board hope that this book will prove to be a valuable piece of knowledge for researchers, students, practitioners and scholars across the globe.

List of Contributors

Otis C. Attucks, Jareer Kassis, Zhenping Zhong, Suparna Gupta, Sam F. Victory, Mustafa Guzel, Dharma Rao Polisetti, Robert Andrews, Adnan M. M. Mjalli and Matthew J. Kostura
TransTech Pharma LLC, High Point, North Carolina, United States of America

Mark Hannink
Biochemistry Department and Life Sciences Center, University of Missouri, Columbia, Missouri, United States of America

Kimberly J. Jasmer
Division of Biological Sciences, University of Missouri, Columbia, Missouri, United States of America

Nagendra K. Kaushik, Neha Kaushik, Daehoon Park and Eun H. Choi
Plasma Bioscience Research Center, Kwangwoon University, Seoul, Korea

Yuh-Mou Sue
Department of Nephrology, Taipei Medical University-Wan Fang Hospital, Taipei, Taiwan

Hsiu-Chu Chou
Department of Anatomy, School of Medicine, College of Medicine, Taipei
Medical University, Taipei, Taiwan

Chih-Cheng Chang, Nian-Jie Yang, Ying Chou and Shu-Hui Juan
Graduate Institute of Medical Sciences, Taipei Medical University, Taipei, Taiwan
Department of Physiology, School of Medicine, College of Medicine, Taipei Medical University, Taipei, Taiwan

Beng Fye Lau, Noorlidah Abdullah and Vikineswary Sabaratnam
Mushroom Research Centre and Institute of Biological Sciences, Faculty of Science, University of Malaya, Kuala Lumpur, Malaysia

Norhaniza Aminudin
Mushroom Research Centre and Institute of Biological Sciences, Faculty of Science, University of Malaya, Kuala Lumpur, Malaysia
University of Malaya Centre for Proteomics Research (UMCPR), Faculty of Medicine, University of Malaya, Kuala Lumpur, Malaysia

Hong Boon Lee
Drug Discovery Laboratory, Cancer Research Initiatives Foundation
(CARIF), Subang Jaya, Selangor, Malaysia
Department of Pharmacy, Faculty of Medicine, University of Malaya, Kuala Lumpur, Malaysia

Ken Choy Yap
Advanced Chemistry Solutions, Kuala Lumpur, Malaysia

Yujuan Zhang, Dongmei Wang, Lina Yang, Dan Zhou and Jingfang Zhang
College of Forestry, Northwest A & F University, Yangling, China

Bidya Dhar Sahu, Meghana Koneru, Madhusudana Kuncha and Ramakrishna Sistla
Medicinal Chemistry and Pharmacology Division, CSIR-Indian Institute of Chemical Technology (IICT), Hyderabad, Andhra Pradesh, India

Anil Kumar Kalvala
Department of Pharmacology and Toxicology, National Institute of Pharmaceutical Education and Research (NIPER), Hyderabad, Andhra Pradesh, India

Jerald Mahesh Kumar
Animal House, CSIR-Centre for Cellular and Molecular Biology (CCMB), Hyderabad, Andhra Pradesh, India

Shyam Sunder Rachamalla
Faculty of Pharmacy, Osmania University, Hyderabad, Andhra Pradesh, India

Jian-bo Yu, Jia Shi, Li-rong Gong, Shu-an Dong, Yan Xu, Yuan Zhang, Xin shun Cao and Li-li Wu
Department of Anesthesiology, Tianjin Nankai Hospital, Tianjin Medical University, Tianjin, China

Ana Stevanovic, Mark Coburn, Rolf Rossaint, Gereon Schälte and
Thilo Werker
Department of Anaesthesiology, University Hospital of the RWTH Aachen, Aachen, Germany

Christian Stoppe
Department of Anaesthesiology, University Hospital of the RWTH Aachen, Aachen, Germany
Department of Thoracic, Cardiac and Vascular Surgery, University Hospital, RWTH Aachen, Aachen, Germany

Institute of Biochemistry and Molecular Cell Biology, RWTH Aachen University, Aachen, Germany

Ares Menon and Andreas Goetzenich
Department of Thoracic, Cardiac and Vascular Surgery, University Hospital, RWTH Aachen, Aachen, Germany

Daren Heyland
Kingston General Hospital, Kingston, Ontario, Canada,

Willibald Wonisch
Institute of Physiological Chemistry, Centre for Physiological Medicine, Medical University of Graz, Graz, Austria
Clinical Institute of Medical and Chemical Laboratory Diagnostics, Medical University of Graz, Graz, Austria

Michael Kiehntopf
Institute of Clinical Chemistry, Friedrich-Schiller University, Jena, Germany

Steffen Rex
Department of Anaesthesiology and Department of Cardiovascular Sciences, University Hospitals Leuven, KU Leuven, Belgium

Kalathookunnel Antony Antu, Mariam Philip Riya, Chandrasekharan K. Chandrakanth and K. Gopalan Raghu
Agroprocessing and Natural Products Division, Council of Scientific and Industrial Research-National Institute for Interdisciplinary Science and Technology (CSIR-NIIST), Thiruvananthapuram, Kerala, India

Arvind Mishra and Arvind K. Srivastava
Division of Biochemistry, Council of Scientific and Industrial Research-Central Drug Research Institute (CSIR-CDRI), Lucknow, Uttar
Pradesh, India

Karunakaran S. Anilkumar
Medicinal Chemistry Division, CSIR-CDRI, Lucknow, Uttar Pradesh, India

Akhilesh K. Tamrakar
Division of Pharmacology, CSIR-CDRI, Lucknow, Uttar Pradesh, India

Kátya Karine Nery Carneiro Lins Perazzo, Anderson Carlos de Vasconcelos Conceição, Juliana CaribéPires dos Santos, Carolina Oliveira Souza and Janice Izabel Druzian
Federal University of Bahia, College of Pharmacy, Department Food Science, Ondina, Salvador, BA, Brazil

Denilson de Jesus Assis
Federal University of Bahia, Department of Chemical Engineering, Federacˌão, Salvador, BA, Brazil

Stefania Briganti, Enrica Flori, Barbara Bellei and Mauro Picardo
Laboratory of Cutaneous Physiopathology, San Gallicano Dermatologic Institute, Istituto di Ricovero e Cura a Carattere Scientifico, Rome, Italy

Xiong Li, Yunqiang Yang and Huaming Lin
Key Laboratory for Plant Biodiversity and Biogeography of East Asia, Kunming Institute of Botany, Chinese Academy of Sciences, Kunming, China
Plant Germplasm and Genomics Center, The Germplasm Bank of Wild Species, Kunming Institute of Botany, Chinese Academy of Sciences, Kunming, China
University of Chinese Academy of Sciences, Beijing, China

Xudong Sun, Xiangyang Hu and Yongping Yang
Key Laboratory for Plant Biodiversity and Biogeography of East Asia, Kunming Institute of Botany, Chinese Academy of Sciences, Kunming, China
Plant Germplasm and Genomics Center, The Germplasm Bank of Wild Species, Kunming Institute of Botany, Chinese Academy of Sciences, Kunming, China

Jinhui Chen
Key Laboratory of Forest Genetics & Biotechnology, Nanjing Forestry University, Nanjing, China

Jian Ren
Department of Grassland Science, Yunnan Agricultural University, Kunming, China

Rafael Medina-Navarro
Department of Experimental Metabolism, Center for Biomedical Research of Michoacán (CIBIMI-IMSS), Morelia, Michoacán, México

Itzia Corona-Candelas and Saúl Barajas-González
Department of Nephrology, General Regional Hospital Nu 1, IMSS, Morelia, Michoacán, Mexico

Margarita Díaz-Flores and Genoveva Durán-Reyes
Biochemistry Medical Research Unit, National Medical Center, IMSS, México City, México

Min Liu, Rui Gao, Qingwei Meng, Yuanyuan Zhang, Chongpeng Bi and Anshan Shan
Institute of Animal Nutrition, Northeast Agricultural University, Harbin, P. R. China

Pabitra Bikash Pal, Krishnendu Sinha and Parames C. Sil
Division of Molecular Medicine, Bose Institute, Kolkata, India

Hui-Xi Zou1. Nan Li, Yan-Qing Lin, Lu-Min Li, Qin-Qin Wu and Xiu-Feng Yan
Zhejiang Provincial Key Lab for Subtropical Water Environment and Marine Biological Resources Protection, College of Life and Environmental Science, Wenzhou University, Wenzhou, People's Republic of China

Qiu-Ying Pang, Li-Dong Lin and Ai-Qin Zhang
Key Laboratory of Saline-Alkali Vegetation Ecology Restoration in Oil Field, Northeast Forest University, Harbin, People's Republic of China

Xiang Hou, Jingfei Zhang, Hussain Ahmad, Hao Zhang and Tian Wang
College of Animal Science and Technology, Nanjing Agricultural University, Nanjing, Jiangsu, P.R. China

Ziwei Xu
College of Animal Science and Technology, Nanjing Agricultural University, Nanjing, Jiangsu, P.R. China Institute of Animal Husbandry and Veterinary Science, Zhejiang Academy of Agricultural Sciences, Hangzhou, Zhejiang, P.R. China

Index